MARINE MAMMAL PHYSIOLOGY

Requisites for Ocean Living

CRC

MARINE BIOLOGY

SERIES

The late Peter L. Lutz, Founding Editor
David H. Evans and Stephen Bortone, Series Editors

PUBLISHED TITLES

Biology of Marine Birds
E.A. Schreiber and Joanna Burger

Biology of the Spotted Seatrout
Stephen A. Bortone

Early Stages of Atlantic Fishes: An Identification Guide for the Western Central North Atlantic
William J. Richards

Biology of the Southern Ocean, Second Edition
George A. Knox

Biology of the Three-Spined Stickleback
Sara Östlund-Nilsson, Ian Mayer, and Felicity Anne Huntingford

Biology and Management of the World Tarpon and Bonefish Fisheries
Jerald S. Ault

Methods in Reproductive Aquaculture: Marine and Freshwater Species
Elsa Cabrita, Vanesa Robles, and Paz Herráez

Sharks and Their Relatives II: Biodiversity, Adaptive Physiology, and Conservation
Jeffrey C. Carrier, John A. Musick, and Michael R. Heithaus

Artificial Reefs in Fisheries Management
Stephen A. Bortone, Frederico Pereira Brandini, Gianna Fabi, and Shinya Otake

Biology of Sharks and Their Relatives, Second Edition
Jeffrey C. Carrier, John A. Musick, and Michael R. Heithaus

The Biology of Sea Turtles, Volume III
Jeanette Wyneken, Kenneth J. Lohmann, and John A. Musick

The Physiology of Fishes, Fourth Edition
David H. Evans, James B. Claiborne, and Suzanne Currie

Interrelationships Between Coral Reefs and Fisheries
Stephen A. Bortone

Impacts of Oil Spill Disasters on Marine Habitats and Fisheries in North America
J. Brian Alford, PhD, Mark S. Peterson, and Christopher C. Green

Hagfish Biology
Susan L. Edwards and Gregory G. Goss

Marine Mammal Physiology: Requisites for Ocean Living
Michael A. Castellini and Jo-Ann Mellish

MARINE MAMMAL PHYSIOLOGY

Requisites for Ocean Living

EDITED BY

MICHAEL A. CASTELLINI

Graduate School, University of Alaska, Fairbanks, Alaska, USA

JO-ANN MELLISH

North Pacific Research Board, Anchorage, Alaska, USA

CRC Press
Taylor & Francis Group
Boca Raton London New York

CRC Press is an imprint of the
Taylor & Francis Group, an **informa** business

CRC Press
Taylor & Francis Group
6000 Broken Sound Parkway NW, Suite 300
Boca Raton, FL 33487-2742

Printed on acid-free paper
Version Date: 20150603

International Standard Book Number-13: 978-1-4822-4267-6 (Hardback)

Visit the Taylor & Francis Web site at
http://www.taylorandfrancis.com

and the CRC Press Web site at
http://www.crcpress.com

Dedication

This book is dedicated to our teachers and colleagues, both past and present, who forged new and exciting areas in marine mammal physiology and biochemistry.

There are some giants in this field who have left us, but who we still refer to almost daily in our work: Laurence Irving and Per Scholander started the modern work on the physiology of diving in the 1930s. It is hard to imagine our field without their historical and groundbreaking work.

Dr. Knut Schmidt-Nielsen really did "write the book" on comparative physiology, Animal Physiology: Adaptation and Environment (Cambridge University Press, 1975), *that was the core from which many of us were both taught and that we used in our own classes as teachers. Of course, Irving, Scholander, and Schmidt-Neilsen knew each other and worked together.*

Dr. Ted Hammel broke ground on the modern aspects of thermoregulation in mammals, including the marine species. In keeping with the pattern, he also knew and worked with Irving, Scholander, and Schmidt-Nielsen.

Where would we be on the complexity of diving biochemistry without Dr. Peter Hochachka? His ability to perceive and understand biochemical pathways and to explore the new world of comparative biochemical adaptation to the environment was second to none. By now, you can probably guess that he also knew all of the scientists we have already mentioned.

The fascinating world of the neural control of the "diving reflex" was explored by Dr. David Jones in a multitude of species, from rabbits to whales, to ducks. He and Peter Hochachka were great friends, in the same Zoology Department at the University of British Columbia.

Carl Hubbs, from Scripps, who identified many species of marine mammals, was a contemporary with Irving and Scholander; Ken Norris taught many of us about the ecological physiology of marine mammals, "virtually created the field of cetology," and received his doctorate from Scripps as Hubbs's PhD student.

All of these mentors have left us, but their impact to the field of marine mammal physiology and biochemistry is fundamental. Many of their works are referenced in these chapters.

Of our current colleagues, Dr. Bob Elsner is certainly the grandfather of them all. A world-known diving physiologist, he studied with Irving, Scholander, Schmidt-Neilsen, and most of the others. Dr. Elsner is over 90 years old and just as we were writing this dedication, he provided us with a copy of his newly published book on diving physiology entitled Diving Seals and Meditating Yogis (University of Chicago Press, 2015).

Dr. Elsner inspired Dr. Gerry Kooyman, also from Scripps, who broke open the field of modern diving recorders and has been recognized around the world for his leadership in diving physiology and Antarctic expeditions. A simple "Google" search on Kooyman and Diving *will result in an astounding number of "hits." Dr. Kooyman defined the "aerobic diving limit," which you will read about in this book.*

Dr. Pat Butler is probably the world's expert on the comparative neural control of diving and has written a number of reviews on diving physiology. He and Dr. Jones were long-standing friends and colleagues.

Dr. George Somero continues to be at the forefront of "comparative biochemistry" and he and Peter Hochachka were the closest of friends. They published the groundbreaking book Strategies of Biochemical Adaptation (W.B. Saunders Company, 1973) *that many of us use in teaching. Their research on high-pressure biochemistry forms the core of much of that work in marine mammals.*

Needless to say, Elsner, Kooyman, Butler, and Somero are all friends and have worked together for years.

Many studied under those noted above and they, in turn, formed many of the author committees, chairs, and thesis projects in a large academic family tree, with many interweaving branches. What was once a small group of researchers has grown into an international field of thousands. The editors thank in particular those founding members of their marine mammal careers, Gerry Kooyman, George Somero, and Peter Hochachka (Castellini), and Sara Iverson (Mellish).

This work is also dedicated to our many current colleagues and students who have shared years of laboratory, classroom, and field time working with marine mammals around the world. From exciting to dangerous, the long nights observing animals, to shared time in camps, on boats, sea ice, and beaches, working with marine mammals is certainly a group effort.

Finally, to our own families and children for their endless supply of support and willingness to tolerate long times without us, to read manuscripts, and listen to our tales of field debacles and near misses, animal antics, and scientific mysteries, we thank you. We could not do this without you.

Michael A. Castellini
Fairbanks, Alaska

Jo-Ann Mellish
Anchorage, Alaska

Contents

Section III: Reproduction

Section IV: Sensory systems

Section V: Environmental interactions

Editors

Michael A. Castellini, PhD, earned his PhD from the Scripps Institution of Oceanography and has been a faculty member at the University of Alaska–Fairbanks (UAF) since 1989. He is the founding science director for the Alaska SeaLife Center, the director of the Institute of Marine Science at UAF, associate dean and then dean for the School of Fisheries and Ocean Sciences. He is now associate dean for the UAF Graduate School, senior faculty in the Center for Arctic Policy Studies, and director of the Center for Global Change. Dr. Castellini's research focuses on how marine mammals have adapted to life in the sea, including their biochemical, physiological, and behavioral adaptation for deep and long-duration diving, extended fasting, exercise physiology, hydrodynamics, and even sleeping patterns. In Alaska, his work has extended into issues of population health, contaminant chemistry, reproductive chemistry, and digestive physiology. Dr. Castellini's graduate students have worked from Alaska to Antarctica on these issues. He has written more than 100 scientific papers and chapters on his work and is involved in local, state, and national panels and committees dealing with policy issues related to marine mammals, ecosystem management, and polar concerns.

Jo-Ann Mellish, PhD, earned her PhD from Dalhousie University in Canada in 1999. She has been on the research faculty at the University of Alaska–Fairbanks with a joint appointment at the Alaska SeaLife Center since 2001. Most recently, she has taken on the role of program manager at the North Pacific Research Board. Although trained in comparative animal physiology, she has yet to run out of research with marine mammals given their enormous diversity. In a similar path to Dr. Castellini, her dean and mentor for many years, her studies on energetics, health, and conditions have taken her from pole to pole. In Alaska, she has focused on innovative approaches to studying physiology in endangered species, and she has worked to raise the bar on thoughtful assessments of the research procedures themselves. For the past decade, Dr. Mellish has been the lead investigator of a highly collaborative team of experts working in Antarctica that melds physiology, ecology, and modeling to better understand phenotypic plasticity and the potential impacts of ecosystem change in high-latitude species. Her work has resulted in 40 scientific papers.

Contributors

Daniel E. Crocker
Department of Biology
Sonoma State University
Rohnert Park, California

Guido Dehnhardt
Institute for Biosciences
Marine Science Center
University of Rostock
Rostock, Germany

Andreas Fahlman
Department of Life Sciences
Texas A&M, Corpus Christi
Corpus Christi, Texas

Frank E. Fish
Department of Biology
West Chester University
West Chester, Pennsylvania

Jeremy A. Goldbogen
Department of Biology
Hopkins Marine Station
Stanford University
Pacific Grove, California

Frederike D. Hanke
Institute for Biosciences
Marine Science Center
University of Rostock
Rostock, Germany

John R. Harley
Department of Chemistry and Biochemistry
University of Alaska–Fairbanks
Fairbanks, Alaska

Mark A. Hindell
Institute for Marine and Antarctic Studies
University of Tasmania
Hobart, Tasmania, Australia

Allyson Hindle
Massachusetts General Hospital
Harvard Medical School
Boston, Massachusetts

Sascha K. Hooker
Sea Mammal Research Unit
Scottish Oceans Institute
University of St. Andrews
St. Andrews, Scotland

Dorian Houser
National Marine Mammal Foundation
San Diego, California

Shawn P. Johnson
Department of Veterinary Science
The Marine Mammal Center
Sausalito, California

Jennifer L. Maresh
Department of Ecology and Evolutionary Biology
University of California, Santa Cruz
Santa Cruz, California

Christopher D. Marshall
Department of Marine Biology
and
Department of Wildlife and Fisheries Biology
Texas A&M University
Galveston, Texas

Birgitte I. McDonald
Moss Landing Marine Laboratories
Moss Landing, California

Jason Mulsow
National Marine Mammal Foundation
San Diego, California

Todd M. O'Hara
Department of Veterinary Medicine
University of Alaska–Fairbanks
Fairbanks, Alaska

Rudy M. Ortiz
School of Natural Sciences
University of California, Merced
Merced, California

Paul J. Ponganis
Center for Marine Biotechnology and Biomedicine
Scripps Institution of Oceanography
University of California, San Diego
La Jolla, California

Jean Potvin
Department of Physics
Saint Louis University
St. Louis, Missouri

David Rosen
Marine Mammal Research Unit
University of British Columbia
Vancouver, British Columbia, Canada

Claire A. Simeone
Department of Veterinary Science
The Marine Mammal Center
Sausalito, California

Miwa Suzuki
Department of Marine Science and Resources College of Bioresource Science
Nihon University
Fujisawa, Kanagawa, Japan

Andrea Walters
Institute for Marine and Antarctic Studies
University of Tasmania
Hobart, Tasmania, Australia

Cassondra L. Williams
Department of Ecology and Evolutionary Biology
University of California, Irvine
Irvine, California

Terrie M. Williams
Department of Ecology and Evolutionary Biology
University of California, Santa Cruz
Santa Cruz, California

Introduction

I.1 What makes a marine mammal different from terrestrial mammals, beyond just their environment?

When confronted for the first time with studying the biology of marine mammals, many students are initially drawn in by the legends, lore, and intrigue of marine species. They certainly are "charismatic megafauna," and their photos and stories adorn numerous websites and books. Their high visibility on display at aquaria gives the public a small glimpse into their lives, but also may feed a tendency to produce polarizing opinions on issues. These sometimes high-profile issues are numerous, from subsistence and commercial hunting, classification and handling of endangered species, local and international conflicts with fisheries, to sport boating and military operations. It is no wonder that many people from around the world have some knowledge of whales, seals, walruses, otters, and more. There are many publications and books that cover these species from almost all perspectives and we include some of those reference works for your consideration.

This textbook, however, takes a step back, to focus on the physiological and biochemical characteristics that have allowed these mammals as a group to effectively exploit the marine environment that is so hostile to humans. While there are thousands of individual scientific papers and research-focused books that have asked this question in extreme detail, we have compiled a series of larger discussions aimed at the student interested in the general aspects of physiology and biochemical adaptation of marine mammal species. The authors have focused on the key challenges that a marine mammal must overcome by living in the ocean, and how they have evolved and adapted biological systems to successfully live where no other mammals can follow.

Our contributing authors come from a wide spectrum of academia, government, industry, and veterinary science, however, all are experienced teachers in their respective fields, and this book is written with that focus in mind. We hope to appeal to a broad range of students, from those that are new to marine mammals, in general, to those that are well into a focused graduate program and need a broader frame of reference. Our author pairings are specifically tailored to have mentors and up-and-coming researchers working together to provide a perspective that is relevant to multiple generations of scientists. The approach was intentionally international, such that our expertise reflected this group of mammals with global distribution.

We cover the fundamental physiological, biochemical, and anatomical "tricks" that have evolved in many species of marine mammals. We discuss variations in those adaptations, through a range of species. For example, the water balance issues facing a newly weaned, fasting elephant seal pup on a California beach are not the same as those facing a whale on a long migration, yet they approach it in a remarkably consistent fashion. The hydrodynamic design of a swimming walrus is not very similar to that of a high-speed dolphin, but they are both constrained by the forces of drag, lift, and buoyancy. Each chapter has a "Toolbox" section where the authors discuss many of the newest methods for working on the physiology of marine mammals.

In short, we are asking you to consider the following questions: If you were designing a marine mammal, what would you need to think about to allow it to live in the ocean? How would you keep it warm? What would you design to allow it to dive for very long periods, to extreme depths? Where would it find water to drink? How would you minimize the cost of swimming and how would it find its prey in the deep and dark? These questions and more are throughout this book. As with all research, we expect that for every question answered, you will ask several more.

I.2 What are the requisites for ocean living?

Our chapter topics are grouped into major themes: diving and locomotion, nutrition and energetics, reproduction, sensory systems, and environmental interactions.

We begin with one of the most fundamental aspects of these species, diving and locomotion. Marine mammals need to be able to swim and dive and to do it very well. Humans can swim and dive too, but it is clear that we cannot hold our breath for over an hour, or dive to a 600 m depth. This group of four chapters (Section I) begins with an analysis of marine mammal hydrodynamics by Jeremy A. Goldbogen, Frank E. Fish, and Jean Potvin (Chapter 1). They explore many of the complications of moving a body through water that must be resolved, including larger questions of drag, swimming at various depths, and more unique aspects such as what happens when a whale swimming at high speeds opens its mouth to catch krill on its baleen.

A marine mammal must be able to hold its breath for a very long time to be able to dive and swim. As mammals, they do not have gills or specialized organs to extract oxygen from the seawater. Where do they carry the oxygen they need, how do they utilize that oxygen, and what happens if they stay underwater for longer than the amount of oxygen they have available? Paul J. Ponganis and Cassondra L. Williams (Chapter 2) discuss how these species can remain underwater for so much longer than terrestrial mammals.

Next, Terrie M. Williams and Jennifer L. Maresh (Chapter 3) discuss the energetic costs of swimming and exercise. Even the most hydrodynamic body shapes incur energy costs to move through the water. How does this efficiency vary among the wide range of marine mammals? Does it cost the same to move a large whale through the water as it does to move a small porpoise? What is the optimal speed to swim? How would you even begin to consider effective measurements of the cost of swimming?

Finally, if the animals dive deeply, how do they withstand the tremendous pressures at depth? Why do they not get "the bends?" What are the physiological and biochemical adaptations that allow their tissues and cells to survive such an insult? There are many marine mammal species that collapse their lungs dozens of times a day, yet this is a critical injury in a human. Sascha K. Hooker and Andreas Falhman (Chapter 4) explore those adaptations that allow marine mammals to dive far deeper than freely diving humans have ever even approached.

The next set of five chapters (Section II) examines how marine mammals are able to feed, obtain energy, fast for very long periods of time, live without freshwater, and manage to maintain their body temperatures in extreme environments.

Christopher D. Marshall and Jeremy A. Goldbogen (Chapter 5) begin Section II by examining the feeding mechanisms that have been developed by marine species. How do dugongs and manatees digest their vegetarian diet? Are the teeth of marine mammals any different than terrestrial species? How do baleen whales obtain their food compared to polar bears?

Mark A. Hindell and Andrea Walters (Chapter 6) continue with questions on diet and nutrition. They examine the ranges of diets in many species, looking at common dietary characteristics. Many whale species filter feed, but this is primarily in polar and high-productivity regions in the ocean. Did you know that some seals also filter feed? Marine mammals are incredibly diverse in their diets, ranging from tiny krill and copepods, all the way to the 6 ft Antarctic cod and, in some cases, other marine mammals.

It is well known that humans cannot survive on the ocean without access to freshwater. Miwa Suzuki and Rudy M. Ortiz (Chapter 7) explore the fascinating question of water balance in marine species. You will discover that marine mammal kidneys look very different from those of a terrestrial mammal, yet have many physiological and biochemical similarities. If available, will a marine mammal drink freshwater or ignore it?

Many marine mammals fast for very long periods either during migration or when on land for breeding purposes. While Suzuki and Ortiz (Chapter 7) looked at this from water balance requirements, David Rosen and Allyson Hindle (Chapter 8) ask how some marine mammals can live for months with little or no access to food? They explore the sources of energy during these periods on a biochemical and physiological basis.

Finally, Michael A. Castellini and Jo-Ann Mellish (Chapter 9) address how marine mammals stay warm in a cold sea, yet do not overheat when on land. They look at the normal body temperature of a marine mammal, and how their fur or blubber helps to maintain that temperature. What about species that live on the frozen sea ice and give birth to pups at extremely cold temperatures? Will some of the predicted changes in water temperature through climate change impact marine mammals?

In the next section of the book (Section III), the authors focus on the amazing physiology of giving birth, and how these mammals have adapted to support their calves/pups in a marine environment. Daniel E. Crocker and Birgitte I. McDonald (Chapter 10) take you onto elephant seal breeding beaches where nursing mothers support rapidly growing newborn pups, yet do not eat or drink for many weeks. They explore the complex biochemical pathways that allow milk production during caloric restriction. They address the support of newborn whale calves and overwintering polar bears alike.

Sensory systems must clearly be very different in marine versus terrestrial mammals, as they have to be able to hear and see underwater with very high resolution. Do they have other senses that we only partially understand?

Dorian Houser and Jason Mulsow (Chapter 11) begin with a focus on the acoustic aspects of the physiology and anatomy of hearing underwater. It is well known that marine mammals can produce a large range of acoustic calls, many of which are not in the audible range of humans. What is the nature of their sound production? What are the differences in the acoustic biology of whales and sea otters? Increasing background noise in the ocean, and how that may influence marine mammals, is another point of discussion in this chapter.

Marine mammals must be able to see both in the air at the surface and deep underwater. The darkness of the ocean at several hundred meters and variable light under the sea ice are not conditions that we as humans are built to contend with. Frederike D. Hanke and Guido Dehnhardt (Chapter 12) consider the difference in the neural structures of the vision system across marine mammals. They look at how they might sense particle flow fields from plankton and other small debris floating in the water. They also examine the extremely sensitive whiskers (vibrissae) of seals, and how they might be used to sense disturbance in the water after a fish has swum nearby.

In the last section of the book (Section V), two sets of authors consider how to assess the health of marine mammals in the environment. First, Shawn P. Johnson and Claire A. Simeone (Chapter 13) explore the world of disease and mortality events in marine mammals. They ask if the respiratory system, which is highly adapted for exchanging large amounts of air very quickly, could be a weakness by trapping foreign particles. They explore possible diseases that could impact seals and sea lions when they haul out on beaches. They examine a wide range of potential sources of disease that could impact these species at all stages of their life cycle.

John R. Harley and Todd M. O'Hara provide the final large chapter of this section (Chapter 14), where they explore potential "poison and toxic" elements in the environment and how they relate to the health of marine mammals. From organic pollutants to heavy metals, marine mammals must deal with a wide range of potential contaminants in the ocean and in their food. Given that we are still in the early stages of how many of these compounds impact human health, there are numerous questions on how they may impact marine mammals. Referring back to the physiology of newborn calves, they talk about how a suite of lipid-soluble contaminants can be transported to the calf through the high-fat milk of the mother. There are also a considerable number of natural toxins in the marine environment, including harmful algal blooms.

In the final chapter (Chapter 15), editor Michael A. Castellini talks about some of the many other fields of physiology and biochemistry that could not be covered in detail in this book. For example, there are many components of pre-partum physiology that also intrigue marine mammal biologists: Do the feeding conditions for the mother in the months preceding birth impact the birth rate of the greater population? This leads into the question of ecological physiology: That is, the biochemical and physiological "boxes" that have been defined in this book provide a reasonable view of the limits of the capabilities of marine mammals. But, we know they do not dive to their maximum breath-hold capacity on each dive, nor swim as fast as possible, nor dive as deep as possible. What is it about the ecology of where they are diving at any given moment that places them within those biochemical and physiological limits?

We conclude by exploring many of the questions we have examined throughout this book, but in terms of what they mean to the reader. From teaching many years of physiology and biochemistry, we have seen that there are some questions that come up repeatedly. Did this book help you to consider those questions and how you might go about studying these particular phenomena? As we were writing this book, even the communication between the authors indicated that "they didn't know that." Despite the fact that the authors have all worked on this group of animals for many years, there is always room to expand our knowledge of the basic principles and our understanding of how they integrate with one another. We hope you share our enjoyment in pulling the diverse expertise of our authors into a single compendium on a vast group of species.

Given the tools you have in this book, how would you design your marine mammal?

section one

Diving and locomotion

chapter one

Hydrodynamics

Jeremy A. Goldbogen, Frank E. Fish, and Jean Potvin

Contents

1.1 Introduction

High-performance locomotion at a low energetic cost is critical for many life functions including capturing food, avoiding predators, and long-distance migration. For marine mammals, aquatic transport is a necessity and thus demands efficient locomotor capacity. Because marine mammals evolved from terrestrial ancestors, these animals exhibit secondary adaptations to the physical challenges associated with life in water. The most fundamental of these challenges are resistive forces, generally known as *drag*, that limit the body's movement through water. Drag arises because water has mass and viscosity, so the motions of animals through the water are not free from the perspective of energy economics. These resistive forces are complex and are largely dependent on shape, scale, and speed. Marine mammals exhibit a wide range of morphologies that reflect different locomotor demands and ecological functions. However, across taxa, most marine mammals are unified by a convergent, fusiform body profile that decreases resistive forces and increases the efficiency of locomotion.

One major factor influencing the morphological design of marine mammals is how much time they spend in water, and what they do while in it. Some animals have evolved a body shape and propulsion system that not only minimizes drag, but also optimizes thrust generation (i.e., whales and dolphins). Not surprisingly, these animals happen to travel over very large distances in the water or chase very fast prey. Other animals have not optimized their hydrodynamics but rather other anatomical characteristics that are more important for other critical life functions, many of which occur on land, at the expense of increased locomotor costs in water. In effect, many marine mammals are faced with conflicting physiological demands of an amphibious lifestyle (i.e., pinnipeds), while others have evolved fully aquatic niches (i.e., cetaceans). In this chapter, we explore the mechanisms that marine mammals use to achieve different levels of swimming performance. Moreover, we will review specific anatomical and behavioral adaptations in a comparative context to understand the evolution of locomotor traits that characterize this extraordinary guild of oceanic predators.

1.2 Hydrodynamic forces at play during locomotion: Drag, lift, and thrust

1.2.1 Flow structures around a body moving underwater

As seen from a distance, the flow of water moving past a marine mammal *coasting* or *gliding* in water appears streamlined and orderly (Figure 1.1a). The incoming flows—or *freestream*—part ways laterally and dorsoventrally at the nose (or rostrum), to accelerate to speeds higher than the freestream's until the body's widest section is reached. Past this point the flows begin to decelerate, and then rejoin to the speed of the freestream past the body. At smaller scales near the animal's body (millimeters to decimeters, Figure 1.1bc) there is a thin layer of fluid "tucked" under the nearest of the fluid streamlines. Typically moving at significantly slower speeds, this layer is called the *boundary layer* (Vogel 2003). Such a thin sheet of water may also appear streamlined, or laminar, like the flows above it; but in the right conditions, it will also appear *turbulent*, that is, populated with swirls or eddies of all sizes spinning in all directions and at varying rates. The thickness of the boundary layer generally increases posteriorly; along the body of a blue whale, for example, it is estimated to be only a few millimeters thick at the rostrum and about 0.2 m at the end of the tail. Whether laminar or turbulent, the boundary layer at the tail of a gliding animal then yields the *turbulent wake* past the tail. This wake is "wrapped" within the streamlines of fluid that are rejoined behind the body (Figure 1.1c).

In contrast, the flow structure about actively swimming mammals is more complicated (Fish 1993). Here, the animal's appendages accelerate portions of the surrounding flows, to add large-scale tornado-like vortical structures that may persist long after the animal has passed by. These vortices can be seen on the surface of the water when paddling in a kayak or canoe. Undulating hydrofoils—or flukes and fins—produce them as well. Generally vortices, and their small-scale eddy cousins within the boundary layer, tend to form whenever a layer of fast fluid moves past a slow-moving layer, or one moving in the opposite direction.

The drag generated by marine mammals is related to the interaction between boundary-layer flows and large-scale flows. Such a link is crucial, as painfully realized by the aircraft aerodynamicists of the early 1900s who, without knowing about, or in some cases not acknowledging the existence of the boundary layer, kept calculating the wrong drag (Bloor 2011). The boundary layer arises from two basic forces: (1) shear forces in-between adjacent layers of fluid are caused by *viscosity*, which tend to resist the layers' relative motion as well as induce

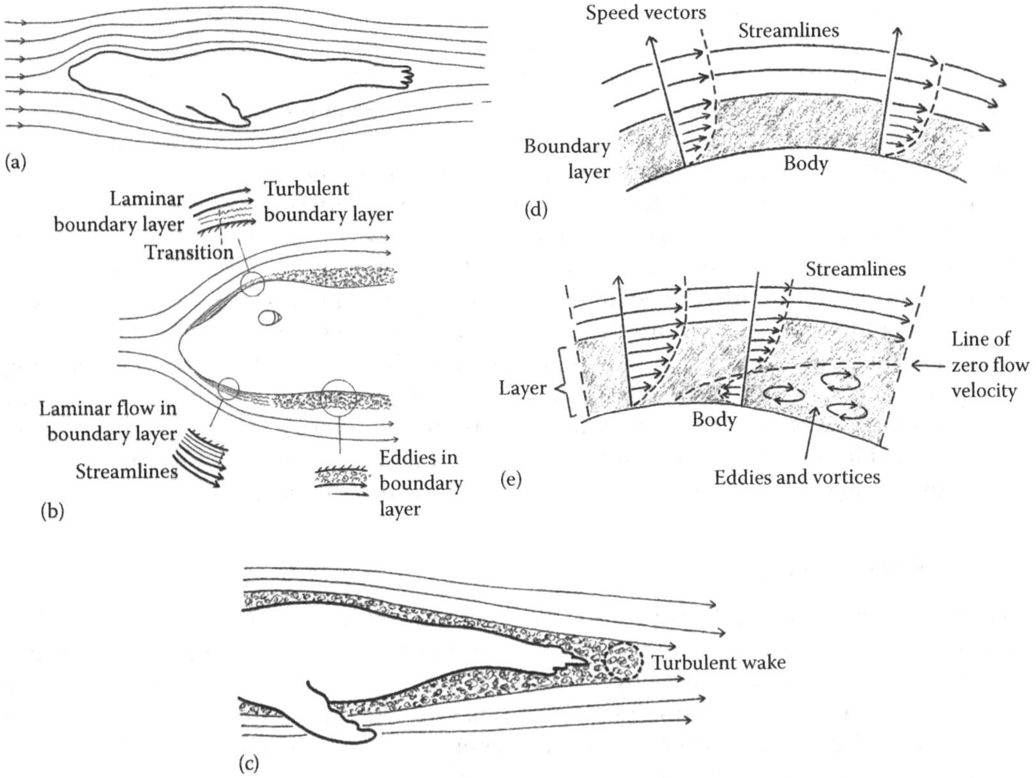

Figure 1.1 Fluid motions around a swimming marine mammal (the motions are shown relative to the body): (a) Flow streamlines showing the path of the fluid particles; (b) laminar and turbulent boundary layers generated along the surface of the body; (c) boundary layer aft of the body, terminating into the turbulent wake; (d) close-up view of the boundary layer, showing the mean speeds of the fluid particles within it (again relative to the animal): increasing from near-zero just above the body surface, up to the speed of the streamline nearest to the top of the layer; (e) boundary layer structure at the threshold of the separated flow genesis.

rotation in each layer, and (2) the molecular forces at the body's surface which prevent the fluid from moving past it—a phenomenon called the *no-slip* condition (Figure 1.1d). The body's surface atoms attract, and come into contact with, the nearest atoms of the liquid; the latter then slow down the motions of the water molecules above it, which in turn slow down those of the layer further up; and so on. The boundary layer is a fluid structure that exists wherever there is flow past a solid surface, such as above the seafloor: there too, the water near the seafloor's surface is always moving at a slower speed than water above it.

The boundary layer is more dynamic and has more structure than the streamlined flows above it. It begins as laminar and its thickness increases posteriorly. But a laminar boundary layer is an unstable structure that is easily disrupted when encountering surface roughness and random perturbations from the flows above it. The result is often a laminar boundary layer changing into a turbulent boundary layer (Figure 1.1b). On a well-streamlined, fusiform animal, and depending on the swim speed, this laminar-to-turbulent transition occurs approximately halfway along the body (Fish and Rohr 1999). Where the transition occurs is dynamic and difficult to predict because it depends on many factors, as well as on the body (and limb) postures adopted at any point in time.

The structure of the boundary layer, that is, how much of it is laminar and how much is turbulent, largely determines how much drag is produced.

Another important feature of the boundary layer is *flow separation*, which occurs when the flows within are decelerated to a near zero speed when approaching an area of greater pressure, as usually occurs along the posterior half of a tapered body. If these adverse pressure gradients are strong enough, the boundary layer flows may reverse direction relative to the streamlined flows above the layer. The interaction between the two creates a large zone of turbulent flow that keeps the streamline flow away from the surface—effectively "separating" the streamlined flow away from the body (Figure 1.1e). The creation of both the turbulent wake and turbulent zone underneath the separated flows lead to more drag. Note that turbulent boundary layers are less prone to adverse pressure gradients than laminar layers and thus less conducive to flow separation and the ensuing drastic increase in drag. "Tripping" or inducing the boundary layer to transition from laminar to turbulent, for example, by dimples (as on a golf ball) or sharp protrusions (by barnacles), can help reduce drag. Note also that in some forms of undulatory swimming where more adverse pressure gradients appear along the body, boundary layer control becomes crucial for efficient propulsion.

1.2.2 *Resistive forces*

In order to move through the water, marine mammals must apply a force to part the fluid around their body. In reaction to this force, the fluid applies a force (drag) onto the animal (i.e., Newton's third law of motion). From the point of view of the swimmer, drag is an energy-dissipating force because it transfers kinetic energy away from the swimmer and into the water. The animal never recovers this energy—instead, it gets dissipated into heat.

There are different types of drag depending on the mechanism that generates the resistive force. The two most important are *friction drag*, due to the shear force (or sliding "friction") among the sub-layers of the boundary layer; and *pressure drag*, due to the existence of the turbulent wake and/or zone under separated flows (when present). Separated flows sustain pressures that are low enough to effectively "suck" the body back into it, thus resulting in drag. Generally, laminar boundary layers produce less friction drag than turbulent boundary layers do. Turbulent wakes of smaller width ("girth") (Figure 1.1c) produce less pressure drag. With non-streamlined or *bluff* bodies, pressure drag dominates over friction drag, and the opposite occurs with streamlined bodies.

Another source of drag appears when a body is accelerated or decelerated, namely, *the acceleration reaction*, which arises because of the necessity to accelerate—along with the animal's body—a fluid mass roughly equivalent to that of the fluid displaced by the body (Daniel 1984). This added mass is important, for example, with sea lions because of their accelerating–decelerating swimming style (Feldkamp 1987a; Stelle et al. 2000). Added mass drag can be reduced by body streamlining, as spheres tow-along far more added mass than javelins. For fusiform body shapes exhibited by marine mammals, the added mass is approximately 5%–10% that of the displaced water (Williams 1987).

Some marine mammals use winglike flippers and flukes to generate lift and produce thrust. Although lift is produced, these surfaces simultaneously generate a fourth type of drag called *induced drag*. Induced drag appears because winglike structures produce a large-scale vortex at the tip of the wing that extends far behind. Like the other vortices and eddies produced in boundary layers and in turbulent wakes, tip vortices are tornado-like structures of spinning fluid whose rotation is driven by the wing's own kinetic energy. As a general rule, induced drag is minimal when the wing is tapered, as is the case with most flukes and flippers. Induced drag is a major energy sink with aircraft, relative to the friction and pressure drag generated by the fuselage. This is because very large wings are required to generate enough lift to support the heavy payload. Because marine mammals are close to neutral buoyancy, they do not require as much lift to "fly" above the seafloor. Thus, marine mammal lifting surfaces are much smaller in relation to the rest of the body relative to aircraft, but they are still important for generating forces to effect rolls, turns, or propulsive forces.

Marine mammals that swim at the sea surface experience additional resistive forces which can increase the cost of swimming (Vogel 1994), often referred to as ventilation drag (Ahlborn 2004) and wave drag (Hertel 1966; Fish 1993) (Figure 1.2). The former happens

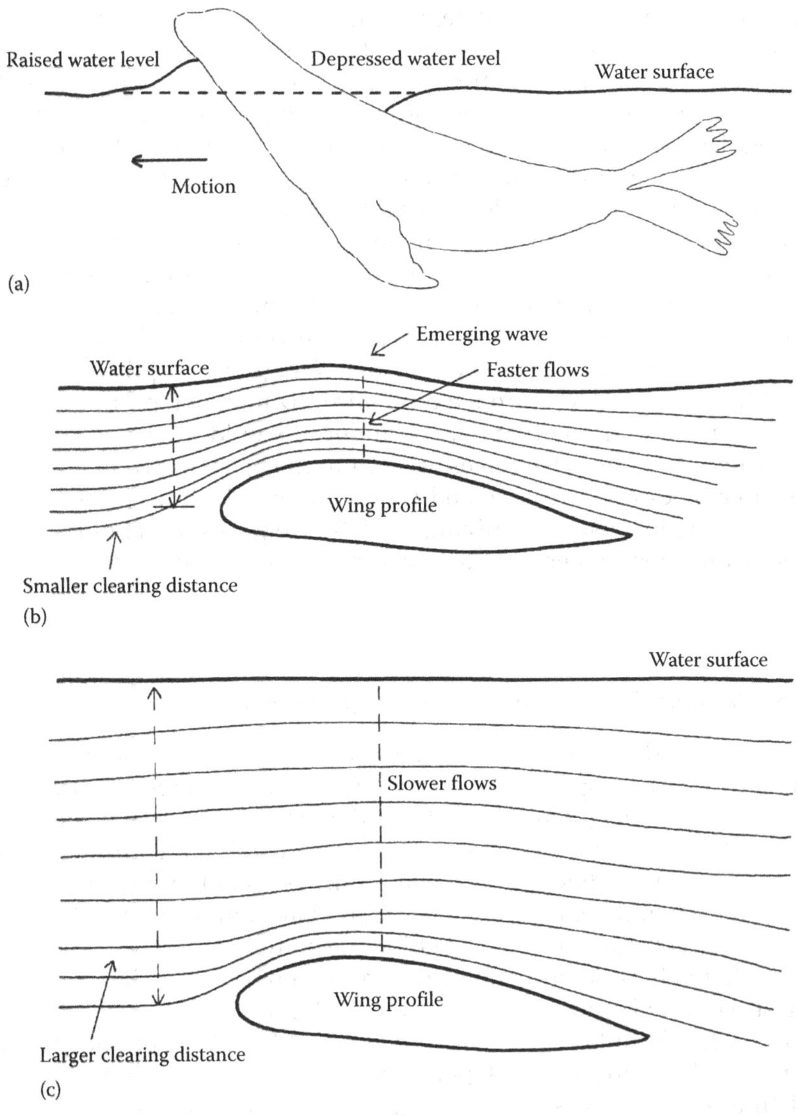

Figure 1.2 Hydrodynamics of a marine mammal near the sea surface: (a) When breaking the surface, (b) just below, and (c) in the process of creating a wave. Note that at a given depth, the height and location of the wave depend on the object's shape and orientation with respect to the surface. The height of the wave shown in (b) has been exaggerated, and in some cases, the back side of the wave may also dip below the mean height of the sea surface.

whenever portions of the body break the water surface, such as with a seal's head or a cetacean's blowhole. *Ventilation drag* arises because of the mass of water that is elevated on the leading side of the exposed body feature; and also because of the depressed wake that follows behind (Figure 1.2a). "Drag" appears because its takes a force to push water upward above the water surface as well as downward below that surface. Ventilation drag can be minimized through streamlining and reduction of the surface-breaking body feature.

Wave drag—a combined form of "wave-making resistance" or "area blockage" drag—arises from the necessity to move the dorsal portion of the freestream over the smaller volume in comparison to the same at depth (Figure 1.2b). As with moving water through a funnel, the fluid has to accelerate through the narrower section in order to pass the same amount of fluid mass per unit time as through the wider section (by Venturi effect). Furthermore, increasing the speed of flow means higher friction drag and (possibly) pressure drag. This is the effect routinely observed in aircraft wind tunnel testing. When the body is close enough to the surface, the accelerating fluid begins forming a wave at the surface above the body. At a depth less than three times its maximum girth, area blockage and wave-making drag effects increase the overall drag. At a depth of half the animal's girth, this increase in drag can be up to ~5 times that experienced at depth. Beyond that point, drag decreases somewhat, to ~3 times the "at depth" value when the dorsal portion of the body is about to break the surface (Hertel 1966). These physical principles suggest that with all else being equal, foraging at depth is always more efficient than at the surface (at least from a drag minimization perspective).

Note that added mass, ventilation, and induced would exist even in a zero viscosity world because these resistive forces arise from the water's own mass. This contrasts with wave (area blockage), pressure and friction drag, which need the boundary layer (and thus viscosity) to exist. Minimizing the latter requires morphology (specifically a forward–backward body taper) that controls the boundary layer and size of the turbulent wake.

1.2.3 *How to limit drag: Forward and backward taper*

As a general rule and during non-foraging travel, the most important contribution to marine mammal body drag is the sum total of friction and pressure drag. In this context, the body *fineness ratio* (body length/body diameter) is a crucial determinant of drag (Figure 1.3). With the so-called bluff shapes where (lateral) body girth far exceeds body length in the direction of flow (i.e., very low fineness ratio), pressure drag is far more important than friction drag. Here, the oncoming flow at maximum girth is tripped and separated into a sizable turbulent wake behind the body (Figure 1.3a).

Animals that exhibit long-distance migration, on the other hand, are never bluff but rather streamlined into a fusiform shape. Recall that the size of the turbulent wake behind a fusiform body depends on the width of the boundary layer posteriorly (Figure 1.1); therefore, a more tapered posterior region yields a smaller turbulent wake and thus smaller pressure drag (Figure 1.3b). Additionally, to reduce friction drag, one also needs high taper over the anterior portion of the body to get the longest extension of the laminar boundary layer. It would thus follow that marine mammals ought to be shaped like javelins with fineness ratios exceeding 20; however, most species have a fineness ratio less than 8 (Fish and Rohr 1999; Ahlborn et al. 2009). Friction drag increases with the body length, thus fineness ratios exceeding 20 would entail

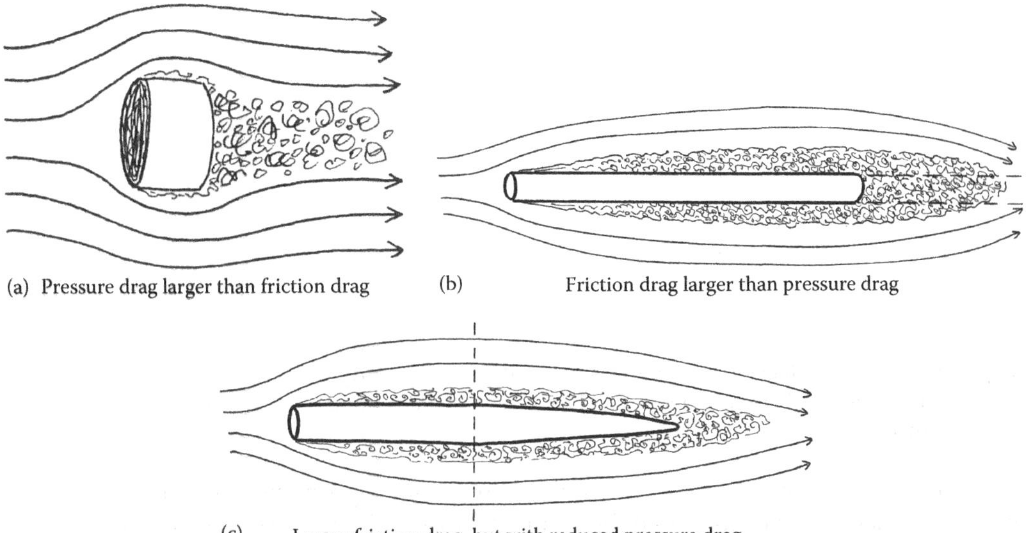

(a) Pressure drag larger than friction drag (b) Friction drag larger than pressure drag

(c) Larger friction drag, but with reduced pressure drag

Figure 1.3 Friction versus pressure drag on bodies of varying aspect ratios and anterior taper: Bluff with a wide turbulent wake (a), streamlined with a narrow turbulent wake (b), and streamlined and tapered with an even narrower turbulent wake (c).

significant friction drag. Moreover, pressure drag is not eliminated by extreme fineness since the boundary layer—and hence the turbulent wake—would also increase. In order to achieve minimum total body drag, rather than just pressure or friction drag, body fineness is limited to a range of 5.5–7.2, depending on body mass (in the 10^3–10^5 kg range) (Ahlborn et al. 2009).

1.2.4 *Lift and propulsion*

As further explored in the next section, paddling involves using pressure drag as a means to generate propulsion. This is a low-speed approach to swimming, however, which can only work for travel over relatively short distances. For long-range swimmers, thrust is generated more efficiently using flippers and flukes that act as lift-producing *hydrofoils*. These hydrodynamic lifting surfaces generate propulsion when oscillated or raised and lowered in rhythmic fashion orthogonal to direction of transport. More specifically, thrust is produced via the combination of the *lift* and *drag* forces that the hydrofoils produce. While drag always points in a direction opposite to the body or appendage in motion, lift is directed perpendicularly to that motion. Using a simplistic description, the lift generated by a wing or hydrofoil arises from the combination of the low-pressure region existing over the upper surface of the wing, in comparison to the higher pressure under the wing. This is the so-called "Bernoulli lift," created when the flow over the wing is faster than the flow underneath. But lift is also the result of the downward deflection of the air (or water) imparted by *both* the vacuum over the upper surface and the deflecting action of the solid bottom surface (Figure 1.4a).

Aircraft wings typically have upper surfaces that are more cambered (curved) than the lower surface because one needs upward-directed lift to compensate for the aircraft's weight. Flippers and flukes, on the other hand, have near-equal camber on both surfaces

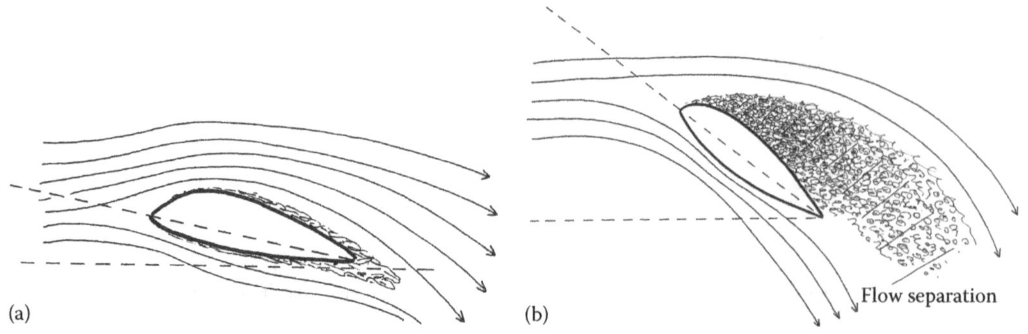

Figure 1.4 (a) The effect of the angle of attack (AOA) on hydrodynamic performance of hydrofoil-like marine mammal appendages. (b) The effect of appendage flexibility on hydrodynamic performance.

(i.e., are more symmetrical)—because they are used to produce forward-directed propulsion, which involves the use of both downward- and upward-directed lift.

The basic design of a lift-producing winglike surface hinges on minimizing drag and maximizing lift. In general, both lift and drag increase with increasing hydrofoil *angle of attack* (AOA). AOA characterizes the angle between the direction of the flow relative to the surface of the foil, and orientation of its mean chord line (Figure 1.4a). In order to produce the most lift at the lowest energetic (drag) cost, marine mammals must optimize the lift-to-drag ratio. At low AOA, drag is the result of friction, pressure, and induced drag. In this regime, the "best" AOA for optimal lift-to-drag is at about 15°, depending on the morphological design of the hydrofoil. This is why dolphins, which use flukes as hydrofoils to produce lift, appear to swim almost effortlessly by barely moving their flukes at low AOA.

When a wing is tilted beyond a certain "critical" AOA of ~18° (depending on specific wing design), drag increases and lift decreases resulting in *stall*. During stall the boundary layer has separated all the way to the wing's leading edge, leaving a large zone of turbulent air—and lots of pressure drag—over the wing (Figure 1.4b). The ratio of lift to drag decreases significantly and this diminishes the flipper or fluke's ability to produce the beneficial force (lift) for control and propulsion. In the average wing example above, 15° is not that far from 18° and so, from the point of view of design, one needs wing profile that keeps the "best" AOA as far away from the "critical" AOA. Because wing friction drag dominates pressure drag at low AOA, it is advantageous to have a wing over which the boundary layer is largely laminar. Marine mammals achieve this by having flexible flippers and flukes (Fish et al. 2006) and/or integration with structures that further control the flow, exemplified by the tubercles on humpback whale flippers (Miklosovic et al. 2004; Fish and Lauder 2006).

Fluke or flipper (hydrofoil) thrust along the line of the body arises from the vertical motions of these appendages. In combination with the forward movement of the body, this motion effectively changes the direction of the flow relative to the hydrofoil (and corresponding AOA), in a manner to tilt forward the lift force relative to the animal's body, and with enough magnitude to cancel the rearward action of the drag (likewise tilted) (see p. 268 in Vogel 2003). It is here that the flukes' flexibility becomes advantageous, as it puts the AOA closer to the "best" AOA and also further away from the critical regime of stall. Many marine mammals that have flippers use them primarily for maneuvering, that is, to effect turns and rolls. Here, the flippers may be rotated to adjust AOA and the magnitude of lift along the vertical.

1.3 Evolutionary biomechanics of marine mammal locomotion: How different marine mammals achieve different levels of locomotor performance

Thrust for swimming is generated from the interaction of the water with the movement of a propulsive surface (e.g., paddles, flippers, flukes). Propulsors, therefore, are large in span and area to increase the volume of water accelerated by the excursion of the propulsor (Blake 1981; Fish 1993a). The propulsive movements can be classified broadly as drag-based oscillatory, lift-based oscillatory, and undulatory (Webb and Blake 1985). Oscillatory propulsion (both drag- and lift-based) uses the motion of the paired appendages (e.g., feet) or a highly modified lunate tail (e.g., flukes), whereas undulatory propulsion uses the movements of the body and tail.

1.3.1 Mechanics of drag-based paddling

Drag-based propulsion is used by semi-aquatic (e.g., muskrat, beaver, platypus, otter) and fully aquatic mammals (e.g., manatee, humpback whale) for swimming and maneuvering (Howell 1930; Fish 1996). Propulsion by drag-based oscillation is produced by the motion of various combinations of the paired appendages (quadrupedal, pectoral, pelvic) either alternately or simultaneously and oriented in either the parasagittal or horizontal planes (Howell 1930; Fish 1996). The stroke cycle includes the power and recovery phases (Fish 1984; Fish and Baudinette 1999). In the power phase, the posterior sweep of the limb generates a large pressure drag, which provides an anterior thrust.

Maximum thrust is generated with a broad paddle area that is configured as a circle or triangle with a constriction at the attachment point with the body (Fish 2004). The constriction minimizes interference drag with the body and provides a long lever arm to increase the velocity of the paddle during the power stroke. Paddle area is increased by abduction of the digits and by interdigital webbing or fringe hairs (Howell 1930; Fish 1984; Thewissen and Fish 1997). The increased paddle area allows for the production of a high-pressure drag on the paddle as it is swept posteriorly. The reaction force to the pressure drag is the thrust that moves the paddling animal forward. The size of the paddle accelerates a large mass of fluid to a low velocity, which is more efficient than accelerating a small mass of fluid to a high velocity (Fish 2004). The recovery phase repositions the limb, incurring a non-thrust generating drag. To limit the reduction in thrust during the recovery phase, drag on the appendage is reduced by adducting the digits or rotating the appendage to reduce the paddle area and by changing the timing of movement to reduce the relative velocity (Fish 1984).

Drag-based oscillation has a low propulsive efficiency (thrust power/total mechanical power output) of ≤ 0.33 (Fish 1984, 1992). This low efficiency occurs because thrust is generated through only half of the stroke cycle (Fish 1984). Energy is lost to increased resistive drag as the foot is repositioned during the recovery phase. In addition, approximately 40%–50% of the total energy expended through the stroke is lost in acceleration of the mass of the limb and the water entrained to the foot (Fish 2000). Propulsive efficiency for the drag-based oscillation is highest at low speeds (Vogel 1994).

1.3.2 Mechanics of lift-based oscillation

Lift-based oscillation is a high-performance swimming mode (Lighthill 1969; Webb 1975; Fish 1996). Several marine mammal taxa exhibit lift-based swimming modes including cetaceans and sirenians (caudal flukes), otariid seals (pectoral flippers), and phocid seals and walrus (pelvic flippers) (Fish 1993a, 1998a). This swimming mode, also described as thunniform or carangiform with lunate tail (Lighthill 1969; Webb 1975), is similar to propulsion by certain fast swimming fish. Lift-based swimming produces forces up to five times greater than drag-based propulsion (Weihs 1989).

An oscillating hydrofoil (i.e., flukes, flippers) generates lift at a controlled angle of attack (Lighthill 1969; Feldkamp 1987a; Fish 1998a,b; Vogel 2003). The angle of attack should be small to avoid separation of the flow from the hydrofoil surface, which reduces lift and increases hydrodynamic drag. Lift is directed perpendicular to the path traversed by the hydrofoil, so it has a component that can be resolved as an anteriorly directed thrust force (Weihs and Webb 1983; Fish 1993a,b; 1998a,b). Thrust is generated almost continuously throughout a stroke cycle. Although the hydrofoil produces some drag, it is small relative to the lift (high lift-to-drag ratio) (Vogel 2003).

Lift-based oscillation is characterized by high propulsive efficiencies. Propulsive efficiencies for cetaceans are 0.75–0.9 (Fish and Rohr 1999), where maximum efficiencies are achieved within the range of normal cruising speeds (0.8–1.5 body lengths/s). For pinnipeds, oscillation of the fore flippers by *Zalophus* and hind flippers by *Pusa* and *Pagophilus* provide maximum efficiencies of 0.8 and 0.88, respectively (Feldkamp 1987b; Fish et al. 1988).

Propulsive flippers and flukes have a planar geometry with a high aspect ratio, where aspect ratio is defined as the square of the span over the planar area of the hydrofoil (Webb 1975; Feldkamp 1987b; Fish 1993a,b; 1998a,b). A high aspect ratio reduces drag while maximizing thrust. An aspect ratio works in concert with the sweep of the hydrofoil geometry. Sweep is the rearward inclination of the leading edge. A combination of low sweep with a high aspect ratio allows for high-efficiency swimming, whereas high sweep may compensate for the reduced lift production of low aspect ratio hydrofoils (Liu and Bose 1993).

The cross-sectional geometry is similar to symmetrical engineered foil sections with an elongated teardrop design. This design has a rounded leading edge that increases to a maximum thickness at approximately 24%–36% of the chord (i.e., the linear distance from the leading to the trailing edges) (Fish 2004). From the maximum thickness, the hydrofoil slowly tapers to the trailing edge. This configuration prevents stalling due to boundary layer separation as the hydrofoil is oscillated.

The typical planform of cetacean flukes has a sweptback, winglike shape with tapering tips. The aspect ratio varies from 2.0 for the Amazon River dolphin (*Inia geoffrensis*) to 6.1 and 6.2 for the fin whale (*Balaenoptera physalus*) and false killer whale (*Pseudorca crassidens*), respectively (Fish and Rohr 1999). The flukes of fast-swimming cetaceans have higher aspect ratios than slow swimmers. The sweepback in flukes ranges from 4.4° for the killer whale (*Orcinus orca*) to 47.4° for the white-sided dolphin (*Lagenorhynchus acutus*) (Fish and Rohr 1999). The flukes of mature male narwhals (*Monodon monoceros*) have a slightly concave leading edge without a sweepback (Hay and Mansfield 1989).

Pinnipeds utilize foreflippers (Otariidae) and hind flippers (Phocidae, Odobenidae) as propulsive hydrofoils. The foreflippers of *Zalophus* have a high aspect ratio of 7.9 and are flapped in a manner reminiscent of underwater flight (Feldkamp 1987b). The foreflippers generate thrust throughout almost the entire stroke cycle. The stroke cycle is divided into a forceful downstroke of the flippers during a power phase that terminates with a paddling phase before the flippers are lifted in a recovery phase (Feldkamp 1987a). The hind flippers of phocids and walrus are oscillated laterally. The two flippers are alternated with the digits of the trailing flipper are fully abducted to generate thrust as the digits of the leading flipper adducted (folded) (Fish et al. 1988). Aspect ratio for the hind flippers of phocids is 3.4–4.0.

1.3.3 *Undulatory swimming*

In terms of performance (i.e., speed, efficiency), undulatory swimming is intermediate between drag-based oscillation and lift-based oscillation. In undulatory swimming, the body and tail are bent into a wave that travels backward at a velocity faster than the animal is moving forward (Webb 1975). For otters and manatees, the undulatory wave is generated by flexion and extension of the spine. As each section of the body accelerates vertically, the wave faces caudally at an angle to the mean motion of the body. Fluid adjacent to the accelerated section produces a reaction force with a component in the direction of thrust (Lighthill 1971). As the traveling moves along the body, its amplitude increases to a maximum at the tip of the tail, where the velocity of the tip is high. For river otters (*Lontra canadensis*), the tail tapers to a point reducing its effectiveness in thrust generation (Fish 1994). The giant river otter (*Pteroneura brasiliensis*) has a laterally expanded tail that is undulated during simultaneous paddling of the hind feet (Fish 2001). The manatee (*Trichechus* sp.) possesses a broad paddle-like tail, which is undulated in the vertical plane (Kojeszewski and Fish 2007). The propulsive efficiency of the manatee is 0.67–0.81 with the highest value at a swimming speed of 0.4 body lengths/s (Kojeszewski and Fish 2007).

1.3.4 *Transition from drag-based to lift-based locomotion*

The evolution of highly derived aquatic mammals (pinnipeds, cetaceans, and sirenians) represents the culmination of a sequence of transitional stages extending from terrestrial quadrupeds to fully aquatic mammals capable of high-performance propulsion (Figure 1.5) (Howell 1930; Gingerich et al. 1990; Fish 1998a; Thewissen 2014). Semi-aquatic mammals are in an evolutionarily precarious position, being unspecialized for locomotor performance in either terrestrial or aquatic environments. The energetic cost of being semi-aquatic is higher than being adaptated for land or water (Williams 1999). Specialized lift-based swimming modes that use oscillation of flippers or flukes have low minimum costs of transport (i.e., metabolic energy consumed to move a unit mass a unit distance) for aquatic mammals, whereas paddling has the highest minimum costs of transport and undulation is intermediate (Fish 2000).

Selective pressures would have been high in the transition to a more aquatic lifestyle. The evolutionary changes would have been directed by increases in swimming speed, propulsive efficiency, energy economy, and dive time, while constrained by morphology and neuromotor patterns. In recent years, fossil species have been discovered, which have added to our knowledge of the transitional stages that evolved into the highly derived aquatic mammals (Thewissen 2014). However, it has only been through observation and experimentation on living analogs of the transitional forms that the selective pressures have been determined directing the course to the pinnipeds, cetaceans, and sirenians.

As mammals became semi-aquatic, the earliest species, such as the pakicetids, would have used a modified terrestrial, quadrupedal gait to move on along the bottom substrate (e.g., hippopotamus; Coughlin and Fish 2009) or swim in a manner similar to the "dog paddle" (e.g., mink, *Mustela vison*; Williams 1983). The primitive quadrupedal gait would eventually be replaced with bipedal paddling, with alternating motions of either the pectoral or pelvic extremities as exemplified by the polar bear (*Ursus maritimus*) and muskrat (*Ondatra zibethicus*), respectively (Flyger and Townsend 1968; Fish 1984). Otters (*L. canadensis*) are able to swim quadrupedally, but they generally swim with 2 feet (Tarasoff et al. 1972; Fish 1994).

Bipedal paddling avoids mechanical and hydrodynamic interference between the ipsolateral limbs, increasing propulsive efficiency. This mode of swimming frees one set

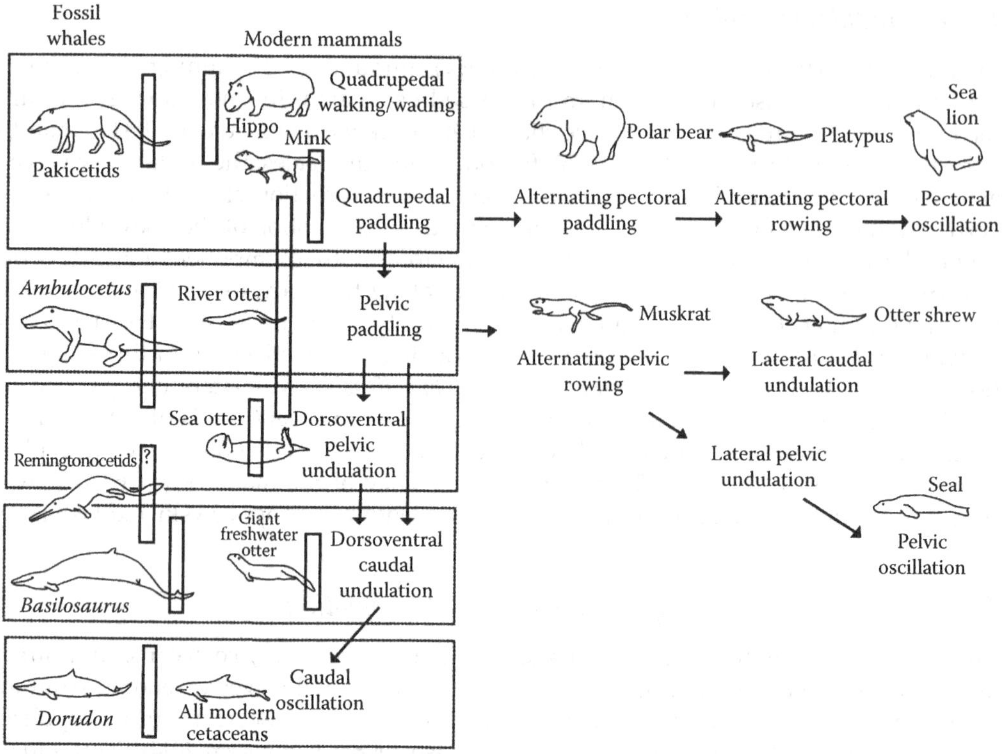

Figure 1.5 The evolution of swimming modes in mammals with examples of extant and extinct aquatic taxa. The transition is shown from quadrupedal walking, through drag-based paddling modes to highly derived lift-based oscillatory swimming. Arrows show how swimming modes are related. The boxes contain fossil and modern species that swim similarly. The bars are indicative that some of the species employ more than one swimming mode. (Reprinted from Thewissen, J.G.M., *The Walking Whales*, University of California Press, Berkeley, CA, 2014, Figure 20. With permission.)

of limbs for stabilization, maneuvering, tactile reception, or food capture and processing (Dagg and Windsor 1972; Fish 2001). With the development of pectoral or pelvic paddling, specialization of the limbs to increase thrust and improve stability would be selected for as a greater commitment to life in water. These specializations include (1) the development of a short, robust humerus and femur, (2) the elongation of digits, (3) an increase in the propulsive surface by the addition of interdigital webbing or fringe hairs, and (4) an increase in the bone density for buoyancy control (Fish 1996).

Paddling modes work well for surface swimming, where the orientation of the limbs helps to maintain stability (Fish and Stein 1991) and their propulsive forces are applied under the body and away from the water's surface. Buoyancy provided by lungs and a non-wettable fur keeps the body in trim and at the surface. However, swimming at the surface is energetically demanding. Wave drag (see Section 1.2.2 and Figure 1.2) occurs when an animal swims at or near the surface (Hertel 1966). Kinetic energy from the animal's motion is transformed into potential energy in the upward displacement of water, forming surface waves, and also lost due to friction and pressure drag. This energy loss can be substantial, up to a maximum of five times the frictional drag (Hertel 1966). The cost of transport in surface paddling is 69% greater than that associated with submerged swimming (Williams 1989). Furthermore, the wave patterns produced at the bow and stern of the surface swimmer constructively interfere and limit maximum swimming speed by effectively trapping the animal between two wave crests (Fish 1982; Fish and Baudinette 1997).

In submerged swimming for underwater foraging, wave drag is negated and swimming speed can be higher than surface swimming (Williams 1999). The platypus (*Ornithorhynchus anatinus*) and muskrat swim underwater by reorienting the limbs laterally and using a rowing motion (Fish 1996, 2000). Rowing can generate thrust, while simultaneously exerting a downward force to counter buoyancy. A further transition from rowing to pectoral oscillation requires only elongation of the forelimb, with a power phase of the stroke directed posteriorly and ventrally. Employing simultaneous strokes of the winglike forelimbs could lead to the lift-based pectoral oscillatory stroke of the sea lion *Zalophus* (Feldkamp 1987a,b; Fish 2000). Alternatively, rowing of the pelvic limbs could have lead to other pinniped swimming mode lateral pelvic oscillation displayed by phocids and *Odobenus*. By control of the angle of attack, the hind flippers can act as a hydrofoil for lift-based propulsion (Fish et al. 1988). The lumbar vertebrae in an early fossil pinniped, *Enaliarctos mealsi*, allowed swimming by lateral movements of the trunk along with propulsion by the limbs (Berta et al. 1989).

Simultaneous pelvic paddling is another swimming mode used for submerged swimming by otters (Fish et al. 1997). This mode transitioned to undulatory swimming as the body and tail are thrown into waves due to movement of pelvis (Fish 1996). The combination of limbed and axial propulsion would allow the powerful back muscles to aid in powering strokes of the hindlimbs. The undulations of the tail contributes thrust production and helps to compensate for the periodic increased drag as both hind limbs are in the recovery phase. Although this mode of swimming is observed in modern otters (Tarasoff et al. 1972; Fish 1994), the fossil quadrupedal cetaceans *Ambulocetus* and *Rhodhocetus*, and the quadrupedal sirenian *Pezosiren* might have used this transitional mode (Gingerich et al. 1990; Thewissen and Fish 1997; Domning 2001).

Further increases in performance would occur in adoption of fully undulatory swimming in conjunction with abandonment of limbs and distal expansion of the tail (Fish 1996). A reduction in the limbs would reduce the drag on the animal. The increased surface area and large amplitude of oscillation at the tip of the tail work against more water, increasing total momentum, as traveling waves move along the body with increasing velocities (Webb 1975). An oscillatory mechanism is made possible by progressively restricting the propulsive wave in the tail and establishing a controlled angle of attack at a pivot point toward the tail tip. This leads to the evolution of a high-efficiency caudal hydrofoil, with a lift-based oscillatory motion like that employed by cetaceans and sirenians (Fish 1996). With the change from surface to subsurface and drag-based to lift-based swimming in the evolution of a fully aquatic lifestyle, buoyancy control could be improved by abandoning fur for blubber.

1.3.5 *Surface versus submerged and buoyancy control*

The evolution from a terrestrial to fully aquatic lifestyle in mammals required the development of buoyancy control mechanisms for stabilization in water (Stein 1989; Fish and Stein 1991). Buoyancy control has major implications on locomotor energetics with respect to the ability to float at the water surface and dive and surface easily (Johansen 1962). Such activities are associated with foraging and escaping predation. For semi-aquatic and aquatic mammals, a distinct division exists in use of fur and blubber for buoyancy control. This division is associated also with the insulatory capacity of fur and blubber in the different environments in which they operate.

Semi-aquatic mammals operate primarily at the water surface. Possession of large, low-density air-filled spaces provides positive buoyancy and decreases the effort needed to float. Sea otters maintain buoyancy in part with lungs that are two times larger than that of other similarly sized mammals (Kooyman 1973; Tarasoff and Kooyman 1973). The non-wettable fur of the sea otter is extremely dense (1008–1573 hairs/mm^2) entrapping a large volume of air for buoyancy (Kenyon 1969; Tarasoff 1974; Williams et al. 1992a). The density of hairs in the pelage of semi-aquatic mammals is high in comparison with terrestrial and aquatic mammals (Tarasoff 1974; Sokolov 1982; Fish et al. 2002). However, the amount of buoyancy afforded by fur becomes proportionately smaller as the body size increases (Fish et al. 2002).

Although the positive buoyancy afforded by fur will reduce energy costs at the surface, buoyancy is a major determinant of locomotor costs when diving (Lovvorn and Jones 1991). To submerge, a positively buoyant animal can expend at least 95% of its total mechanical energy to work against the buoyant force (Stephenson et al. 1989). The buoyancy of a diving sea otter can be 20 times greater than the drag on the body (Lovvorn and Jones 1991). At depth, fur has serious limitations. The air in the fur will compress reducing buoyancy at depth and reduce the insulation value of the fur. Reduction of insulation performance is also attributed to water infiltration. Furthermore, maintenance of the air layer requires a large fraction of the daily energy budget devoted to grooming (Kenyon 1969; Williams 1989). However, specialized fur structures on some areas of the body may enhance insulation as well as drag reduction (Erdsack et al. 2015).

Blubber offers a structure for aquatic mammals, which requires lower maintenance costs than fur, is not prone to fouling, is an effective thermal barrier in water, can be used as an energy reserve, facilitates streamlining, has spring-like properties to reduce locomotor effort, and provides buoyancy (Lang 1966; Kooyman 1973; Brodie 1975; Pabst 1996). Buoyancy from blubber is not depth-sensitive. The lipid composition of blubber makes it essentially incompressible relative to air, but not as buoyant (Lovvorn and Jones 1991). Without large lungs (Kooyman 1973; Tarasoff and Kooyman 1973), aquatic mammals offset the high density of the body tissues (i.e., bone, muscle) with a thick layer of blubber. Approximately 20%–30% of the total mass of marine mammals is blubber (Kooyman 1973).

The distribution of buoyancy is associated with the maintenance of trim (i.e., fore and aft angle of a vessel). Having longitudinal trim provides streamlining to reduce drag. For sea otters, the elongate shape of the lungs can keep the body in trim on the water surface. The diaphragm is oriented diagonally in pinnipeds and longitudinally in the mantee to extend the lungs and their buoyancy of a large portion of the body to maintain trim. Mysticete whales of the family Balaenopteridae feed by engulfing large volumes of water and prey in a distended throat pouch (Pivorunas 1979; Shadwick et al. 2013). This engulfment would produce increased drag and a torque that would pitch the whale downward when the mouth is opened. The flippers of the minke whale (*Balaenoptera acutorostrata*) are canted at an angle to produce a lift that would produce an opposing torque to trim the body during the feeding maneuver (Cooper et al. 2008).

1.3.6 Maneuverability

The ability to maneuver (i.e., turn) with speed becomes imperative in the acquisition and capture of prey. Maneuverability is defined as the space required to execute a turn, and agility is defined as the rapidity that direction can be changed and is measured as the rate of turn (Norberg 1990; Walker 2000). Small animals have an advantage with respect to maneuverability and agility compared to large animals because turn radius increases directly with body mass (Howland 1974; Weihs and Webb 1984). Escape by small prey animals is possible as they are able to turn in smaller radii and with higher angular velocities than the larger predators, such as marine mammals (Domenici 2001). However, marine mammals hold an advantage in that their absolute swimming speed is substantially greater than the speed of the prey.

Marine mammals exhibit divergent body designs that suggest differences in performance regarding stability and maneuverability. A body design adapted for stability when swimming would aid in minimizing energy expenditure and increase propulsive efficiency. In addition, a stable body design would reduce transverse movements of the body that could interfere with effective use of sensory systems. To understand how variation in the morphology of animals can affect maneuverability, consideration should be given to parameters associated with stability. Morphological characters that deviate from those of a stable design (i.e., like an arrow) are expected to enhance maneuvering performance is dependent on the location and design of control surfaces (i.e., flippers, fins, flukes) relative to the center of gravity (CG), and rigidity of the body. Most analyses to date have been performed on cetaceans and sea lions (Fish 2002; Fish et al. 2003) with comparisons to a self-stabilizing design like an arrow (Harris 1936; Wegener 1991; Fish 2002). The arrow has a rigid body, the CG is positioned anterior on the body, and control surfaces exhibiting sweep and dihedral that are located far posterior of CG.

1.3.7 *Maneuverability in cetaceans*

The placement and design of control surfaces of cetaceans indicates a relatively stable (e.g., like an arrow) configuration (Fish 2002), although there are marked differences between species. For example, the CG for the common dolphin *Tursiops truncatus* is located at a position of 41% of body length (Fish 2002). Although this position appears to enhance stability, it also appears to be nearly coincident with the center of buoyancy (Slijper 1979; Weihs 1993; Fish 2002). As a result, delphinids can be unstable with respect to roll and can side-swim, swim upside down, and barrel-roll (Layne and Caldwell 1964; Klima et al. 1987). However, the dorsal fin, when present, is located approximately over the center of gravity and is immobile (Fish and Rohr 1999). This position limits the dorsal fin's effectiveness in developing a turning moment but allows the fin to prevent side-slip and oppose rolling. Alternatively, the beluga whale, *Delphinapterus leucas*, which lacks a dorsal fin, rolls during turns (Fish 2002).

The mobile control surfaces of cetaceans are located at a distance from the CG and provide the major percentage of area for control (Slijper 1961; Aleyev 1977; Edel and Winn 1978; Fish and Battle 1995). The mobility of flippers in dolphins capable of rapid sprints and fast cruising appears to be more constrained when compared to the flippers of slow-swimming, highly maneuverable animals (Howell 1930; Pilleri et al. 1976; Klima et al. 1987). For example, the shoulder musculature of *I. geoffrensis* is highly differentiated in contrast to the faster swimming *Lagenorhynchus albirostris, Phocoena phocoena,* and *T. truncatus* (Klima et al. 1987).

Flexibility in the body of cetaceans is generally constrained (Long et al. 1997). The highly compressed cervical vertebrae and streamlined body form restrict bending in the anterior region of the body, although some species have un-fused cervical vertebrae that facilitate flexion of the neck (Ridgway and Harrison 1985; Narita and Kuratani 2005). Turns are initiated anteriorly with lateral flexion of the head, and adduction and rotation of the flippers into the turn (Fish 1997). When not actively fluking, there is substantial lateral flexion of the peduncle in concert with twisting at the base of the flukes (Fish 2002). Such unpowered turns for cetaceans have smaller minimum radii than turns in which the animal is actively swimming (Fish 2002). The increased flexibility of the body during unpowered turns in conjunction with mobility of the flippers and twisting of the flukes permits small turn radii. Minimum radii for unpowered turns by cetaceans were reported to range from 0.10 to 0.15 body lengths (Fish 2002). When scaled to the body length, cetaceans generally demonstrate unpowered turning radii of <50% of the body length with minimum radii ranging from 11% to 17% of body length. The extremely flexible body and mobile flippers of the river dolphin *I. geoffrensis* enables some of the smallest radius turns (Fish 2002). Both *Delphinapterus* and *Inia* inhabit structurally complex habitats (i.e., pack ice, flooded forest), where increased flexibility for enhanced maneuverability is necessary.

Differences in turning performance between species are associated with swimming speed, size, and habitat (Fish 2002). *Inia* and *Delphinapterus* produce low-speed, small radius turns. Faster speed but larger radius turns are performed by pelagic delphinids. Most turning maneuvers by cetaceans are performed at <200°/s and <1.5 *g*, although turns of 453.3°/s and 3.6 *g* have been measured in fast-swimming *Lagenorhynchus obliquidens* (Fish 2002). Humpback whales, *Megaptera novaeangliae*, use long, mobile flippers to effect highly aquabatic maneuvers (Edel and Winn 1978; Fish and Battle 1995; Fish et al. 2011). These flippers have an aspect ratio of 6.1 with a streamlined cross-sectional profile similar to engineered foil sections (Fish and Battle 1995). The flippers generate lift that produces a centripetal force for banked turns (Howland 1974; Weihs 1981; Fish and Battle 1995). Therefore, more lift is required to produce a tighter turn. Enhanced lift production is achieved by increasing the angle of attack of the winglike flippers. However, at too high an angle of attack, the flipper could stall and lose lift by instead generating too much drag.

The rounded tubercles along the leading edge of humpback whale flippers delay stall to higher angles of attack by modifying the flow over the flippers surface. As the water flow impacts the leading edge in the troughs between two adjacent tubercles, it was deflected into the center of the trough producing a pair of vortices with opposite spins. Each vortex that is immediately flanking the flow over the tubercle has a spin, which is in the same direction as the flow. Sandwiched between two vortices, the flow over the tubercle is energized and accelerated to avoid separation from the wing surface and prevent stall (Fish et al. 2011). Humpback whales use this advanced hydrodynamic feature of their flippers during bubble-net feeding maneuvers, which consist of underwater exhalations from the blowhole produce bubble clouds or columns, which completely encircle and concentrate the prey in a spiral (Hain et al. 1982; Wiley et al. 2011).

1.3.8 Maneuverability in pinnipeds

The placement of the flippers of sea lions (*Zalophus californicus*) is dynamically unstable. The roots of the large pectoral flippers are located near the center of gravity. The flippers provide little rotational dampening about the yaw and pitch axes, although they could retard rotational and translational motion in regard to roll and heave, respectively. The smaller pelvic flippers are in the preferred location to develop sufficient torque to act like an airplane stabilizer or ship rudder and resist rotational instabilities.

The attitude of the *Zalophus* flippers are highly variable, because of the high mobility of the pectoral and pelvic flippers (English 1976; Godfrey 1985). The ability of the sea lion to adduct the pectoral flippers against the body and also adduct the pelvic flippers can effectively produce a condition where the animal is devoid of control surfaces and potentially susceptible to all instabilities. The mobility of the pectoral and pelvic flippers also permits dynamic production of lift that can induce torques around CG to promote instabilities. The location of the pectoral flippers close to CG would not produce large torques and would be less effective in rapidly inducing turns. However, the pectoral flippers are used for propulsion (Feldkamp 1987a,b), and propulsors arranged around CG are postulated to promote maneuverability (Webb et al. 1996). The highly flexible body of *Zalophus* also enhances maneuverability. Bending of the body and neck is an integral component of turning in conjunction with the flippers of pinnipeds (Godfrey 1985; Fish et al. 2011). Dorsal bending of the spine allows the body to curve smoothly, maintaining a streamlined appearance throughout the turn.

Zalophus generally can turn in small radii and at faster rates than cetaceans of similar size. Minimum turn radii are 0.09–0.16 body lengths (Fish 2002; Fish et al. 2011). The maximum turning rate of *Zalophus* is 690°/s and the maximum centripetal acceleration is 5.13 *g*. Such performance is superior to turning rates for cetaceans.

1.3.9 Energy capture from the external environment

As the energy cost of swimming can be large in the marine environment, marine mammals have adopted a variety of behavioral, morphological, and physiological mechanisms to swim economically. Such mechanisms rely upon energy management by capturing energy from external and internal sources. External sources of energy can be utilized from the prevailing physics of the environment (i.e., gravity, hydrodynamics, waves). In contrast, kinetic energy from muscular contraction that would typically be lost can be recycled internally from the elastic properties of the connective tissues of the body (Pabst 1996). In both cases, the available energy to perform the work of swimming is augmented or conserved to lower metabolic power consumption, increase dive time, and increase speed.

The capture of external forms of energy to add to the total energy budget for movement by marine mammals is known as *free-riding*. The simplest type of free-riding behavior is gliding while diving. The density of the body increases with depth due to increased hydrostatic pressure. When diving deeply (>20 m), lung collapse reduces the net buoyant force causing the animal to sink (Ridgway et al. 1969; Ridgway and Howard 1979; Kooyman and Ponganis 1998; Moore et al. 2011). The animal can glide deeper as gravity now supplies the motile force. The gliding configuration of the body minimizes drag and reduces the metabolic cost of swimming. Whales, dolphins, and seals intermittently switch between active swimming and gliding depending on the dive depth and the net buoyancy of the body (Williams et al. 2000; Williams 2001). Exhalation before diving by pinnipeds has been considered a mechanism to prevent decompression sickness, however, this behavior may effectively reduce buoyancy to decrease the energy cost of swimming during the initial descent (Kooyman 1973). During deep dives, dolphins can reduce energy costs by approximately 20% when transiting to the bottom by using intermittent swimming behaviors (Williams et al. 1996; Williams 2001). During ascent, the reverse occurs and the animal accelerates by actively swimming until its lungs re-inflate sufficiently to provide positive buoyancy (Skrovan et al. 1999).

The occurrence of highly organized formations by cetaceans has been suggested as an adaptation for energy economy (Kelly 1959). Formation swimmers are able to capture energy from the vortex patterns in the wakes of conspecifics and decrease drag with a concomitant decrease in overall energy cost of locomotion (Weihs 1973). In addition, when two bodies are in close proximity, the water flow between them is accelerated resulting in an attractive force due to the Bernoulli effect (Kelly 1959; Fish et al. 2013).

Groups of whales and dolphins swim in side-by-side and echelon formations to draft (Weihs 2004; Fish et al. 2013). Small dolphins often position themselves beside and slightly behind the maximum diameter of a larger animal (Tavolga and Essapian 1957; Norris and Prescott 1961; Reid et al. 1995; Marino and Stowe 1997; Noren 2008). While the larger dolphin will experience increased drag, the smaller gains an energetic benefit (Weihs 2004; Noren 2008; Noren et al. 2008). This effect is beneficial particularly for young whales in order to maintain speed with their mothers. A neonatal dolphin could use this mechanism to gain up to 90% of the thrust needed to move alongside its mother.

Dolphins are able to reduce energy costs by riding the bow waves generated by large whales and boats (Fish and Hui 1991; Williams et al. 1992b). Williams et al. (1992b) found that wave-riding dolphins could swim at a higher speed while reducing or maintaining metabolic rate, heart rate, lactate production, and respiratory rate. Dolphins can either ride bow waves like a surfer (Caldwell and Fields 1959), or they make use of the pressure front created by the boat (Scholander 1959; Fejer and Backus 1960). This behavior is complex with any energy savings to the dolphin related to bow design, swimming depth, body orientation, and distance from the ship (Fish and Hui 1991).

Wind-wave riding and surf-wave riding can use gravity to reduce the energy cost of swimming (Caldwell and Fields 1959). These wave riding behaviors differ from bow-wave riding because they use the interaction of the dolphin's weight and slope of the wave front to produce movement analogous to human surfers (Hayes 1953; Fejer and Backus 1960; Perry et al. 1961). Dolphins and sea lions have been observed to surf on inshore waves (Norris and Prescott 1961; Riedman 1990). Dolphins ride waves with a forward slope of 10°–18° at velocities of 5–6 m/s (Hertel 1969). In the open sea, the flukes of large whales can absorb energy from ocean waves (Bose and Lien 1990). Whales absorb 25% of their propulsive power from head seas and 33% from following seas, by synchronizing the motion of the wave with the motion of the flukes. This energetic advantage is not generally available to dolphins, due to the difference in length between the animal and the wavelength of oceanic waves (Curren 1992).

1.4 *Tools and methods for hydrodynamics research*

There are a variety of tools and methodologies available to researchers to study marine mammal hydrodynamics that include both theoretical and empirical approaches. As we have shown, body shape and mechanical design are critically important aspects that influence hydrodynamic performance. Traditional techniques for quantifying morphology include direct measurement and photogrammetry (Bose et al. 1990; Fish and Battle 1995; Ginter et al. 2012). Surface and internal morphology can be quantified using advanced bioimaging techniques such as computed tomography (CT) (Fish et al. 2006, 2007; Weber et al. 2009a,b, 2014).

How bodies and appendages move relative to flow, both actively and passively, is also a fundamental aspect of hydrodynamic research (Fish and Lauder 2006). The kinematics of bodies and appendages can be quantified using high-speed video cameras (Feldkamp 1987a; Fish 1993). Multiple cameras at orthogonal axes can provide movement data in three dimensions (Friedman and Leftwich 2014), although the size and large-scale trajectories of most marine mammal movements may often preclude this approach. However, the recent advent of animal-borne tags equipped with movement sensors can provide some information on the kinematics of swimming and maneuvering in free-ranging animals (Goldbogen et al. 2006). Common sensors within contemporary tags include tri-axial accelerometers, magnetometers, and more recently, gyroscopes. Data from these tags can be used to determine swimming gaits, speed, and body orientation dynamics (Miller et al. 2004; Watanabe et al. 2011; Goldbogen et al. 2013). Although animal-borne tag data rarely inform hydrodynamic phenomena directly, they are important tools for assessing a variety swimming metrics that indirectly reflect hydrodynamic performance and its greater physiological relevance (Potvin et al. 2009, 2012; Sato et al. 2011; Adachi et al. 2014).

Arguably the most difficult aspect of hydrodynamic research on marine mammals is the quantification of the flows and the forces they generate. Recently, the flows around the fluke of an actively swimming dolphin have been quantified using digital particle image velocimetry (DPIV) to measure propulsive forces (Fish et al. 2014). DPIV is an experimental technique that enables the quantification of flow velocities and momentum changes through the tracking of individual particles suspended in the fluid (Fish and Lauder 2006). Such an approach is a major step forward for visualizing flows and understanding marine mammal hydrodynamics, which in the past has been limited to opportunistic observations with bioluminescent organisms suspended in the water (Rohr et al. 1998). When experimental approaches are not available, computer modeling (or simulation) of the flows about a swimming animal, a technique known as *computational fluid dynamics* (CFD), can provide important insights into how morphology can influence hydrodynamic performance (Fish et al. 2008; Weber et al. 2009a).

CFD approaches exploit the capacity of modern computer processing power to calculate the motions of fluid particles down to scales of approximately one hundredth of a millimeter, over body lengths spanning those up to the largest whales. This computational technique enables the visualization of fluid structures that cannot be resolved in the laboratory due to their extreme scale or ephemeral nature. An important example is the boundary layer found very close to the body, or the trail of vortices produced by the sweeping motions of a fluke. One can also simulate, although imperfectly, the turbulence trailing the body. These data allow the estimation of the force applied on and resulting motions by a swimmer. Tag design has also involved the use of CFD, to identify the flow characteristics moving past the sensors and the added drag associated with its attachment (Hazekamp et al. 2010; Shorter et al. 2014). Such data can help inform tag placement-dependent hydrodynamic effects on the energetic cost of swimming (Pavlov et al. 2007; Pavlov and Rashad 2012; van der Hoop et al. 2014).

1.5 Lingering mysteries and future challenges

Due to the logistical difficulties associated with research on marine mammals, there remain extensive challenges and lingering mysteries related to hydrodynamic performance. Despite the advent of animal-borne tags and significant advances in comparative scaling analyses across taxa (Sato et al. 2007; Watanabe et al. 2011), there is still a dearth of information on the maneuvering performance envelope in the wild. Moreover, at the largest scale it is difficult to obtain key data such as the precise length and mass of the tagged animal. However, the emerging use of aerial vehicles equipped with cameras should make this possible in the future (Koski et al. 2013). Body size data are essential for understanding both the biomechanics and biology of swimmers, as well as the scaling of drag, swimming speeds, and ultimately, of the metabolic costs of foraging and living in water (Potvin et al. 2012).

Hydrodynamical modeling is also facing new challenges. In particular, modeling approaches require more realistic body shapes, especially in the case of the very large whales that cannot be photographed from all angles in the laboratory or in the field. Again, the use of drones, together with image and fluid-lensing analysis that remove the distortion optics of water could bring quantum leaps of improvement (Chirayath et al. 2015).

Another challenge relates to the flow simulations of the boundary layer of animals covered with fur or with complex skin structures such as dermal ridges (Erdsack et al. 2015), although CFD could be used here to provide some insight (Oeffner and Lauder 2012). However, CFD is not an exact simulation of the fluid at play because of its approximations of small-scale turbulence, which is difficult to describe computationally. An alternative could be a technique known as *direct numerical simulations* (DNS), which in principle could allow a more exact treatment of the turbulence happening in-between hair strands or even within dermal ridges, down to the smallest turbulence scale possible, the so-called *Kolmogorov scale*. Finally, the *Holy Grail* of marine mammal hydrodynamics would involve using fluid–structure interactions (FSIs) for computer simulations of cetacean swimming. The primary challenge of this approach is the difficulty of characterizing all relevant input parameters despite the physics being well understood.

Acknowledgment

We acknowledge Deborah J. Albert for illustrating Figures 1.1, 1.2, 1.3, and 1.4.

References

Adachi, T., J.L. Maresh, P.W. Robinson et al. 2014. The foraging benefits of being fat in a highly migratory marine mammal. *Proceedings of the Royal Society B—Biological Sciences* 281(1797):20142120.

Ahlborn, B.K. 2004. *Zoological Physics*, 2nd edn. Berlin, Germany: Springer-Verlag.

Ahlborn, B.K., R.W. Blake, and K.H.S. Chan. 2009. Optimal fineness ratio for minimum drag in large whales. *Canadian Journal of Zoology—Revue Canadienne De Zoologie* 87(2):124–131.

Alex Shorter, K., M.M. Murray, M. Johnson, M. Moore, and L.E. Howle. 2014. Drag of suction cup tags on swimming animals: Modeling and measurement. *Marine Mammal Science* 30(2):726–746.

Aleyev, Y.G. 1977. *Nekton*. The Hague, the Netherlands: Junk.

Berta, A., C.E. Ray, and A.R. Wyss. 1989. Skeleton of the oldest known pinniped *Enaliarctos mealsi*. *Science* 244:60–62.

Blake, R.W. 1981. Influence of pectoral fin shape on thrust and drag in labriform locomotion. *Journal of Zoology* 194:53–66.

Bloor, D. 2011. *The Enigma of the Aerofoil, Rival Theories in Aerodynamics, 1909–1930*. Chicago, IL: University of Chicago Press.

Bose, N. and J. Lien. 1990. Energy absorption from ocean waves: A free ride for cetaceans. *Proceedings of the Royal Society B—Biological Sciences* 240:591–605.

Bose, N., J. Lien, and J. Ahia. 1990. Measurements of the bodies and flukes of several cetacean species. *Proceedings of the Royal Society of London Series B—Biological Sciences* 242(1305):163–173.

Brodie, P.F. 1975. Cetacean energetics, an overview of intraspecific size variation. *Ecology* 56(1):152–161.

Caldwell, D.K. and H.M. Fields. 1959. Surf-riding by Atlantic bottle-nosed dolphins. *Journal of Mammalogy* 40:454–455.

Chirayath, V., O. Galvan-Lopez, and R. Instrella. 2015. Blind wave field characterization from fluid lensing. https://stacks.stanford.edu/file/druid:cg133bt2261/Chirayath_Lopez_Instrella_Blind_Wave_Field_Characterization_from_Fluid_Lensing.pdf.

Cooper, L.N., N. Sedano, S. Johansson et al. 2008. Hydrodynamic performance of the minke whale (*Balaenoptera acutorostrata*) flipper. *Journal of Experimental Biology* 211(12):1859–1867.

Coughlin, B.L. and F.E. Fish. 2009. Underwater locomotion of the hippopotamus: Reduced gravity movements for a massive mammal. *Journal of Mammalogy* 90:675–679.

Curren, K.C. 1992. Designs for swimming: Morphometrics and swimming dynamics of several cetacean species. Memorial University of Newfoundland, St. John's, Newfoundland, Canada.

Dagg, A.I. and D.E. Windsor. 1972. Swimming in northern terrestrial mammals. *Canadian Journal of Zoology* 50:117–130.

Daniel, T.L. 1984. Unsteady aspects of aquatic locomotion. *American Zoologist* 24:121–134.

Domenici, P. 2001. The scaling of locomotor performance in predator–prey encounters: From fish to killer whales. *Comparative Biochemistry and Physiology A—Molecular and Integrative Physiology* 131(1):169–182.
Domning, D.P. 2001. The earliest known fully quadrupedal sirenian. *Nature* 413:625–627.
Edel, R.K. and H.E. Winn. 1978. Observations on underwater locomotion and flipper movement of the humpback whale *Megaptera novaeangliae*. *Marine Biology* 48:279–287.
English, A.W. 1976. Limb movements and locomotor function in the California sea lion (*Zalophus californianus*). *Journal of Zoology* 178:341–364.
Erdsack, N., G. Dehnhardt, M. Witt, A. Wree, U. Siebert, and W. Hanke. 2015. Unique fur and skin structure in harbour seals (*Phoca vitulina*)—Thermal insulation, drag reduction, or both? 12.
Fejer, A.A. and R.H. Backus. 1960. Porpoises and the bow-riding of ships under way. *Nature* 188:700–703.
Feldkamp, S.D. 1987a. Foreflipper propulsion in the California sea lion, *Zalophus californianus*. *Journal of Zoology* 212:43–57.
Feldkamp, S.D. 1987b. Swimming in the California sea lion: Morphometrics, drag and energetics. *Journal of Experimental Biology* 131:117–135.
Fish, F.E. 1982. Aerobic energetics of surface swimming in the muskrat (*Ondatra zibethicus*). *Physiological Zoology* 55:180–189.
Fish, F.E. 1984. Mechanics, power output, and efficiency of the swimming muskrat (*Ondatra zibethicus*). *Journal of Experimental Biology* 110:183–201.
Fish, F.E. 1992. Aquatic locomotion. In *Mammalian Energetics: Interdisciplinary Views of Metabolism and Reproduction*, eds. T. Tomasi and T. Horton. Ithaca, NY: Cornell University Press.
Fish, F.E. 1993a. Influence of hydrodynamic design and propulsive mode on mammalian swimming energetics. *Australian Journal of Zoology* 42:79–101.
Fish, F.E. 1993b. Power output and propuslive efficiency of swimming bottlenose dolphins (*Tursiops truncatus*). *Journal of Experimental Biology* 185:179–193.
Fish, F.E. 1994. Association of propulsive mode with behavior by swimming river otters (*Lutra canadensis*). *Journal of Mammalogy* 75(4):989–997.
Fish, F.E. 1996. Transitions from drag-based to lift-based propulsion in mammalian aquatic swimming. *American Zoologist* 36(5):628–641.
Fish, F.E. 1998a. Biomechanical perspective on the origin of cetacean flukes. In *The Emergence of Whales: Evolutionary Patterns in the Origin of Cetacea*, ed. J.G.M. Thewissen. New York: Plenum Press.
Fish, F.E. 1998b. Comparative kinematics and hydrodynamics of odontocete cetaceans: Morphological and ecological correlates with swimming performance. *Journal of Experimental Biology* 201(20):2867–2877.
Fish, F.E. 2000. Biomechanics and energetics in aquatic and semiaquatic mammals: Platypus to whale. *Physiological and Biochemical Zoology* 73(6):683–698.
Fish, F.E. 2001. Mechanism for evolutionary transition in swimming mode by mammals. In *Secondary Adaptation of Tetrapods to Life in Water*, eds. J.M. Mazin, P. Vignaud, and V. de Buffrénil. München, Germany: Verlag Dr. Friedrich Pfeil.
Fish, F.E. 2002. Balancing requirements for stability and maneuverability in cetaceans. *Integrative and Comparative Biology* 42(1):85–93.
Fish, F.E. 2004. Structure and mechanics of nonpiscine control surfaces. *IEEE Journal of Oceanic Engineering* 29(3):605–621.
Fish, F.E. and J.M. Battle. 1995. Hydrodynamic design of the humpback whale flipper. *Journal of Morphology* 225(1):51–60.
Fish, F.E. and R.V. Baudinette. 1999. Energetics of locomotion by the Australian water rat (*Hydromys chrysogaster*): Comparison of swimming and running in a semiaquatic mammal. *Journal of Experimental Biology* 202(4):353–363.
Fish, F.E., R.V. Baudinette, P. Frappell, and M. Sarre. 1997. Energetics of swimming by the platypus (*Ornithorhynchus anatinus*): Metabolic effort associated with rowing. *Journal of Experimental Biology* 200:2647–2652.
Fish, F.E., J.T. Beneski, and D.R. Ketten. 2007. Examination of the three-dimensional geometry of cetacean flukes using computed tomography scans: Hydrodynamic implications. *Anatomical Record—Advances in Integrative Anatomy and Evolutionary Biology* 290(6):614–623.

Fish, F.E., K.T. Goetz, D.J. Rugh, and L.V. Brattstrom. 2013. Hydrodynamic patterns associated with echelon formation swimming by feeding bowhead whales (*Balaena mysticetus*). *Marine Mammal Science* 29(4):E498–E507.
Fish, F.E., L.E. Howle, and M.M. Murray. 2008. Hydrodynamic flow control in marine mammals. *Integrative and Comparative Biology* 48(6):788–800.
Fish, F.E. and C.A. Hui. 1991. Dolphin swimming: A review. *Mammal Review* 21:181–196.
Fish, F.E., J. Hurley, and D.P. Costa. 2003. Maneuverability by the sea lion *Zalophus californianus*: Turning performance of an unstable body design. *Journal of Experimental Biology* 206(4):667–674.
Fish, F.E., S. Innes, and K. Ronald. 1988. Kinematics and estimated thrust production of swimming harp and ringed seals. *Journal of Experimental Biology* 137(1):157–173.
Fish, F.E. and G.V. Lauder. 2006. Passive and active flow control by swimming fishes and mammals. *Annual Review of Fluid Mechanics* 38:193–224.
Fish, F.E., P. Legac, T.M. Williams, and T. Wei. 2014. Measurement of hydrodynamic force generation by swimming dolphins using bubble DPIV. *Journal of Experimental Biology* 217(2):252–260.
Fish, F.E., M.K. Nusbaum, J.T. Beneski, and D.R. Ketten. 2006. Passive cambering and flexible propulsors: Cetacean flukes. *Bioinspiration & Biomimetics* 1(4):S42–S48.
Fish, F.E. and J.J. Rohr. 1999. Review of dolphin hydrodynamics and swimming performance. Technical report 1801, SPAWARS System Center, San Diego, CA.
Fish, F.E., J. Smelstoys, R.V. Baudinette, and P.S. Reynolds. 2002. Fur doesn't fly, it floats: Buoyancy of pelage in semi-aquatic mammals. *Aquatic Mammals* 28(2):103–112.
Fish, F.E. and B.R. Stein. 1991. Functional correlates of differences in bone density among terrestrial and aquatic genera in the family Mustelidae (Mammalia). *Zoomorphology* 110(6):339–345.
Fish, F.E., P.W. Weber, M.M. Murray, and L.E. Howle. 2011. The humpback whale's flipper: Application of bio-inspired tubercle technology. *Integrative and Comparative Biology* 51:203–213.
Flyger, V. and M.R. Townsend. 1968. The migration of polar bears. *Scientific American* 218:108–116.
Friedman, C. and M.C. Leftwich. 2014. The kinematics of the California sea lion foreflipper during forward swimming. *Bioinspiration & Biomimetics* 9(4):046010.
Gingerich, P.D., B.H. Smith, and E.L. Simons. 1990. Hind limbs of Eocene Basilosaurus: Evidence of feet in whales. *Science* 249(13):154–157.
Ginter, C.C., T.J. DeWitt, F.E. Fish, and C.D. Marshall. 2012. Fused traditional and geometric morphometrics demonstrate pinniped whisker diversity. *PLoS One* 7(4):e34481.
Godfrey, S.J. 1985. Additional observations of subaqueous locomotion in the California sea lion. *Aquatic Mammals* 11(2):53–57.
Goldbogen, J.A., J. Calambokidis, A.S. Friedlaender et al. 2013. Underwater acrobatics by the world's largest predator: 360 degrees rolling manoeuvres by lunge-feeding blue whales. *Biology Letters* 9:20120986.
Goldbogen, J.A., J. Calambokidis, R.E. Shadwick, E.M. Oleson, M.A. McDonald, and J.A. Hildebrand. 2006. Kinematics of foraging dives and lunge-feeding in fin whales. *Journal of Experimental Biology* 209(7):1231–1244.
Hain, J.H.W., G.R. Carter, S.D. Kraus, C.A. Mayo, and H.E. Winn. 1982. Feeding behavior of the humpback whale, *Megaptera novaeangliae*, in the western North Atlantic. *Fisheries Bulletin* 80:259–268.
Harris, J.E. 1936. The role of the fins in the equilibrium of the swimming fish. I. Wind-tunnel tests on a model of *Mustelus canis* (Mitchill). *Journal of Experimental Biology* 13(4):476–493.
Hay, K. and A. Mansfield. 1989. Narwhal *Monodon monoceros* Linnaeus, 1758. In *Handbook of Marine Mammals*, Vol. 4, eds. S.H. Ridgway and R. Harrison. San Diego, CA: Academic Press.
Hayes, W.D. 1953. Wave riding of dolphins. *Nature* 172:1060.
Hazekamp, A.A.H., R. Mayer, and N. Osinga. 2010. Flow simulation along a seal: The impact of an external device. *European Journal of Wildlife Research* 56(2):131–140.
Hertel, H. 1966. *Structure, Form, Movement*. New York: Reinhold.
Hertel, H. 1969. Hydrodynamics of swimming and wave-riding dolphins. In *The Biology of Marine Mammals*, eds. H.T. Andersen. New York: Academic Press, pp. 31–63.
Howell, A.B. 1930. *Aquatic Mammals*. Springfield, IL: Charles C Thomas.
Howland, H.C. 1974. Optimal strategies for predator avoidance: Relative importance of speed and maneuverability. *Journal of Theoretical Biology* 47(2):333–350.

Johansen, K. 1962. Buoyancy and insulation in the muskrat. *Journal of Mammalogy* 43:64–68.

Kelly, H.R. 1959. A two-body problem in the echelon-formation swimming of porpoise. U.S. Naval Ordinance Test Station, China Lake, CA.

Kenyon, K.W. 1969. The sea otter in the eastern Pacific Ocean. *North American Fauna* 68:1–352.

Klima, M., H.A. Oelschläger, and D. Wünsch. 1987. Morphology of the pectoral girdle in the Amazon dolphin *Inia geoffrensis* with special reference to the shoulder joint and the movements of the flippers. *Zeitschrift für Säugetierkunde* 45:288–309.

Kojeszewski, T. and F.E. Fish. 2007. Swimming kinematics of the Florida manatee (*Trichechus manatus latirostris*): Hydrodynamic analysis of an undulatory mammalian swimmer. *Journal of Experimental Biology* 210:2411–2418.

Kooyman, G.L. 1973. Respiratory adaptations in marine mammals. *American Zoologist* 13:457–468.

Kooyman, G.L. and P.J. Ponganis. 1998. The physiological basis of diving to depth: Birds and mammals. *Annual Review of Physiology* 60:19–32.

Koski, W.R., C.Q. da-Silva, J. Zeh, and R.R. Reeves. 2013. Evaluation of the potential to use capture-recapture analyses of photographs to estimate the size of the Eastern Canada–West Greenland bowhead whale (*Balaena mysticetus*) population. *Canadian Wildlife Biology and Management* 2(1):23–35.

Lang, T.G. 1966. Hydrodynamic analysis of cetacean performance. In *Whales, Dolphins and Porpoises*, ed. K.S. Norris. Berkeley, CA: University of California Press.

Layne, J.N. and D.K. Caldwell. 1964. Behavior of the Amazon Dolphin, *Inia geoffrensis* (Blainville), in captivity. *Zoologica* 49:81–111.

Lighthill, J. 1969. Hydrodynamics of aquatic animal propulsion—A survey. *Annual Review of Fluid Mechanics* 1:413–446.

Lighthill, J. 1971. Large-amplitude elongated-body theory of fish locomotion. *Proceedings of the Royal Society B—Biological Sciences* 179:125–138.

Liu, P. and N. Bose. 1993. Propulsive performance of three naturally occurring oscillating propeller planforms. *Ocean Engineering* 20:57–75.

Long, J.H. Jr., D.A. Pabst, W.R. Shepherd, and W.A. McLellan. 1997. Locomotor design of dolphin vertebral columns: Bending mechanics and morphology of *Delphinus delphis*. *Journal of Experimental Biology* 200:65–81.

Lovvorn, J.R. and D.R. Jones. 1991. Effect of body size, body fat, and change in pressure with depth on buoyancy and costs of diving in ducks. *Canadian Journal of Zoology* 69:2879–2887.

Marino, L. and J. Stowe. 1997. Lateralized behavior in two captive bottlenose dolphins (*Tursiops truncatus*). *Zoo Biology* 16:173–177.

Miklosovic, D.S., M.M. Murray, L.E. Howle, and F.E. Fish. 2004. Leading-edge tubercles delay stall on humpback whale (*Megaptera novaeangliae*) flippers. *Physics of Fluids* 16(5):L39–L42.

Miller, P.J.O., M.P. Johnson, P.L. Tyack, and E.A. Terray. 2004. Swimming gaits, passive drag and buoyancy of diving sperm whales *Physeter macrocephalus*. *Journal of Experimental Biology* 207(11):1953–1967.

Moore, M.J., T. Hammar, J. Arruda et al. 2011. Hyperbaric computed tomographic measurement of lung compression in seals and dolphins. *Journal of Experimental Biology* 214(14):2390–2397.

Narita, Y. and S. Kuratani. 2005. Evolution of the vertebral formulae in mammals: A perspective on developmental constraints. *Journal of Experimental Zoology Part B: Molecular and Developmental Evolution* 304(2):91–106.

Norberg, U.M. 1990. *Vertebrate Flight: Mechanics, Physiology, Morphology, Ecology and Evolution*. Berlin, Germany: Springer-Verlag.

Noren, S.R. 2008. Infant carrying behaviour in dolphins: Costly parental care in an aquatic environment. *Functional Ecology* 22(2):284–288.

Noren, S.R., G. Biedenbach, J.V. Redfern, and E.F. Edwards. 2008. Hitching a ride: The formation locomotion strategy of dolphin calves. *Functional Ecology* 22(2):278–283.

Norris, K.S. and J.H. Prescott. 1961. Observations on Pacific cetaceans of California and Mexican waters. *University of California Publications in Zoology* 63:291–401.

Oeffner, J. and G.V. Lauder. 2012. The hydrodynamic function of shark skin and two biomimetic applications. *Journal of Experimental Biology* 215(5):785–795.

Pabst, D.A. 1996. Morphology of the subdermal connective tissue sheath of dolphins: A new fibre-wound, thin-walled, pressurized cylinder model for swimming vertebrates. *Journal of Zoology* 238:35–52.

Pavlov, V.V. and A.M. Rashad. 2012. A non-invasive dolphin telemetry tag: Computer design and numerical flow simulation. *Marine Mammal Science* 28(1):E16–E27.

Pavlov, V.V., R.P. Wilson, and K. Lucke. 2007. A new approach to tag design in dolphin telemetry: Computer simulations to minimise deleterious effects. *Deep Sea Research Part II: Topical Studies in Oceanography* 54(3–4):404–414.

Perry, B., A.J. Acosta, and T. Kiceniuk. 1961. Simulated wave-riding dolphins. *Nature* 192:148–150.

Pilleri, G., M. Gihr, P.E. Purves, K. Zbinden, and C. Kraus. 1976. On the behaviour, bioacoustics and functional morphology of the Indus River dolphin (*Platanista indi* Blyth, 1859). *Investigations on Cetacea* 6:11–141.

Pivorunas, A. 1979. Feeding mechanisms of baleen whales. *American Scientist* 67(4):432–440.

Potvin, J., J.A. Goldbogen, and R.E. Shadwick. 2009. Passive versus active engulfment: Verdict from trajectory simulations of lunge-feeding fin whales *Balaenoptera physalus*. *Journal of the Royal Society Interface* 6:1005–1025.

Potvin, J., J.A. Goldbogen, and R.E. Shadwick. 2012. Metabolic expenditures of lunge feeding rorquals across scale: Implications for the evolution of filter feeding and the limits of maximum body size. *PLoS One* 7(9):e44854. Doi:10.1371/journal.pone.0044854.

Reid, K., J. Mann, J.R. Weiner, and N. Hecker. 1995. Infant development in two aquarium bottlenose dolphins. *Zoo Biology* 14:135–147.

Ridgway, S.H. and R. Harrison. 1985. *Handbook of Marine Mammals*, Vol. 3: *The Sirenians and Baleen Whales*. London, UK: Academic Press.

Ridgway, S.H. and R. Howard. 1979. Dolphin lung collapse and intramuscular circulation during free diving: Evidence from nitrogen washout. *Science* 206:1182–1183.

Ridgway, S.H., B.L. Scronce, and J. Kanwishe. 1969. Respiration and deep diving in bottlenose porpoise. *Science* 166(3913):1651–1654.

Riedman, M. 1990. *The Pinnipeds: Seals, Sea Lions, and Walruses*. Berkeley, CA: University of California Press.

Rohr, J., M.I. Latz, S. Fallon, J.C. Nauen, and E. Hendricks. 1998. Experimental approaches towards interpreting dolphin-stimulated bioluminescence. *Journal of Experimental Biology* 201(9):1447–1460.

Sato, K., K. Shiomi, G. Marshall, G.L. Kooyman, and P.L. Ponganis. 2011. Stroke rates and diving air volumes of emperor penguins: Implications for dive performance. *Journal of Experimental Biology* 214(17):2854–2863.

Sato, K., Y. Watanuki, A. Takahashi et al. 2007. Stroke frequency, but not swimming speed, is related to body size in free-ranging seabirds, pinnipeds and cetaceans. *Proceedings of the Royal Society B—Biological Sciences* 274(1609):471–477.

Scholander, P.F. 1959. Wave-riding dolphins: How do they do it? *Science* 129:1085–1087.

Shadwick, R.E., J.A. Goldbogen, J. Potvin, N.D. Pyenson, and A. Wayne Vogl. 2013. Novel muscle and connective tissue design enables high extensibility and controls engulfment volume in lunge-feeding rorqual whales. *Journal of Experimental Biology* 216(14):2691–2701.

Skrovan, R.C., T.M. Williams, P.S. Berry, P.W. Moore, and R.W. Davis. 1999. The diving physiology of bottlenose dolphins (*Tursiops truncatus*). II. Biomechanics and changes in buoyancy at depth. *Journal of Experimental Biology* 202(20):2749–2761.

Slijper, E.J. 1961. Locomotion and locomotory organs in whales and dolphins (Cetacea). *Symposium of the Zoological Society of London* 5:77–94.

Slijper, E.J. 1979. *Whales*, 2nd English edn. London, UK: Cornell University Press.

Sokolov, V.E. 1982. *Mammalian Skin*. Berkeley, CA: University of California Press.

Stein, B.R. 1989. Bone density and adaptation in semiaquatic mammals. *Journal of Mammalogy* 70:467–476.

Stelle, L.L., R.W. Blake, and A.W. Trites. 2000. Hydrodynamic drag in Steller sea lions (*Eumetopias jubatus*). *Journal of Experimental Biology* 203:1915–1923.

Stephenson, R., J.R. Lovvorn, M.R.A. Heieis, D.R. Jones, and R.W. Blake. 1989. A hydromechanical estimate of the power requirements of diving and surface swimming in lesser scaup (*Aythya affinis*). *Journal of Experimental Biology* 147:507–519.

Tarasoff, F.J. 1974. Anatomical adaptations in the river otter, sea otter and harp seal with reference to thermal regulation. In *Functional Anatomy of Marine Mammals*, Vol. 2, ed. R.J. Harrison. London, UK: Academic Press.

Tarasoff, F.J. and G.L. Kooyman. 1973. Observations on the anatomy of the respiratory system of the river otter, sea otter, and harp seal. I. The topography, weight, and measurements of the lungs. *Canadian Journal of Zoology* 51:163–170.

Tavolga, M.C. and F.S. Essapian. 1957. The behavior of the bottle-nosed dolphin (*Tursiops truncatus*): Mating, pregnancy, parturition and mother-infant behavior. *Zoologica* 42:11–31.

Thewissen, J.G.M. 2014. *The Walking Whales*. Berkeley, CA: University of California Press.

Thewissen, J.G.M. and F.E. Fish. 1997. Locomotor evolution in the earliest cetaceans: Functional model, modern analogues, and paleontological evidence. *Paleobiology* 23(4):482–490.

van der Hoop, J.M., A. Fahlman, T. Hurst et al. 2014. Bottlenose dolphins modify behavior to reduce metabolic effect of tag attachment. *Journal of Experimental Biology* 217(23):4229–4236.

Vogel, S. 1994. *Life in Moving Fluids: The Physical Biology of Flow*, 2nd edn. Princeton, NJ: Princeton University Press.

Vogel, S. 2003. *Comparative Biomechanics: Life's Physical World*. Princeton, NJ: Princeton University Press.

Walker, J.A. 2000. Does a rigid body limit maneuverability? *Journal of Experimental Biology* 203:3391–3396.

Watanabe, Y.Y., K. Sato, Y. Watanuki et al. 2011. Scaling of swim speed in breath-hold divers. *Journal of Animal Ecology* 80(1):57–68.

Webb, P.W. 1975. Hydrodynamics and energetics of fish propulsion. *Bulletin of the Fisheries Research Board of Canada* 190:1–158.

Webb, P.W. and R.W. Blake. 1985. Swimming. In *Functional Vertebrate Morphology*, eds. M. Hildebrand, D.M. Bramble, K.F. Liem, and D.B. Wake. Cambridge, MA: Harvard University Press.

Weber, P.W., L.E. Howle, M.M. Murray, and F.E. Fish. 2009a. Lift and drag performance of odontocete cetacean flippers. *Journal of Experimental Biology* 212(14):2149–2158.

Weber, P.W., L.E. Howle, M.M. Murray, J.S. Reidenberg, and F.E. Fish. 2014. Hydrodynamic performance of the flippers of large-bodied cetaceans in relation to locomotor ecology. *Marine Mammal Science* 30(2):413–432.

Webb, P.W., G.D. LaLiberte, and A.J. Schrank. 1996. Does body and fin form affect the maneuverability of fish traversing vertical and horizontal slits? *Environmental Biology of Fishes* 46(1):7–14.

Weber, P.W., M.M. Murray, L.E. Howle, and F.E. Fish. 2009b. Comparison of real and idealized cetacean flippers. *Bioinspiration & Biomimetics* 4(4):046001.

Wegener, P.P. 1991. *What Makes Airplanes Fly?* New York: Springer-Verlag.

Weihs, D. 1973. Hydromechanics of fish schooling. *Nature* 241:290–291.

Weihs, D. 1981. Effect of swimming path curvature on the energetics of fish. *Fisheries Bulletin* 79:171–176.

Weihs, D. 1989. Design features and mechanics of axial locomotion in fish. *American Zoologist* 29:151–160.

Weihs, D. 1993. Stability of aquatic animal locomotion. *Contemporary Mathematics* 141:443–461.

Weihs, D. 2004. The hydrodynamics of dolphin drafting. *Journal of Biology* 3:8.

Weihs, D. and P.W. Webb. 1983. Optimization of locomotion. In *Fish Biomechanics*, eds. P.W. Webb and D. Weihs. New York: Praeger.

Weihs, D. and P.W. Webb. 1984. Optimal avoidance and evasion tactics in predator-prey interactions. *Journal of Theoretical Biology* 106(2):189–206.

Wiley, D., C. Ware, A. Bocconcelli et al. 2011. Underwater components of humpback whale bubble-net feeding behaviour. *Behaviour* 148(5–6):575–602.

Williams, T.D., D.D. Allen, J.M. Groff, and R.L. Glass. 1992a. An analysis of California sea otter (*Enhydra lutris*) pelage and integument. *Marine Mammal Science* 8:1–18.

Williams, T.M. 1983. Locomotion in the North American mink, a semi-aquatic mammal. I. Swimming energetics and body drag. *Journal of Experimental Biology* 103:155–168.

Williams, T.M. 1987. Approaches for the study of exercise physiology and hydrodynamics in marine mammals. In *Approaches to Marine Mammals Energetics*, eds. A.C. Huntley, D.P. Costa, G.A.J. Worthy, and M.A. Castellini. Lawrence, KS: Society for Marine Mammology.
Williams, T.M. 1989. Swimming by sea otters: Adaptations for low energetic cost locomotion. *Journal of Comparative Physiology A: Sensory Neural and Behavioral Physiology* 164(6):815–824.
Williams, T.M. 1999. The evolution of cost efficient swimming in marine mammals: Limits to energetic optimization. *Philosophical Transactions Royal Society of London B Biological Sciences* 353:1–9.
Williams, T.M. 2001. Intermittent swimming by mammals: A strategy for increasing energetic efficiency during diving. *American Zoologist* 41(2):166–176.
Williams, T.M., R.W. Davis, L.A. Fuiman et al. 2000. Sink or swim: Strategies for cost-efficient diving by marine mammals. *Science* 288(5463):133–136.
Williams, T.M., W.A. Friedl, M.L. Fong, R.M. Yamada, P. Sedivy, and J.E. Haun. 1992b. Travel at low energetic cost by swimming and wave-riding bottle-nosed dolphins. *Nature* 355(6363):821–823.
Williams, T.M., S.F. Shippee, and M.J. Rothe. 1996. Strategies for reducing foraging costs in dolphins. In *Aquatic Predators and Their Prey*, eds. S.P.R. Greenstreet and M.L. Tasker. Oxford, UK: Fishing News Books.

chapter two

Oxygen stores and diving

Paul J. Ponganis and Cassondra L. Williams

Contents

2.1 *The big picture challenge and summary*

Marine mammals are able to dive for incredibly long durations. One of the central questions in marine mammal physiology is how these air-breathing vertebrates are able to remain underwater so long on a single breath (Butler and Jones 1997; Ponganis 2015). Elephant seals and beaked whales can dive for as long as 2 hours. And many species make routine, frequent dives to depth with only very short surface periods and little rest between dives. The key to such dive patterns and long dive durations is found both in the amount of oxygen (O_2) marine mammals take down with them and in the rate of consumption of that O_2 during diving. In general, marine mammals store much greater amounts of O_2 than terrestrial mammals. But enhanced O_2 storage is not enough to explain the routine, let alone, the longest dive durations of these animals. Regulation of the depletion rate of O_2 is essential to diving ability. Such O_2 store management is achieved through the *dive response* (decreased heart rate or *bradycardia* and redistribution of blood flow to tissues) as well as through efficient locomotory patterns. Finally, while terrestrial mammals are not able to withstand low levels of blood O_2, some marine mammals have demonstrated the ability to dive with very little O_2 remaining in the blood (*hypoxemic tolerance*). Thus, marine mammals can extend dive durations by (1) bringing more O_2 with them, (2) using it more efficiently, and (3) depleting it to very low levels during dives.

Many marine mammals dive repetitively with very short surface intervals. Due to the cost efficiency of aerobic metabolism for ATP production, such diving patterns in many species have led to the concept that most dive are aerobic and less than an *aerobic dive limit* (ADL, the dive duration associated with the onset of post-dive blood lactate accumulation).

Our understanding of O_2 storage, rates of O_2 depletion, the dive response, and aerobic dive limits in marine mammals is limited by the difficulty of measuring these parameters in freely diving animals. Early work on the diving physiology of seals involved *forced submersions*, an approach that allowed for measurements under extreme conditions (Scholander 1940). Advances in microprocessor technology have led to the development of small physiological loggers that can record some of these parameters in freely diving animals (Ponganis 2015). Much of this recent work has been performed on pinnipeds, due to the ability to capture these animals on land. Since cetaceans are not easily accessible and dive in the open ocean, they present the greatest challenge in measuring these variables.

Measurement of O_2 stores, recording of heart rate during diving, and determination of ADLs each has its own challenges. Body O_2 stores, located in the lungs, blood, and muscle, are dependent on many different variables: diving lung volume, blood volume, hemoglobin (Hb) concentration, muscle mass, and myoglobin (Mb) concentration. Accurate recording of heart rate during dives as deep as 500 m is not a simple matter. And documentation of post-dive blood lactate concentrations has been achieved in only a few species under special circumstances. This chapter will review (1) the magnitude and distribution of O_2 stores in marine mammals, (2) the dive response, (3) the ADL concept, and (4) limitations in the measurement and documentation of these parameters and variables.

2.2 *What is known*

2.2.1 *Magnitude and distribution of total body O_2 stores*

How much O_2 is stored and where it is stored varies among species; the best divers tend to store more O_2 in the muscle and blood than in the lungs. In comparison to humans ("a non-diver"), mass-specific body O_2 stores are elevated 1.5- to almost 5-fold in all marine mammals except the manatee (Table 2.1). The largest mass-specific O_2 stores are found in the phocid seals; the northern elephant seal (*Mirounga angustirostris*), with some of the longest routine dive durations of any pinniped, has the highest mass-specific O_2 store at 94 ml O_2 kg^{-1} (Simpson et al. 1970; Bryden 1972; Thorson and Le Boeuf 1994). The manatee (*Trichechus manutus*), as the shallowest diver, has the lowest O_2 store at 21 ml O_2 kg^{-1} (Lenfant et al. 1970; Gallivan et al. 1986). In general, diving mammals with greater diving capacities have significantly larger O_2 stores.

As a percentage of the total body O_2 store in diving mammals, the blood and muscle O_2 stores range from 54% in the sea otter (*Enhdyra lutris*) to 97% in the northern elephant seal (Table 2.1). In deeper divers, there appears to be decreased dependence on the respiratory

Table 2.1 Total body O_2 stores in humans and marine mammals with percentages of contribution from each of the three O_2 stores and common dive durations

Species	Total O_2 store (ml O_2 kg^{-1})	Lung (%)	Blood (%)	Muscle (%)	Common dive durations (min)
Human (*Homo sapiens*)	24	42	44	14	1–2
Bottlenose dolphin (*Tursiops truncatus*)	34	27	33	40	1
Narwhal (*Monodon monoceros*)	75	12	38	50	1–15
Sperm whale (*Physeter macrocephalus*)	81	5	64	30	40–60
Manatee (*Trichecus manutus*)	21	33	60	7	2–3
Sea otter (*Enhydra lutris*)	69	45	33	21	1–3
Northern fur seal (*Callorhinus ursinus*)	42	24	43	33	2
California sea lion (*Zalophus californianus*)	55	13	39	48	2
Walrus (*Odobenus rosmarus*)	38	24	50	26	4–6
Crabeater seal (*Lobodon carcinophagus*)	43	12	67	21	5
Weddell seal (*Leptonychotes weddellii*)	89	4	66	30	10–15
Hooded seal (*Cystophora cristata*)	90	7	51	42	5–25
Northern elephant seal (*Mirounga angustirostris*)	94	3	71	26	15–30

Sources: Data from Lenfant et al. (1970); Simpson et al. (1970); Bryden (1972); Sleet et al. (1981); Gallivan et al. (1986); Ponganis et al. (1993); Thorson and Le Boeuf (1994); Miller et al. (2004); Burns et al. (2007); Weise and Costa (2007); Ponganis et al. (2011); Williams et al. (2011); Noren et al. (2012); Ponganis (2015); Thometz et al. (2015).

O_2 store. This is especially evident in deep-diving mammals such as the sperm whale, elephant seal, hooded seal, and Weddell seal. With relatively little O_2 available in the lungs, the need for blood O_2 uptake from the lung is decreased. Less dependence on the pulmonary O_2 store decreases the need for gas exchange, thus allowing for *lung collapse* and the bradycardia of diving to decrease nitrogen absorption and the risks of decompression sickness (Chapter 4) without increasing the risk of lower blood O_2 levels (*hypoxemia*) in these deep divers.

2.2.1.1 Respiratory O_2 stores

In general, cetaceans are considered to dive at full lung capacity, and pinnipeds at 50% total lung volume (Gentry and Kooyman 1986; Kooyman 1989; Ponganis 2011). However, pre-dive lung volumes in pinnipeds are probably variable. More recent evidence suggests that sea lions inspire deeper and have larger diving air volumes for deeper dives (McDonald and Ponganis 2012). Manatees are considered to dive on inspiration with full lung volumes, and sea otters with 60% total lung capacity (Ponganis et al. 2003).

Lung volumes of diving mammals are in the general range of terrestrial mammals (Kooyman 1973; Fahlman et al. 2011; Piscitelli et al. 2013). Notable exceptions are the small lungs of the deep-diving whales and the large lungs of the shallow-diving sea otter. Bottlenose whales (*Hyperoodon ampullatus*), sperm whales (*Physeter macrocephalus*), and pygmy and dwarf sperm whales (*Kogia breviceps, Kogia sima*) have lung volumes of 21–28 ml kg^{-1} (Scholander 1940; Miller et al. 2004; Piscitelli et al. 2010). Allometric analyses of lung mass in cetaceans revealed that kogiids, physeterids, ziphiids, and mysticetes all had relative lung masses similar to terrestrial mammals, while delphinids, phocoenids, and monodonts had relatively larger lung masses (Piscitelli et al. 2010, 2013). Sea otter lung volume has been measured at 345 ml kg^{-1} (Lenfant et al. 1970). The high lung volume in the otter presumably contributes to its buoyancy at the surface, where it feeds, grooms, and cares for its young. Such buoyancy in the otter also elevates more of the body out of the water while the animal is at the surface; this should reduce body heat loss due to conduction in water (Chapter 9).

The percentage of the total body O_2 store provided by the respiratory store is shown in Table 2.1. While total lung capacity is often necessary to calculate the respiratory O_2 store, the respiratory store is not equivalent to the quantity of O_2 in the total lung capacity. This is because many marine mammals dive with lung volumes that are much less than the total lung capacity (see Section 2.3.1).

2.2.1.2 Blood O_2 stores

The blood O_2 store is the largest O_2 store in many marine mammals. Almost all pinnipeds store 50% or more of their O_2 in the blood (Table 2.1). The blood O_2 store is also highest in both walruses and manatees; however, in sea otters the respiratory O_2 store is the highest at 45%. In Cetacea, blood is the largest O_2 store in deep, long duration divers, such as the sperm whale (*P. macrocephalus*) (Table 2.1).

Blood O_2 stores are calculated using three parameters: (1) blood volume (one-third is assumed to be arterial and two-thirds venous), (2) hemoglobin (Hb) concentration, and (3) the amount of oxygen extracted from hemoglobin during a dive (net Hb desaturation) (Lenfant et al. 1970; Kooyman 1989) (see Section 2.3.2). Many marine mammals have elevated values of these parameters when compared to terrestrial mammals. The typical human Hb concentration of 15 g dl^{-1} and blood volume of 70 ml kg^{-1} are 50%–70% lower than Hb concentrations and blood volumes of marine mammals. The greatest elevations in both Hb and blood volume are in the longest-duration divers and in highly active species (Ridgway and Johnston 1966; Ponganis 2011). This is exemplified in the phocid seals. The northern elephant seal and the gray seal (*Halichoerus grypus*) have blood volumes of 216 and 213 ml kg^{-1} and Hb concentrations of 25 and 20 g dl^{-1}, respectively. In general, Hb and blood volume are not as high in otariids as in phocid seals, but some otariid species have hemoglobin concentrations up to 23 g dl^{-1} and blood volumes up to 186 ml kg^{-1}. Hb concentration and blood volume span a wide range in cetaceans, from the shallow-diving bottlenose dolphin with values similar to standard human values to the sperm whale, with an Hb concentration of 22 g dl^{-1} and a blood volume of 200 ml kg^{-1} (Sleet et al. 1981; Ridgway 1986). Consequently, the bottlenose dolphin only stores 33% of total body O_2 in the blood store, while the sperm whale's blood store accounts for 64% of the total O_2 store (Table 2.1). Manatee and walrus values are not particularly high, but the sea otter's values are within the otariid range at 19 g dl^{-1} and 174 ml kg^{-1} (Thometz et al. 2015).

2.2.1.3 Muscle O_2 stores

In all marine mammals except the manatee, muscle is a significant component of the total O_2 store in comparison to their non-diving counterparts (Table 2.1). A large muscle

O_2 store, and especially a high muscle O_2 content concentrated in the primary underwater locomotory muscles, will have implications for the nature of cardiovascular responses required for effective utilization of O_2 stores during dives.

The muscle O_2 store is based on the concentration of Mb and the muscle mass. In marine mammals, Mb concentrations vary almost 100-fold, from 0.1 g 100 g^{-1} muscle in manatees to 9.5 g 100 g^{-1} muscle in hooded seals. The Mb concentration in marine mammals is often indicative of their diving capacity and dive patterns. The Mb concentration is highest in the deep, long duration divers such as the hooded, northern elephant, ribbon, harp, and Weddell seals (with Mb concentrations from 5.4 to 9.5 g 100 g^{-1} muscle) (Lenfant et al. 1970; Ponganis et al. 1993; Burns et al. 2007; Hassrick et al. 2010). Among otariids, the Galapagos (*Zalophus wollebaeki*) and California (*Zalophus californianus*) sea lions have some of the highest Mb concentrations (5.3–5.4 g 100 g^{-1} muscle) (Weise and Costa 2007; Villegas-Amtmann and Costa 2010). Mb content in the sperm whale, bottlenose whale, and the narwhal (*Monodon monoceros*) is also high, ranging from 5.4 to 7.9 g 100 g^{-1} muscle (Scholander 1940; Williams, Noren, and Glenn 2011).

2.2.2 *The dive response*

The dive response forms the basis for the management of O_2 stores, and, indeed, for all of diving physiology. As exemplified in the forced submersion experiments of Scholander and Irving, the heart rate decreases quickly and dramatically in these extreme situations, often to heart rates of less than 10 beats per min in seals (Scholander 1940; Scholander et al. 1942). Under these conditions, cardiac output is severely reduced, and there is widespread *peripheral vasoconstriction* of the arterial vessels to maintain blood pressure. The reduced blood flow is directed away from most organs, and is essentially distributed to the brain and heart (Blix et al. 1983; Zapol 1987). This reduction in blood flow decreases blood O_2 consumption by most organs and isolates the muscle from the circulation, so that muscle metabolism is dependent on the enhanced myoglobin-bound O_2 store and eventually on anaerobic glycolysis. This severe reduction in heart rate and peripheral blood flow, the so-called *dive reflex* (see Figure 2.1), results in a very slow rate of blood O_2 depletion with conservation of the blood O_2 store for metabolism of the brain and heart (Elsner et al. 1966; Kerem and Elsner 1973).

In free dives and spontaneous breath-holds, however, the reduction in heart rate and peripheral blood flow (the dive response) is often variable and not as great as during forced submersion (Elsner 1965; Thompson and Fedak 1993; Andrews et al. 1997). This was well illustrated in Elsner's studies of trained breath-holds of seals in the 1960s (Figure 2.1). In free dives, although diving heart rates are variable, heart rates can be in the range of 20–40 beats min^{-1} (about 2–5 times those during forced submersions); gastrointestinal blood flow, hepatic blood flow, renal blood flow, and even some muscle blood flow all appear to be maintained in short-duration dives of Weddell seals (Davis et al. 1983; Guppy et al. 1986; Hill et al. 1987; Guyton et al. 1995). At times, however, the heart rate can be quite low even during free dives; during a 14 min dive of a gray seal, average dive heart rate was near 4 beats min^{-1} (Thompson and Fedak 1993). And, during deep dives of California sea lions (Figure 2.2), the heart rate can be less than 10 beats per min at maximum depth (McDonald and Ponganis 2014). Thus, the degrees of bradycardia and tissue blood flow reduction during free dives are variable and probably dependent on the nature and circumstances of a given dive. Due to this variable dive response, the depletion rate of the blood O_2 store and the duration of aerobic metabolism are variable in different dives. Similarly, the depletion rate of the muscle O_2 store will also be variable, depending on locomotory requirements during a given dive (Chapters 1 and 3) as well as any blood O_2 supplementation of muscle

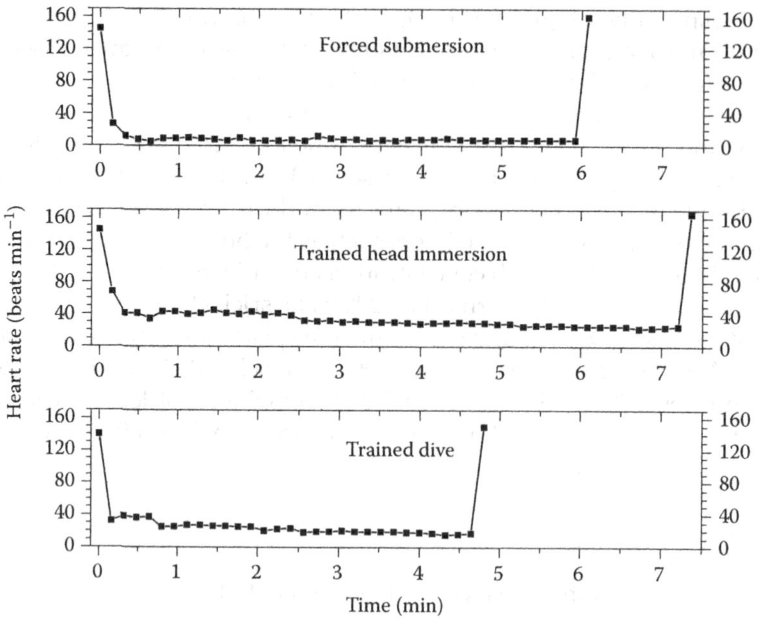

Figure 2.1 The heart rate profiles during forced submersion, trained head immersion, and a trained dive of a young harbor seal (*Phoca vitulina*) illustrates the differences in the dive response under these three conditions. During forced submersion, the heart rates were 5–10 beats per min (the classic dive reflex with extreme bradycardia). In contrast, the heart rates were more variable and higher during the voluntary head immersion and dive, with the initial heart rates near 40 beats per min that later decreased to about 20 beats per min. The breath-holds started at 0 min, and were approximately 6, 7.5, and 4.7 min in duration. The heart rate data were at 10 s intervals. (Adapted from Elsner, R., *Hvalradets Skrifter*, 48, 24, 1965.)

during a dive. The dive response and rates of O_2 store depletion are further detailed in recent reviews (Ponganis et al. 2011; Davis 2014; Elsner 2015; Ponganis 2015). The decrease in cardiac output and the distribution of the blood flow to tissues will also affect the magnitude and distribution of nitrogen uptake from the lungs at depth (Chapter 4). The range and complexity of the dive response during free dives is illustrated for shallow versus deep dives of a California sea lion in Figure 2.2.

2.2.3 *Aerobic dive limits*

The enhanced O_2 stores, the dive response, and the workload of muscle all combine to contribute to the duration of aerobic metabolism during a dive and an animal's aerobic dive limit (ADL). Originally, an aerobic dive limit (Figure 2.3) was measured by determining post-dive blood lactate levels in Weddell seals (Kooyman et al. 1980, 1983). Importantly, it was noted that most free dives (90%–97%) of Weddell seals were less than this limit, and that dives beyond that limit were associated with longer surface intervals. Hence, the concept developed that most dives were aerobic, and that it was the efficiency of aerobic metabolism that allowed animals to dive frequently.

Although an ADL has only been determined in a few other marine mammals (Ponganis et al. 1997a,c; Shaffer et al. 1997; Williams et al. 1999), the concept of an ADL and aerobic diving has been applied to multiple species and has become fundamental in the interpretation

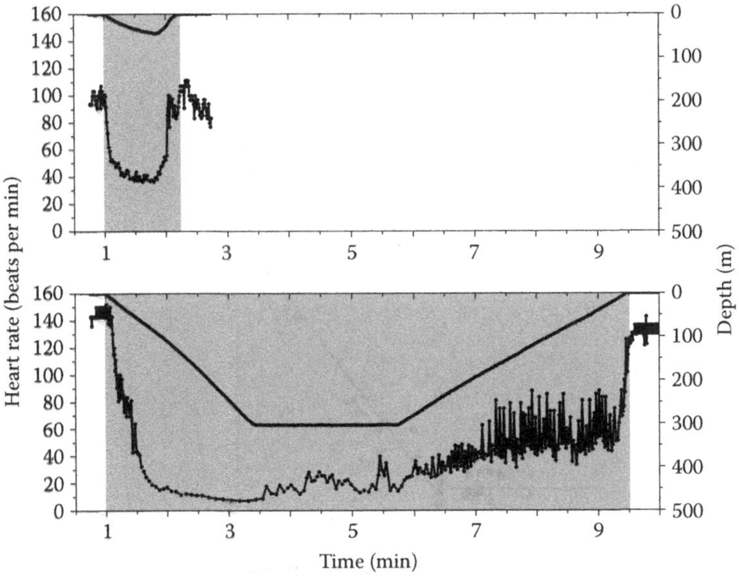

Figure 2.2 The heart rate profiles of a California sea lion during a 1.3 min dive to 45 m and an 8.5 min to 305 m illustrate the range and variability of the dive response during different types of dives. Note that pre- and post-dive heart rates were higher for the deep dive and that the minimum heart rates during the deep dive were near 10 beats per min, in the same range as observed during forced submersions. Although the heart rates during the short dive were higher, the minimum heart rates were still less than that at rest on land (54 beats per min). The gray-shaded area indicates the dive period; in each panel, upper trace is depth, lower trace is heart rate. (Adapted from McDonald, B.I. and Ponganis, P.J., *J. Exp. Biol.*, 217, 1525, 2014.)

of diving behavior and foraging ecology. This widespread application of the ADL concept has been based on two observations from Kooyman's original papers. First, those authors noted that about 95% of dives were less than the blood lactate-determined ADL of the Weddell seal. Consequently, a *behavioral ADL* (ADL_b) has often been determined on the basis of dive duration distribution with a 90%–95% cutoff threshold for the ADL_b.

Kooyman and co-workers also noted that the blood lactate-determined ADL could be estimated by dividing O_2 stores by a diving metabolic rate (a *calculated aerobic dive limit*, ADL_c). The diving metabolic rate was the average metabolic rate measured from O_2 consumption measurements made when the seal surfaced and breathed under a metabolic dome (Castellini et al. 1992; Ponganis et al. 1993). That metabolic rate was calculated as the total O_2 consumed during the surface interval divided by the sum of the dive duration and the surface interval. Thus, the metabolic rate in the calculation was not the actual O_2 consumption rate during a dive, but rather the O_2 consumption rate of the entire dive event (dive + post-dive surface interval). As emphasized in many reviews (Kooyman and Ponganis 1998; Ponganis et al. 2011; Ponganis 2015), this calculation does not reflect the actual status of the body O_2 store at the ADL. It is a simple prediction of the dive duration associated with the onset of post-dive blood lactate accumulation. The body O_2 store is not completely depleted at the ADL. The ADL is considered primarily secondary to muscle O_2 depletion in the primary locomotory muscles, with subsequent glycolysis and muscle lactate accumulation during the dive, and lactate washout into blood after the dive. This has been most completely demonstrated in an avian diver, the emperor penguin (*Aptenodytes forsteri*) (Ponganis et al. 1997b; Meir and Ponganis 2009; Williams et al. 2011).

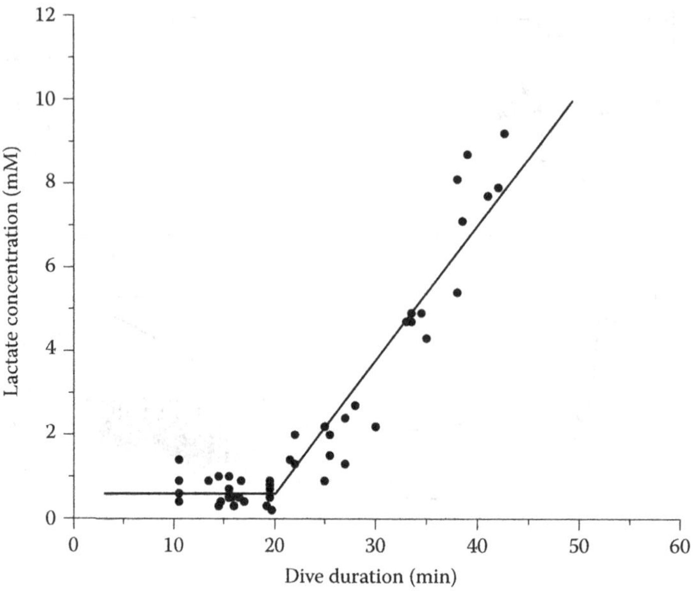

Figure 2.3 Aerobic dive limit of Weddell seals. The graph shows lactate concentrations measured from blood samples taken from seals after dives of various durations. Initially, post-dive blood lactate values remain near resting values and then begin to increase at a dive duration near 20 min, the ADL. (Adapted from Kooyman, G.L. et al., *J. Compar. Physiol.*, 138, 335, 1980.)

2.3 *Toolbox: How these parameters are measured*

The measurement of these physiological parameters is challenging in a number of ways. The calculation of O_2 stores requires a number of anatomical or physiological measurements which can be made using a variety of methods. However, these measurements or calculations often require a number of assumptions which have not been verified in diving marine mammals. As a result, accurate measurement of each O_2 store has its own limitations. There are also significant challenges in heart rate measurements and the determination of ADLs. As mentioned earlier, the results from early studies of forcibly submerged animals can be quite different from how animals respond while freely diving. Yet, making similar measurements in freely diving animals is much more difficult.

Recent work on the physiological responses in freely diving animals has been assisted by technological advances, which have paved the way for small microprocessor-based data recorders that can be attached to free-diving animals. However, while the ability to record physiological data has improved, the need to recapture the animal and remove the physiological sensors is still a major impediment in increasing our understanding of the diving physiology of many marine mammals.

This section will address current methods for determining: (1) O_2 stores; (2) heart rate; and (3) the ADL in marine mammals.

2.3.1 *Measuring the respiratory O_2 store*

The respiratory O_2 store is not the maximum amount of O_2 that can be stored in the lungs. There are two primary factors in measuring the respiratory O_2 store: (1) the diving air volume and (2) the net extraction of O_2 from the diving air volume. The diving air volume is not necessarily the same as the total lung capacity. The total lung capacity is the maximum amount of air in the lungs after maximum inspiration. Total lung capacity has been measured in a number of marine mammals, using a variety of methods including inflation of excised lungs, helium dilution techniques, nitrogen washout, tidal volume measurements, and inspiratory capacity (for further review, see Ponganis 2015). Most total lung capacity measurements have been made by inflation of excised lungs, including bottlenose whales (*H. ampullatus*), pygmy and dwarf sperm whales (*K. breviceps, K. sima*), and sea otters (*Enhydra lutris*) (Piscitelli et al. 2010, 2013).

The diving air volume is the volume of air in the lungs at the start of a dive. Some marine mammals, such as many pinnipeds, dive after exhalation. As a result, diving lung volume is far less than the total lung capacity. In simulated dives in pressure chambers, the diving lung volumes of phocid seals and sea lions have been estimated to be about 50% total lung capacity (Kooyman et al. 1973; Kooyman and Sinnett 1982). Such values, obtained from restrained animals, may not be representative of the initial lung volumes of free-diving animals. However, these values have been used extensively in calculations of O_2 stores. Cetaceans, in contrast, dive on inspiration. Thus, it is often assumed that their diving lung volume is near total lung capacity. Although measurements on cetaceans are not common, diving lung volumes determined by buoyancy-swim velocity calculations are similar to total lung capacity measured in excised lungs of some deep-diving whales (Scholander 1940, Miller et al. 2004).

The true, usable, respiratory O_2 store is based on the amount of O_2 extracted from the diving lung volume. It is unlikely that the O_2 in the lungs can be completely depleted. Thus, the net extraction of O_2 is key to estimating the usable respiratory O_2 store. The net O_2 extraction is the difference between the O_2 fraction at the start of the dive (initial O_2 fraction) and the lowest possible O_2 at the end of dives. This difference is assumed to be 15% in diving mammals (Kooyman 1989). At the start of dives, O_2 fractions are likely near 20%, or slightly less for pinnipeds which exhale before diving. The expiratory O_2 fractions measured at the end of forced submersion and forced dives in seals and other marine mammals were approximately 2%–4%, consistent with the assumption of 15% net O_2 fraction (Scholander 1940; Ridgway et al. 1969; Kooyman et al. 1973; Ponganis et al. 1993).

2.3.2 *Measuring the blood O_2 store*

To calculate the total blood O_2 store, a number of parameters must be measured: blood volume (BV), mass, Hb concentration, Hb carrying capacity (1.34 ml O_2 g Hb^{-1}), arterial and venous Hb saturation (SaO_2 and SvO_2), and the assumption that one-third of the blood volume is arterial and two-thirds venous (Lenfant et al. 1970; Kooyman 1989). The amount of O_2 in the total blood volume can be calculated by the following equation:

$$\begin{aligned}\text{Blood } O_2 = {} & (0.33\text{BV (dl kg}^{-1}) \times \text{Mass (kg)} \times \text{Hb (g dl}^{-1}) \times 1.34\ (\text{ml } O_2 \text{ g Hb}^{-1}) \times \text{SaO}_2\ (\%)) \\ & + (0.67\text{BV (dl kg}^{-1}) \times \text{Mass (kg)} \times \text{Hb (g dl}^{-1}) \times 1.34\ (\text{ml } O_2 \text{ g Hb}^{-1}) \times \text{SvO}_2\ (\%))\end{aligned} \tag{2.1}$$

Note that BV is in dl kg^{-1}, not in ml kg^{-1}.

The usable blood O_2 store, however, is dependent both on the initial Hb saturation of blood (initial SvO_2 and SaO_2) and on how much O_2 can be extracted from the blood on both the arterial side and venous side during a dive. Typically, it is assumed that the SaO_2 can decrease during a dive from 95% to 20%, and that the venous O_2, with an initial venous content of 5 ml O_2 dl^{-1} less than the arterial value, can be completely depleted. This has been estimated in most cases, but has been calculated as the difference between the initial O_2 value at the beginning of the dive and the lowest O_2 value obtained in any dive (or the difference between the beginning and end values for SaO_2 and SvO_2) (Lenfant et al. 1970; Kooyman 1989). See Section 2.3.2.3 for further review.

2.3.2.1 *Blood volume*

The most common way to determine the blood volume in marine mammals is to measure the plasma volume and the hematocrit. Then, the blood volume is calculated by the formula:

$$\text{Total blood volume} = \frac{\text{Plasma volume}}{(1 - \text{hematocrit})} \tag{2.2}$$

Plasma volume is typically measured using Evans blue dye (T-1824) and the indicator-dilution principle (El-Sayed et al. 1995). Hematocrit (or packed cell volume) is measured by centrifugation in a calibrated tube. There are a number of challenges in accurately measuring these variables, especially in animals with large blood volumes. The hematocrit can be variable due to splenic relaxation/contraction under different conditions (e.g., under anesthesia versus during a dive) (Turner and Hodgetts 1959; Qvist et al. 1986; Cross et al. 1988; Ponganis et al. 1993).

2.3.2.2 *Hb concentration*

Standard commercial kits, using spectrophotometric methods, are available to measure Hb concentrations from blood samples. Blood samples obtained during dives or as soon as possible after dives are best for accurate measurements of Hb concentration because, as reviewed above, Hb concentrations may vary due to the state of splenic contraction/relaxation.

2.3.2.3 *Hb saturation (SaO_2 and SvO_2)*

Hb saturation refers to the percentage of hemoglobin bound with O_2. O_2 reversibly binds to hemoglobin based on the partial pressure of O_2 (P_{O_2}) in the blood. The relationship between the P_{O_2} and percent of bound Hb is calculated from the O_2–Hb dissociation curve. Thus, to calculate Hb saturation at the beginning and end of dives, the P_{O_2} and the *O_2–Hb dissociation curve* must be known. This relationship is explained in more detail in most basic physiology textbooks and reviewed and illustrated in Ponganis (2015).

However, since pre- and end-of-dive measurements of P_{O_2} are difficult to obtain, assumptions of initial and final arterial and venous Hb saturation have frequently been used. Arterial blood is typically assumed to be 95% saturated (Lenfant et al. 1970). The venous store is more difficult to estimate. In humans, venous Hb saturation is 60%–80%. However, in some marine mammals, it is now known venous Hb saturations and P_{O_2} can be elevated above the expected resting values. This has been observed both prior to and during dives of California sea lions, and during dives of elephant seals (Meir et al. 2009; Ponganis et al. 2011; McDonald and Ponganis 2013). This increase in venous O_2 content is called *arterialization* of venous blood. In sea lions, the venous blood in some dives was increased to above 95% saturation (McDonald and Ponganis 2013). Since very few mea surements of blood O_2 during diving have been made, it is unknown if arterialization of venous blood occurs in other marine mammals. However, as a mechanism to increase O_2 storage, and ultimately enhance breath-hold capacity, it would seem to be a highly beneficial mechanism.

Finally, in most mammals, the blood O_2 store cannot be completely depleted. Thus to calculate the usable blood O_2 store, it is necessary to determine the minimum blood O_2 level during dives. However, there have been few studies of blood O_2 store depletion during diving. In marine mammals, the lowest arterial saturation is generally assumed to be 20%. Forced submersion studies demonstrated that seals were tolerant of arterial Hb saturation values down to about 20% (Elsner et al. 1970; Kerem and Elsner 1973). Tolerance of low levels of O_2 in the blood is called *hypoxemic tolerance*. Northern elephant seals have extreme hypoxemic tolerance with arterial Hb saturation often reaching values less than 5% (Meir et al. 2009). On the venous side, values near zero are often assumed as forced submersion studies in seals have shown blood O_2 venous store can be completely depleted (Kerem and Elsner 1973). In more recent studies on freely diving northern elephant seals and California sea lions, venous Hb saturation values were near zero at times (Meir et al. 2009; McDonald and Ponganis 2013).

2.3.3 *Measuring the muscle O_2 store*

The total muscle O_2 store is calculated from muscle mass, Mb concentration, and the Mb O_2 carrying capacity (1.34 ml O_2 g^{-1} Mb).

$$\text{Muscle } O_2 \text{ store} = \text{Muscle mass (kg)} \times \text{Mb concentration (g Mb kg}^{-1}) \times 1.34 \text{ (ml } O_2 \text{ g Mb}^{-1}) \quad (2.3)$$

As with the other O_2 store calculations, there are potential sources of error in this calculation, including muscle mass data, inhomogeneous Mb concentrations, and potential issues with the technique to determine Mb concentration.

2.3.3.1 *Muscle mass*

Most measurements of muscle mass have been made by anatomical dissection and thus, muscle mass is not known for many species. In marine mammal species where muscle mass has been measured, it is close to 30% (Ponganis 2015). Thus, muscle mass is often assumed to be 30%. However, this assumption does not hold for the balaenopterid whales, in which the muscle mass is 45%–62% of body mass (Lockyer 1976).

2.3.3.2 *Mb concentration analysis and potential problems*

Myoglobin concentrations have most often been determined using a spectrophotometric method (Reynafarje 1963). This technique involves determining the difference in tissue absorbances at two wavelengths (538 and 568 nm) and dividing by the Mb extinction coefficient. Although Hb and Mb have similar absorbance spectra, the effect of Hb from blood-perfused tissue on the Mb measurement is eliminated due to the assumed identical extinction coefficients of Hb at those wavelengths. A recent study suggests this assumption may not always be correct and extinction coefficients can vary in different species (Masuda et al. 2008). However, in marine mammals with very high Mb concentrations, the effect of the error may be minimal. For example, in the sperm whale (*P. macrocephalus*), the use of the Reynafarje method would lead to a less than 4% error in Mb concentration.

The assumption that Mb concentration is constant throughout a muscle or among different muscles may also be incorrect. In several studies of seals, Mb concentrations have been shown to differ: (1) between individual muscles, (2) in the location within a single muscle, and (3) with the season of the year (Neshumova et al. 1983; Neshumova and Cherepanova 1984; Petrov and Shoshenko 1987; Polasek and Davis 2001; Polasek et al. 2006). However, at least in the Baikal seal (*Phoca sibirica*), the mean Mb concentration of all muscles was not that different from the Mb content of the primary locomotory muscle (Neshumova and Cherepanova 1984). Thus, the use of Mb concentration measurements from a single locomotory muscle should provide a reasonable estimate of muscle O_2 stores.

2.3.4 Measuring heart rate

Heart rates of diving animals are usually determined from the electrocardiogram (ECG) signal. The typical ECG signal that is associated with a heartbeat consists of a P wave (atrial contraction), QRS complex (ventricular contraction), and T wave (ventricular relaxation). Beat-to-beat heart rates are usually calculated from the time intervals between successive R waves (the R–R interval). There have been two common types of recorders used in heart rate studies of marine mammals (Ponganis 2007). The first technique consists of an electronic R wave detector that recognizes the R wave based on its height, width, and polarity. The second technique is the actual recording of the ECG signal.

The R-wave detector is usually programmed to count the number of detected R waves over a given time interval, and then store that value in memory. Such recorders provide a record of heart rate at fixed intervals throughout the dive. The advantages of such recorders are that signal processing is automatic, memory usage is minimal, and heart rates can be recorded over long time periods. Disadvantages are that potential error due to artifact is unrecognized and beat-to-beat heart rates are not recorded.

In contrast, an ECG recorder provides a continuous ECG record, from which each R–R interval can be calculated so that the instantaneous heart rate profiles can be constructed. The advantages of this technique are that accurate R wave detection can be verified, beat-to-beat changes in heart rate are recorded, and variability in heart rate profiles can be assessed. The disadvantages include memory storage (typically the ECG is recorded at 50–100 Hz), the need for peak detection programs to recognize the R waves and calculate the heart rate, and the time to process the ECG records with such programs.

The challenge in collecting heart rate data is secure placement of the ECG electrodes in the proper position so that a good signal is detected and there is no muscle or movement artifact in the record. This varies with species. Both surface and subcutaneous electrodes have been used in marine mammals (Hill et al. 1987; Williams et al. 1992; Thompson and Fedak 1993; Andrews et al. 1997; Hindle et al. 2010; Davis and Williams 2012; McDonald and Ponganis 2014). Surface electrodes have usually been attached with epoxy glues in pinnipeds and with suction cups in cetaceans.

2.3.5 Measuring blood lactate to determine the ADL

Documentation of blood lactate concentrations to determine the ADL of a given species has required vascular access to the animal to obtain blood samples and the subsequent analysis of those samples for lactate content. The ability to reliably obtain blood samples in the post-dive period without disturbing the animal has been most critical and difficult to achieve. Such studies have been carried out under two conditions—with trained or captive animals, or with wild animals at an isolated dive hole to which they must return in order to breathe. Isolated dive hole studies have been conducted only in McMurdo Sound, Antarctica—with Weddell seals and emperor penguins (Ponganis et al. 1997b; Kooyman et al. 1980, 1983). The studies with captive or trained animals have included bottlenose dolphins, beluga whales (*Delphinapterus leucas*), Baikal seals, and California sea lions (Ponganis et al. 1997a; Williams et al. 1999). Blood samples have been obtained through indwelling venous or arterial catheters in pinnipeds, and through needle venipuncture of tail fluke vessels in the cetaceans.

Blood or plasma lactate concentrations were originally measured through standard spectrophotometric enzyme assays. Lactate analyzers are now available commercially; these devices are based on the enzymatically linked generation of an electrical current. A typical reaction involves the generation of an electrical current on a platinum electrode by hydrogen peroxide, the production of which is linked to lactate by the enzyme, lactate oxidase. Although such analyzers simplify and streamline the analysis process, the technical bottleneck in lactate determinations of ADLs remains obtaining the blood sample in the post-dive period.

If all reactions that consume or generate lactate could be immediately inhibited in a blood sample, it is possible that a remote blood sampling device (Hill 1986; Ponganis et al. 2009) could collect samples during short trips to sea. Lactate determinations could then potentially be performed days later. The limitation with this approach is the size of the device, the limited number of samples obtained, and the need for a guaranteed quick return of the animal (both for sample analysis and the removal of the device and catheter).

Intravascular blood lactate sensors have been developed and used under experimental conditions in other animals (Baker and Gough 1995). This specific approach has involved the detection of lactate with an oxygen electrode enveloped by an outer membrane containing lactate oxidase. The resultant reaction consumes O_2 inside the sensor envelope, decreasing the partial pressure of O_2 within the outer membrane and decreasing the current output of the O_2 electrode. A second O_2 electrode (with no lactate oxidase) provides a reference O_2 current, thus providing a difference between the two electrode currents that is proportional to the blood lactate concentration. Such an approach holds promise to measure blood lactate concentrations and determine ADLs in more species of marine mammals. However, there are many technical difficulties to be evaluated, including prevention of thrombus formation, pressure and temperature effects on the reactions, the lifespan of the embedded lactate oxidase, and the sensitivity and accuracy of the technique under conditions of low and changing blood P_{O_2} during and after a dive. Despite the testing and development needed for this approach in marine mammals, it may hold the most promise for providing blood lactate concentration profiles and subsequent determination of an ADL.

2.3.6 Summary

It should now be clear that there are a number of assumptions and potential sources of error in determining the O_2 storage of any particular species. The measurement of heart rate and the ADL also have significant challenges. It should come as no surprise that this area of diving physiology is still an active area of research as new measurement techniques are developed and new measurements are made. Consequently, what we know about O_2 storage in marine mammals is subject to frequent revision and our understanding of how marine mammals can make extended dives continues to improve.

2.4 Unanswered questions

The primary challenge in understanding the physiology underlying extended dive durations remains making measurements in freely diving mammals. There are significant differences in the physiological responses to mammals undergoing forced submersion experiments and animals freely diving in their natural environment. Measurements in free-diving cetaceans are especially difficult due to their completely aquatic lifestyle and to the difficulties of recorder attachment.

A number of the physiological parameters discussed above may also vary depending on the type of dive or even within dives; these include diving lung volume, O_2 extraction from lung or blood, hematocrit, and initial/final blood O_2 saturations. How these parameters change during or between dives remains an unanswered question.

As more physiological data are collected in freely diving marine mammals, we are learning how the dive response and O_2 store management vary during different types of dives and between species. We now know that heart rates and O_2 depletion rates can vary considerably depending on the nature and circumstances of a given dive. Yet, the how and why these parameters vary remains a challenge for future studies and new technological advances. Finally, the ADL has been defined by the accumulation of blood lactate during dives, yet we still have not conquered the challenge of making continuous blood lactate measurements. Such measurements are essential to interpretations of diving/foraging behavior and to evaluation of the ADL hypothesis that most dives are aerobic.

Glossary

Aerobic dive limit (ADL): The shortest dive duration associated with an elevation in post-dive blood lactate concentration. This is determined by lactate measurements in post-dive blood samples. The ADL is important because it indicates an increased reliance on anaerobic metabolism, the result of which is lactate accumulation. In the Weddell seal, the accumulation of lactate has been associated with increased time at the surface, which results in less time spent underwater for important activities, such as foraging.

Arterialization: The arterialization of venous blood means that more O_2 than normal is found in the venous circulation. For a diving mammal, such an increase in pre-dive venous O_2 saturation (arterialization) would result in an increased blood O_2 store, allowing these mammals to dive aerobically for longer periods.

Behavioral aerobic dive limit (ADL_b): The behavioral ADL is often considered the dive duration below which are 90%–97% of all dives of a given animal. The behavioral ADL has also been estimated as the dive duration after which the time spent at the surface (post-dive surface interval) is increased significantly.

Bradycardia: A bradycardia is a heart rate that is below resting levels. In diving mammals, a bradycardia during diving is part of the dive response and is considered to reduce O_2 consumption during the dive.

Calculated aerobic dive limit (ADL_c): The total sum of all three oxygen stores in a species divided by the diving metabolic rate. The ADL_c is subject to many assumptions and is often used when the aerobic dive limit cannot be determined.

Dive response: The cardiovascular response during a breath-hold or dive. This response involves both a decrease in heart rate and an increase in peripheral vasoconstriction via activation of the parasympathetic and sympathetic nervous systems. *Dive reflex* usually refers to a most extreme dive response as elicited by forced submersion.

Forced submersion: An experimental technique in which an animal is constrained and held underwater for a duration unknown to it. This approach was utilized in early studies on diving physiology.

Hypoxemia: A low blood O_2 level; in humans, arterial hemoglobin saturations less than 90% are considered hypoxemic; *hypoxemic tolerance* refers to the ability of an animal or tissue to tolerate and survive such low O_2 levels.

Lung collapse: The process of alveolar collapse and lack of gas exchange at depth secondary to the increase in ambient pressure during a dive; see Chapter 4.

O_2–Hb dissociation curve: The curve that describes the relationship between hemoglobin saturation and the partial pressure of O_2 (P_{O_2}).

Peripheral vasoconstriction: The constriction of arterial blood vessels. Such constriction maintains blood pressure when heart rate and cardiac output decrease.

References

Andrews, R.D., D.R. Jones, J.D. Williams et al. 1997. Heart rates of northern elephant seals diving at sea and resting on the beach. *Journal of Experimental Biology* 200:2083–2095.

Baker, D.A. and D.A. Gough. 1995. A continuous, implantable lactate sensor. *Analytical Chemistry* 67:1536–1540.

Blix, A.S., R.W. Elsner, and J.K. Kjekhus. 1983. Cardiac output and its distribution through capillaries and A-V shunts in diving seals. *Acta Physiologica Scandanavica* 118:109–116.

Bryden, M.M. 1972. Body size and composition of elephant seals (*Mirounga leonina*): Absolute measurements and estimates from bone dimensions. *Journal of Zoology (London)* 167:265–276.

Burns, J.M., K.C. Lestyk, M.O. Hammill, L.P. Folkow, and A.S. Blix. 2007. Size and distribution of oxygen stores in harp and hooded seals from birth to maturity. *Journal of Comparative Physiology B* 177:687–700.

Butler, P.J. and D.R. Jones. 1997. The physiology of diving of birds and mammals. *Physiological Reviews* 77:837–899.

Castellini, M.A., G.L. Kooyman, and P.J. Ponganis. 1992. Metabolic rates of freely diving Weddell seals: Correlations with oxygen stores, swim velocity, and diving duration. *Journal of Experimental Biology* 165:181–194.

Cross, J.P., C.G. Mackintosh, and J.F.T. Griffin. 1988. Effect of physical restraint and xylazine sedation of haemotological values in red deer (*Cervus elaphus*). *Research in Veterinary Science* 45:281–286.

Davis, R.W. 2014. A review of the multi-level adaptations for maximizing aerobic dive duration in marine mammals: From biochemistry to behavior. *Journal of Comparative Physiology B* 184:23–53.

Davis, R.W., M.A. Castellini, G.L. Kooyman, and R. Maue. 1983. Renal GFR and hepatic blood flow during voluntary diving in Weddell seals. *American Journal of Physiology* 245:R743–R748.

Davis, R.W. and T.M. Williams. 2012. The marine mammal dive response is exercise modulated to maximize aerobic dive duration. *Journal of Comparative Physiology A* 198:583–591.

El-Sayed, H., S.R. Goodall, and R. Hainsworth. 1995. Re-evaluation of Evans Blue dye dilution method of plasma volume measurements. *Clinical and Laboratory Haematology* 17:189–194.

Elsner, R. 1965. Heart rate response in forced versus trained experimental dives of pinnipeds. *Hvalradets Skrifter* 48:24–29.

Elsner, R. 2015. *Diving Seals and Meditating Yogis: Strategic Metabolic Retreats.* Chicago, IL: Universtiy of Chicago Press.

Elsner, R., D.L. Franklin, R.L. Van Citters, and D.W. Kenney. 1966. Cardiovascular defense against asphyxia. *Science* 153:941–949.

Elsner, R., J.T. Shurley, D.D. Hammond, and R.E. Brooks. 1970. Cerebral tolerance to hypoxemia in asphyxiated Weddell seals. *Respiration Physiology* 9:287–297.

Fahlman, A., S.H. Loring, M. Ferrigno et al. 2011. Static inflation and deflation pressure–volume curves from excised lungs of marine mammals. *Journal of Experimental Biology* 214:3822–3828.

Gallivan, G.J., J.W. Kanwisher, and R.C. Best. 1986. Heart rates and gas exchange in the Amazonian manatee (*Trichecus manutus*) in relation to diving. *Journal of Comparative Physiology B* 156:415–423.

Gentry, R.L. and G.L. Kooyman, eds. 1986. *Fur Seals: Maternal Strategies on Land and at Sea.* Princeton, NJ: Princeton University Press.

Guppy, M., R.D. Hill, G.C. Liggins, W.M. Zapol, and P.W. Hochachka. 1986. Micro-computer assisted metabolic studies of voluntary diving of Weddell seals. *American Journal of Physiology* 250:175–187.

Guyton, G.P., K.S. Stanek, R.C. Schneider et al. 1995. Myoglobin-saturation in free-diving Weddell seals. *Journal of Applied Physiology* 79:1148–1155.

Hassrick, J.L., D.E. Crocker, N.M. Teutschel et al. 2010. Condition and mass impact oxygen stores and dive duration in adult female northern elephant seals. *Journal of Experimental Biology* 213(4):585–592.

Hill, R.D. 1986. Microcomputer monitor and blood sampler for free-diving Weddell seals Leptonychotes weddelli. *Journal of Applied Physiology* 61:1570–1576.

Hill, R.D., R.C. Schneider, G.C. Liggins et al. 1987. Heart rate and body temperature during free diving of Weddell seals. *American Journal of Physiology* 253:R344–R351.

Hindle, A.G., B.L. Young, D.A.S. Rosen, M. Haulena, and A.W. Trites. 2010. Dive response differs between shallow- and deep-diving Steller sea lions. *Journal of Expermental Marine Biology and Ecology* 394:141–148.

Kerem, D. and R. Elsner. 1973. Cerebral tolerance to asphyxial hypoxia in the harbor seal. *Respiration Physiology* 19:188–200.

Kooyman, G.L. 1973. Respiratory adaptations in marine mammals. *American Zoologist* 13:457–468.

Kooyman, G.L. 1989. Diverse divers physiology and behavior. In: D.S. Farner, ed., *Zoophysiology*, Vol. 23. Berlin, Germany: Springer-Verlag.

Kooyman, G.L., M.A. Castellini, R.W. Davis, and R.A. Maue. 1983. Aerobic dive limits in immature Weddell seals. *Journal of Comparative Physiology* 151:171–174.

Kooyman, G.L., D.H. Kerem, W.B. Campbell, and J.J. Wright. 1973. Pulmonary gas exchange in freely diving Weddell seals (*Leptonychotes weddelli*). *Respiration Physiology* 17:283–290.

Kooyman, G.L. and P.J. Ponganis. 1998. The physiological basis of diving to depth: Birds and mammals. *Annual Review of Physiology* 60:19–32.

Kooyman, G.L. and E.E. Sinnett. 1982. Pulmonary shunts in harbor seals and sea lions during simulated dives to depth. *Physiological Zoology* 55:105–111.

Kooyman, G.L., E.A. Wahrenbrock, M.A. Castellini, R.W. Davis, and E.E. Sinnett. 1980. Aerobic and anaerobic metabolism during diving in Weddell seals: Evidence of preferred pathways from blood chemistry and behavior. *Journal of Comparative Physiology* 138:335–346.

Lenfant, C., K. Johansen, and J.D. Torrance. 1970. Gas transport and oxygen storage capacity in some pinnipeds and the sea otter. *Respiration Physiology* 9:277–286.

Lockyer, C. 1976. Body weights of some species of large whales. *Journal du Conseil/Conseil Permanent International pour l'Exploration de la Mer* 36:259–273.

Masuda, K., T. Kent, L. Ping-Chang et al. 2008. Determination of myoglobin concentration in blood perfused tissue. *European Journal of Applied Physiology* 104:41–48.

McDonald, B.I. and P.J. Ponganis. 2012. Lung collapse in the diving sea lion: Hold the nitrogen and save the oxygen. *Biology Letters* 8(6):1047–1049.

McDonald, B.I. and P.J. Ponganis. 2013. Insights from venous oxygen profiles: Oxygen utilization and management in diving California sea lions. *Journal of Experimental Biology* 216:3332–3341.

McDonald, B.I. and P.J. Ponganis. 2014. Deep-diving sea lions exhibit extreme bradycardia in long-duration dives. *Journal of Experimental Biology* 217:1525–1534.

Meir, J.U., C.D. Champagne, D.P. Costa, C.L. Williams, and P.J. Ponganis. 2009. Extreme hypoxemic tolerance and blood oxygen depletion in diving elephant seals. *American Journal of Physiology Regulatory, Integrative and Comparative Physiology* 297(4):R927–R939.

Meir, J.U. and P.J. Ponganis. 2009. High-affinity hemoglobin and blood oxygen saturation in diving emperor penguins. *Journal of Experimental Biology* 212(20):3330–3338.

Miller, P.J., M.P. Johnson, P.L. Tyack, and E.A. Terray. 2004. Swimming gait, passive drag and buoyancy of diving sperm whales. *Journal of Experimental Biology* 207:1953–1967.

Neshumova, T.V. and V.A. Cherepanova. 1984. Blood supply and myoglobin stocks in muscles of the seal *Pusa siberica* and muskrat *Ondatra zibethica*. *Journal of Evolutionary Biochemistry and Physiology* 20:282–287.

Neshumova, T.V., V.A. Cherapanova, and E.A. Petrov. 1983. Myoglobin concentration in muscles of the seal *Pusa sibirica*. *Journal of Evolutionary Biochemistry and Physiology* 19:93–95.

Noren, S., T. Williams, K. Ramirez, J. Boehm, M. Glenn, and L. Cornell. 2012. Changes in partial pressures of respiratory gases during submerged voluntary breath hold across odontocetes: Is body mass important? *Journal of Comparative Physiology B: Biochemical, Systemic, and Environmental Physiology* 182(2):299–309.

Petrov, E.A. and K.A. Shoshenko. 1987. Total store of oxygen and duration of diving of the Nerpa. In: G.I. Galazii, ed., *Morphology and Ecology of Fish*. Novbosibirsk, Russia: Academy of Sciences of Russia, Siberian Division, pp. 110–128.

Piscitelli, M., W. McLellan, S. Rommel, J. Blum, S. Barco, and D.A. Pabst. 2010. Lung size and thoracic morphology in shallow (*Tursiops truncatus*) and deep (*Kogia* spp.) diving cetaceans. *Journal of Morphology* 271:654–673.

Piscitelli, M.A., S.A. Raverty, M.A. Lillie, and R.E. Shadwick. 2013. A review of cetacean lung morphology and mechanics. *Journal of Morphology* 274(12):1425–1440.

Polasek, L. and R.W. Davis. 2001. Heterogeneity of myoglobin distirbution in the locomotory muscles of five cetacean species. *Journal of Experimental Biology* 204:209–215.

Polasek, L., K.A. Dickson, and R.W. Davis. 2006. Metabolic indicators in the skeletal muscles of harbor seals (*Phoca vitulina*). *American Journal of Physiology* 290:R1720–R1727.

Ponganis, P.J. 2007. Bio-logging of physiological parameters in higher marine vertebrates. *Deep-Sea Research II* 54:183–192.

Ponganis, P.J. 2011. Diving mammals. In: R. Terjung, ed., *Comprehensive Physiology*. Hoboken, NJ: John Wiley & Sons, Inc., pp. 447–465.

Ponganis, P.J. 2015. *Diving Physiology of Marine Mammals and Seabirds*. Cambridge, UK: Cambridge University Press.

Ponganis, P.J., G.L. Kooyman, E.A. Baronov, P.H. Thorson, and B.S. Stewart. 1997a. The aerobic submersion limit of Baikal seals, *Phoca sibirica*. *Canadian Journal of Zoology* 75:1323–1327.

Ponganis, P.J., G.L. Kooyman, and M.A. Castellini. 1993. Determinants of the aerobic dive limit of Weddell seals: Analysis of diving metabolic rates, post-dive end tidal P_{O2}'s, and blood and muscle oxygen stores. *Physiological Zoology* 66:732–749.

Ponganis, P.J., G.L. Kooyman, and S.H. Ridgway. 2003. Comparative diving physiology. In: A.O. Brubakk and T.S. Neuman, eds., *Physiology and Medicine of Diving*. Edinburgh, UK: Saunders, pp. 211–226.

Ponganis, P.J., G.L. Kooyman, L.N. Starke, C.A. Kooyman, and T.G. Kooyman. 1997b. Post-dive blood lactate concentrations in emperor penguins, *Aptenodytes forsteri*. *Journal of Experimental Biology* 200:1623–1626.

Ponganis, P.J., G.L. Kooyman, L.M. Winter, and L.N. Starke. 1997c. Heart rate and plasma lactate responses during submerged swimming and diving in California sea lions (*Zalophus californianus*). *Journal of Comparative Physiology B* 167:9–16.

Ponganis, P.J., J.U. Meir, and C.L Williams. 2011. In pursuit of Irving and Scholander: A review of oxygen store management in seals and penguins. *Journal of Experimental Biology* 214:3325–3339.

Ponganis, P.J., T.K. Stockard, J.U. Meir, C.L. Williams, K.V. Ponganis, and R. Howard. 2009. O_2 store management in diving emperor penguins. *Journal of Experimental Biology* 212:217–224.

Qvist, J., R.D. Hill, R.C. Schneider et al. 1986. Hemoglobin concentrations and blood gas tensions of free-diving Weddell seals. *Journal of Applied Physiology* 61:1560–1569.

Reynafarje, B. 1963. Simplified method for the determination of myoglobin. *Journal of Laboratory and Clinical Medicine* 61:139–145.

Ridgway, S.H. 1986. Diving by cetaceans. In: A.O. Brubakk, J.W. Kanwisher, and G. Sundnes, eds., *Diving in Animals and Man*. Trondheim, Norway: Royal Norwegian Society of Science and Letters, pp. 33–62.

Ridgway, S.H. and D.G. Johnston. 1966. Blood oxygen and ecology of porpoises of three genera. *Science* 151:456–458.

Ridgway, S.H., B.L. Scronce, and J. Kanwisher. 1969. Respiration and deep diving in the bottlenose porpoise. *Science* 166:1651–1654.

Scholander, P.F. 1940. Experimental investigations on the respiratory function in diving mammals and birds. *Hvalradets Skrifter* 22:1–131.

Scholander, P.F., L. Irving, and S.W. Grinnell. 1942. Aerobic and anaerobic changes in seal muscle during diving. *Journal of Biological Chemistry* 142:431–440.

Shaffer, S.A., D.P. Costa, T.M. Williams, and S.H. Ridgway. 1997. Diving and swimming performance of white whales, *Delphinapterus leucas*: An assessment of plasma lactate and blood gas levels and respiratory rates. *Journal of Experimental Biology* 200:3091–3099.

Simpson, J.G., W.G. Gilmartin, and S.H. Ridgway. 1970. Blood volume and other hematologic values in young elephant seals (*Mirounga angustirostris*). *American Journal of Veterinary Research* 31:14449–14452.

Sleet, R.B., J.L. Sumich, and L.J. Weber. 1981. Estimates of total blood volume and total body weight of sperm whale (*Physeter catodon*). *Canadian Journal of Zoology* 59:567–570.

Thometz, N.M., M. Murray, and T.M. Williams. 2015. Ontogeny of oxygen storage capacity and diving ability in the southern sea otter (*Enhydra lutris* nereis): Costs and benefits of large lungs. *Physiological and Biochemical Zoology* 88:311–327.

Thompson, D. and M.A. Fedak. 1993. Cardiac responses of grey seals during diving at sea. *Journal of Experimental Biology* 174:139–164.

Thorson, P.H. and B.J. Le Boeuf. 1994. Developmental aspects of diving in northern elephant seal pups. In: B.J. Le Boeuf and R.M. Laws, eds., *Elephant Seals: Population Ecology, Behavior, and Physiology*. Berkeley, CA: University of California Press.

Turner, A.W. and V.E. Hodgetts. 1959. The dynamic red cell storage function of the spleen in sheep. I. Relationship to fluctuations of jugular hematocrit. *Australian Journal of Experimental Biology* 37:399–420.

Villegas-Amtmann, S. and D.P. Costa. 2010. Oxygen stores plasticity linked to foraging behaviour and pregnancy in a diving predator, the Galapagos sea lion. *Functional Ecology* 24(4):785–795.

Weise, M.J. and D.P. Costa. 2007. Total body oxygen stores and physiological diving capacity of California sea lions as a function of sex and age. *Journal of Experimental Biology* 210:278–289.

Williams, C.L., J.U. Meir, and P.J. Ponganis. 2011. What triggers the aerobic dive limit? Muscle oxygen depletion during dives of emperor penguins. *Journal of Experimental Biology* 214:1801–1812.

Williams, T.M., W.A. Friedl, M.L. Fong, R.M. Yamada, P. Sedivy, and J.E. Haun. 1992. Travel at low energetic cost by swimming and wave-riding bottlenose dolphins. *Nature* 355:821–823.

Williams, T.M., J.E. Haun, and W.A. Friedl. 1999. The diving physiology of bottlenose dolphins (*Tursiops truncatus*). I. Balancing the demands of exercise for energy conservation at depth. *Journal of Experimental Biology* 202:2739–2748.

Williams, T.M., S.R. Noren, and M. Glenn. 2011. Extreme physiological adaptations as predictors of climate-change sensitivity in the narwhal, *Monodon monoceros*. *Marine Mammal Science* 27(2):334–349.

Zapol, W.M. 1987. Diving adaptations of the Weddell seal. *Scientific American* 256(6):100–105.

chapter three

Exercise energetics

Terrie M. Williams and Jennifer L. Maresh

Contents

3.1 Introduction: The challenge of exercising while diving

The question of how cetaceans, pinnipeds, and other marine mammal groups exercise while holding their breath has been debated by scientists for many decades (Castellini et al. 1985; Butler 1988; Davis et al. 2004; Davis and Williams 2012). In the previous chapters, we found that the solution to this problem has involved external changes such as body streamlining and efficient swimming styles to reduce energetic costs (Chapter 1), as well as internal changes, most notably the enhancement of on-board oxygen stores to support aerobic processes while underwater (Chapter 2). Here, we examine how the efficient use of those oxygen stores, termed *energetics*, enables marine mammals to breath-hold for periods far beyond those of terrestrial mammals, and consequently allows this group of mammals to dive to remarkable depths.

As air-breathing vertebrates, all marine mammals are obligated to periodically return to the water surface to breathe. This interruption in aquatic activities, one of the most important being foraging, is due to an evolutionary history that involved the re-invasion of the oceans by the terrestrial ancestors of cetaceans (Thewissen et al. 1994), pinnipeds (Berta 2012), sea otters (Riedman and Estes 1990), and sirenians (Barnes et al. 1985). As a result, the internal building blocks required for exercising while diving by ancestral marine mammals included tolerance or modification of morphological, physiological, biochemical, and molecular mechanisms originally intended for locomotion on land (Williams 1999).

The paradox concerning exercise while diving stems from two seemingly conflicting physiological responses that occur when a marine mammal submerges. The ensuing dive response is characterized by the cessation of breathing (apnea), a slowing of the heart (bradycardia), peripheral vasoconstriction, and metabolic downregulation of noncritical tissues (Chapter 2). Conversely, the exercise response of mammals promotes an increase in metabolism, heart rate (tachycardia), and respiratory rates. How can both responses occur when a submerged marine mammal is actively swimming, migrating, or pursing underwater prey? We might expect that limited on-board oxygen stores would constrain energetically expensive behaviors in diving mammals. Instead, we find that many species of marine mammal that chase and consume prey while holding their breath often demonstrate exceptional speeds and complex, energetically costly maneuvers during foraging (Aguilar de Soto et al. 2008; Goldbogen et al. 2008).

Studies that simultaneously measured swimming performance, stroke frequency, and heart rate in diving mammals provide clues about the balance between diving and exercise responses in cetaceans and pinnipeds (Davis and Williams 2012; Noren et al. 2012; Williams et al. 2015b). Rather than a single dominant response, there is an interplay between exercise and dive responses in which both the depth of the dive and the intensity of exercise alter the level of bradycardia. This variability impacts the energetic and physiological costs associated with a dive. The deeper the dive by cetaceans and pinnipeds, the lower the minimum heart rate achieved (Figure 3.1a–c). Instead of maintaining a single level of bradycardia when submerged, marine mammals show considerable variability in diving heart rate. Swimming exercise plays an important role in dictating this variability. Periods of high stroke frequency show a relaxation in bradycardia (higher heart rates) while low stroke frequency swimming during a dive is associated with the most intense levels of bradycardia (and lowest heart rates). Such responses have been demonstrated for a wide range of species including bottlenose dolphins (*Tursiops truncatus*), Weddell seals (*Leptonychotes weddellii*), gray seals (*Halichoerus grypus*), California sea lions (*Zalophus californianus*), and human breath-hold divers. Importantly, because the heart rate of marine mammals, like terrestrial mammals, correlates directly with oxygen consumption (Figure 3.1d), cardiac variability will influence the movement and utilization of oxygen and carbon dioxide during a dive.

The physical separation of two critical resources required for survival by diving mammals, air at the water surface and prey at depth, should not be considered a simple biological constraint. Instead, the physiological challenge associated with this dichotomy may have led to a unique selection pressure for locomotor efficiency in pinnipeds, cetaceans, and other mammalian groups that hunt while submerged (Williams et al. 2015a). In the following section, we examine the energetics of marine living mammals and the costs related to resting, swimming, and diving activities. These costs represent the maintenance energetic requirements of the animal as well as the additional energy that must be expended to move though water. It is important to recognize that depending on the species, age, and reproductive status, the total energetic balance of an individual marine mammal must also account for the energy required for thermoregulation, growth, reproduction, and the assimilation of food, which will be addressed in subsequent chapters.

3.2 *Energetic costs of marine mammals*

How many fish must a marine mammal eat to survive or to successfully reproduce? Answering such questions requires an understanding of how energetic costs are partitioned across time by an animal. Often, metabolic rate is used as the common metric for evaluating these costs, as it provides an index of the instantaneous and long-term energy requirements, fuel utilization, and heat production of an animal. For marine mammals, the rate of metabolism also serves as an important indicator of the consumption rate of oxygen stores within the lungs, blood, and muscle that represent a major determinant of breath-hold duration (Chapter 2). Thus, the metabolic rates from rest to high activity, and the associated energetic costs ultimately dictate the diving capacity and survival of individual marine mammal species.

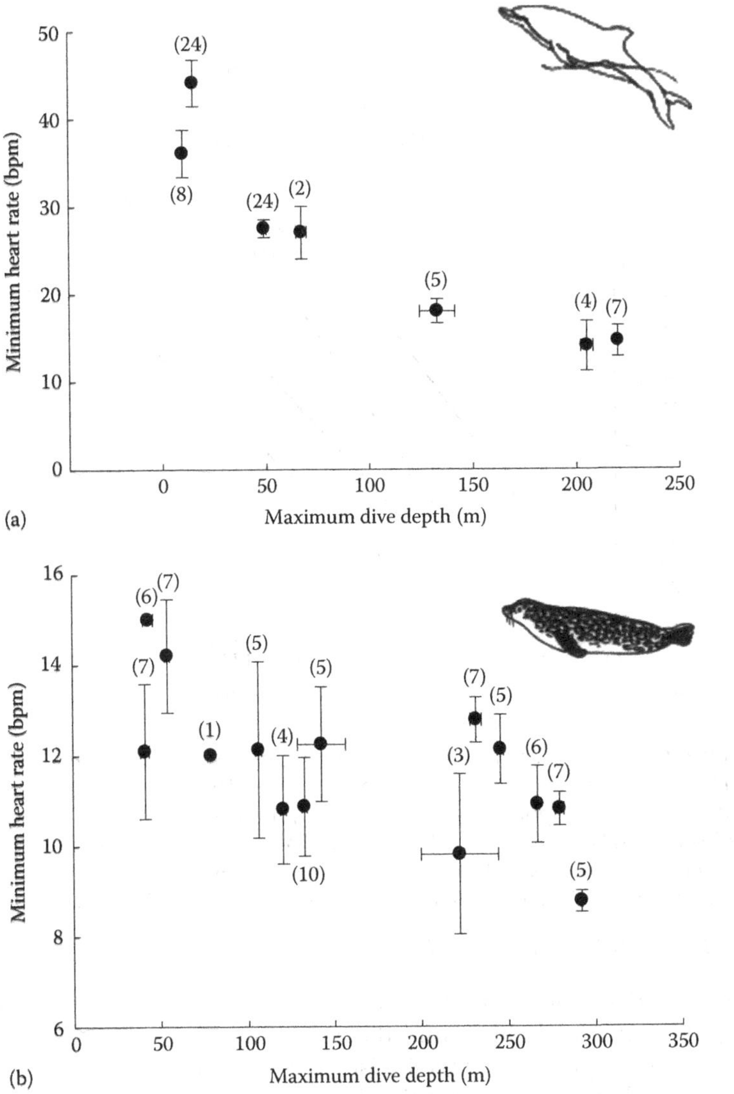

Figure 3.1 Cardiac responses during diving and exercise in marine mammals. The minimum heart rate (level of bradycardia achieved) is plotted in relation to the maximum dive depth for bottlenose dolphins (a) and Weddell seals (b).

(Continued)

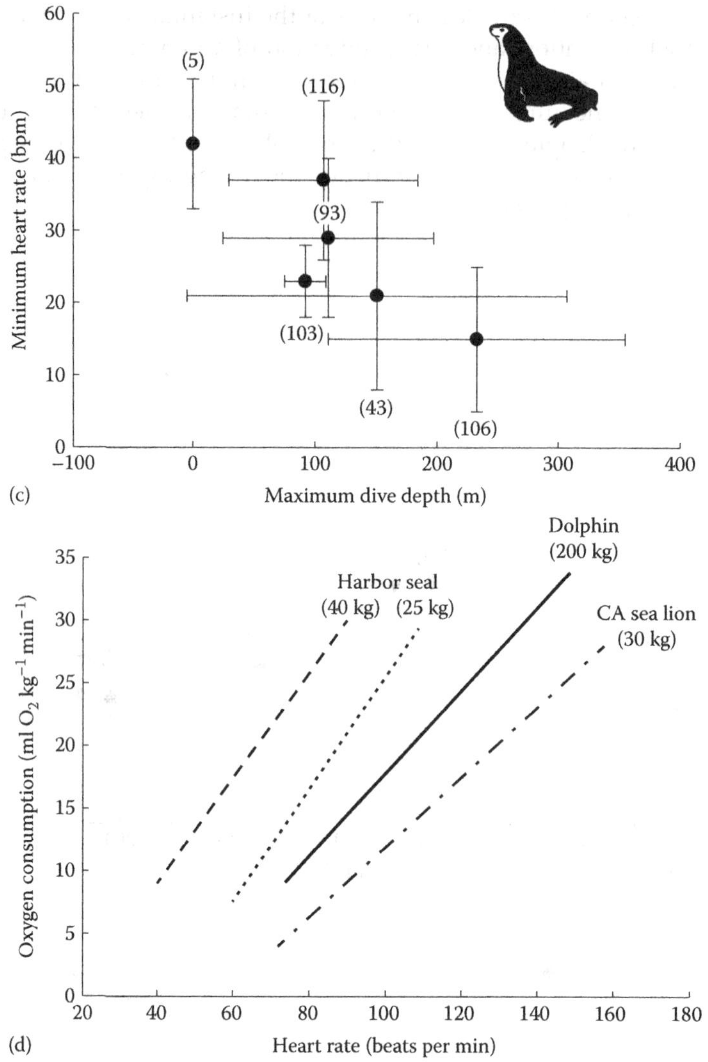

***Figure 3.1* (*Continued*)** Cardiac responses during diving and exercise in marine mammals. California sea lions (c). The number of dives is indicated in parentheses. Panel (d) shows the relationship between heart rate and oxygen consumption during exercise for pinnipeds swimming in a flume and dolphins pushing on a load cell. (Data are from Williams, T.M. et al., *J. Compar. Physiol. B*, 160(6), 637, 1991; Butler, P.J. et al., *Funct. Ecol.*, 18(2):168, 2004. Williams, T.M. et al., *J. Exp. Biol.*, 179(1), 31, 1993; McDonald, B.I. and Ponganis, P.J., *J. Exp. Biol.*, 217(9), 1525, 2014; Williams, T.M. et al., *Nat. Commun.*, 6, 6055, 2015b.)

3.2.1 *Resting metabolic rates and costs*

Resting Metabolic Rate, sometimes termed basal metabolic rate, is a measure of the minimum energy requirements for an animal that is awake and alert but otherwise not engaged in any activity that might elevate metabolism above baseline maintenance levels (Kleiber 1975). In this physiological state, the animal is a non-reproductive adult (e.g., no growing, pregnancy, or lactation costs), inactive, fasted (e.g., post-prandial), relaxed but not asleep, and in a thermally neutral environment. For accurate inter-species comparisons, it is critical that these so-called *Kleiber standards* are met during metabolic measurements.

Because the Kleiber standards were originally developed for terrestrial mammals measured in air, applying them to semi- or fully aquatic mammals has proved challenging. This has led to differing opinions as to what represents *resting* in a marine mammal. It is unclear whether resting metabolic rates of sea otters, pinnipeds and cetaceans measured in air is comparable to that measured in water. The dive response, described earlier and in Chapter 2, complicates measurements as it can affect metabolic rate through cardiovascular changes within individual tissues (Zapol et al. 1979). For some species, face immersion is enough to trigger this response (Ridgway et al. 1975). Thus, metabolic rates measured during submersion are considered by some to be confounded by the dive response; others believe these represent true baseline energetic costs in marine mammals. The definition of maintenance metabolism is especially complicated for species such as cetaceans and sirenians, for whom measurements on land are neither feasible nor physiologically relevant. For deep-diving species, some non-essential tissues may enter a hypometabolic state which reduces the rate of oxygen consumption while diving (Maresh et al. in review) as well as provides a critical seasonal strategy for conserving energy while fasting (Tift et al. 2013). For these reasons, comparisons of the baseline resting costs between the various marine mammal groups or between marine mammals and terrestrial mammals are not always straightforward.

Despite the limitations, some general trends in resting metabolism are apparent for the major marine mammal groups (Figure 3.2a). The relationship between body mass and Resting Metabolic Rate (RMR_{MM}) for marine mammals is best described by

$$RMR_{MM} = 581 mass^{0.68} \quad (n = 12 \text{ species}, r^2 = 0.66) \tag{3.1}$$

where

metabolic rate is in kJ day^{-1}

body mass is in kg (Maresh 2014)

From this relationship, it appears that with the exception of the sirenians and the tropical Hawaiian monk seal (*Monachus schauinslandi*, Williams et al. 2011), marine mammals have approximately twice the resting metabolic costs of similarly sized terrestrial mammals. This trend is most consistent in otariids, cetaceans and sea otters, but more variable in phocid seals. Notably, this allometric regression includes marine mammals ranging in size from a 25 kg sea otter (*Enhydra lutris*) to a 5300 kg killer whale (*Orcinus orca*). Adult mysticete whales far exceed this mass range (e.g., an adult blue whale, *Balaenoptera musculus*, is >180,000 kg; Folkens et al. 2002), and the Resting Metabolic Rates of the largest of the mysticetes have never been directly measured. It is unknown how such extreme body size will affect the allometric relationship for Resting Metabolic Rate in marine mammals.

Many explanations have been proposed for the observed elevation in Resting Metabolic Rates of marine mammals compared to terrestrial mammals predicted by Kleiber (1975). Currently, the most widely accepted hypothesis suggests that an elevation in metabolic rate is necessary to offset the high cost of endothermy when in water (e.g., Speakman and Król 2010; Hudson et al. 2013). Because thermoregulation may not be a problem for all but the smallest marine mammals (Boyd 2002; Porter and Kearney 2009), many investigators have proposed that other traits in marine mammals set metabolism (e.g., Liwanag et al. 2012; Heim et al. 2015). In these studies, high metabolic rates are considered an exaptation rather than an adaptation per se, whereby the elevated metabolism of ancestral carnivores that secondarily invaded the marine environment provided a competitive advantage over

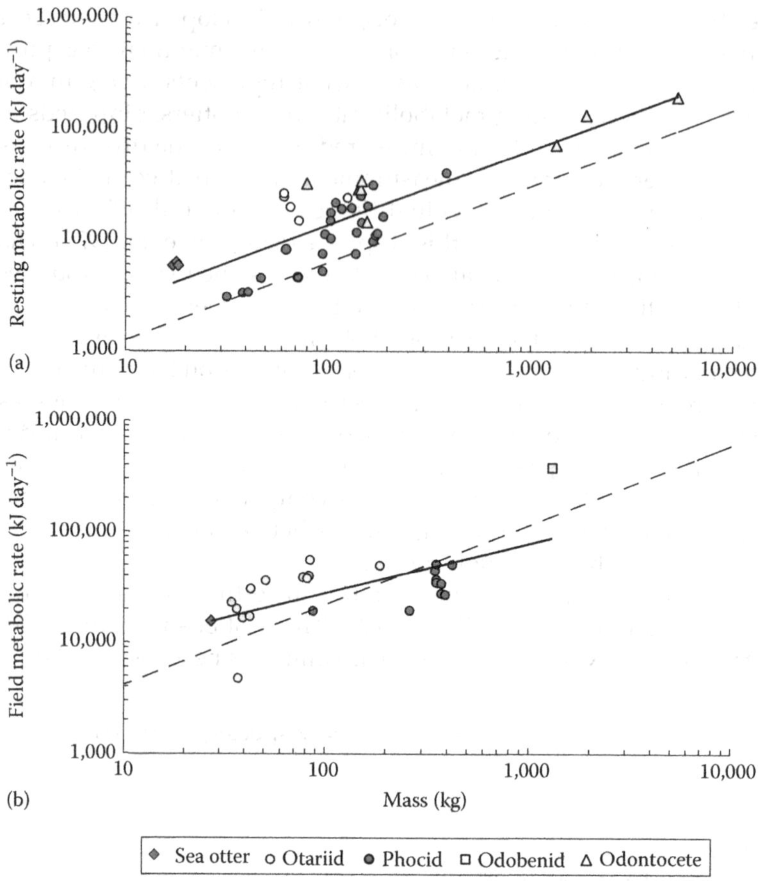

Figure 3.2 Resting (a) and field (b) metabolic rates of marine mammals (solid lines) compared to terrestrial mammals (dashed lines). Each point denotes a separate metabolic measurement for an individual marine mammal. Symbols representing the major marine mammal groups are shown on the bottom. (Data from Maresh, J.L., Bioenergetics of marine mammals: The influence of body size, reproductive status, locomotion and phylogeny on metabolism, PhD thesis, University of California, Santa Cruz, CA, 2014.)

other aquatic species with low metabolic rates. Until additional data that adhere to standard Kleiber measurement conditions are available for a wider range of marine mammals, some confusion regarding what physiological state represents baseline or resting conditions in aquatic mammals will remain.

3.2.2 *Metabolic costs of swimming and diving*

As for any active animal, the metabolic rate of marine mammals increases in response to the level of physical exertion. However, the energy expended for swimming by marine mammals differs from the pattern observed for running by terrestrial mammals and reflects the unique physical forces that must be overcome to move through water. In Chapter 1, we learned that the body drag of swimmers increases exponentially with locomotor speed. This has a profound effect on the cost of swimming, especially at higher speeds. For runners, the rate of oxygen consumption ($\dot{V}O_2$) typically increases linearly with speed until aerobic limits are reached (Taylor et al. 1982). In contrast, the $\dot{V}O_2$ of phocid seals (Davis et al. 1985; Fedak 1986), sea lions (Feldkamp 1987), dolphins (Williams et al. 1992), and killer whales (Kreite 1994) increases curvilinearly as the animals swim faster in flumes or in open water (Figure 3.3a).

The shape of this relationship has two key effects on the swimming behaviors of marine mammals. First, slow or routine speeds can be performed for prolonged periods as they often result in relatively small physiological changes from rest. This facilitates prolonged performance such as migrations or lengthy movements between foraging areas that may last from hours to months. For example, the metabolic rate, respiration rate, heart rate, and levels of blood lactate of oceanic bottlenose dolphins swimming at routine speeds of 2.0 m s^{-1} are only slightly elevated from levels measured during rest in water (Williams et al. 1992). By comparison, the

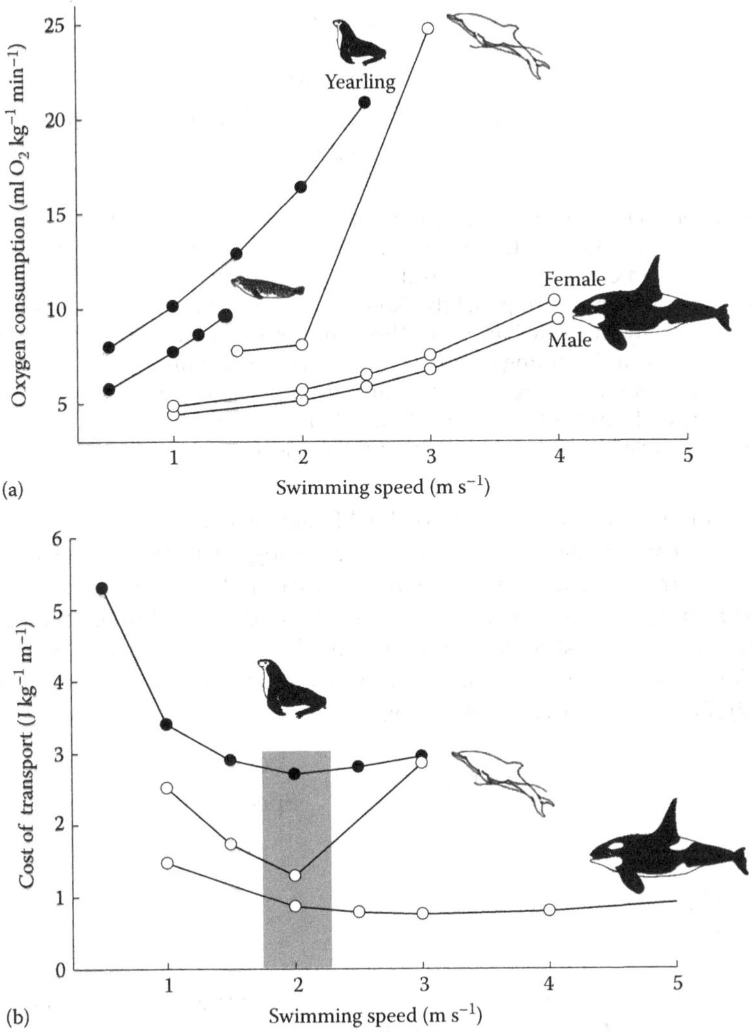

Figure 3.3 Oxygen consumption (a) and cost of transport (b, c) for swimming in marine mammals. Panel (a) shows the curvilinear increase in oxygen consumption with the swimming speed in pinnipeds and cetaceans. This results in a nonlinear relationship between the cost of transport (COT) and speed as shown in panel (b). Note that the minimum COT occurs in the mid-range of speeds as denoted by the gray bar. *(Continued)*

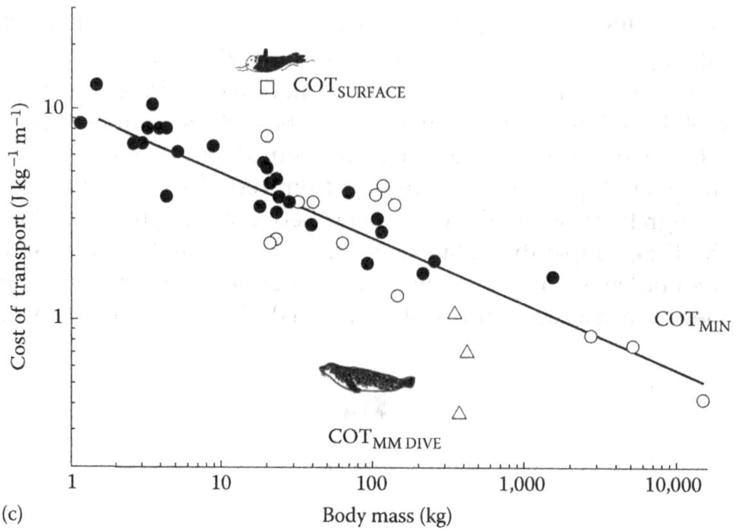

(c)

Figure 3.3 (Continued) Oxygen consumption (a) and cost of transport (b, c) for swimming in marine mammals. Panel (a) shows the curvilinear increase in oxygen consumption with swimming speed in pinnipeds and cetaceans. This results in a nonlinear relationship between cost of transport (COT) and speed as shown in panel (b). Note that the minimum COT occurs in the mid-range of speeds as denoted by the gray bar. In (c), these minimum values are compared for swimming marine (open symbols) and running terrestrial (filled symbols) mammals. Costs of surface (square) and submerged (circle) swimming sea otters and diving Weddell seals (triangles) are provided for comparison. (Data from Williams, T.M., *Philos. Trans. Roy. Soc. B: Biol. Sci.*, 354(1380), 193, 1999; Williams, T.M. et al., *Integr. Compar. Biol.*, doi:10.1093/icb/icv025, 2015a.)

2012 Olympic gold medal sprint pace of Michael Phelps was 1.9 m s^{-1} for an 800 m swim with an obvious, significant increase in respiratory rate during the post-race period.

The second effect of the curvilinear relationship between oxygen consumption and speed is that high speed swimming is enormously expensive. Small changes in speed as marine mammals move past preferred ranges result in a progressive increase in the cost of transport (COT), that is the energetic cost to move a unit of body mass a meter (Schmidt-Nielsen 1972) (Figure 3.3b,c), calculated from

$$\text{COT (J kg}^{-1}\text{ m}^{-1}) = \frac{\dot{V}O_2}{\text{Speed}} \tag{3.2}$$

where

$\dot{V}O_2$ is in J kg^{-1} s^{-1}
speed is in m s^{-1}

The resulting COT is an important metric for comparing the energy expended for locomotion by different animals. The increase in COT with speed for swimming marine mammals is in marked contrast to the progressive decline in COT with speed that occurs for running mammals (Taylor et al. 1970). As a result, high speed swimming by most marine mammals, even the small, athletic dolphins, are either of comparatively short duration or involve behavioral maneuvers that enable the animals to circumvent the exceptionally high energetic transport costs of fast swimming (see Section 3.2.4). Even short-finned pilot whales (*Globicephala macrorhynchus*), named "the cheetahs of the deep sea," can sustain their unusual 9.0 m s^{-1} sprints chasing prey for only 20–80 s (Aguilar de Soto et al. 2008).

Together these changes in the amount of energy expended per meter moved result in a characteristic U-shaped relationship between COT and swimming speed (Figure 3.3b). Importantly, for each swimmer and species there is a minimum COT that defines the speed or range of speeds that enable the animals to move the furthest for the least investment of energy (Schmidt-Nielsen 1972). Equivalent to the miles per gallon (MPG) fuel economy rating of automobiles, the minimum COT (COT_{MIN}) provides an index of the energetic efficiency of marine mammals. From Figure 3.3b, it is obvious that COT_{MIN} occurs in the trough of the U-shaped curve that relates COT to swimming speed, and that it decreases with body mass. Interestingly, for many of the marine mammals measured to date this minimum occurs at approximately 1.5–2.5 m s^{-1} regardless of the size of the animal. Not surprisingly, the preferred or routine speeds of many marine mammals fall within or near this cost-efficient range of swimming speeds (see Chapter 1).

Like Resting Metabolic Rate, the minimum cost of transport for marine mammals varies predictably with body mass (Table 3.1) according to

$$COT_{MIN} = 7.79\ \text{mass}^{-0.29} \quad (n = 6\ \text{species},\ r^2 = 0.83) \tag{3.3}$$

where

COT_{MIN} is the minimum cost of transport in J kg^{-1} m^{-1}
body mass is in kg (Williams 1999)

Remarkably, this relationship is indistinguishable from that describing the total cost of transport for running mammals and attests to the shared ancestral lineages of highly active mammalian specialists. Terrestrial mammals, phocid seals, otariids, large and small odontocetes, and an estimate for a mysticete, the gray whale (*Eschrichtius robustus*), follow the same regression with a few notable exceptions (Figure 3.3c). Phocid seals tend to show lower transport costs compared to other marine mammals, especially when diving. This difference has been attributed to metabolic changes associated with the dive response as well as to the energy savings associated with incorporation of extended periods of gliding during deep dives (see Section 3.2.4). Conversely, sea otters demonstrate COT_{MIN} levels that are two or three times predicted for other similarly sized marine mammals, depending on whether the otter is swimming submerged or on the water surface, respectively (Williams 1989). The inefficient paddling swimming style of sea otters and a surface swimming position instigate these high costs, and demonstrate the challenges that ancestral marine mammals must have encountered when transitioning from land to sea (Williams 1999).

3.2.3 *Field metabolic rates*

As might be expected, measuring the energetic costs of free-ranging marine mammals that spend more than 90% of their lives submerged is challenging and has involved a wide variety of methods (see Section 3.3). Because energy expenditure is related to the body size for wild mammals (Nagy 2005), the field metabolic rate of marine mammals (FMR_{MM} in kJ day^{-1}) can be estimated by using the equation:

$$FMR_{MM} = 3511\ \text{mass}^{0.45} \quad (n = 10\ \text{species},\ r^2 = 0.43) \tag{3.4}$$

where body mass is in kg (Maresh 2014). Based on this equation developed from measurements of field energetic costs for marine mammals ranging in body mass from 27 kg sea

Table 3.1 Energetic costs and swimming speeds of marine mammals

Species	Body mass (kg)	$\dot{V}O_{2rest}$ (ml O_2 kg^{-1} min^{-1})	$\dot{V}O_{2swim}$ (ml O_2 kg^{-1} min^{-1})	COT_{MIN} (J kg^{-1} m^{-1})	Speed (m s^{-1})	Method
Sea otter (surface)	20	13.5	29.6	12.6	0.8	Flume
Sea otter (submerged)	20	13.5	17.6	7.4	0.8	Flume
California sea lion	21		13.7	2.3	2	Flume
	23	6.3	22	2.8	2.6	Flume
	23	6.6	13	2.4	1.8	Flume
Harbor seal	32		23.6	3.6	2.2	Flume
	33	5.1	15.2	3.6	1.4	Flume
	63	4.6	9.6	2.3	1.4	Flume
Gray seal	104	7.7	15	3.9	1.3	Flume
Steller sea lion	116		24.9	4.3	2	Flume
	139		22.6	3.5	2	Flume
Bottlenose dolphin	145	4.6	8.1	1.3	2.1	Ocean swim
Killer whale	2,738			0.84	3.1	Field respiration rate
	5,153			0.75	3.1	Field respiration rate
Gray whale	15,000			0.4	2.1	Field respiration rate

Sources: Data from Kreite, B., Bioenergetics of the killer whale, *Orcinus orca*, PhD thesis, University of British Columbia, Vancouver, British Columbia, Canada, 1994; Williams, T.M., *Philos. Trans. Roy. Soc. B: Biol. Sci.*, 354(1380), 193, 1999; Rosen, D.A.S. and Trites, A.W., *Marine Mammal. Sci.*, 18(2), 513, 2002.

Notes: Oxygen consumption at rest ($\dot{V}O_{2rest}$) was determined for animals resting on the water surface prior to exercise. Swimming values ($\dot{V}O_{2swim}$) were determined during exercise in flumes or open water as indicated. $\dot{V}O_2$ in ml O_2 kg^{-1} min^{-1} was converted to metabolic energy (joules, J) assuming a caloric equivalent of 4.8 kcal L^{-1} of O_2 and a conversion factor of 4.187×10^3 J kcal^{-1}.

otters to 1310 kg walruses (*Odobenus rosmarus*), we find that field energy expenditure in the smallest marine mammals is higher than would be predicted for a terrestrial mammal. Conversely, FMR is lower than would be predicted in the largest marine mammals (Figure 3.2b). When the comparison is made between marine mammals and only terrestrial carnivores, the differences in field metabolic rates often disappear (Maresh 2014). Although the intercept for the FMR to body mass relationships are not significantly different between the groups, the slope as indicated by the exponent is shallower for marine mammals (exponent = 0.45) than that for terrestrial mammals (exponent = 0.72). Based on this, it appears that mass-specific field metabolic rates decline with increasing body size at a faster rate in marine mammals than in terrestrial mammals, as proposed by Boyd (2002). The implication is that energy expenditure in marine mammals will be elevated above that of a terrestrial mammal up to approximately 250 kg, above which a marine mammal's field metabolic rate becomes more economical than that of a similarly sized terrestrial mammal. The diversity of marine mammal groups comprising the FMR relationship may be a critical underlying factor driving this trend. Thus, Equation 3.4, while useful as a first approximation of field metabolic rate, should be considered an estimate of the diverse energy needs of the many marine species that make up this mammalian group.

Based on their taxonomic and ecological diversity, we might expect different metabolic adaptations for aquatic living by mustelids, otariids, phocids, odontocetes, and mysticetes. Indeed, phylogeny does seem to explain much of the residual variation of FMR after body size is taken into account. Among marine mammals, phylogeny is highly correlated with activity levels, reproductive strategies, locomotor mechanics, and other physiological and ecological drivers of overall energy expenditure. For example, among the Pinnipedia, phocid seals tend to swim at more economical speeds and engage in less energetically costly acrobatic maneuvers than otariids (Fish 1994; Chapter 1) which is reflected in their FMR. Both athletic otariids and odonotocetes demonstrate higher field metabolic rates relative to the general mammalian trend. Mysticete whales are more likely to swim at low-cost cruising speeds rather than engage in costly sprints typical of small odontocetes. To date, field metabolic rates have not been directly measured in adult mysticetes. However, based on shared economical swimming patterns, it is reasonable to predict that the FMR of this cetacean group may follow the trends of phocid seals rather than those of otariids. Not surprisingly, sea otters, as small-bodied surface swimmers, show higher than predicted field metabolic rates among the marine mammals (Thometz et al. 2014). Using the same logic, we might also predict that Hawaiian monk seals and sirenians, as moderately sized, slow, tropical swimmers, will have low field metabolic rates compared to other marine mammal groups; this remains to be tested.

Because body size and phylogeny are correlated, it is possible to assign the major marine mammal taxonomic groups to one of two distinct *pace of life* categories based on these metrics. Species with a fast pace of life tend to be relatively small, highly active swimmers, and shallow diving, with a reproductive strategy that involves long lactation periods and little-to-no separation of foraging from lactation (i.e., income breeders). Membership in this group includes otariid seals, most odontocetes, and sea otters. Species with a slow pace of life tend to be relatively large, economical swimmers, and deep diving, with a reproductive strategy that involves short lactation periods and geographical and temporal separation of foraging from lactation (i.e., capital breeders). Membership in this group includes phocid seals, and possibly some of the larger but inaccessible odontocetes and mysticetes. As the only herbivorous marine mammal group, sirenians represent a special case of a taxonomic group with exceptionally low energy costs below those of any similarly sized marine or terrestrial mammal. Tropical living, as exemplified by the Hawaiian monk seal, may also contribute to the slower life pace of these marine mammals.

3.2.4 *Behavioral strategies to reduce the energetic cost of swimming and diving*

In contrast to the constant stroking of swimming humans, marine mammals display a wide range of unsteady swimming behaviors when moving through water (Williams et al. 2015a). Both pinnipeds and cetaceans use extended glides, burst-and-glide swimming, wave-riding, and for some species a roller-coaster pattern of movements while transiting (Davis et al. 2003; Davis and Weihs 2007) and foraging (Williams et al. 2015b) that reduce the energetic costs associated with locomotion. The obvious benefit of these behaviors is the conservation of oxygen stores and prolongation of aerobic dives.

Two key factors, buoyancy control and drag reduction through body streamlining, facilitate these energy-saving strategies by marine mammals. As discussed in Chapter 1, the streamlined shape and submerged swimming position of marine mammals reduce the overall hydrodynamic drag encountered, and the associated energy required for moving through the water. Once again this is illustrated by the sea otter, a unique species of marine mammal that can swim for extended periods either on the water surface or submerged. The oxygen consumption of sea otters swimming submerged at 0.80 m s^{-1} is 17.55 ml O_2 kg^{-1} min^{-1}, which is reduced by more than 40% if the animal simply changes to a submerged mode of swimming at the same speed (Williams 1989). As discussed above, this difference impacts the minimum cost of transport of sea otters, which is 69% higher for the animals swimming on the water surface than when submerged (Figure 3.3c). In view of this, it is not surprising that marine mammals spend 90%–95% of their time below the water surface when migrating across ocean basins, moving between prey patches, or even transiting short distances.

The necessity to breathe requires that marine mammals periodically surface, and even here we find several behaviors that enable the animals to reduce the energetic cost of swimming. High speed swimming near the water surface is exceptionally expensive due to additional drag associated with the generation of waves (Chapter 1). Consequently, surface intervals tend to be brief for marine mammals, ranging from <1 s for fast swimming dolphins and otariids (Hui 1989) to 3–5 s and longer for harbor seals (*Phoca vitulina*) (Williams et al. 1991) and large whales.

Body streamlining facilitates other cost-efficient behaviors when fast-moving marine mammals swim near the water surface. These include wave-riding and porpoising. As the latter name implies these behaviors are most often equated with small cetaceans; however, sea lions, harbor seals, and fur seals have also been observed to occasionally perform both energy-saving strategies when swimming at high speed.

Porpoising enables fast-swimming mammals to avoid the problem of elevated drag on the water surface by leaping into the air. By comparing the theoretical energetic costs of dolphins swimming on the water surface to the costs associated with swimming submerged and leaping, Au and Weihs (1980) and Blake (1983) predicted the most economical swimming positions based on the speed of forward movement. For all, speeds swimming more than three body diameters below the water surface were the most economical position for a dolphin. When surfacing to breathe, the optimum swimming position changed with speed. Theoretically, the cost of leaping is greater than the energy expended to overcome surface drag at slow speeds, so slow swimming dolphins should remain in the water when breathing at the surface. At high speeds, the relative energetic costs are reversed as wave drag increases exponentially. As a result, the most energetically efficient strategy is to take to the air and *porpoise* when taking a breath.

Wave-riding by dolphins has been described by observers aboard ships since the time of Greek mythology. Wild dolphins appear to surf effortlessly by positioning their streamlined bodies in the bow or stern wakes of boats, matching the speed of the vessel without propulsive movements by their flukes (Scholander 1959). Indeed, the physiological responses of bottlenose dolphins trained to ride in the wake of a research boat demonstrated a significant energetic benefit to wave-riding (Williams et al. 1992). The heart rate, respiration rate, and by inference metabolic rate were reduced when dolphins changed from active swimming near the water surface to wave-riding. The resulting minimum cost of transport for wave-riding at 3.8 m s^{-1} was nearly identical to that recorded for the same dolphins freely swimming at 2.1 m s^{-1} outside of the wake zone of the boat. In this case, wave-riding enabled the dolphins to move twice as fast for the same energetic expenditure.

Interestingly, marine mammals will take advantage of waves generated from a variety of sources including wind, currents, and even larger whales (Woodcock and McBride 1951; Caldwell and Fields 1959; Würsig and Würsig 1979) in addition to the wake of ships in order to save energy by surfing.

Marine mammals also rely on an economy of movement to reduce energetic costs when submerged. For a wide variety of species, stroke frequency may be decoupled from speed during the descent phase of a dive. This is due primarily to the incorporation of prolonged (>12 s) periods of gliding (Figure 3.4; Williams et al. 2000). The ability to *turn the motor off* has been attributed to buoyancy changes with hydrostatically induced lung compression at depth and allows many species of marine mammal including pinnipeds and cetaceans to maintain forward motion without the energetic cost of active stroking (Skrovan et al. 1999). For one elite diver, the Weddell seal, over 78% of the dive descent may be spent gliding rather than actively stroking. By incorporating this intermittent mode of swimming during the dive, Weddell seals realize a 9.2%–59.6% reduction in diving energetic costs depending on depth. For exceptionally buoyant species, an opposite pattern may occur. This has been reported for sea otters which have proportionately large lungs compared to other marine mammals (Thometz et al. 2015), and for the unusually buoyant right whale (*Eubalaena glacialis*) which tends to glide on the ascent rather than the descent (Nowacek et al. 2001). Regardless of the direction, the energetic savings associated with decreased stroking translates into increased aerobic dive duration, and hence foraging time, when submerged.

The interaction between buoyancy, body condition (e.g., the proportion of fat to lean mass), and hydrostatic pressure as marine mammals move through the water column affords many opportunities for saving energy during transit swimming and diving. Passive descents of diving seals have been likened to *drifting leaves* that provide low cost rest periods during migratory movements (Mitani et al. 2009) or *falling rocks* that support low cost hunting tactics (Williams et al. 2000). Furthermore, marked changes in buoyancy

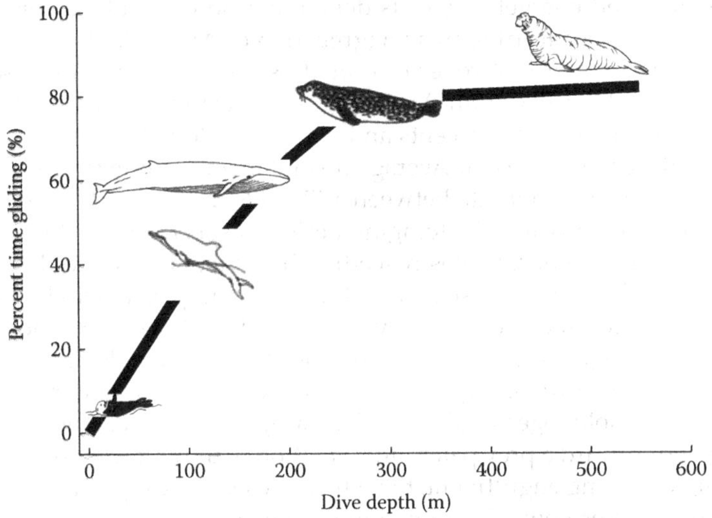

Figure 3.4 Percentage glide time during descent in relation to dive depth for marine mammals. Each animal symbol represents an individual species. The data were described by the nonlinear function, percentage glide time = 85.9 – (2820.3/depth). Except for the dolphins, the range of depths was determined by the free-ranging behavior of instrumented animals. (Redrawn from Williams, T.M. et al., *Integr. Compar. Biol.*, doi:10.1093/icb/icv025, 2015a.)

that occur as the percentage of body fat changes with season (e.g., Adachi et al. 2014), pregnancy (e.g., Dunkin et al. 2010), lactation and fasting (e.g., Costa et al. 1986; Crocker et al. 1997), will alter the ability of marine mammals to take advantage of these energy-saving behaviors.

3.2.5 *The high cost of foraging activities*

For active predators like most marine mammals, behaviors associated with foraging represent a major component of daily energy costs. In particular, high speed chases, prey handling, and prey consumption can entail high stroke frequencies and a large investment of energy when submerged (Williams et al. 2004; Aguilar de Soto et al. 2008; Maresh 2014). To mitigate these costs, marine mammals often switch between several swimming modes during a foraging dive. Typically, a dive may begin with a series of quick, high amplitude swimming strokes, followed by long periods of gliding on descent. Many deep-diving pinnipeds including Weddell seals (Williams et al. 2004) and elephant seals (*Mirounga angustirostris*, Adachi et al. 2014), as well as diving cetaceans including bottlenose dolphins and blue whales (Williams et al. 2000), beaked whales (*Ziphius cavirostris* and *Mesoplodon densirostris*) (Tyack et al. 2006), sperm whales (*Physeter macrocephalus*) (Miller et al. 2004), and pilot whales (Aguilar de Soto et al. 2008), save energy on the ascent by using burst-and-glide unsteady swimming, and by limiting continuous stroking periods to the short interval on the initial turn around at depth.

Such variability and use of unsteady swimming modes are typical of swimming vertebrates from sharks to pinnipeds and have been shown to provide an energetic advantage over continuous locomotion (Williams et al. 2000; Gleiss et al. 2011). In addition to foraging dives, both large-scale migrations of birds and mammals (Davis and Weihs 2007; Gleiss et al. 2011; Bishop et al. 2015) and the intra-dive foraging periods of pinnipeds often incorporate a roller-coaster series of powered and non-powered phases that result in performance, behavioral, and energetic benefits depending on the context. For example, active foraging by Weddell seals feeding in an aggregation of Antarctic silverfish (*Pleuragramma antarcticum*) involves a series of roller-coaster dips and rises that are associated with a low-frequency (7.2 ± 0.7 strokes min^{-1}) stroking descents followed by moderate stroke frequency (28.5 ± 0.8 strokes min^{-1}) ascents and fish encounters (Williams et al. 2015b). Only rarely do the seals feed on descent. Average instantaneous energetic costs are dictated by the stroking patterns and alternate between 17.2 ± 1.6 and 68.2 ± 2.0 J kg^{-1} min^{-1} on each gliding dip and powered rise of the foraging period, respectively (Williams et al. 2015a).

One of the most energetically costly feeding behaviors is displayed by lunge-feeding rorqual whales (balaenopterids such as blue and humpback whales) and has been described as "the largest biomechanical event on Earth and one of the most extreme feeding methods among aquatic vertebrates" (Brodie 1993). Balaenids are bulk filter feeders that capture prey by engulfing large volumes of water containing dense aggregations of plankton or nekton (Goldbogen et al. 2011). This lunge feeding behavior requires acceleration to high speeds toward a prey patch, and inflation of the accordion-like buccal cavity to an 80° gape angle. During engulfment, the whale presents the equivalent of a massive, flat plate to oncoming water flow. The animal must overcome exceptionally high drag forces due to a reduction in body streamlining and engulfment drag that rapidly decelerates the whale to a near halt (Chapter 1; Goldbogen et al. 2007). The exceptionally high level of biomechanical work required to lunge feed underlies the relatively short maximum dive durations observed for foraging balaenids (Goldbogen et al. 2012), which are often shorter than predicted based on allometric calculations of aerobic dive limits (Chapter 2).

Despite accounting for upwards of 60% of total foraging costs, the energetic payoff of the lunge feeding foraging strategy is high, with each lunge providing 6–237 times more energy consumed than expended during foraging by blue whales (Goldbogen et al. 2011).

3.3 Toolbox

The methods for assessing energetic costs in active marine mammals are changing quickly as advances in microprocessor technology afford new opportunities and finer scale monitoring of free-ranging animals. Combined with traditional methods of direct and indirect calorimetry, it is now possible to evaluate the energetics of marine mammals on many scales from instantaneous to prolonged intervals that range from seconds to daily, seasonal, and annual periods of time.

3.3.1 Measuring oxygen consumption

One of the most common methods that sets the foundation for evaluating the energetic cost of exercise in marine mammals is the determination of the rate of oxygen consumption ($\dot{V}O_2$) using open-flow respirometry. Details of the methods and calibrations are provided in Fedak et al. (1981) and Davis et al. (1985) for marine mammals as modified from Withers (1977) who developed many of the techniques for open-flow respirometry on terrestrial species. $\dot{V}O_2$ during rest, swimming, and diving has been directly measured for a wide variety of marine mammals using this method (Table 3.1) by training or directing animals to breathe into an appropriately-sized metabolic chamber to capture exhalations. A Plexiglas skylight floating on the water surface often suffices as a respiratory chamber and has been used to measure $\dot{V}O_2$ of sea otters to killer whales. Creative placement of the metabolic chamber above load cells (Williams et al. 1993), pools (Worthy 1987; Thometz et al. 2014), flumes (Davis et al. 1985; Fedak 1986; Feldkamp 1987; Rosen and Trites 2002), diving towers (Yeates et al. 2007), open water (Fahlman et al. 2008), and even breathing holes in the polar ice (Castellini et al. 1992; Williams et al. 2004) have allowed measurement of the metabolic rates of resting and exercising marine mammals.

The technique involves drawing air through the metabolic chamber with a vacuum pump at flow rates that maintain the fractional concentration of oxygen in the metabolic chamber above 0.2000 to avoid hypoxic conditions for the animals. Samples of expired air from the exhaust port of the chamber are usually dried with Drierite and scrubbed of carbon dioxide with Sodasorb before entering an oxygen analyzer (e.g., Sable Systems International, Inc., Las Vegas, Nevada). $\dot{V}O_2$ is calculated based on the airflow rate ($\dot{V}I$), and the difference between the fractional concentration of gas entering the chamber (FIO_2) and in the expired air (FEO_2) using equations from Fedak et al. (1981) and an assumed respiratory quotient (RQ) of 0.77 (Williams et al. 2004). A typical equation modified from Equation 4b in Withers (1977) for determining $\dot{V}O_2$ of marine mammals from open-flow respirometry is

$$\dot{V}O_2 = \frac{(\dot{V}I)(FIO_2 - FEO_2)}{(1 - FIO_2) + RQ(FIO_2 - FEO_2)} \qquad (3.5)$$

Note that the specific equation used as well as the interpretation of the results will depend on the exact experimental setup and the physiological status of the animals.

3.3.2 *Indirect methods for determining energetic costs*

Except in a few unique circumstances, it is not possible to directly measure the oxygen consumption of wild marine mammals. However, by pairing direct measures of metabolism in controlled circumstances with other physiological or biomechanical parameters, it is possible to estimate nearly instantaneous to long-term energetic costs for a wide variety of behaviors. For these measurements calibrated instrumentation is deployed on wild marine mammals, and records respiration rates, heart rates, acceleration, swimming stroke rates, or speed. Because each of these parameters is correlated to the rate of oxygen consumption (e.g., Figures 3.1 and 3.3), they can be used to determine free-ranging energetic costs as the animals move through the environment. The energetics of swimming, diving, migrating, and foraging of pinnipeds (Boyd et al. 1999; Butler et al. 2004; Williams et al. 2004) as well as small (Williams et al. 1992, 2015b) and large (Goldbogen et al. 2008, 2011, 2012) cetaceans have been estimated using this indirect method.

Recently, the use of tri-axial accelerometers has enabled energetics to be measured on individual behaviors, and even on an instantaneous basis. Accelerometry-based methods rely on a quantified relationship between movement in three dimensions and the metabolic power required to fuel that movement. One method, termed overall dynamic body acceleration (ODBA), integrates the dynamic acceleration of the animal's body in each of the three movement vectors. Higher levels of acceleration indicate movements requiring more metabolic power (e.g., Fahlman et al. 2008; Skinner et al. 2014). A second method involving accelerometry uses information from only one movement axis to identify individual swim strokes, the total number of which can be used to estimate overall energy expenditure on a cost-per-stroke basis (e.g., Williams et al. 2004; Maresh et al. 2014). To be successful, both of these methods require species-specific calibration of the acceleration–energetics relationship, for which respirometry or doubly labeled water can be useful. While there remains some uncertainty regarding the use of accelerometry to measure the field energetics of free-ranging marine mammals (e.g., Dalton et al. 2014), these methods are increasingly popular as they can provide data at a high temporal and behavioral resolution over long periods of time (Fahlman et al. 2008).

3.3.3 *Measuring field metabolic rates*

While the measurement of oxygen consumption makes respirometry the *gold standard* for estimating metabolic rates and energetic costs, it is generally limited to captive settings. Of the various methods available for measuring FMR, the doubly labeled water (DLW) method has been used most often in free-ranging, foraging pinnipeds (Boyd 2002). Like respirometry, the DLW method is based on the principle that the bodies of living animals are always in a state of change, constantly exchanging materials such as oxygen, carbon dioxide, fuels, and water with the external environment. The rate at which these various materials turnover in the body is proportional to the animal's metabolic rate (Speakman 1997). DLW specifically measures the rate of carbon dioxide (CO_2) produced in expired gases as a proxy for metabolism and energy expenditure. In this method, water labeled with heavy isotopes of oxygen (O^{18}) and hydrogen (deuterium D^2 or tritium H^3) is injected into the animal, and the rates of isotope dilution in the body over time are used to determine CO_2 production. Doubly labeled water performs best over prolonged periods, from hours to days, depending on the size of the animal, and ultimately provides a single value that represents energy expenditure summed over the entire measurement period (Costa 1987). Researchers interested in measuring the energy costs of specific, discreet behaviors should consider other methods.

For marine mammals, the use of doubly labeled water has three major limitations. First, it is prohibitively expensive for exceptionally large animals, most notably mysticete whales. For example, as of this writing, an average-sized, adult blue whale would require a 30 L injection of D^2O^{18}, at a minimum cost of over $13 million! Second, because the DLW method requires blood samples at both the beginning and end of the measurement period, it is of limited utility when the chance of recapture is low, as is the case for many far-ranging and elusive marine mammals, particularly cetaceans. Finally, calculations of energy expenditure using the DLW method are very sensitive to estimates of the animal's mass; this limits the utility of the method to instances where precise body mass measurements are possible—a rare luxury for biologists studying free-ranging marine mammals, and impossible for those studying mysticetes. Despite these limitations, DLW is still considered one of the best methods for estimating field metabolic rate when logistically feasible (Speakman 1997; Sparling et al. 2008).

3.4 *Unsolved mysteries and future directions*

Estimates of energy expenditure and, by extension, the prey-energy requirements of free-ranging marine mammals are fundamental to many questions surrounding the biology, management, and conservation of populations. For each member of any population, survival and successful reproduction depend on achieving positive energy balance, whereby individuals acquire enough energy from foraging to fuel critical life processes. In gray whales, for example, a sexually mature female may not successfully produce a calf if her energy intake is reduced by just 4% during pregnancy. Her own survival is predicted to be impacted if energy intake is reduced by 42% during this same time (Villegas-Amtmann et al. 2015). Similarly, the high energetic demands of raising a pup superimposed on the daily energetic costs of foraging and activity by female sea otters can push the mother over the energetic cliff and into a state of starvation. The potential for pup abandonment is escalated under these energetic circumstances (Thometz et al. 2014). Thus, energy balance in individuals has consequences for reproductive fitness, survival and, ultimately, demography.

Strong evolutionary pressures are in place for energy minimizing mechanisms and behaviors that help balance the trade-offs between the costs and benefits of all activities. Today, cumulative anthropogenic disturbances of marine mammals, especially during foraging, are expected to challenge energetic balance by instigating avoidance behaviors that increase movement costs. Consequently, the movement energetics of marine mammals has received growing attention in the scientific community (e.g., New et al. 2013; Braithwaite et al. 2015).

Overall, exercise and movement are important life-history behaviors with marked consequences for fecundity and fitness. In general, the temporal and spatial scales over which marine mammals move and the consequent balance of energy budgets are correlated with body size (Boyd 2002). *Movement* in these groups can refer to the relatively short distances associated with daily territorial patrols in male sea otters, to extraordinary annual, ocean basin scale foraging migrations of mysticete whales and phocid seals. *Movement* can also refer to the depths to which marine mammals dive. Although motivations and cues vary, one constant unifies the movements of all active animals: it comes at a high energetic cost. One can only imagine the total energetic costs that supported the remarkable movements of a female gray whale that migrated 22,511 km (nearly 14,000 miles) roundtrip between Russia and Baja, Mexico to set the record for the longest migratory movement of any mammal (Mate et al. 2015).

We find that marine mammals are increasingly altering their activity patterns in response to man-made influences. Individuals are traveling farther to find prey in response to climate change. Some species are taking longer, more circuitous transit routes to avoid acute anthropogenic disturbances, and others are altering their diving behaviors. Predictive models incorporating movement energetics and behavioral flexibility can be used to examine the cumulative impacts of these disturbances on individuals and ultimately, to predict the vulnerability of marine mammal populations to rapidly changing environments.

Under normal circumstances, behavioral control of exercise at depth, as discussed in this chapter, enables marine mammals to perform deep-sea activities within the physiological limits imposed by the mammalian cardio-respiratory system. The physiological and energetic repercussions of the flight responses of diving marine mammals are currently unknown. However, these responses may override behavioral safeguards that enable extreme dives by marine mammals and should be considered when assessing the potential impacts of man-made disturbances on these oceanic mammals (Williams et al. 2015b).

As evident from this chapter, our understanding of energetics in marine mammals is based primarily on a few well-studied species, most notably, harbor, gray, elephant and Weddell seals, California sea lions and fur seals, sea otters, and bottlenose dolphins. Much less is known about the energetic costs of other species, particularly the cetaceans. With over 120 species of marine mammals worldwide, living in tropical to polar habitats, there are clearly many questions regarding the unique energetic costs and resource demands of this group that need to be explored.

References

Adachi, T., J.L. Maresh, P.W. Robinson et al. 2014. The foraging benefits of being fat in a highly migratory marine mammal. *Proceedings of the Royal Society B: Biological Sciences* 281(1797): 20142120.

Aguilar de Soto, N., M.P. Johnson, P.T. Madsen et al. 2008. Cheetahs of the deep sea: Deep foraging sprints in short-finned pilot whales off Tenerife (Canary Islands). *Journal of Animal Ecology* 77(5): 936–947.

Au, D. and D. Weihs. 1980. At high speeds dolphins save energy by leaping. *Nature* 284(5756): 548–550.

Barnes, L.G., D.P. Domning, and C.E. Ray. 1985. Status of studies on fossil marine mammals. *Marine Mammal Science* 1(1): 15–53.

Berta, A. 2012. *Return to the Sea: The Life and Evolutionary Times of Marine Mammals.* University of California Press, Berkeley, CA.

Bishop, C.M., R.J. Spivey, L.A. Hawkes et al. 2015. The roller coaster flight strategy of bar-headed geese conserves energy during Himalayan migrations. *Science* 347(6219): 250–254.

Blake, R.W. 1983. Energetics of leaping in dolphins and other aquatic animals. *Journal of the Marine Biological Association of the United Kingdom* 63(01): 61–70.

Boyd, I.L. 2002. Energetics: Consequences for fitness. In: *Marine Mammal Biology–An Evolutionary Approach*, ed. A. Rus Hoelzel. Oxford, UK: Blackwell Science Ltd., pp. 247–277.

Boyd, I.L., R.M. Bevan, A.J. Woakes, and P.J. Butler. 1999. Heart rate and behavior of fur seals: Implications for measurement of field energetics. *American Journal of Physiology—Heart and Circulatory Physiology* 276(3): H844–H857.

Braithwaite, J.E., J.J. Meeuwig, and M.R. Hipsey. 2015. Optimal migration energetics of humpback whales and the implications of disturbance. *Conservation Physiology* 3(1): 1–15.

Brodie, P.F. 1993. Noise generated by the jaw actions of feeding fin whales. *Canadian Journal of Zoology* 71(12): 2546–2550.

Butler, P.J. 1988. The exercise response and the "classical" diving response during natural submersion in birds and mammals. *Canadian Journal of Zoology* 66(1): 29–39.

Butler, P.J., J.A. Green, I.L., Boyd, and J.R. Speakman. 2004. Measuring metabolic rate in the field: The pros and cons of the doubly labelled water and heart rate methods. *Functional Ecology* 18(2):168–183.
Caldwell, D.K. and H.M. Fields. 1959. Surf-riding by Atlantic bottle-nosed dolphins. *Journal of Mammalogy* 40:454–455.
Castellini, M.A., G.L. Kooyman, and P.J. Ponganis. 1992. Metabolic rates of freely diving Weddell seals: Correlations with oxygen stores, swim velocity and diving duration. *Journal of Experimental Biology* 165(1): 181–194.
Castellini, M.A., B.J. Murphy, M.A. Fedak, K. Ronald, N. Gofton, and P.W. Hochachka. 1985. Potentially conflicting metabolic demands of diving and exercise in seals. *Journal of Applied Physiology* 58(2): 392–399.
Costa, D.P. 1987. Isotopic methods for quantifying material and energy intake of free-ranging marine mammals. *Approaches to Marine Mammal Energetics* 1: 43–66.
Costa, D.P., B.J. Le Boeuf, A.C. Huntley, and C.L. Ortiz. 1986. The energetics of lactation in the northern elephant seal, *Mirounga angustirostris*. *Journal of Zoology* 209(1): 21–33.
Crocker, D.E., B.J. Le Boeuf, and D.P. Costa. 1997. Drift diving in female northern elephant seals: Implications for food processing. *Canadian Journal of Zoology* 75(1): 27–39.
Dalton, A.J.M., D.A.S. Rosen, and A.W. Trites. 2014. Season and time of day affect the ability of accelerometry and the doubly labeled water methods to measure energy expenditure in northern fur seals (*Callorhinus ursinus*). *Journal of Experimental Marine Biology and Ecology* 452: 125–136.
Davis, R.W., L.A. Fuiman, T.M. Williams, M. Horning, and W. Hagey. 2003. Classification of Weddell seal dives based on 3-dimensional movements and video-recorded observations. *Marine Ecology Progress Series* 264: 109–122.
Davis, R.W., L. Polasek, R. Watson, A. Fuson, T.M. Williams, and S.B. Kanatous. 2004. The diving paradox: New insights into the role of the dive response in air-breathing vertebrates. *Comparative Biochemistry and Physiology Part A: Molecular & Integrative Physiology* 138(3): 263–268.
Davis, R.W. and D. Weihs. 2007. Locomotion in diving elephant seals: Physical and physiological constraints. *Philosophical Transactions of the Royal Society B: Biological Sciences* 362(1487): 2141–2150.
Davis, R.W. and T.M. Williams. 2012. The marine mammal dive response is exercise modulated to maximize aerobic dive duration. *Journal of Comparative Physiology A* 198(8): 583–591.
Davis, R.W., T.M. Williams, and G.L. Kooyman. 1985. Swimming metabolism of yearling and adult harbor seals *Phoca vitulina*. *Physiological Zoology* 58:590–596.
Dunkin, R.C., W.A. McLellan, J.E. Blum, and D.A. Pabst. 2010. The buoyancy of the integument of Atlantic bottlenose dolphins (*Tursiops truncatus*): Effects of growth, reproduction, and nutritional state. *Marine Mammal Science* 26(3): 573–587.
Fahlman, A., R. Wilson, C. Svärd, D.A. Rosen, and A.W. Trites. 2008. Activity and diving metabolism correlate in Steller sea lion *Eumetopias jubatus*. *Aquatic Biology* 2: 75–84.
Fedak, M.A. 1986. Diving and exercise in seals: A benthic perspective. In: *Diving in Animals and Man*, eds. A.D. Brubakk, J.W. Kanwisher, and G. Sundnes. Trondheim, Norway: Tapir Publishers.
Fedak, M.A., L. Rome, and H.J. Seeherman. 1981. One-step N_2-dilution technique for calibrating open-circuit VO_2 measuring systems. *Journal of Applied Physiology* 51(3): 772–776.
Feldkamp, S.D. 1987. Swimming in the California sea lion: Morphometrics, drag and energetics. *Journal of Experimental Biology* 131(1): 117–135.
Fish, F.E. 1994. Influence of hydrodynamic-design and propulsive mode on mammalian swimming energetics. *Australian Journal of Zoology* 42(1): 79–101.
Folkens, P.A., R.R. Reeves, B.S. Stewart, P.J. Clapham, and J.A. Powell. 2002. *Guide to Marine Mammals of the World*. New York: AA Knopf.
Gleiss, A.C., R.P. Wilson, and E.L. Shepard. 2011. Making overall dynamic body acceleration work: On the theory of acceleration as a proxy for energy expenditure. *Methods in Ecology and Evolution* 2(1): 23–33.
Goldbogen, J.A., J. Calambokidis, D.A. Croll et al. 2008. Foraging behavior of humpback whales: Kinematic and respiratory patterns suggest a high cost for a lunge. *Journal of Experimental Biology* 211(23): 3712–3719.
Goldbogen, J.A., J. Calambokidis, D.A. Croll et al. 2012. Scaling of lunge-feeding performance in rorqual whales: Mass-specific energy expenditure increases with body size and progressively limits diving capacity. *Functional Ecology* 26(1): 216–226.

Goldbogen, J.A., J. Calambokidis, E. Oleson et al. 2011. Mechanics, hydrodynamics and energetics of blue whale lunge feeding: Efficiency dependence on krill density. *Journal of Experimental Biology* 214(1): 131–146.

Goldbogen, J.A., N.D. Pyenson, and R.E. Shadwick. 2007. Big gulps require high drag for fin whale lunge feeding. *Marine Ecology Progress Series* 349: 289–301.

Heim, N.A., M.L. Knope, E.K. Schaal, S.C. Wang, and J.L. Payne. 2015. Cope's rule in the evolution of marine animals. *Science* 347(6224): 867–870.

Hudson, L.N., N.J. Isaac, and D.C. Reuman. 2013. The relationship between body mass and field metabolic rate among individual birds and mammals. *Journal of Animal Ecology* 82(5): 1009–1020.

Hui, C.A. 1989. Surfacing behavior and ventilation in free-ranging dolphins. *Journal of Mammalogy*: 833–835.

Kleiber, M. 1975. *The Fire of Life: An Introduction to Animal Energetics*, 2nd edn. New York: Kreiger.

Kreite, B. 1994. Bioenergetics of the killer whale, *Orcinus orca*. PhD thesis, University of British Columbia, Vancouver, British Columbia, Canada.

Liwanag, H.E., A. Berta, D.P. Costa, M. Abney, and T.M. Williams. 2012. Morphological and thermal properties of mammalian insulation: The evolution of fur for aquatic living. *Biological Journal of the Linnean Society* 106(4): 926–939.

Maresh, J.L. 2014. Bioenergetics of marine mammals: The influence of body size, reproductive status, locomotion and phylogeny on metabolism. PhD thesis, University of California, Santa Cruz, CA.

Maresh, J.L., T. Adachi, A. Takahashi et al. in review. Summing the strokes: Energy economy in northern elephant seals during large-scale foraging migrations. *Movement Ecology*.

Maresh, J.L., S.E. Simmons, D.E. Crocker, B.I. McDonald, T.M. Williams, and D.P. Costa. 2014. Free-swimming northern elephant seals have low field metabolic rates that are sensitive to an increased cost of transport. *Journal of Experimental Biology* 217(9): 1485–1495.

Mate, B.R., V.Y. Ilyashenko, A.L. Bradford, V.V. Vertyankin, G.A. Tsidulko, V.V. Rozhnov, and L.M. Irvine. 2015. Critically endangered western gray whales migrate to the eastern North Pacific. *Biology Letters*. 11(4):20150071. doi: 10.1098/rsbl.2015.0071.

McDonald, B.I. and P.J. Ponganis. 2014. Deep-diving sea lions exhibit extreme bradycardia in long-duration dives. *Journal of Experimental Biology* 217(9): 1525–1534.

Miller, P.J., M.P. Johnson, P.L. Tyack, and E.A. Terray. 2004. Swimming gaits, passive drag and buoyancy of diving sperm whales *Physeter macrocephalus*. *Journal of Experimental Biology* 207(11): 1953–1967.

Mitani, Y., R.D. Andrews, K. Sato, A. Kato, Y. Naito, and D.P. Costa. 2009. Three-dimensional resting behaviour of northern elephant seals: Drifting like a falling leaf. *Biology Letters*. 6: 163–166. rsbl20090719.

Nagy, K.A. 2005. Field metabolic rate and body size. *Journal of Experimental Biology* 208(9): 1621–1625.

New, L.F., J. Harwood, L. Thomas et al. 2013. Modelling the biological significance of behavioural change in coastal bottlenose dolphins in response to disturbance. *Functional Ecology* 27(2): 314–322.

Noren, S.R., T. Kendall, V. Cuccurullo, and T.M. Williams. 2012. The dive response redefined: Underwater behavior influences cardiac variability in freely diving dolphins. *Journal of Experimental Biology* 215(16): 2735–2741.

Nowacek, D.P., M.P. Johnson, P.L. Tyack, K.A. Shorter, W.A. McLellan, and D.A. Pabst. 2001. Buoyant balaenids: The ups and downs of buoyancy in right whales. *Proceedings of the Royal Society of London B: Biological Sciences* 268(1478): 1811–1816.

Porter, W.P. and M. Kearney. 2009. Size, shape, and the thermal niche of endotherms. *Proceedings of the National Academy of Sciences* 106(Supplement 2): 19666–19672.

Ridgway, S.H., D.A. Carder, and W. Clark. 1975. Conditioned bradycardia in the sea lion *Zalophus californianus*. *Nature* 256: 37–38.

Riedman, M. and J.A. Estes. 1990. The sea otter (*Enhydra lutris*): Behavior, ecology, and natural history. Biological Report (USA), No. 90(14). Biological Report of the U.S. Fish & Wildlife Service, 90, 126pp.

Rosen, D.A.S. and A.W. Trites. 2002. Cost of transport in Steller sea lions, *Eumetopias jubatus*. *Marine Mammal Science* 18(2): 513–524.

Schmidt-Nielsen, K. 1972. Locomotion: Energy cost of swimming, flying, and running. *Science* 177(4045): 222–228.

Scholander, P.F. 1959. Wave-riding dolphins: How do they do it? At present only the dolphin knows the answer to this free-for-all in hydrodynamics. *Science* 129(3356): 1085–1087.

Skinner, J.P., Y. Mitani, V.N. Burkanov, and R.D. Andrews. 2014. Proxies of food intake and energy expenditure for estimating the time–energy budgets of lactating northern fur seals *Callorhinus ursinus*. *Journal of Experimental Marine Biology and Ecology* 461: 107–115.

Skrovan, R.C., T.M. Williams, P.S. Berry, P.W. Moore, and R.W. Davis. 1999. The diving physiology of bottlenose dolphins (*Tursiops truncatus*). II. Biomechanics and changes in buoyancy at depth. *Journal of Experimental Biology* 202(20): 2749–2761.

Sparling, C.E., D. Thompson, M.A. Fedak, S.L. Gallon, and J.R. Speakman. 2008. Estimating field metabolic rates of pinnipeds: Doubly labelled water gets the seal of approval. *Functional Ecology* 22(2): 245–254.

Speakman, J.R. 1997. *Doubly Labelled Water: Theory and Practice*. Chapman & Hall, London, UK.

Speakman, J.R. and E. Król. 2010. Maximal heat dissipation capacity and hyperthermia risk: Neglected key factors in the ecology of endotherms. *Journal of Animal Ecology* 79(4): 726–746.

Taylor, C.R., N.C. Heglund, and G.M. Maloiy. 1982. Energetics and mechanics of terrestrial locomotion. I. Metabolic energy consumption as a function of speed and body size in birds and mammals. *Journal of Experimental Biology* 97(1): 1–21.

Taylor, C.R., K. Schmidt-Nielsen, and J.L. Raab. 1970. Scaling of energetic cost of running to body size in mammals. *American Journal of Physiology–Legacy Content* 219(4): 1104–1107.

Thewissen, J.G.M., S.T. Hussain, and M. Arif. 1994. Fossil evidence for the origin of aquatic locomotion in archaeocete whales. *Science* 263(5144): 210–212.

Thometz, N.M., M.J. Murray, and T.M. Williams. 2015. Ontogeny of oxygen storage capacity and diving ability in the Southern sea otter (*Enhydra lutris nereis*): Costs and benefits of large lungs. *Physiological and Biochemical Zoology* 88(3):311–327. doi: 10.1086/681019.

Thometz, N.M., M.T. Tinker, M.M. Staedler, K.A. Mayer, and T.M. Williams. 2014. Energetic demands of immature sea otters from birth to weaning: Implications for maternal costs, reproductive behavior and population-level trends. *Journal of Experimental Biology* 217(12): 2053–2061.

Tift, M.S., E.C. Ranalli, D.S. Houser, R.M. Ortiz, and D.E. Crocker. 2013. Development enhances hypometabolism in northern elephant seal pups (*Mirounga angustirostris*). *Functional Ecology* 27(5): 1155–1165.

Tyack, P.L., M.P. Johnson, N.A. Soto, A. Sturlese, and P.T. Madsen. 2006. Extreme diving of beaked whales. *Journal of Experimental Biology* 209(21): 4238–4253.

Villegas-Amtmann, S., L.K. Schwarz, J.L. Sumich, and D.P. Costa. 2015. Population consequences of lost foraging opportunity in eastern female gray whales. *Ecosphere* (in press).

Williams, T.M. 1989. Swimming by sea otters: Adaptations for low energetic cost locomotion. *Journal of Comparative Physiology A: Neuroethology, Sensory, Neural, and Behavioral Physiology* 164(6): 815–824.

Williams, T.M.. 1999. The evolution of cost efficient swimming in marine mammals: Limits to energetic optimization. *Philosophical Transactions of the Royal Society B: Biological Sciences* 354(1380): 193–201.

Williams, T.M., R.W. Davis, L.A. Fuiman et al. 2000. Sink or swim: Strategies for cost-efficient diving by marine mammals. *Science* 288(5463): 133–136.

Williams, T.M., W.A. Friedl, M.L. Fong, R.M. Yamada, P. Sedivy, and J.E. Haun. 1992. Travel at low energetic cost by swimming and wave-riding bottlenose dolphins. *Nature* 355(6363): 821–823.

Williams, T.M., W.A. Friedl, and J.E. Haun. 1993. The physiology of bottlenose dolphins (*Tursiops truncatus*): Heart rate, metabolic rate and plasma lactate concentration during exercise. *Journal of Experimental Biology* 179(1): 31–46.

Williams, T.M., L.A. Fuiman, and R.W. Davis. 2015a. Locomotion and the cost of hunting in large, stealthy marine carnivores. *Integrative and Comparative Biology*. doi:10.1093/icb/icv025.

Williams, T.M., L.A. Fuiman, M. Horning, and R.W. Davis. 2004. The cost of foraging by a marine predator, the Weddell seal *Leptonychotes weddellii*: Pricing by the stroke. *Journal of Experimental Biology* 207(6): 973–982.

Williams, T.M., L.A. Fuiman, T.L. Kendall et al. 2015b. Exercise at depth alters bradycardia and incidence of cardiac anomalies in deep-diving marine mammals. *Nature Communications* 6:6055.

Williams, T.M., G.L. Kooyman, and D.A. Croll. 1991. The effect of submergence on heart rate and oxygen consumption of swimming seals and sea lions. *Journal of Comparative Physiology B* 160(6): 637–644.

Williams, T.M., B. Richter, T. Kendall, and R. Dunkin. 2011. Metabolic Demands of a Tropical Marine Carnivore, the Hawaiian Monk Seal (*Monachus schauinslandi*): Implications for fisheries competition. *Aquatic Mammals* 37(3): 372–376. doi: 10.1578/AM.37.3.2011.372.

Withers, P.C. 1977. Measurement of VO_2, VCO_2, and evaporative water loss with a flow-through mask. *Journal of Applied Physiology* 42(1): 120–123.

Woodcock, A.H. and A.F. McBride. 1951. Wave-riding dolphins. *Journal of Experimental Biology* 28(2): 215–217.

Worthy, G.A.J. 1987. Metabolism and growth of young harp and grey seals. *Canadian Journal of Zoology* 65(6): 1377–1382.

Würsig, B. and M. Würsig. 1979. Behavior and ecology of the bottlenose dolphin, *Tursiops truncatus*, in the South Atlantic. *Fishery Bulletin* 77(2): 399–412.

Yeates, L.C., T.M. Williams, and T.L. Fink. 2007. Diving and foraging energetics of the smallest marine mammal, the sea otter (*Enhydra lutris*). *Journal of Experimental Biology* 210(11): 1960–1970.

Zapol, W.M., G.C. Liggins, R.C. Schneider et al. 1979. Regional blood flow during simulated diving in the conscious Weddell seal. *Journal of Applied Physiology* 47(5): 968–973.

chapter four

Pressure regulation

Sascha K. Hooker and Andreas Fahlman

Contents

4.1 The big picture challenge and summary

Marine mammals dive to find their food but have to return to the surface to breathe. Some species occupy relatively shallow foraging niches, while others have adapted to target prey deep in the ocean. Diving records for elephant seals exceed 2000 m, and beaked whales have been recorded to dive as deep as 3000 m. Although management of available O_2 has been the focus of diving research for over 70 years, relatively little effort has been dedicated toward understanding the effects of elevated pressure on marine mammals. Pressure is an important thermodynamic variable that, like temperature, can have significant physiological implications. Marine mammals are thought to have various anatomical and biochemical adaptations to help them to withstand the pressure at depth, in particular, reducing any body gas-filled spaces, and allowing their chest and respiratory system to compress to the limit of collapse.

However, they still have to cope with changes in pressure many times per day, and rapidly and repeatedly recruit their alveoli each time they surface. How do they avoid other problems associated with pressure, such as *atelectasis*, the "bends" or decompression sickness (DCS), high pressure nervous syndrome (HPNS), shallow-water blackout, or N_2 narcosis?

4.2 Current knowledge

4.2.1 General abilities of marine mammals

Different marine mammal species show very different diving patterns, driven by their habitats and prey (Table 4.1, Figure 4.1). Some species, such as elephant seals (Le Boeuf et al. 1986, 1988) and sperm whales (Miller et al. 2004; Watkins et al. 1993), dive repeatedly

Table 4.1 Overview of marine mammal families and their approximate diving depths

Order Suborder	Family	Common name (number of species)	Approximate diving depths[a]	Shown in Figure 4.1
Cetacea				
Mysticeti	Balaenidae	Right and bowhead whales (4)	**Shallow/ moderate**	
	Neobalaenidae	Pygmy right whale (1)	Shallow	
	Balaenopteridae	Rorquals (8)	**Shallow/ moderate**	
	Eschrichtiidae	Gray whale (1)	**Shallow**	
Odontoceti	Physeteridae	Sperm whale (1)	**Very deep**	Physeter
	Kogiidae	Pygmy and dwarf sperm (2)	Very deep	
	Monodontidae	Narwhal and beluga (2)	**Deep**	
	Ziphiidae	Beaked whales (21)	**Very deep**	Ziphius
	Delphinidae	Oceanic dolphins (36)	**Moderate/deep**	Globicephala
	Phocoenidae	Porpoises (6)	**Shallow/ moderate**	Phocoena
	Platanistidae	South Asia river dolphin (1)	Shallow	
	Iniidae	Boto (1)	Shallow	
	Lipotidae	Baiji (extinct)	Shallow	
	Pontoporiidae	Franciscana (1)	Shallow	
Sirenia	Trichechidae	Manatees (3)	Shallow	
	Dugongidae	Dugong (1)	**Shallow**	
Carnivora	Mustelidae	Otters (2)	**Shallow**	
	Ursidae	Polar bear (1)	Shallow	
Pinnipedia	Orariidae	Fur seals and sea lions (16)	**Shallow/ moderate**	Arctocephalus, Zalophus
	Phocidae	True seals (18)	**Moderate/deep/ very deep**	Mirounga
	Odobenidae	Walrus (1)	**Moderate**	

Notes: Shallow <200 m; moderate 200–500 m; deep 500–1000 m; very deep >1000 m, those shown in bold have been confirmed by time–depth recorders, those not in bold are speculated.

[a] We suggest the online Penguinness database for updates on diving depths listed by species (Ropert-Coudert and Kato 2012).

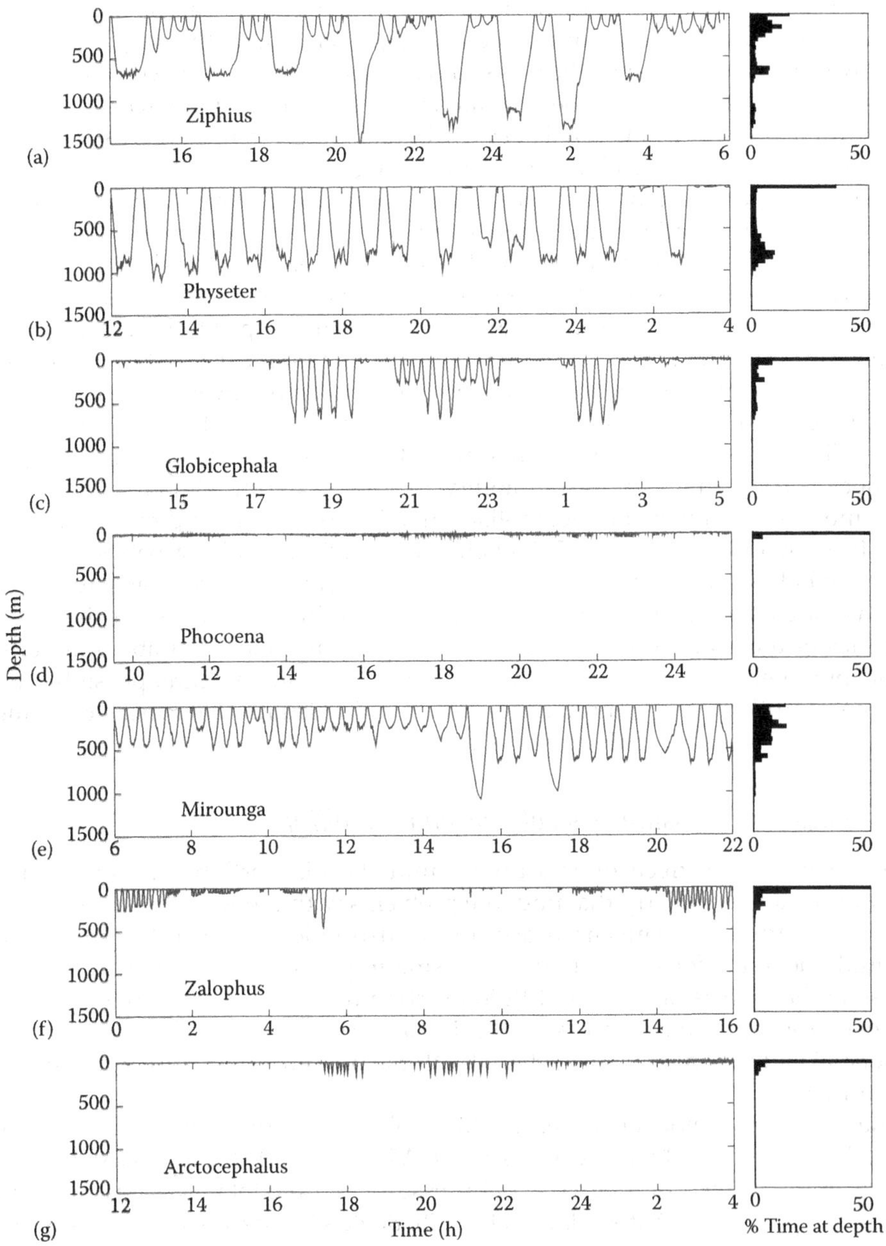

Figure 4.1 **(See color insert.)** Dive traces (left) and frequency histograms (right) showing time at depth (in 50 m depth intervals) for various marine mammal species. Dive traces are plotted to identical scales: 1500 m depth over a 16 h time period, to illustrate the differences in the use of depth and patterns of diving between species. (a) Cuvier's beaked whale (Ziphiidae); Mediterranean, September 2003; (b) sperm whale (Physeteridae), Azores, August 2010; (c) short-finned pilot whale (Delphinidae), Canary Islands, October 2004; (d) harbor porpoise (Phocoenidae), Jutland Peninsula, Denmark, October 2012; (e) northern elephant seal (Phocidae), eastern Pacific, March 2014; (f) California sea lion (Otariidae), San Nicholas Island, California, November 2006; and (g) Antarctic fur seal (Otariidae), South Georgia, December 2001. (Data courtesy of (a) M. Johnson, (b) C. Oliveira, (c) N. Aguilar and M. Johnson, (d) D. Wisniewska (e) D. Costa, (f) D. Costa, (g) S. Hooker.)

to great depths, while others, such as dugongs, spend much of their lives in relatively shallow sea-grass environments (Chilvers et al. 2004). Even among the deep divers, some are relatively sedate divers, while others descend and ascend at high speeds (Figure 4.1). With the advent of sophisticated microelectronics, we are building up knowledge of the diving behavior of many marine mammal species although many gaps remain, especially in our understanding of how physiology may limit diving.

There is a general tendency for greater body size to favor greater dive capacity because of the differential scaling between body O_2 stores and metabolic rate (Halsey et al. 2006). However, other factors may also play a role, such as the ability to reduce body temperature (Boyd and Croxall 1996) or the costs associated with engulfment capacity (Goldbogen et al. 2012). Given constraints on swimming speed, there is also a general relationship between dive time and dive depth, such that the deepest dives are necessarily the longest. Thus, the deepest divers also tend to be relatively large (Table 4.1). This relationship has also led to predictions that as animals grow, their diving capacity will increase. There is some support for this, and, at least among pinniped species, it appears that there is some ontogeny of the ability to dive to depth, although this is possibly driven more by O_2 stores and metabolism than by adaptation to pressure (Fowler et al. 2006; Horning and Trillmich 1997; McLellan et al. 2002). Similarly, babysitting behavior in sperm whales is thought necessary because young animals are unable to dive as deep or for as long as adults (Whitehead 1996). Some preliminary work on the mechanical properties of the lung in pinnipeds showed higher and more variable lung *compliance* in wild pinnipeds compared with those raised under human care, possibly indicating that lung conditioning is important for diving animals and that repeated diving helps protect against *lung squeeze* (Fahlman et al. 2014a).

4.2.2 Physics of pressure, gas volume, and solubility

Before looking at the effects of pressure on animals, it is worth briefly considering some of the physics and chemistry that underlie problems with pressure. Pressure is a measure of the force exerted over a given surface area. Hydrostatic pressure is the pressure exerted by a fluid due to the force of gravity, increasing in proportion to depth from the surface because of the increasing weight of fluid exerting downward force from above. Thus in the ocean, from atmospheric pressure (1 ATA) at the surface, pressure increases by 1 ATA for every 10 m descended. So, at 1000 m depth the pressure is 100 times greater than that at the surface.

The volume of 1 mole of gas (6.02×10^{23} molecules) without water vapor (dry) under standard temperature (0°C), and pressure (1 ATA) is 22.7 L for all gases. In a mixture, each gas exerts its own pressure (partial pressure) and the total gas pressure is the sum of all partial pressures (Dalton's law; Table 4.2). Boyle's law describes the inverse relationship between gas volume and pressure of a gas phase (Table 4.2). A popular experiment to show this is to attach a styrofoam cup to deep-ocean sampling equipment. Such cups return from depth thimble-sized as the pressure has compressed the air in them, forcing this structural change. Airspaces in animals similarly become compressed at depth, potentially causing damage to any surrounding rigid structures, causing, for example, *barotrauma* or *lung squeeze*.

While liquids are essentially incompressible, increasing pressure does have an effect on the solubility of dissolved gases (Henry's law, Table 4.2). This is well illustrated by the gas dissolved in carbonated drink, which is "released" from solution as the pressure drops (as the lid is opened). Gases dissolved in liquids are measured in terms of the

Table 4.2 Gas laws relating to pressure

Boyle's law	$P_1 \cdot V_1 = P_2 \cdot V_2$	Volume will decrease in inverse proportion to the increase in pressure.
Dalton's law	$p(1) \alpha P$	Total gas pressure is the sum of all partial pressures; all gases will stay in the same proportions.
Henry's law	$p = k_H M$	The solubility of gas in a liquid is directly proportional to the partial pressure of the gas above the liquid. The Henry's law constant, k_H, varies for different gases, for example, N_2 is less soluble (k_H =1640 L atm/mol) than O_2 (k_H = 770 L atm/mol) or CO_2 (k_H = 29 L atm/mol).

Notes: V, volume; P, pressure; p, partial pressure; M, molar concentration.

corresponding partial pressure of that gas in gas phase, referred to as the gas tension. However, a variation in solubility can have a significant effect on this. Some gases, such as N_2 and O_2 have relatively low solubility, while CO_2 dissociates into carbonic acid and has high solubility. Consequently, the molar concentration dissolved will be related to both the partial pressure and the solubility, and the molar concentration of a given change in gas tension will differ considerably for different gases.

During descent, the partial pressure of the gases in the lung increases in proportion with the pressure (Dalton's law). As long as the alveoli are open and there is gas exchange, the gas tension in the *arterial* blood increases as gas diffuses from high to low pressure. It is this increased partial pressure of different gases which leads to many of the gas diseases discussed in Section 4.2.6.

4.2.3 *Pressure, biochemistry, and blood flow*

The effects of pressure on the body can be caused via its effect on gas spaces and solubility, or can be caused directly in terms of biochemical reactions and cellular structure. The deepest point in the ocean lies at a depth of 11.8 km (i.e., pressures of >1000 ATA). However, most marine mammals live within the upper waters of the ocean at pressures of 1–300 ATA. Although protein denaturation is only caused at extremely high pressures (>4000 ATA) beyond those found in the ocean, pressures in the ocean can nevertheless have an effect at the biochemical and cellular level. During any reaction, pressure can have an effect via the changes in volume caused by the reaction. When a process occurs with an increase in volume, pressure inhibits that process; and when a process occurs with a decrease in volume, pressure enhances that process (Somero 1992).

Comparative studies have shown that the pressure sensitivities of enzymes, structural proteins, and membrane-based systems differ markedly between shallow- and deep-living species. Some fish that live as shallow as 500 m show biochemical adaptations that allow their enzymatic reactions to be pressure-tolerant (Siebenaller and Garrett 2002; Somero 1992).

Although many marine mammal species stay in relatively shallow waters for much of the time, others visit depths well beyond these (Table 4.1). What is perhaps most remarkable is that whereas many marine species have a relatively narrow range of pressure at which they function, marine mammals must function over a remarkable range of diving depths. There have been only a few studies of the biochemical tolerance to pressure found in marine mammals. In general, these have shown that marine mammal tissue enzymes and living red blood cells appear to be adapted to pressure, either showing no reaction to pressure changes, or even functioning better under pressure conditions (Castellini et al. 2001, 2002; Croll et al. 1992; Williams et al. 2001). Recent work has looked at immune response with pressure exposure, and showed reduced response in belugas compared to humans following pressure exposure (Thompson and Romano 2015). Studies in rats and rabbits have shown that animals that have been immune compromised have a reduced risk of DCS (Kayar et al. 1997; Ward et al. 1990). This could potentially be an adaptation protecting against dive-related pathologies (Thompson and Romano 2015). In addition, platelets from elephant seals responded less to agonists which was suggested to be an adaptation toward repeated pressure changes (Field et al. 2001). Cetaceans appear to lack a number of clotting factors, for example, Hageman factor, common to terrestrial mammals, leading to their more *hypocoagulable* blood (Robinson et al. 1969). Their absence may help to improve microcirculation at depth and/or to reduce *venous* thrombosis, which has been suggested important in human DCS (Montcalm-Smith et al. 2008; Pontier et al. 2011).

Pressure in the body can have a detrimental effect on blood pressure and blood flow to the brain. Marine mammals possess extensive *venous plexuses* (Costidis and Rommel 2012). Some, such as the retia mirabilia found in cetaceans and sirenians, might be related to diving ability (Vogl and Fisher 1982). In cetaceans, these are a series of vascular networks of densely looped *arteries* primarily located along the base of the brain case, along and within the vertebral column, and retro-pleurally lining the ventral aspect of the rib arches. It is the only path of *arterial* blood to the brain in adult cetaceans and may be an adaptation allowing effective drainage of blood from the central nervous system during periods of elevated pressure (D. Garcia-Parraga, 2015, pers. comm.). Alternatively, the *venous* portion of these *rete* may allow intra-thoracic and vascular engorgement to prevent *lung squeeze*, similar to that seen in humans (Brown and Butler 2000; Craig 1968) while the *arterial rete* may act as a filter for *arterial* gas *emboli* preventing DCS (neuroprotective effect) (Nagel et al. 1968; Ponganis et al. 2003; Scholander 1940).

4.2.4 *Pressure and airspaces*

It is the airspaces that cause the most problem to diving animals, and for this reason many of the airspaces present in terrestrial mammals have been lost in the evolution of marine mammals. There are three major airspaces that are a liability for divers—the lung, the facial *sinuses*, and the middle ear.

Marine mammals have lost their facial *sinuses* and so avoid problems with these. The middle ear is an air-filled rigid cavity with little or no compressibility. In pinnipeds, a pressure differential is prevented by a complex vascular *sinus* lining the wall of the middle ear cavity (Odend'hal and Poulter 1966; Stenfors et al. 2001). The negative pressure that develops in the middle ear during diving pulls blood into the *venous sinus* and helps fill the internal volume. In cetaceans, a similar mechanism is found in the pterygoid and peribullar *sinuses* possessing elaborate plexiform *veins*. More elaborate and voluminous *sinus* vasculature is found in deep divers (e.g., physeterids, kogiids, ziphiids) compared to shallow-diving delphinids (Fraser and Purves 1960).

4.2.5 *Lungs and diving lung collapse*

The major airspace that is affected by pressure during diving is the respiratory system. One problem is *lung squeeze*, as pressure reduces the air volume within an incompressible rib cage. Human divers face this problem with their largely incompressible ribs and chest wall (Ferretti 2001; Lundgren and Miller 1999). Scholander (1940) suggested that marine mammals are able to circumvent this problem by having a highly compliant chest wall. In the pinniped, volume–pressure curves (*compliance*) showed that the chest wall provides little resistance to compression (Fahlman et al. 2014a; Leith 1976). Whether cetaceans also show high chest *compliance* is not known, but studies in the pilot whale (Olsen et al. 1969), beluga whale, and bottlenose dolphin (Fahlman, unpub. obs.) indicate that the chest recoils to low lung volumes. Thus, whereas for humans, pulmonary vascular engorgement is required to fill the space left by the compressed lungs, the pinniped, and possibly the cetacean, chest offers little resistance toward compression regardless of the depth to which they dive. The *venous* thoracic *rete* that exists in many cetaceans may help pool blood in the thoracic cavity and prevent excessive negative intra-thoracic pressures from developing (Hui 1975; Vogl and Fisher 1982). In addition, dolphins and sperm whales have a tracheal lumen lined with transitional epithelia (Cozzi et al. 2005; Leith 1989), and it has been suggested that this structure may engorge with blood and so offset the limited compression of a stiff trachea. Thus, both the chest and trachea may resist compression in some species, but blood engorgement may help fill the space caused by reduced gas volume and thereby prevent *barotrauma*.

The second problem relates to gas uptake by the pressurized lung. As depth increases, the lung volume reduces and the pulmonary pressure and the partial pressure of the gases in the respiratory system increase (Boyle's and Dalton's laws). This increases the solubility of these gases within the blood (Henry's law), potentially causing a problem with uptake of N_2 and the risk of forming inert gas bubbles during ascent and decompression (i.e., DCS symptoms).

Marine mammals show modifications to their lungs compared to terrestrial mammals. In general, marine mammals have reinforced upper airways and a lack of smaller respiratory bronchii compared to terrestrial mammals (Kooyman 1973). However, there is great variation in tracheal stiffness between species and some appear not very different from terrestrial mammals (Moore et al. 2014). Dolphins show the most extreme modifications including the presence of a series of bronchial *sphincter* muscles found in the terminal segments of the airways (Kooyman 1973). The function of these is largely unknown, but it is hypothesized to relate to management of lung air (Belanger 1940; Ninomiya et al. 2004). In combination with chest wall musculature, compression/expansion of the chest could be used to help control volume (and thus buoyancy) without the need to exhale (Garcia-Parraga, pers. comm.)

Otariids (fur seals and sea lions) have robust cartilaginous airway reinforcement extending to the alveolar sac, whereas phocids have no cartilage in the terminal airway, but the walls are thickened by connective tissue and smooth muscle (Kooyman 1973). Thus for all marine mammals, the thin-walled alveoli compress under increasing hydrostatic pressure causing a graded decrease in the amount of respiratory gases absorbed by the blood stream as the depth of diving increases. Effective gas exchange between lungs and blood ceases when all alveoli are collapsed. This reinforced lung structure may also facilitate high ventilation rates at the surface (Denison and Kooyman 1973; Denison et al. 1971).

The mechanism leading to alveolar collapse (previously called *lung collapse*) was first proposed by Scholander (1940): "It seems almost evident that the compression of such a system would begin with the alveols and end with the most rigid parts." In this hypothetical situation, compression would result in a gradual reduction of the alveolar surface area and increasing alveolar thickness reducing diffusion (i.e., a pulmonary shunt) and eventually leading to a termination of gas exchange (Kooyman and Sinnett 1982; McDonald and Ponganis 2012). Progressive compression of alveoli was thus thought to reduce gas uptake by the blood up to some critical depth of alveolar collapse at which gas uptake would cease (Figure 4.2). Early empirical work supported this, suggesting alveolar collapse in bottlenose dolphins at 70 m depth would lead to the muscle N_2 washout rates observed (Ridgway and Howard 1979). Similarly, use of an *arterial* N_2 blood sampler during Weddell seal dives suggested alveolar collapse at approximately 30 m (Falke et al. 1985). More recently, continuous *arterial* partial pressure of O_2 during diving has shown suggested depth of alveolar

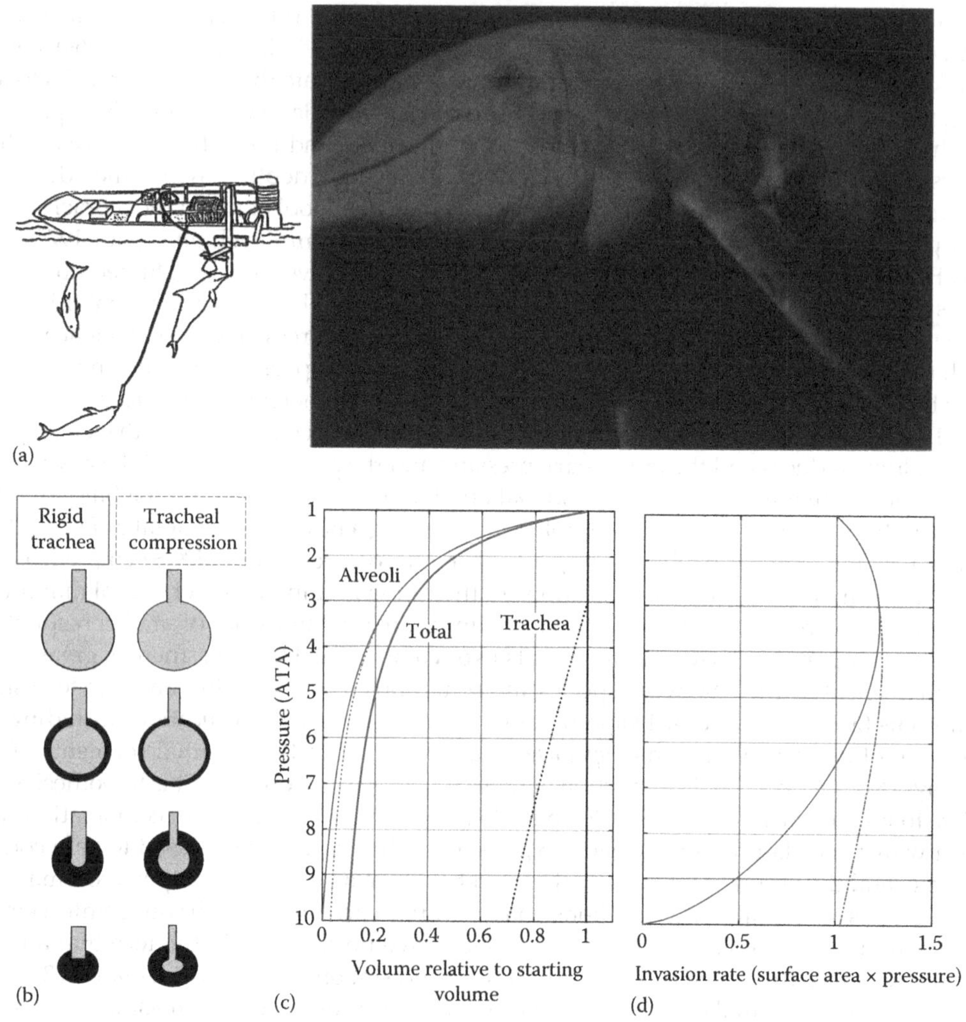

Figure 4.2 **(See color insert.)** Compression of lung and thorax results in changes in invasion/diffusion rate. (a) Chest compression of a bottlenose dolphin (Tuffy), photographed at 300 m depth, with experimental setup shown on the left. Thoracic collapse is visible behind the left flipper. (Image copyright U.S. Navy; details published in Ridgway et al. 1969). (b) Graded lung compression showing illustrations of two models: (i) rigid trachea (solid) and (ii) tracheal compression at depths less than alveolar collapse (dotted). (c) Modeled relative volume changes of trachea and alveoli based on these models. Total volume (trachea + alveoli = 1 l) reduces according to Boyle's law (thick gray line). Volume changes are shown based on a rigid trachea (solid black line, 0.1 l), and assuming that the tracheal volume decreases exponentially after 3 ATA (dotted black line), with resulting changes to the alveolar volume from 0.9 l at the surface, decreasing more quickly with a rigid trachea (gray solid line) than with a compressing trachea (dotted gray line). (d) The resulting invasion rate (alveolar surface area × pressure) (sensu Scholander 1940). The invasion rate is plotted from 1 at the surface breath-hold, to 0 at the depth of alveolar collapse. Fick's law states that the diffusion rate is proportional to surface area × pressure/membrane thickness, thus the invasion rate divided by the membrane thickness would give the diffusion rate (i.e., this would further reduce the invasion rate). It can be seen that near the surface, the invasion rate increases with a pressure increase, but that at deeper depths this reduces to the depth of alveolar collapse. The effect of tracheal compression at depths shallower than alveolar collapse effects deeper alveolar collapse and increases the range of depths which have a high invasion (and diffusion) rate.

collapse of 225 m for California sea lions (McDonald and Ponganis 2012). Experimental work showed deeper alveolar collapse depths closer to 170 m for sea lions in a *hyperbaric* chamber (Kooyman and Sinnett 1982).

However, experimental studies of the tracheal structure under pressure (using *hyperbaric* chambers) have shown that the trachea is not incompressible (Kooyman et al. 1970). There is also considerable interspecific variation in tracheal *compliance* in the excised conducting airways of several phocid seals and odontocetes (Moore et al. 2014), perhaps related to variation in life history and diving abilities. Furthermore, the pressure–volume curves of excised lungs diverged during deflation (i.e., exhalation) versus compression (Denison et al. 1971; Fahlman et al. 2011; Kooyman and Sinnett 1982; Piscitelli et al. 2010).

Thus, if the trachea were to begin to compress prior to full alveolar collapse, it would cause alveolar collapse depths to be deeper than initially predicted by Scholander's balloon-pipe model (Figure 4.2) (Bostrom et al. 2008; Fitz-Clarke 2007). In fact, a model with a slightly compliant trachea suggested alveolar collapse depths of 110 m for bottlenose dolphins based on the Ridgway and Howard (1979) study and suggested that peak *arterial* N_2 seen at 30 m depth by Falke et al. (1985) might, rather than showing alveolar collapse depth, correspond to a depth-dependent pulmonary shunt that would affect the diffusion rate, that is, the alveoli might not collapse until greater depths (Bostrom et al. 2008). That said, theoretical models are highly dependent on appropriate assumptions, and even though some studies have tested model output against empirical data for dolphins, seals and sea lions (Fahlman et al. 2009; Hooker et al. 2009), there is scant information about the physiological responses of most marine mammal species. Thus, model results should be viewed with care but are useful in defining important areas for further research and providing alternative explanations to earlier experimental work.

Whether breath-hold diving marine mammals experience alveolar collapse is not disputed, but there is probably considerable variability in how gas exchange is managed both within and between species. This variability would arise from differences in respiratory anatomy (Fahlman et al. 2014a; Moore et al. 2011; Moore et al. 2014; Piscitelli et al. 2010) and behavior, for example, shallow versus deep divers. Our current understanding suggests that the inherent anatomy of each species may limit deep diving. For example, deep divers seem to have smaller lungs as compared with shallow-diving species (Piscitelli et al. 2010). Still, there is evidence that there is limited plasticity in altering the structural properties to help prevent *barotrauma* (Fahlman et al. 2014a). Interestingly, lung conditioning appears to help increase vital capacity in humans (Johansson and Schagatay 2012) and recent work has suggested that lung growth may occur even in adults (Butler et al. 2012). Thus, repeated *atelectasis* and alveolar recruitment may be important for healthy lung function. However, a more important factor on a dive-to-dive basis for determining the depth of alveolar collapse is the diving lung volume, that is, the air volume at the outset of the dive. Different species also have different tactics in terms of inspired air volume. Although phocid seals are thought to exhale prior to diving, the important question is to what extent does this occur, that is, what is the diving lung volume? Studies show that they still dive with as much as 60% of their inspiratory volume (Kooyman et al. 1971). Fur seals, sea lions and cetaceans are thought to dive on inhalation (Kerem et al. 1975; Kooyman 1973).

Whatever the depth of alveolar collapse, many marine mammal species will reach depths sufficient to cause a period of *atelectasis* followed by recruitment. That they do this repeatedly and without any apparent side effects as the animal resurfaces is remarkable (Denison and Kooyman 1973). The lungs produce a pulmonary *surfactant* that lines the alveolar air–water interface, and varies the surface tension with lung volume to reduce the work of breathing, alter *compliance*, and prevent adhesion of respiratory surfaces. An anti-adhesive *surfactant* with greater fluidity and rapid expansion capabilities to cope with repeated collapse and reinflation has been reported in pinnipeds. These functional adaptations appear to be supported by molecular modifications in key protein and lipid compositional changes, as well as adaptations in the secretory mechanisms of the cells (Foot et al. 2006). Thus, the *surfactant* compositions in pinnipeds may have been selected to help recruitment of closed alveoli following deep dives. *Surfactant* production may also be triggered by diving, supporting the idea that pressure is the driving force behind observed differences in *surfactant* levels (Foot et al. 2006). While selection at the molecular level has been investigated between different marine mammal groups, analysis of surfactant composition has been primarily conducted on samples from pinnipeds and much less is known about cetacean surfactants. Preliminary data suggest that the fluidizing phospholipids are not increased in odontocetes (Gutierrez et al. 2015).

4.2.6 *Diving diseases*

There are several diving diseases brought about by the effect of pressure on the body (Figure 4.3). The two diseases which are most likely to be problem for marine mammals are decompression sickness and shallow-water blackout.

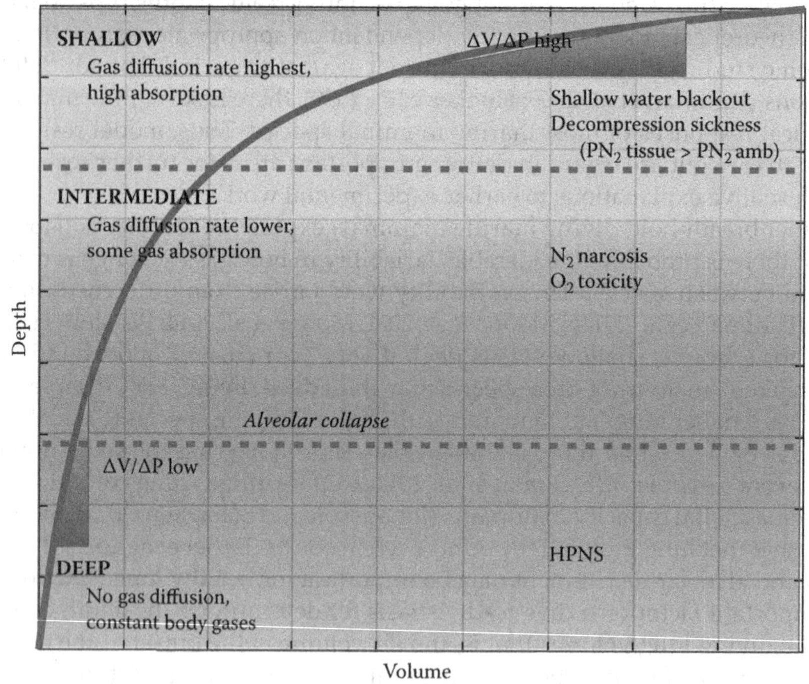

Figure 4.3 **(See color insert.)** Risks of different diving-related problems are related to pressure and gas diffusion. The water column can be divided into a shallow, intermediate, and deep region. In the shallow region, the rate of change of volume (ΔV/ΔP) is high, and there is high gas diffusion (see Figure 4.2). In this region, gases are exchanged and animals may be at risk of gas bubble disease when blood and tissue PN_2 exceeds ambient pressure. In addition, shallow-water blackout may also occur in this region due to the rapid changes in volume. In the intermediate region, a reduction in the alveolar surface area and thickening of the alveolar membrane reduces gas exchange. The N_2 and O_2 taken up may cause nitrogen narcosis and increase the risk for O_2 toxicity. Once the alveoli collapse in the deep region, no further gas is exchanged and as pressure increases, animals may be more at risk of HPNS.

1. *High pressure nervous syndrome* (HPNS) is caused by the hydrostatic pressure inducing tremors and convulsions.
2. *Nitrogen narcosis* is caused by the anesthetic effect of certain gases at high pressure ultimately leading to loss of consciousness at deeper depths.
3. *Decompression sickness* (DCS) is caused by the dissolved gas coming out of solution and forming bubbles during a reduction in pressure. Bubbles may cause lesions or *embolize* blood vessels leading to ischemic damage.
4. *Shallow-water blackout* is the loss of consciousness due to cerebral *hypoxia* caused by depletion of blood O_2 associated with a rapid drop in partial pressure of lung O_2 toward the end of a breath-hold dive.
5. *Oxygen toxicity* is caused by the high partial pressure of O_2 with short exposure causing central nervous system toxicity, longer exposures causing pulmonary or ocular toxicity.

We are familiar with these syndromes due to their effect on human divers, and primarily scuba divers. While scuba divers breathe compressed air and are, therefore, exposed to the deleterious effects of increased blood gas concentrations (i.e., at 30 m depth they breathe four times the gas concentrations as would someone breathing at the surface), breath-hold divers bring down only what is contained in the lungs, which mixes with gases previously absorbed and only causes problems if it is not completely removed during the surface interval. Consequently, scuba divers absorb significantly more N_2 as compared to breath-hold divers, which may have deleterious effects. This was well demonstrated by Scholander (1940) who pressurized a container with two frogs, one held underwater and one breathing under pressure. The frog breathing under pressure died while that held underwater (breath-holding) survived.

Thus until recently, it has been widely believed that free-divers, and particularly those with shallow depths of alveolar collapse (such as incurred by exhalation prior to diving), would be relatively immune from diving diseases. Alveolar collapse at shallow depths (Figure 4.2) would reduce the amount of inert gas absorbed and minimize the likelihood of supersaturation and bubble formation during ascent. However, recent work looking at human free-divers has suggested otherwise (Lemaitre et al. 2009). In fact, rapid, repetitive breath-hold diving in humans can result in decompression sickness (Schipke et al. 2006) and modeling work suggests that it may even be possible to develop DCS after a single deep breath-hold dive (Fitz-Clarke 2009). It seems that in free-diving animals, tissues can become highly saturated under certain circumstances depending on the iterative process of loading during diving and washout at the surface (Paulev 1967); and there is ample evidence that marine mammals are living with blood and tissue N_2 tensions that exceed ambient levels (de Quiros et al. 2013b; Falke et al. 1985; Moore et al. 2009; Ridgway and Howard 1979).

4.2.6.1 *High pressure nervous syndrome and nitrogen narcosis*

In humans, the symptoms of HPNS appear at pressures exceeding 11 ATA (Halsey 1982; Jain 1994), although individual differences make it difficult to give an absolute pressure (or depth) at which the symptoms first appear (Bennett and Rostain 2003; Brauer et al. 1975). Although species show differences in their susceptibility to elevated pressures, this variability appears to be related to the complexity of the CNS, and organisms with a less complex CNS seem to have a higher tolerance (Hunter and Bennett 1974; Rostain et al. 1983). Many marine mammals would therefore appear to be at risk of HPNS since they have fast descent rates (1–2 m/s), and spend time at pressures far exceeding 11 ATA.

However, we still know very little about why marine mammals are apparently unaffected by HPNS, and whether they possess specific neuroanatomical and physiological adaptations to protect them. The mechanism of HPNS is currently unknown, but an opinion currently held by many researchers is that HPNS is caused by compression of the neural membranes. The compression is thought to change the structure and function of the neural membranes, with secondary effects resulting in HPNS. The secondary effects could involve changes in axonal conduction, synaptic transmission, or changes in the release of neurotransmitters. Generally, membranes of the nervous system function over only a small range of pressures—so, the deep-diving marine mammals are all the more remarkable for their tolerance to a high range of pressures.

Interestingly, anesthetic gases appear to ameliorate symptoms and provide increased pressure tolerance in terrestrial mammals (Hunter and Bennett 1974). As N_2 behaves like an anesthetic gas at elevated pressure, it has been suggested that high N_2 levels could ameliorate HPNS symptoms. The myelin sheath is made up of lipids, and with a higher solubility for N_2 in lipids, this might alter the "flexibility" for lipid chains at pressure, allowing for more effective neurotransmission. Why marine mammals do not suffer from the narcotic effect of N_2 is unknown, but the tissue PN_2 levels of deep-diving species may be their primary way to avoid HPNS. The critical PN_2 for prevention of HPNS is less than that causing N_2 narcosis in humans (Halsey 1982). So it is possible that regulation of N_2 is to obtain levels sufficient to prevent HPNS but low enough to avoid N_2 narcosis.

4.2.6.2 Oxygen toxicity

In terrestrial mammals, there is a relationship between O_2 tension (PO_2), exposure time, and O_2 toxicity (Harabin et al. 1995). During breath-holding, however, O_2 toxicity is unlikely to be a problem as the PO_2 will only transiently reach high pressures and the continuous uptake of O_2 will not last for an extended period of time (McDonald and Ponganis 2013; Qvist et al. 1986).

4.2.6.3 Decompression sickness

Theoretical modeling attempts have been made to estimate blood and tissue levels of N_2 (Fahlman et al. 2006; Houser et al. 2001; Zimmer and Tyack 2007), O_2, and CO_2 (Fahlman et al. 2007) in diving vertebrates. When combined with a refined description of Scholander's model of lung compression (Bostrom et al. 2008; Fitz-Clarke 2007) most model results suggest that accumulation of both N_2 and CO_2 may reach levels that would cause DCS symptoms in similar-sized terrestrial mammals (Fahlman et al. 2007, 2014b; Hooker et al. 2009, 2012; Kvadsheim et al. 2012).

Decompression sickness has been suggested as a potential explanation for lesions coincident with intravascular and major organ gas *emboli* in beaked whales mass stranded in conjunction with military exercises deploying sonar (Fernandez et al. 2005; Jepson et al. 2003). There is some controversy about the proximate cause of gas *emboli* (Hooker et al. 2012) although it is widely agreed that it appeared to be linked to anthropogenic disturbance. These types of lesions have also been reported in some single-stranded cetaceans for which they do not appear to have been immediately fatal (Jepson et al. 2005). Differences in N_2 solubility between species and between the body blubber and acoustic fats may align with these observations (Koopman and Westgate 2012).

Osteonecrosis-type surface lesions have been reported in sperm whales (Moore and Early 2004). These were hypothesized to have been caused by repetitive formation of asymptomatic N_2 *emboli* over time and suggest that sperm whales live with sub-lethal decompression-induced bubbles on a regular basis, but with long-term impacts on bone health. However, experimental work using a captive bottlenose dolphin undergoing a dive schedule designed to induce bubble formation was unable to detect bubbles (Houser et al. 2010). In contrast, a low level of bubble incidence has been detected in stranded (common and white-sided) dolphins using B-mode ultrasound. Furthermore, these dolphins showed normal behavior upon release and did not restrand, suggesting some tolerance to bubble formation (Dennison et al. 2012). Bubbles have also been observed from marine mammals bycaught in fishing nets, which died at depth (Moore et al. 2009). These bubbles suggest the animals' tissues were supersaturated sufficiently to cause bubble formation when depressurized (as nets were hauled). However, whether tissue and blood N_2 levels represented the routine load at the time of entrapment or whether these levels were elevated as animals struggled in the nets is not clear. Two recent studies have reported neural deficiencies in California sea lions admitted to a rehabilitation facility (Van Bonn et al. 2011, 2013). In both instances, cerebral gas lesions were observed following magnetic resonance imaging (MRI). While the etiology cannot be determined, it was postulated that *barotrauma* would be one plausible explanation.

These cases highlight a growing body of evidence that indicates that our understanding how marine vertebrates manage gases during diving is very rudimentary, and that they may often experience blood and tissue tensions that are higher than previously thought (Hooker et al. 2012), and that may cause bubbles to form (de Quiros et al. 2013b; Dennison et al. 2012; Garcia-Parraga et al. 2014; Moore et al. 2009; Moore and Early 2004).

4.2.6.4 *Shallow-water blackout*

While shallow-water blackout has not been observed in marine mammals, it has been documented for human breath-hold divers and, therefore, would be expected to be a potential problem for marine mammals (Figure 4.4). In humans, it is often linked with hyperventilation before the dive, which helps reduce the vascular CO_2 tension (PCO_2) and thus the urge to breathe. During the dive, the *arterial* PO_2 (PaO_2) drops as O_2 is consumed

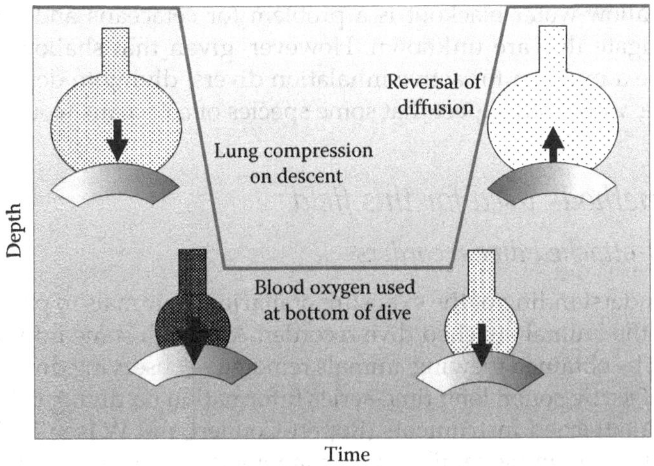

Figure 4.4 **(See color insert.)** Mechanism underlying shallow-water blackout. Alveolar O_2 pressure increases with ambient pressure on descent, increasing diffusion into blood. Arterial O_2 is increased at the bottom of a shallow dive, and then decreases (due to metabolism) over the course of the dive. During ascent, alveolar O_2 pressure decreases as the lung volume increases with decreasing ambient pressure. This can cause a reversal of the diffusion gradient, pulling O_2 from the blood into the lungs, causing a transient decrease in arterial O_2 and leading to blackout. Darker shading of the alveoli indicates higher partial pressure. Darker vessel shading indicates higher blood gas tension. The arrows show direction of net diffusion and the greater arrow thickness indicates greater diffusion rate.

(Figure 4.4; McDonald and Ponganis 2012; Meir et al. 2009). On ascent, the reduction in pressure causes the lung volume to increase, reducing the partial pressure of O_2, and potentially reversing the O_2 gradient across the lung, causing a rapid drop in alveolar PO_2 and resulting *hypoxemia* (Figure 4.4; McDonald and Ponganis 2012; Meir et al. 2009). Marine mammals hyperventilate before and after deep dives (Kooyman 1989). Such behavior could increase susceptibility to shallow water blackout if they dive on *hypocapnic* blood PO_2 levels.

Some phocid species, for example, the Weddell seal, harbor seal, and elephant seal have been documented to be extremely *hypoxia* tolerant, which may be an adaptation to this problem (Elsner et al. 1970; Kerem and Elsner 1973; Meir et al. 2009). Unlike phocids, otariid seals are thought to be more reliant on their lung O_2 stores (Kooyman 1973). Continuous *arterial* PO_2 loggers attached to California sea lions have shown a relatively deep depth of alveolar collapse (225 m). Alveolar collapse potentially mitigates the potential for shallow-water blackout as cessation of gas exchange preserved a reservoir of O_2 that supplemented blood O_2 during ascent (McDonald and Ponganis 2012). In contrast, during the relatively shallow (less than 160 m) dives of Antarctic fur seals, animals were found to exhale during the ascent portion of dives (Hooker et al. 2005). This behavior appeared to be associated with maintenance of lung compression. Since these dives were shallower than the likely depth of alveolar collapse, they could potentially result in extreme lung O_2 depletion and severe *hypoxemia*. Exhaling on ascent was suggested to mitigate the risk of shallow-water blackout by maintaining compressed alveoli and minimizing diffusion between lungs and blood during ascent (Hooker et al. 2005). A similar effect was previously observed during measurement of blood N_2 levels in a forced dive of a harbor seal. The seal exhaled during decompression from a simulated dive to 130 m in a *hyperbaric* chamber, which resulted in delaying the removal of N_2 from its blood until it surfaced and breathed (Kooyman et al. 1972). Thus, exhalation can maintain blood gas concentrations during ascent and help prevent drops in *arterial* PO_2 that may cause unconsciousness.

Whether shallow-water blackout is a problem for cetaceans and whether they have methods to mitigate this are unknown. However, given that shallow-water blackout is more likely to be a problem for active inhalation divers, diving to depths shallower than alveolar collapse, we would expect that some species of cetaceans would be vulnerable.

4.3 Tools/methods used for this field

4.3.1 Animal-attached dive recorders

Crucial to our understanding of the exposure of marine mammals to pressure has been the development of the animal-attached dive recorder. Although some information about diving behavior can be obtained viewing animals remotely—observing dive times or following animals sub-surface by sonar, long time-series information on diving behavior necessitates the use of animal-attached instruments (Ropert-Coudert and Wilson 2005). The earliest of these instruments were simple capillary tube manometers (Scholander 1940). The tubes were welded closed at one end and dusted internally with dye that was easily soluble in water. The intrusion of water as the air within the tube was compressed showed the maximum pressure achieved (Scholander 1940). In some cases, multiple rings could be identified (Kooyman 1965), or addition of a radioactive bead that followed the meniscus could record time spent at depth (Wilson and Bain 1984). However, such instruments, although benefiting from their small size, were unable to provide the resolution achieved by time-depth recorders.

The earliest time-depth recorders linked measurement of pressure to a moving needle which recorded directly as a trace on a 60 min kitchen timer with a smoked glass disc mounted on it (Kooyman 1965), and later on carbon-coated paper with a quartz motor allowing recording durations of 25 days (Naito et al. 1990). These analog systems were superseded by digital solid-state recorders. Early limitations were in terms of memory capacity (and thus, temporal resolution of data sampling), number of parallel sensors which could be supported, 8-bit resolution (allowing only 255 measurements, e.g., 1000 m depth could only be recorded to the nearest 4 m), and instrument size (Ropert-Coudert and Wilson 2005).

Most of these impediments have become vastly improved, and tags can now combine hydrophone recordings, video recordings, accelerometers, speed sensors, and oceanographic measurements alongside records of animal depth. Thus, these instruments not only log a detailed diary of the behavior and physiology of the animal but can also provide a detailed record of their external environment.

4.3.2 Observation of lung structure and dynamics

For semi-aquatic marine mammals such as pinnipeds, several aspects of the diving response have been effectively studied in lab-based settings. In terms of investigations examining the effect of pressure on diving animals, these have included sub-cellular based studies (Castellini and Castellini 2004) but also whole animal studies (Kooyman et al. 1972; Kooyman and Sinnett 1982) using *hyperbaric* chambers. Such studies have examined anatomical changes, for example, the flexibility of the trachea under pressure (Kooyman et al. 1970), but also changes in pulmonary shunt and its effects on blood gas content under breath-hold and pressure (Kooyman et al. 1972; Kooyman and Sinnett 1982).

The field of medical imaging (ultrasound, CT, and MRI) is becoming ever more refined, and there is great scope for the application of this type of study to investigations of diving animals. Recent work has used a *hyperbaric* chamber inside a CT scanner to obtain 3D images of the changes in lung compression of dead marine mammals as pressure was increased (Moore et al. 2011). There has been some progress on development of a MRI-compatible *hyperbaric* chamber and, combined with trained animals that dive on cue, there is potential for work on the respiratory alterations caused by pressure.

While imaging studies may become feasible in the future, recent work has investigated the structural properties of the respiratory system in anesthetized pinnipeds (Fahlman et al. 2014a). The static *compliance* (pressure–volume relationship) of the lung and chest were estimated in anesthetized individuals during manual ventilation. The results agreed with Scholander's suggestion that the chest provides little resistance to compression and is able to compress to very low volumes. In cetaceans, anesthesia is uncommon and studies on respiratory physiology and lung mechanics have been performed in conscious animals (Fahlman et al. 2015; Kooyman and Cornell 1981; Olsen et al. 1969). Both anatomical studies (Cotten et al. 2008) and work on live animals using a custom-made pneumotachometer (Fahlman et al. 2015) suggest that exhalation during voluntary breaths is passive while inhalation requires substantial work from the respiratory muscles. Thus, the elastic recoil of the chest may help empty the lung to low volumes which would be beneficial to avoid negative pressures (*lung squeeze*) to develop inside the lung during diving. In addition, works on excised lungs (Kooyman and Sinnett 1979) and live dolphins (Fahlman et al. 2015; Kooyman and Cornell 1981) suggest that the respiratory architecture may play a role in their unusual ability to respire. For example, respiratory flow rates are as much as three times higher than those in the terrestrial champion, the horse. The vital capacity can be as much as 80%–90% of the total lung capacity and exchanged in as little as 200–300 ms. The unusual anatomy of the respiratory system of odontocetes may enable rapid and efficient gas exchange during short surface intervals in addition to the proposed mechanism for alveolar collapse (Fahlman et al. 2015; Kooyman and Cornell 1981; Kooyman and Sinnett 1979).

4.3.3 Measurement of blood flow, gases, and bubble formation

The cardiac output and the blood-flow distribution is, in addition to pulmonary gas exchange, possibly the most important variable that diving animals are able to alter to help manage gases. Blood flow measurements are few and difficult to perform. Dye, coronary angiography, or thermal dilution experiments can be made to determine cardiac output, but are all invasive. Trans-thoracic ultrasound provides a non-invasive way to determine cardiac frequency and stroke volume and would be a viable method for most smaller marine mammals (Miedler et al. in press). To study blood-flow distribution, studies have either provided a snapshot of blood flow distribution at one time point using micro-labeled spheres during forced dives (Zapol et al. 1979) or continuous measurements using Doppler probes from a single blood vessel (Ponganis et al. 1990). While the micro-sphere studies provide global distribution of blood flow, these studies are done during forced-diving, which most likely alters the physiology of the animal and questions how valid the response may be as compared with natural dives. Future alternatives may be to use animals trained to dive on command and short-lived radioactive isotopes that can be imaged in a PET-CT, or ultrasound blood flow probes at selected blood vessels. While these require invasive surgery, technological advances may minimize impact and provide an interesting alternative that can be used on wild animals. In addition, ultrasound Doppler probes placed in strategic locations could be used to answer whether diving marine mammals experience intravascular gas bubbles during natural dives.

The measurement of blood gases (O_2 and N_2) has been undertaken in studies on captive animals during forced dives (Kooyman and Sinnett 1982; Kooyman et al. 1972), and on wild animals using a bespoke device that sampled *arterial* gas for later processing in the lab (Falke et al. 1985). One clever study inserted a needle covered in a gas permeable silicone sleeve into the muscle of dolphins following a dive bout. The needle was attached to a mass spectrometer which pulled gases out of the muscle according to the partial pressure gradients, enabling the N_2 washout to be measured (Ridgway and Howard 1979). In another study, blood samples were collected following a dive bout and PN_2 levels measured using a Van Slyke method (Houser et al. 2010). Recent use of intravascular O_2 electrodes has generated some very interesting results in California sea lions and elephant seals. In the latter, the evidence of extreme *hypoxia* tolerance was shown where *arterial* and *venous* PO_2 levels of 15 and 3 mmHg, respectively, were observed during extended dives (Meir et al. 2009). In the sea lion, the *arterial* data showed evidence of alveolar collapse and the depth of cessation of gas exchange was much deeper than formerly believed (McDonald and Ponganis 2012). Together these studies continue to expand the knowledge of physiological plasticity that appears possible in marine mammals.

Beyond blood gas measurements, other studies have shown that, during certain circumstances, blood gas levels may become supersaturated and then result in intravascular bubbles (de Quiros et al. 2012, 2013b; Dennison et al. 2012; Van Bonn et al. 2011, 2013). Whether and when bubbles are formed is of particular interest in establishing preconditions for diving diseases. Ultrasound (audible Doppler and visual transthoracic echo imaging) has been used to study intravascular bubbles in humans. Emerging technologies, such as dual frequency ultrasound, should enable the study of extravascular bubbles. Studies in stranded marine mammals have shown that under certain circumstances animals may experience renal bubbles (Dennison et al. 2012). However, controlled studies in the bottlenose dolphin (Houser et al. 2010) or Steller sea lion (Moore, unpub. obs.) have failed to replicate these findings. One possible reason could be that the dive history of wild-stranded animals is much more extensive than those starting off with near sea surface blood and tissue PN_2 levels. The development of technology to continuously monitor free ranging animals will be needed to test if free-ranging animals do experience gas bubbles while foraging.

4.4 Future of this topic

It is difficult to formulate a list of the lingering questions for this topic, since there remain so many gaps in our knowledge and understanding of how marine mammals cope with repeated exposure to high pressure. This is largely because experimental research, at least on obligate aquatic mammals diving to pressure has been a relatively intractable problem (Kooyman 2006). This continues to be the case, although advances in electronics are beginning to allow the pursuit of new avenues. However, in trying to make more sense of this field, it will be crucial to understand more about the phylogenetic variation in species behavior and physiology. We need to be open-minded in terms of questioning some of the assumptions, speculation, and hypotheses that have been suggested previously.

4.4.1 Phylogenetic variation

The vast majority of studies of the diving physiology of marine mammals have been done on seals, with a handful of studies of trained bottlenose dolphins (Noren et al. 2004, 2012; Ridgway and Howard 1979; Ridgway et al. 1969; Williams et al. 1993, 2015). An illustration of the type of problem that can arise is oft-quoted wisdom that "seals dive on exhalation." In fact, this was initially shown for some phocid seals, but studies of several otariid seals suggest that these seals are inhalation divers (Hooker et al. 2005; Kooyman 1973; McDonald and Ponganis 2012). Thus care is needed in assuming that particular species or species groups are representative of an overall pattern.

Our knowledge of marine mammal adaptations to pressure remains extremely limited. We need a better understanding of both physiology and physiological plasticity particularly in terms of phylogenetic variability. Thus far, we have only a handful of studies suggesting the depth of alveolar collapse, even fewer documenting changes in blood flow with diving and yet these are crucial to our ability to build a framework to understand changes in gas uptake that animals are exposed to during dives.

4.4.2 Form and function

Even in terms of more tractable studies such as the investigation of beach-cast animals to better describe anatomical adaptations, there are often difficulties inferring function from anatomical form. Although we see certain features of form that we speculate are functional in preventing problems due to pressure, we rarely have the opportunity for establishing definitive proof. There are many features that are suggested to be adaptations for diving, such as the reinforced conducting airways, the *rete* mirabilia and bronchial *sphincters*, but for which we can only speculate as to function.

While work with marine mammals under human care is controversial, access to animals that voluntarily participate in studies can allow us to perform experiments under physiologically normal conditions that would be logistically and ethically difficult with wild animals, and that may provide vital information for conservation efforts.

4.4.3 Consideration of alternatives

Finally, perhaps for any science hampered by limited data, there is a tendency toward generalization. In addition to better consideration of phylogenetic variation, we perhaps need to question the assumptions we often make. In terms of understanding how marine mammals avoid decompression sickness, we have a tendency to consider only the role of N_2. However, careful measurement of gas composition of bubbles in stranded cetaceans has suggested a potential role for CO_2 as the initial instigator to help bubbles form and grow (de Quiros et al. 2012, 2013a). Carbon dioxide is produced by aerobic metabolism and may accumulate to high levels in metabolically active tissues. Previous work in terrestrial mammals has shown that rising CO_2 levels were associated with a higher DCS incidence, and it was suggested that the higher diffusion rate of CO_2 could initiate bubble growth (Behnke 1951). Since bubble gas composition changes in a predictable manner in animals with decompression stress as compared with those receiving gas *emboli* from *barotrauma* or from putrefaction gas, gas composition may therefore be a viable way to assess whether stranded animals have experienced decompression stress. Gas composition in bubbles from stranded animals believed to have experienced gas bubble disease had a high level of CO_2 (de Quirós et al. 2012). It was suggested that elevated levels of CO_2 follow a burst of activity, and its much higher diffusion rate, as compared with N_2, could initiate the growth of a bubble which would then grow with diffusion of N_2 into the bubble as supersaturation increases close to the surface (de Quirós et al. 2012; Fahlman et al. 2014b). This hypothesis could explain how sonar-induced changes in dive behavior and the resultant burst of activity could cause increased DCS risk by causing an aversion response that increases anaerobic metabolism and muscle and vascular CO_2 levels to initiate bubble growth (Fahlman et al. 2014b).

In addition to managing several blood gases, animal are managing several physiological stresses simultaneously, such as dealing with buoyancy, thermoregulation, energy balance and the need for muscle oxygenation (Hooker et al. 2012). Thus, diving physiology and the regulation of pressure effects cannot be considered in isolation. New work examining the interaction between these stresses (e.g., between exercise and diving) (Williams et al. 2015) is likely to help illuminate the ways that diving animals manage this balance.

Glossary

Artery/arterial: A blood vessel carrying blood away from the heart. Systemic arteries carry oxygenated blood from the heart to the body and pulmonary arteries carry deoxygenated blood from the heart to the lungs.

ATA: Atmospheres absolute, a pressure unit that includes surface pressure.

Atelectasis: Alveolar closure when no gas exchange occurs.

Barotrauma: Trauma caused by pressure.

Compliance: A measure of the ease of expansion of a structure.

Emboli/Embolism: The lodging of an embolus (blood clot, fat globule, or gas bubble) in a blood vessel.

Hyperbaric: Higher pressure.

Hypocapnia: CO_2 levels lower than normal.

Hypocoagulable: Less prone to coagulate (form clots).

Hypoxemia: Arterial O_2 levels lower than normal.

Hypoxia: Tissue O_2 levels lower than normal.

Lung squeeze: When the chest is exposed to a pressure that reduces the lung volume below the functional residual capacity and a negative pressure develops inside the lung.

Plexus/rete: A network of blood vessels.
Sinus: A cavity within a bone or tissue.
Sphincter: A muscle that can close off a body cavity.
Surfactant: Compound which lowers the surface tension.
Vein/venous: A blood vessel carrying blood toward the heart. Systemic veins carry deoxygenated blood from the body back to the heart and pulmonary veins carry oxygenated blood from the lungs to the heart.

References

Behnke, A.R. 1951. Decompression sickness following exposure to high pressures. In *Decompression Sickness*, ed. J.F. Fulton. Philadelphia, PA: Saunders, 53–89.

Belanger, L.F. 1940. A study of the histological structure of the respiratory portion of the lungs of aquatic mammals. *American Journal of Anatomy* 67:437–461.

Bennett, P.B. and J.C. Rostain. 2003. The high pressure nervous syndrome. In *Bennett and Elliott's Physiology and Medicine of Diving*, eds. A.O. Brubakk and T.S. Neuman. New York: Saunders, 194–237.

Bostrom, B.L., A. Fahlman, and D.R. Jones. 2008. Tracheal compression delays alveolar collapse during deep diving in marine mammals. *Respiratory Physiology & Neurobiology* 161:298–305.

Boyd, I.L. and J.P. Croxall. 1996. Dive durations in pinnipeds and seabirds. *Canadian Journal of Zoology* 74:1696–1705.

Brauer, R.W., R.W. Beaver, W.M. Mansfield, F. Oconnor, and L.W. White. 1975. Rate factors in development of high pressure neurological syndrome. *Journal of Applied Physiology* 38(2):220–227.

Brown, R.E. and J.P. Butler. 2000. The absolute necessity of chest-wall collapse during diving in breath-hold diving mammals. *Aquatic Mammals* 26(1):26–32.

Butler, J.P., S.H. Loring, S. Patz, A. Tsuda, D.A. Yablonskiy, and S.J. Mentzer. 2012. Evidence for adult lung growth in humans. *New England Journal of Medicine* 367(3):244–247.

Castellini, M.A. and J.M. Castellini. 2004. Defining the limits of diving biochemistry in marine mammals. *Comparative Biochemistry and Physiology B* 139:509–518.

Castellini, M.A., J.M. Castellini, and P.M. Rivera. 2001. Adaptations to pressure in the RBC metabolism of diving mammals. *Comparative Biochemistry and Physiology A* 129(4):751–757.

Castellini, M.A., P.M. Rivera, and J.M. Castellini. 2002. Biochemical aspects of pressure tolerance in marine mammals. *Comparative Biochemistry and Physiology A* 133:893–899.

Chilvers, B.L., S. Delean, N.J. Gales et al. 2004. Diving behaviour of dugongs, *Dugong dugon*. *Journal of Experimental Marine Biology and Ecology* 304(2):203–224.

Costidis, A. and S.A. Rommel. 2012. Vascularization of air sinuses and fat bodies in the head of the bottlenose dolphin (*Tursiops truncatus*): Morphological implications on physiology. *Frontiers in Physiology* 3:243.

Cotten, P.B., M.A. Piscitelli, W.A. McLellan, S.A. Rommel, J.L. Dearolf, and D.A. Pabst. 2008. The gross morphology and histochemistry of respiratory muscles in bottlenose dolphins, *Tursiops truncatus*. *Journal of Morphology* 269(12):1520–1538.

Cozzi, B., P. Bagnoli, F. Acocella, and M.L. Constantino. 2005. Structure and biomechanical properties of the trachea of the striped dolphin *Stenella coeruleoalba*: Evidence for evolutionary adaptations to diving. *The Anatomical Record Part A* 284A:500–510.

Craig, A.B. Jr. 1968. Depth limits of breath hold diving (an example of fennology). *Respiration Physiology* 5:14–22.

Croll, D.A., M.K. Nishiguchi, and S. Kaupp. 1992. Pressure and lactate-dehydrogenase function in diving mammals and birds. *Physiological Zoology* 65(5):1022–1027.

de Quirós, Y.B., Ó. González-Díaz, M. Arbelo, E. Sierra, S. Sacchini, and A. Fernández. 2012. Decompression versus decomposition: Distribution, quantity and gas composition of bubbles in stranded marine mammals. *Frontiers in Physiology* 3:177.

de Quiros, Y.B., O. Gonzalez-Diaz, A. Mollerlokken et al. 2013a. Differentiation at autopsy between in vivo gas embolism and putrefaction using gas composition analysis. *International Journal of Legal Medicine* 127(2):437–445.

de Quiros, Y.B., J.S. Seewald, S.P. Sylva et al. 2013b. Compositional discrimination of decompression and decomposition gas bubbles in bycaught seals and dolphins. *PLoS ONE* 8(12):12.

Denison, D.M. and G.L. Kooyman. 1973. The structure and function of the small airways in pinniped and sea otter lungs. *Respiration Physiology* 17:1–10.

Denison, D.M., D.A. Warrell, and J.B. West. 1971. Airway structure and alveolar emptying in the lungs of sea lions and dogs. *Respiration Physiology* 13:253–260.

Dennison, S., M.J. Moore, A. Fahlman et al. 2012. Bubbles in live-stranded dolphins. *Proceedings of the Royal Society of London B* 279:1396–1404.

Elsner, R., J.T. Shurley, D.D. Hammond, and R.E. Brooks. 1970. Cerebral tolerance to hypoxemia in asphyxiated Weddell seals. *Respiration Physiology* 9:287–297.

Fahlman, A., S.K. Hooker, A. Szowka, B.L. Bostrom, and D.R. Jones. 2009. Estimating the effect of lung collapse and pulmonary shunt on gas exchange during breath-hold diving: The Scholander and Kooyman legacy. *Respiratory Physiology & Neurobiology* 165:28–39.

Fahlman, A., S.H. Loring, M. Ferrigno et al. 2011. Static inflation and deflation pressure-volume curves in excised lungs of marine mammals. *Journal of Experimental Biology* 214:3822–3828.

Fahlman, A., S.H. Loring, S.P. Johnson et al. 2014a. Inflation and deflation pressure-volume loops in anesthetized pinnipeds confirms compliant chest and lungs. *Frontiers in Physiology* 5:433.

Fahlman, A., S.H. Loring, G. Levine, J. Rocho-Levine, T. Austin, and M. Brodsky. 2015. Lung mechanics and pulmonary function testing in cetaceans. *Journal of Experimental Biology* 218:2030–2038.

Fahlman, A., A. Olszowka, B. Bostrom, and D.R. Jones. 2006. Deep diving mammals: Dive behavior and circulatory adjustments contribute to bends avoidance. *Respiratory Physiology & Neurobiology* 153:66–77.

Fahlman, A., A. Schmidt, D.R. Jones, B.L. Bostrom, and Y. Handrich. 2007. To what extent might N2 limit dive performance in king penguins? *Journal of Experimental Biology* 210:3344–3355.

Fahlman, A., P.L. Tyack, P.J.O. Miller, and P.H. Kvadsheim. 2014b. How man-made interference might cause gas bubble emboli in deep diving whales. *Frontiers in Physiology* 5:13.

Falke, K.J., R.D. Hill, J. Qvist et al. 1985. Seal lungs collapse during free diving—Evidence from arterial nitrogen tensions. *Science* 229:556–558.

Fernandez, A., J.F. Edwards, F. Rodriguez et al. 2005. "Gas and fat embolic syndrome" involving a mass stranding of beaked whales (family *Ziphiidae*) exposed to anthropogenic sonar signals. *Veterinary Pathology* 42:446–457.

Ferretti, G. 2001. Extreme human breath-hold diving. *European Journal of Applied Physiology* 84(4):254–271.

Field, C.L., N.J. Walker, and F. Tablin. 2001. Northern elephant seal platelets: Analysis of shape change and response to platelet agonists. *Thrombosis Research* 101(4):267–277.

Fitz-Clarke, J.R. 2007. Mechanics of airway and alveolar collapse in human breath-hold diving. *Respiratory Physiology & Neurobiology* 159(2):202–210.

Fitz-Clarke, J.R. 2009. Risk of decompression sickness in extreme human breath-hold diving. *Undersea Hyperbaric Medicine* 36:83–91.

Foot, N.J., S. Orgeig, and C.B. Daniels. 2006. The evolution of a physiological system: The pulmonary surfactant system in diving mammals. *Respiratory Physiology & Neurobiology* 154(1–2):118–138.

Fowler, S.L., D.P. Costa, J.P.Y. Arnould, N.J. Gales, and C.E. Kuhn. 2006. Ontogeny of diving behaviour in the Australian sea lion: Trials of adolescence in a late bloomer. *Journal of Animal Ecology* 75:358–367.

Fraser, F.C., and P.E. Purves. 1960. Hearing in cetaceans—Evolution of the accessory air sacs and the structure and function of the outer and middle ear in recent cetaceans. *Bulletin of the British Museum (Natural History) Zoology* 7(1):1–140.

Garcia-Parraga, D., J.L. Crespo-Picazo, Y. Bernaldo de Quiros et al. 2014. Decompression sickness ("the bends") in sea turtles. *Diseases of aquatic organisms* 111(3):191–205.

Goldbogen, J.A., J. Calambokidis, D.A. Croll et al. 2012. Scaling of lunge-feeding performance in rorqual whales: Mass-specific energy expenditure increases with body size and progressively limits diving capacity. *Functional Ecology* 26(1):216–226.

Gutierrez, D.B., A. Fahlman, M. Gardner et al. 2015. Phosphatidylcholine composition of pulmonary surfactant from terrestrial and marine diving mammals. *Respiratory Physiology & Neurobiology* 211:29–36.

Halsey, L.G., P.J. Butler, and T.M. Blackburn. 2006. A phylogenetic analysis of the allometry of diving. *American Naturalist* 167(2):276–287.

Halsey, M.J. 1982. Effects of high pressure on the central nervous system. *Physiological Reviews* 62(4):1341–1377.

Harabin, A.L., S.S. Survanshi, and L.D. Homer. 1995. A model for predicting central nervous system oxygen toxicity from hyperbaric oxygen exposures in humans. *Toxicology and Applied Pharmacology* 132(1):19–26.

Hooker, S.K., R.W. Baird, and A. Fahlman. 2009. Could beaked whales get the bends? Effect of diving behaviour and physiology on modelled gas exchange for three species: *Ziphius cavirostris, Mesoplodon densirostris* and *Hyperoodon ampullatus. Respiratory Physiology & Neurobiology* 167:235–246.

Hooker, S.K., A. Fahlman, M.J. Moore et al. 2012. Deadly diving? Physiological and behavioural management of decompression stress in diving mammals. *Proceedings of the Royal Society of London B* 279(1731):1041–1050.

Hooker, S.K., P.J.O. Miller, M.P. Johnson, O.P. Cox, and I.L. Boyd. 2005. Ascent exhalations of Antarctic fur seals: A behavioural adaptation for breath-hold diving? *Proceedings of the Royal Society of London B* 272:355–363.

Horning, M. and F. Trillmich. 1997. Ontogeny of diving behaviour in the Galápagos fur seal. *Behaviour* 134:1211–1257.

Houser, D.S., L.A. Dankiewicz-Talmadge, T.K. Stockard, and P.J. Ponganis. 2010. Investigation of the potential for vascular bubble formation in a repetitively diving dolphin. *Journal of Experimental Biology* 213:52–62.

Houser, D.S., R. Howard, and S. Ridgway. 2001. Can diving-induced tissue nitrogen supersaturation increase the chance of acoustically driven bubble growth in marine mammals. *Journal of Theoretical Biology* 213:183–195.

Hui, C.A. 1975. Thoracic collapse as affected by the *retia thoracica* in the dolphin. *Respiration Physiology* 25(1):63–70.

Hunter, W.L. Jr. and P.B. Bennett. 1974. The causes, mechanisms and prevention of the high pressure nervous syndrome. *Undersea Biomedical Research* 1(1):1–28.

Jain, K.K. 1994. High pressure neurological syndrome (HPNS). *Acta Neurologica Scandinavica* 90(1):45–50.

Jepson, P.D., M. Arbelo, R. Deaville et al. 2003. Gas-bubble lesions in stranded cetaceans. *Nature* 425:575–576.

Jepson, P.D., R. Deaville, I.A.P. Patterson et al. 2005. Acute and chronic gas bubble lesions in cetaceans stranded in the United Kingdom. *Veterinary Pathology* 42:291–305.

Johansson, O. and E. Schagatay. 2012. Lung-packing and stretching increases vital capacity in recreational freedivers. Paper read at *European Respiratory Society Annual Congress,* Vienna, Austria, 2012.

Kayar, S.R., E.O. Aukhert, M.J. Axley, L.D. Homer, and A.L. Harabin. 1997. Lower decompression sickness risk in rats by intravenous injection of foreign protein. *Undersea & Hyperbaric Medicine* 24(4):329–335.

Kerem, D., and R. Elsner. 1973. Cerebral tolerance to asphyxial hypoxia in the harbor seal. *Respiration Physiology* 19:188–200.

Kerem, D., J.A. Kylstra, and H.A. Saltzman. 1975. Respiratory flow rates in the sea lion. *Undersea Biomedical Research* 2:20–27.

Koopman, H.N. and A.J. Westgate. 2012. Solubility of nitrogen in marine mammal blubber depends on its lipid composition. *Journal of Experimental Biology* 215:3856–3863.

Kooyman, G.L. 1965. Techniques used in measuring diving capacities of Weddell seals. *Polar Record* 12:391–394.

Kooyman, G.L. 1973. Respiratory adaptations in marine mammals. *American Zoologist* 13:457–468.

Kooyman, G.L. 1989. *Diverse Divers.* Berlin, Germany: Springer-Verlag.

Kooyman, G.L. 2006. Mysteries of adaptation to hypoxia and pressure in marine mammals. The Kenneth S. Norris lifetime achievement award lecture presented on December 12, 2005, San Diego, California. *Marine Mammal Science* 22:507–526.
Kooyman, G.L. and L.H. Cornell. 1981. Flow properties of expiration and inspiration in a trained bottle-nosed porpoise. *Physiological Zoology* 54(1):55–61.
Kooyman, G.L., D.D. Hammond, and J.P. Schroeder. 1970. Bronchograms and tracheograms of seals under pressure. *Science* 169:82–84.
Kooyman, G.L., D.H. Kerem, W.B. Campbell, and J.J. Wright. 1971. Pulmonary function in freely diving Weddell seals, *Leptonychotes weddelli*. *Respiration Physiology* 12:271–282.
Kooyman, G.L., J.P. Schroeder, D.M. Denison, D.D. Hammond, J.M. Wright, and W.P. Bergman. 1972. Blood nitrogen tensions of seals during simulated deep dives. *American Journal of Physiology* 223:1016–20.
Kooyman, G.L. and E.E. Sinnett. 1979. Mechanical properties of the harbor porpoise lung, *Phocoena phocoena*. *Respiration Physiology* 36(3):287–300.
Kooyman, G.L. and E.E. Sinnett. 1982. Pulmonary shunts in harbor seals and sea lions during simulated dives to depth. *Physiological Zoology* 55:105–111.
Kvadsheim, P.H., P.J.O. Miller, P.L. Tyack, L.D. Sivle, F.P.A. Lam, and A. Fahlman. 2012. Estimated tissue and blood N_2 levels and risk of decompression sickness in deep-, intermediate-, and shallow-diving toothed whales during exposure to naval sonar. *Frontiers in Physiology* 3:125.
Le Boeuf, B.J., D.P. Costa, A.C. Huntley, and S.D. Feldkamp. 1988. Continuous, deep diving in female northern elephant seals, *Mirounga angustirostris*. *Canadian Journal of Zoology* 66:446–458.
Le Boeuf, B.J., D.P. Costa, A.C. Huntley, G.L. Kooyman, and R.W. Davis. 1986. Pattern and depth of dives in northern elephant seals, *Mirounga angustirostris*. *Journal of Zoology, London* 208:1–7.
Leith, D. 1976. Comparative mammalian respiratory mechanics. *Physiologist* 19:485–510.
Leith, D.E. 1989. Adaptations to deep breath-hold diving: Respiratory and circulatory mechanics. *Undersea & Hyperbaric Medicine* 16(5):345–353.
Lemaitre, F., A. Fahlman, B. Gardette, and K. Kohshi. 2009. Decompression sickness in breath-hold divers: A review. *Journal of Sports Sciences* 27(14):1519–1534.
Lundgren, C.E.G. and J.N. Miller. 1999. *The Lung at Depth*. New York: Marcel Dekker.
McDonald, B.I. and P.J. Ponganis. 2012. Lung collapse in the diving sea lion: Hold the nitrogen and save the oxygen. *Biology Letters* 8(6):1047–1049.
McDonald, B.I. and P.J. Ponganis. 2013. Insights from venous oxygen profiles: oxygen utilization and management in diving California sea lions. *Journal of Experimental Biology* 216:3332–3341.
McLellan, W.A., H.N. Koopman, S.A. Rommel et al. 2002. Ontogenetic allometry and body composition of harbour porpoises (*Phocoena phocoena*, L.) from the western North Atlantic. *Journal of Zoology* 257:457–471.
Meir, J.U., C.D. Champagne, D.P. Costa, C.L. Williams, and P.J. Ponganis. 2009. Extreme hypoxemic tolerance and blood oxygen depletion in diving elephant seals. *American Journal of Physiology—Regulatory Integrative and Comparative Physiology* 297(4):R927–R939.
Miedler, S., A. Fahlman, M. Valls Torres, T. Alvaro Alvez, and D. Garcia-Parraga. in press. Evaluating cardiac physiology through echocardiography in bottlenose dolphins: using stroke volume and cardiac output to estimate systolic left ventricular function during rest and following exercise. *Journal of Experimental Biology*.
Miller, P.J.O., M.P. Johnson, and P.L. Tyack. 2004. Sperm whale behaviour indicates the use of echolocation click buzzes "creaks" in prey capture. *Proceedings of the Royal Society of London B* 271:2239–2247.
Montcalm-Smith, E.A., A. Fahlman, and S.R. Kayar. 2008. Pharmacological interventions to decompression sickness in rats: Comparison of five agents. *Aviation Space and Environmental Medicine* 79(1):7–13.
Moore, C., M. Moore, S. Trumble et al. 2014. A comparative analysis of marine mammal tracheas. *Journal of Experimental Biology* 217(7):1154–1166.
Moore, M.J., A.L. Bogomolni, S.E. Dennison et al. 2009. Gas bubbles in seals, dolphins, and porpoises entangled and drowned at depth in gillnets. *Veterinary Pathology* 46:536–547.

Moore, M.J. and G.A. Early. 2004. Cumulative sperm whale bone damage and the bends. *Science* 306:2215.
Moore, M.J., T. Hammar, J. Arruda et al. 2011. Hyperbaric computed tomographic measurement of lung compression in seals and dolphins. *Journal of Experimental Biology* 214:2390–2397.
Nagel, E.L., P.J. Morgane, W.L. McFarlane, and R.E. Galliano. 1968. Rete mirabile of dolphin: Its pressure-damping effect on cerebral circulation. *Science* 161(844):898–900.
Naito, Y., T. Asaga, and Y. Ohyama. 1990. Diving behavior of Adelie penguins determined by time-depth recorder. *Condor* 92(3):582–586.
Ninomiya, H., T. Inomata, and H. Shirouzu. 2004. Microvasculature in the terminal air spaces of the lungs of the Baird's beaked whale (*Berardius bairdii*). *Journal of Veterinary Medical Science* 66(12):1491–1495.
Noren, S.R., V. Cuccurullo, and T.M. Williams. 2004. The development of diving bradycardia in bottlenose dolphins (*Tursiops truncatus*). *Journal of Comparative Physiology B* 174:139–147.
Noren, S.R., T. Kendall, V. Cuccurullo, and T.M. Williams. 2012. The dive response redefined: Underwater behavior influences cardiac variability in freely diving dolphins. *Journal of Experimental Biology* 215(16):2735–2741.
Odend'hal, S. and T.C. Poulter. 1966. Pressure regulation in the middle ear cavity of sea lions: A possible mechanism. *Science* 153(737):768–769.
Olsen, C.R., F.C. Hale, and R. Elsner. 1969. Mechanics of ventilation in the pilot whale. *Respiration Physiology* 7(2):137–149.
Paulev, P. 1967. Nitrogen tissue tensions following repeated breath-hold dives. *Journal of Applied Physiology* 22:714–718.
Piscitelli, M.A., W.A. McLellan, S.A. Rommel, J.E. Blum, S.G. Barco, and D.A. Pabst. 2010. Lung size and thoracic morphology in shallow- and deep-diving cetaceans. *Journal of Morphology* 271(6):654–673.
Ponganis, P.J., G.L. Kooyman, and S.H. Ridgway. 2003. Comparative diving physiology. In *Bennett and Elliott's Physiology and Medicine of Diving*, eds. A.O. Brubakk and T.S. Neuman. London, UK: Saunders, Elsevier Science Ltd., 211–226.
Ponganis, P.J., G.L. Kooyman, M.H. Zornow, M.A. Castellini, and D.A. Croll. 1990. Cardiac-output and stroke volume in swimming harbour seals. *Journal of Comparative Physiology B* 160(5):473–482.
Pontier, J.M., N. Vallee, M. Ignatescu, and L. Bourdon. 2011. Pharmacological intervention against bubble-induced platelet aggregation in a rat model of decompression sickness. *Journal of Applied Physiology* 110(3):724–729.
Qvist, J., R.D. Hill, R.C. Schneider et al. 1986. Hemoglobin concentrations and blood-gas tensions of free-diving Weddell seals. *Journal of Applied Physiology* 61(4):1560–1569.
Ridgway, S.H. and R. Howard. 1979. Dolphin lung collapse and intramuscular circulation during free diving: Evidence from nitrogen washout. *Science* 206:1182–1183.
Ridgway, S.H., B.L. Scronce, and J. Kanwisher. 1969. Respiration and deep diving in the bottlenose porpoise. *Science* 166:1651–1654.
Robinson, A.J., M. Kropatkin, and P.M. Aggeler. 1969. Hageman factor (factor XII) deficiency in marine mammals. *Science* 166:1420–1422.
Ropert-Coudert, Y. and A. Kato. 2012. *The Penguiness Book*. World Wide Web electronic publication (http://penguinessbook.scarmarbin.be/), version 2.0, Accessed 10 March, 2015.
Ropert-Coudert, Y. and R.P. Wilson. 2005. Trends and perspectives in animal-attached remote sensing. *Frontiers in Ecology and the Environment* 3:437–444.
Rostain, J.C., C. Lemaire, M.C. Gardettechauffour, J. Doucet, and R. Naquet. 1983. Estimation of human susceptibility to the high pressure nervous syndrome. *Journal of Applied Physiology* 54(4):1063–1070.
Schipke, J.D., E. Gams, and O. Kallweit. 2006. Decompression sickness following breath-hold diving. *Research in Sports Medicine* 14(3):163–178.
Scholander, P.F. 1940. Experimental investigations on the respiratory function in diving mammals and birds. *Hvalradets Skrifter* 22:1–131.

Siebenaller, J.F. and D.J. Garrett. 2002. The effects of the deep-sea environment on transmembrane signaling. *Comparative Biochemistry and Physiology B* 131(4):675–694.

Somero, G.N. 1992. Adaptations to high hydrostatic-pressure. *Annual Review of Physiology* 54:557–577.

Stenfors, L.E., J. Sade, S. Hellstrom, and M. Anniko. 2001. How can the hooded seal dive to a depth of 1000 m without rupturing its tympanic membrane? A morphological and functional study. *Acta Oto-Laryngologica* 121(6):689–695.

Thompson, L.A. and T.A. Romano. 2015. Beluga (*Delphinapterus leucas*) granulocytes and monocytes display variable responses to *in vitro* pressure exposures. *Frontiers in Physiology* 6:128.

Van Bonn, W., S. Dennison, P. Cook, and A. Fahlman. 2013. Gas bubble disease in the brain of a living California sea lion (*Zalophus californianus*). *Frontiers in Physiology* 4:6.

Van Bonn, W., E. Montie, S. Dennison et al. 2011. Evidence of injury caused by gas bubbles in a live marine mammal: Barotrauma in a California sea lion *Zalophus californianus*. *Diseases of Aquatic Organisms* 96:89–96.

Vogl, A.W. and H.D. Fisher. 1982. Arterial retia related to supply of the central nervous system in two small toothed whales-narwhal (*Monodon monoceros*) and beluga (*Delphinapterus leucas*). *Journal of Morphology* 174(1):41–56.

Ward, C.A., D. McCullough, D. Yee, D. Stanga, and W.D. Fraser. 1990. Complement activation involvement in decompression sickness of rabbits. *Undersea Biomedical Research* 17(1):51–66.

Watkins, W.A., M.A. Daher, K.M. Fristrup, T.J. Howald, and G. Notarbartolo-di-Sciara. 1993. Sperm whales tagged with transponders and tracked underwater by sonar. *Marine Mammal Science* 9:55–67.

Whitehead, H. 1996. Babysitting, dive synchrony, and indications of alloparental care in sperm whales. *Behavioural Ecology and Sociobiology* 38:237–244.

Williams, E.E., B.S. Stewart, C.A. Beuchat, G.N. Somero, and J.R. Hazel. 2001. Hydrostatic-pressure and temperature effects on the molecular order of erythrocyte membranes from deep-, shallow-, and non-diving mammals. *Canadian Journal of Zoology* 79(5):888–894.

Williams, T.M., W.A. Friedl, and J.E. Haun. 1993. The physiology of bottlenose dolphins (*Tursiops truncatus*)—Heart-rate, metabolic rate and plasma lactate concentration during exercise. *Journal of Experimental Biology* 179:31–46.

Williams, T.M., L.A. Fuiman, T. Kendall et al. 2015. Exercise at depth alters bradycardia and incidence of cardiac anomalies in deep-diving marine mammals. *Nature Communications* 6:9.

Wilson, R.P. and C.A.R. Bain. 1984. An inexpensive depth gauge for penguins. *Journal of Wildlife Management* 48(4):1077–1084.

Zapol, W.M., G.C. Liggins, R.C. Schneider et al. 1979. Regional blood flow during simulated diving in the conscious Weddell seal. *Journal of Applied Physiology* 47(5):968–973.

Zimmer, W.M.X. and P.L. Tyack. 2007. Repetitive shallow dives pose decompression risk in deep-diving beaked whales. *Marine Mammal Science* 23:888–925.

section two

Nutrition and energetics

chapter five

Feeding mechanisms

Christopher D. Marshall and Jeremy A. Goldbogen

Contents

5.1 Introduction

Feeding is an essential behavior that is required not only to sustain the organism but also to acquire the energy needed for reproduction. Therefore, feeding adaptations are directly tied to fitness and these phenotypes experience strong selection pressures. The physical forces experienced in an aquatic environment impart strong selection pressures upon all vertebrates and have influenced the evolution of foraging strategies. This basic conceptual framework is fundamental in understanding marine mammal feeding mechanisms of a diverse guild of predators and grazers that evolved from terrestrial ancestors. With respect to evolutionary history, *marine mammals* collectively represent a paraphyletic group because several independent mammal lineages made this transition back to aquatic environments at different evolutionary time scales. Mapping traits associated with different feeding mechanisms onto a phylogeny demonstrates a diversity of feeding methods and foraging strategies among marine mammals. The ancestors of the orders Cetacea and Sirenia were among the first mammals to make the transition back to the sea, followed by pinnipeds (walruses, otariids, and phocids), sea otters, and polar bears more recently. The selection pressure for these multiple re-invasions was abundant food. Therefore, evolutionary innovations for specific feeding mechanisms and strategies are driving factors in marine mammal evolution. This assemblage of marine mammal lineages and the timing of their transition to a fully aquatic or semi-aquatic life history manifest itself into a diversity of foraging strategies that we are only now beginning to explore and appreciate.

The investigation of feeding mechanisms among any vertebrate group is an integrative and comparative endeavor that involves morphological, physiological, developmental, behavioral performance, and ecological studies. In addition to understanding the evolution of mammalian feeding in aquatic environments, understanding prey-capture tactics and feeding performance are important considerations for trophic ecological questions since such behavior can determine prey choice due to energetic constraints (Emlin 1966; Schoener 1971; Bowen et al. 2002; Wainwright and Bellwood 2002). Feeding has a direct bearing on the fitness of an organism by determining the behavioral capacity of an animal to exploit its resources (Arnold 1983; Wainwright and Reilly 1994). Performance is an important link between morphology and ecology because morphology is a primary predictor of performance and performance is a predictor of ecology.

Marine mammals represent the most recent vertebrate guild that underwent a major evolutionary transition from land to sea (Pyenson et al. 2014). This macroevolutionary shift from a terrestrial to a primarily aquatic lifestyle required a complex suite of feeding adaptations, which today are exhibited among several diverse lineages of cetaceans (whales and dolphins), sirenians (manatees and dugongs), and pinnipeds (seals, sea lions, and walruses). Collectively, these extant species feed on a wide variety of resources that span all trophic levels including marine algae, aquatic angiosperms, small zooplankton (such as krill and copepods), fish, squid, and even marine amniotes including other marine mammals. Therefore, marine mammals are important consumers in diverse ocean ecosystems worldwide through the use of multiple innovations that enable highly successful feeding mechanisms.

5.2 *Knowledge by order*

5.2.1 *Order Cetacea: Whales and dolphins*

Cetaceans (whales and dolphins) represent a radiation of carnivorous marine mammals descended from terrestrial even-toed ungulates (artiodactyls) that diversified approximately 50 Ma during the establishment of the circum-Antarctic current system in the Southern Ocean (Fordyce and Barnes 1994). Given our current understanding of the Southern Ocean as a whale feeding hotspot (Nowacek et al. 2011), this restructuring of the oceans presumably changed the abundance and diversity of oceanic resources in a dramatic way (Fordyce 1980; Steeman et al. 2009; Marx and Uhen 2010; Slater et al. 2010; Pyenson et al. 2014). Extant cetaceans are represented by two major clades that exhibit very divergent feeding strategies: toothed whales (Odontoceti) and baleen whales (Mysticeti). Odontocetes possess adaptations of the skull (and other systems) for the integrated dual functions of prey capture and echolocation to target single prey items, whereas baleen whales feed in bulk on aggregations of prey using baleen as a filter (e.g., Slijper 1962; Werth 2001, 2004). Both echolocation and bulk filter feeding represent major evolutionary innovations that underlie the ecological success of this adaptive radiation of marine mammals.

5.2.1.1 *Cetacea: Odontoceti*

It is thought that toothed whales evolved echolocation to feed at night in shallow waters on diel migrating cephalopods, and later this physiological adaptation was exapted in many odontocete lineages to exploit deep prey that were also available during the day (Lindberg and Pyenson 2007). The deepest diving toothed whale lineages are indeed largely teuthophagous (Clarke 1996), as exemplified by beaked (Ziphiidae) and sperm whales (Physeteridae), but their diets can also be supplemented with fish in many geographic regions (Slijper 1962; Gaskin 1982). Several other odontocete lineages (oceanic dolphins, Delphinidae; porpoises, Phocoenidae) also feed on both fish and cephalopods, although narwhals and belugas (Monodontidae) may also feed on crustaceans and benthic invertebrates (Dahl et al. 2000). These generalized and perhaps opportunistic feeding preferences reflect the flexibility of feeding strategies afforded by echolocation, a key evolutionary innovation that is a hallmark of toothed whale life history and functional ecology.

Odontocete (and all cetacean) skulls are perhaps the most derived among mammals. The morphology of the rostrum, nares, cranium, ear bones, and mandible and biomechanics of jaw adduction have been drastically modified (Rommel 1990). The unusual morphology of cetacean skulls is due to overlapping and telescoping of bones that shorten the cranium and the elongation of the facial region by lengthening of the maxilla and premaxilla, and mandible. Such modifications of the skull are linked to feeding, respiration, and the generation and reception of sound used for echolocation (Rommel 1990; Marshall 2002). Variation in the ratio of facial length versus cranial length is variable among cetaceans and results in species with very short blunt skulls (e.g., *Kogia* and *Globicephala*) or long narrow rostra (e.g., platanistids; Werth 2006b). Dentition of odontocete jaws varies from virtually edentulous (e.g., ziphiids) to several hundred simple homodont teeth (e.g., platanistids). These traits are related to diet. It is thought that constraint of the genetic component in early dental development has been released in cetaceans (Armfield et al. 2013). Changes in the development of teeth can drive morphological evolution as observed in odontocetes. Concomitantly, their feeding mechanisms are among the most specialized and varied among mammals. They range from ram and raptorial feeding to suction feeding, whereas mysticetes are generally categorized as filter feeders. In odontocetes and many secondarily aquatic tetrapods, there is a dichotomy of cranial morphology associated with feeding mode. Piscivory tends to be associated with long narrow rostra and mandibles, and jaws filled with numerous teeth. This *ecomorph* has evolved independently several times among several aquatic vertebrate groups (e.g., extinct marine reptiles, gharials, and some odontocetes) and is an adaptation for high velocity of the jaw tips to capture elusive prey, at the expense of bite force. On the other extreme, teuthophagy (squid-eating) in toothed whales is associated with short, blunt rostra and mandibles, a reduction in tooth number (or function), expanded basihyoid bone, and the use of suction as the primary feeding mode. However, important exceptions to this dichotomy, as observed in sperm whales (*Physeter macrocephalus*) and beaked whales, exist. Suction in marine mammals is generated by the rapid depression and retraction of the tongue (but in some cases by the addition of fast jaw opening velocities), which results in a rapid increase in buccal volume and concomitant decrease in pressure (Gordon 1984; Werth 2000, 2006a,b, 2007; Marshall et al. 2008, 2014; Marshall 2009). It is thought that the muscles associated with the enlarged basihyoid bone of the hyoid apparatus (Reidenberg and Laitman 1994; Heyning and Mead 1996; Werth 2007) result in a greater force of lingual depression that presumably increases subambient pressure (Heyning and Mead 1996; Werth 2006b). However, orofacial morphology and tongue shape may be just as important in directing the subambient pressure anteriorly (Bloodworth and Marshall 2007; Marshall et al. 2008, 2014, in press; Kane and Marshall 2009). Such traits and correlated performance measures are well known for teleost fish (e.g., Lauder 1985; Wainwright and Day 2007; Wainwright et al. 2007; Van Wassenbergh and Aerts 2009).

To acquire prey, toothed whales use two primary mechanisms: ram and suction. Ram feeding occurs when the whale's attack speed and agility outperform that of targeted prey, thereby resulting in raptorial capture (Weihs and Webb 1984). In contrast, suction can be generated from negative intraoral pressures through the rapid depression of the hyoid and tongue. These two capture mechanisms are not necessarily exclusive, and some toothed whale species can use these strategies either together or in sequence. Suction appears to be an important mechanism for prey capture in odontocetes as demonstrated by the well-developed tongue musculature and increased surface area of the hyoid for attachment of these muscles (Bloodworth and Marshall 2005, 2007; Werth 2007). The hyolingual apparatus exhibits a greater proportion of extrinsic muscle fibers (connects the tongue to other structures), and a lower proportion of intrinsic muscle fibers (connects the tongue to itself), as compared to both terrestrial mammals and other aquatic mammals (Werth 2007).

In general, suction feeders produce negative intraoral pressures through the rapid depression of the hyolingual apparatus (Werth 2000, 2007; Bloodworth and Marshall 2005; Marshall et al. 2008, 2014, in press; Kane and Marshall 2009). This so-called *gular depression* to generate suction during feeding has been experimentally demonstrated (Werth 2006a) in odontocetes of varying head shape and bluntness of the rostrum (common dolphin, *Delphinus delphis*; Atlantic white-sided dolphin, *Lagenorhychus acutus*; and harbor porpoise, *Phocoena phocoena*). The greatest suction capability in this study was found in harbor porpoises, which also possessed the bluntest rostrum in the study (Werth 2006b). In addition, gular depression as a mechanism to produce suction has been empirically demonstrated in several toothed whale species including pygmy sperm whale (*Kogia* sp.), long-finned pilot whales (*Globicephala melas*), belugas (*Delphinapterus leucas*) (Werth 2000; Bloodworth and Marshall 2005, 2007; Kane and Marshall 2009) as well as harbor porpoises (Kastelein et al. 1997). In addition to blunt rostra and wide jaws, odontocetes that can produce a more circular mouth aperture to enhance suction feeding performance (Bloodworth and Marshall 2005; Kane and Marshall 2009). This phenotype, termed amblygnathy (blunt, wide), is well represented in most toothed whale families and is well developed in globicephaline delphinids (Werth 2006b). Toothed whales that exhibit amblygnathy tend to be larger, deep divers, and many species have evolved reduced dentition. The two exceptions to this trend are river dolphins (Platanistoidea) and sperm whales (Physeteridae), which have elongated skulls with teeth. Comparative morphological data for skull and dental traits analyzed in a phylogenetic context suggest that suction feeding evolved once early in evolutionary history, thereby representing the ancestral condition of crown cetaceans, but intermediate ancestral reconstructions may also indicate that many extant odontocetes exhibit secondarily derived suction feeding mechanisms (Werth 2006a,b; Johnston and Berta 2011).

The morphological design of the skull and hyoid bones alone do not completely determine nor limit feeding mode and performance. Therefore, the function of soft tissues must also be considered to fully understand the integrative biology of feeding in cetaceans. For example, beluga whales (*D. leucas*) often exhibit discrete ram and suction components during feeding, but the latter is greatly enhanced by pursing of the lips to occlude the lateral gape (Kane and Marshall 2009). Lip pursing behaviors act to form a small circular aperture to magnify negative intraoral pressures, a mechanism that is convergent with more basal vertebrates (Kane and Marshall 2009). Other odontocetes, such as bottlenose and pacific white-sided dolphins, have limited lip pursing abilities and instead use ram primarily to capture prey, although suction may be used to manipulate prey within the mouth to facilitate deglutition (Bloodworth and Marshall 2005; Kane and Marshall 2009). In other species, ram and suction may be more synchronized, as demonstrated in long-finned pilot whales, where sub-maximal gape angles or soft tissue adaptations effectively occlude lateral gape to enhance suction performance during raptorial capture events (Werth 2000; Bloodworth and Marshall 2007; Kane and Marshall 2009).

5.2.1.2 *Cetacea: Mysticeti*

Baleen whales (Mysticeti) evolved from toothed whales presumably to exploit aggregations of prey, rather than single prey items. Baleen whales are edentulous and exhibit calcified alpha-keratin baleen racks that act as the primary feeding structure (Szewciw et al. 2010). All baleen whales are obligate bulk filter feeders that engulf a volume of prey-laden water that is subsequently filtered out of the mouth using two racks of vertically oriented baleen plates that hang down from the top of the rostrum. Each baleen rack consists of an array of plates that are obliquely angled relative to the whale's long body axis. Each plate consists of a series of tubules that are imbedded in a calcified, keratin matrix (Fudge et al. 2009; Szewciw et al. 2010). The plates are frayed on the lingual side, thereby exposing the core tubules that derive from the palatal epithelium. Because these plates are embedded in a cream white Zwischensubstanz less than 1 cm apart (Pinto and Shadwick 2013), these exposed tubules, or fringes, collectively form a fibrous mat that acts as the filter that separates prey that are suspended in engulfed water. There is a tremendous amount of diversity among mysticetes with respect to plate and fringe morphology, and these variants correlate generally with prey preference and ecological niche. For example, finer fringes are associated with smaller prey, as exemplified by bowhead and right whales that feed primarily on copepods (Werth 2012).

There are different modes of filter feeding represented among different baleen whale families. These modes include suction and ram hydraulic phenomena, the latter of which can be either intermittent or continuous, but these elements manifest in baleen whales in very different ways compared to toothed whales. Mysticetes have very large skulls and most lack soft tissues that completely occlude the lateral gape, so enhanced suction performance may not be possible in most species. However, one monotypic family of gray whales (Eschrichtidae) is known to use suction to feed along the seafloor on benthic invertebrates (Nerini 1984). Gray whales have well developed hyolingual musculature and ventral grooves that facilitate the depression of the hyoid and tongue to generate suction (Werth 2007). However, it is unclear how adequate suction is produced without lateral occlusion of the gape. Gray whales have been observed both in captivity and virtually using digital movement tags to roll onto their lateral sides during feeding (Ray and Schevill 1974; Woodward and Winn 2006). Therefore, the flow of water and prey must enter and exit through the side of the mouth, and relatively short baleen plates may enable this unique filtration mechanism (Werth 2001). In addition, gray whales could use the substrate and the geomorphology of the seafloor to passively increase suction distance. The ability to use the substrate to passively increase the suction distance on benthic prey is known for both ray-finned and chondrichthyan fish (Carroll et al. 2004; Nauwelaerts et al. 2007). Interestingly, other observations suggest that gray whales are generalist filter feeders and may also use ram in combination with suction to feed throughout the water column (Pyenson and Lindberg 2011). Despite several observational and anatomical studies, the filter feeding mechanisms in gray whales remain poorly understood. Even more enigmatic are pygmy right whales (Cetotheriidae/Neobalaeninae; *Caperea marginata*), which share many morphological characters with some mysticetes that exhibit continuous ram feeding (Balaenidae; Bowhead and right whales) and others that are obligate intermittent ram feeders (Balaenopteridae; rorqual whales) (Fordyce and Marx 2012).

The two mysticete families Balaenidae and Balaenopteridae are extremely divergent in both the anatomy of the feeding apparatus and the hydrodynamic mechanisms employed to capture and filter prey from seawater. Broadly, balaenids have large and stiff tongues that direct continuous flow of prey-laden water past long baleen plates, whereas balaenopterids have largely flaccid tongues for inversion and expansion of the large ventral pouch. Balaenids swim at slow, steady speeds (<1 m/s) to continuously drive water into the mouth (Simon et al. 2009). Such a strategy may be required to efficiently filter water with an enlarged mouth aperture that should incur significant drag and thus large energy costs (Werth 2004). Despite this large anterior opening, there are two much smaller posterior openings of the mouth where water is thought to exit the mouth after being filtered through the baleen (Werth 2004). The difference in area between the anterior and posterior openings creates a Venturi effect, or suction, in front of the mouth that should help reduce a bow wave from forming that would deflect prey away from the mouth (Werth 2004). In addition, perioral structures such as the subrostral gap, orolabial sulcus, curvature of baleen, mandibular rotation, and lingual mobility also permit the steady flow of water through the baleen that improves the efficiency of filtration.

In contrast, Balaenopterids (rorquals) exhibit a dynamic process that involves a lunge, or a rapid acceleration to high speed (Goldbogen et al. 2006; Simon et al. 2012; Kot et al. 2014), and the subsequent engulfment of a large volume of prey-laden water (Orton and Brodie 1987). After the target volume of prey and water is engulfed, the mouth closes just enough to leave the baleen exposed for filtration and this so-called lunge filter feeding mechanism is enabled by an integrated suite. The engulfed water is then driven past the baleen plates through the contraction of the expanded ventral pouch. This so-called lunge filter feeding mechanism is enabled by a integrated suite of morphological and mechanical adaptations that facilitates the lunge feeding process: hyper-expandable ventral groove blubber (Shadwick et al. 2013), elongate and curved jaws (Goldbogen et al. 2010; Pyenson et al. 2013), and a sensory organ in the un-fused mandibular symphysis (Pyenson et al. 2012). In many large-bodied rorqual species, the size of the engulfed volume is commensurate with the whale's body size (Goldbogen et al. 2007). This is due in large part to the positive allometry of the skull and ventral pouch, whereby large whales exhibit relatively larger oropharyngeal cavities. The positive allometry of the engulfment apparatus is accompanied by the negative allometry of the caudal peduncle. As a result, larger rorquals have big heads and short tails, which may reflect that whales are investing growth in the anterior region at the expense of the posterior region (Goldbogen et al. 2010). The scaling of pouch allometry appears to have important consequences for rorqual diving physiology, feeding performance, and ecological niche (Goldbogen et al. 2012).

5.2.2 Order Sirenia: Manatees and dugongs

Sirenians were among the first mammal lineages to return to aquatic habitats. The earliest sirenians appear in the fossil record ~50 Ma in the middle to late Eocene (Savage et al. 1994; Domning 2001) and our modern dugongs and manatees first appeared in the middle Eocene and Oligocene, respectively. Our modern assemblage includes three manatee species with the Family Trichechidae: West Indian manatees (*Trichechus manatus*), West African manatees (*Trichechus senegalensis*), Amazonian manatees (*Trichechus inunguis*), and two species within the Family Dugongidae, dugongs (*Dugong dugon*) and the recently extinct Steller's sea cow (*Hydrodamalis gigas*). Sirenians are distinct among marine mammals in that they are herbivorous and possess many adaptations for grasping, excavating, and processing aquatic plants. Although sirenians may be superficially similar in both morphology and diet, they are quite different in that manatees are considered to be generalists, whereas dugongs are benthic specialists. There are several overarching functional themes in sirenian feeding mechanisms that include the degree of deflection of the rostrum, tooth replacement, the vibrissal-muscular complex. Among all sirenians, the skull is dominated by a broad narial basin and a rostrum comprised of enlarged premaxillary bones. The variation of the degree of rostral deflection of these premaxillary bones relative to the palatal plane reflects the location within the water column where feeding occurs most efficiently (Domning 1980, 1982; Domning and Hayek 1986). On one extreme of rostral deflection are Amazonian manatees. They feed primarily at the surface upon natant *floating meadows* of the Amazonian rivers, lakes and floodplains, inundated vegetation of the vàrzea, igapó, and emergent grasses (Poaceae formally Gramineae; Best 1981; Rosas 1994). West African manatees also primarily consume natant vegetation. Both species inhabit turbid aquatic habitats where submerged aquatic plants are not widely supported (Best 1981). Corresponding these trichechids possess the least deflected snouts (~25°–42° and 15°–40° respectively). On the other extreme of rostral deflection are dugongs. Dugongs are benthic-feeding specialists that consume primarily sea grasses. They consume both the above- and belowground biomass. Belowground biomass contains rhizomes, which dugongs often target and are a rich source of carbohydrates. Dugongs possess the greatest rostral deflection (70°). West Indian manatees are the ultimate generalists among sirenians. They inhabit a variety of habitats that span coastal marine, estuarine, and freshwater ecosystems. They consume more than 60 species of aquatic vegetation that are distributed throughout these habitats (Hartman 1979). Much of their diet is comprised of sea grasses, brackish and freshwater submerged aquatic vegetation, but many terrestrial grasses and other types of vegetation are also consumed (Hartman 1979). The locations of these food resources span the water column from benthic, midwater, natant, emergent, and terrestrial. Their intermediate snout deflection (29°–52°) allows them to consume vegetation throughout these various locations (Domning 1980). More recently, it has been demonstrated that such correlations are much more complex and interesting. The tusk size, body size, and degree of snout deflection are among several suites of important ecomorphological traits that determined resource partitioning in extinct dugongid assemblages (Velez-Juarbe et al. 2012).

Unlike many marine mammals, sirenians masticate and process their food. Plant matter is processed using teeth and/or cornified palatal pads. This processing aids in reducing particle size, increasing surface area, and ruptures the tough plant cell walls. Sirenians use hindgut fermentation to digest the cellulose and cell contents of the plants they consume (Reynolds and Rommel 1996). Plant matter, particularly grasses (which contain high levels of silica) can be highly abrasive. This selection pressure has resulted in numerous adaptations by herbivores, terrestrial and aquatic, to resist tooth wear, since the life of an herbivore only lasts as long as their teeth. Classic examples include the hypsodont teeth of grazers such as horses, the open rooted ever-growing incisors of rodents, and the serial replacement of multi-rooted molars of elephants (Unger 2010). Trichechids exhibit a novel mechanism of cheek-tooth replacement, which differs significantly from dugongs (and other herbivores). At any one time 6–8 cheek teeth are erupted and functional. As they wear, these teeth migrate horizontally from the posterior region of the tooth-row to the anterior region (Domning and Hayek 1984). Teeth at the anterior locations have little-to-no crown remaining. Once non-functional, the roots are reabsorbed and the tooth falls out. New molars erupt at the posterior tooth-row and migrate anteriorly as replacements. Manatees apparently have an indeterminate number of teeth that can be replaced in this manner. This is an evolutionary novel solution to coping with an abrasive diet. Dugongs do not possess such a conveyor-belt mechanism to resist an abrasive diet and are thought to be at a disadvantage compared to trichechids when consuming grasses. Instead, dugong teeth are open-rooted, simple peg-like molars that consist of dentin covered by cementum. These molars erupt slowly over their lifetime, through anterior drift, also known as molar progression (Lanyon and Sanson 2006a,b; Unger 2010). Unlike manatees, however, dugongs have a finite number of molars (six). However, some evidence suggests that dugongid cheek teeth may not be functional at all. Instead the enlarged, heavily cornified and rugose palatal pads may function to masticate, process, and transport sea grasses into the buccal cavity. Although trichechids also possess such cornified palatal pads, those of dugongs are much more robust and cover a larger surface area (Marsh et al. 1999; Lanyon and Sanson 2006a,b).

The facial muscles of all sirenians form a muscular hydrostat (Kier and Smith 1985), or short muscular snout, that is capable of highly complex and varied movements (Marshall et al. 1998a, 2003). The complex facial muscles surround a series of six perioral bristle fields, or modified vibrissae, which are located on both the broad and expanded upper and lower lip margins (Reep et al. 1998, 2001; Marshall et al. 2003). Vibrissae are specialized hairs that transmit tactile information from the environment to the central nervous system. Although these bristles are homologous with mystacial vibrissae of other mammals, manatee vibrissae differ in that they are short, thick, and robust. These are also unusual in they function in both motor and sensory roles. When contracted, semicircular facial muscles protrude the largest pair of perioral bristles on the upper (U2 bristle fields) and lower (L1) lips, which are then used to handle and manipulate vegetation (Marshall et al. 2000). How this is accomplished differs between manatees and dugongs (Marshall et al. 1998b, 2003). In manatees, the paired upper bristle fields are protruded anteriorly and then medially in a grasping motion, pushing vegetation in the mouth. This is alternated with a sweeping motion of the lower bristle fields to further push vegetation into the mouth. For dugongs, the upper bristle fields have a slightly more horizontal distribution across the lip margin (oral disk) but are also protruded (as much as 6 cm). However, rather than moving toward the midline, the upper bristles are moved laterally, in a breast-stroke-like motion (Marshall et al. 2003). As dugongs graze along the seafloor, this functions to part the sea grass in front of the animal and introduce plant material into the side of the mouth. As in manatees, the bristle fields on the lower jaw alternate and sweep vegetation further into the mouth. When feeding upon small species of sea grasses, this mechanism can be used to literally excavate the root system (belowground biomass), with rhizomes, from the benthic substrate. This action can be seen as a signature *feeding trail* on the seafloor, which creates substantial bioturbation. Feeding trails span the width of the dugong rostrum, can be as deep as 5 cm, and as long as 10 m (Anderson and Birtles 1978). Up to 90% of the vegetation can be removed from these feeding trails (Preen 1995).

5.2.3 Order Carnivora: Pinnipedia

Pinnipeds are a monophyletic lineage of carnivores that include sea lions (Otariidae), seals (Phocidae), and walruses (Odobenidae). Each family has successfully transitioned back to the aquatic environment, albeit at different time scales. Extant pinnipeds consume a diversity of prey that include fish, cephalopods, bivalves, crustaceans, invertebrates, and large amniote prey (penguins, seabirds, other marine mammals) (King 1983; Riedman 1990; Pauly et al. 1998). Although pinnipeds are a major carnivoran lineage, we know less regarding their aquatic feeding mechanisms than other marine mammal groups (i.e., cetaceans and sirenians). Our knowledge of how pinnipeds feed is largely descriptive, and lacks the detailed functional and phylogenetic analyses found for basal aquatic vertebrates (e.g., actinopterygian fish; but see Jones and Goswami 2010; Jones et al. 2013). Three themes in the feeding function of pinnipeds are (1) loss of mastication, (2) a reduced and simplified dentition, and (3) importance of suction (Fay 1982; King 1983; Kastelein et al. 1994; Werth 2006b; Marshall et al. 2008, 2014). Based on craniodental morphology and some functional studies, pinniped feeding has been generally characterized as raptorial biting, suction, grip-and-tear, or filter feeding (Adam and Berta 2002) (Figure 5.1).

Functional studies of terrestrial carnivores demonstrate that craniodental and mandibular morphology are good predictors of feeding performance and diet (e.g., Van Valkenburgh 1989; Biknevicius and Van Valkenburgh 1996; Sacco and Van Valkenburgh 2004). It is also well known that key biomechanical cranial elements of mammals are associated with specific prey types (Radinsky 1981a,b; Kiltie 1982; Weijs 1994; Perez-Barberia and Gordon 1999). Capture of specific prey types would subsequently suggest that craniodental characteristics are associated with specific prey capture modes in pinnipeds. However, only a handful of feeding kinematic and performance studies on pinnipeds have been conducted (Fay 1982; Kastelein et al. 1994; Marshall et al. 2008, 2014, in press; Hocking et al. 2013, 2014).

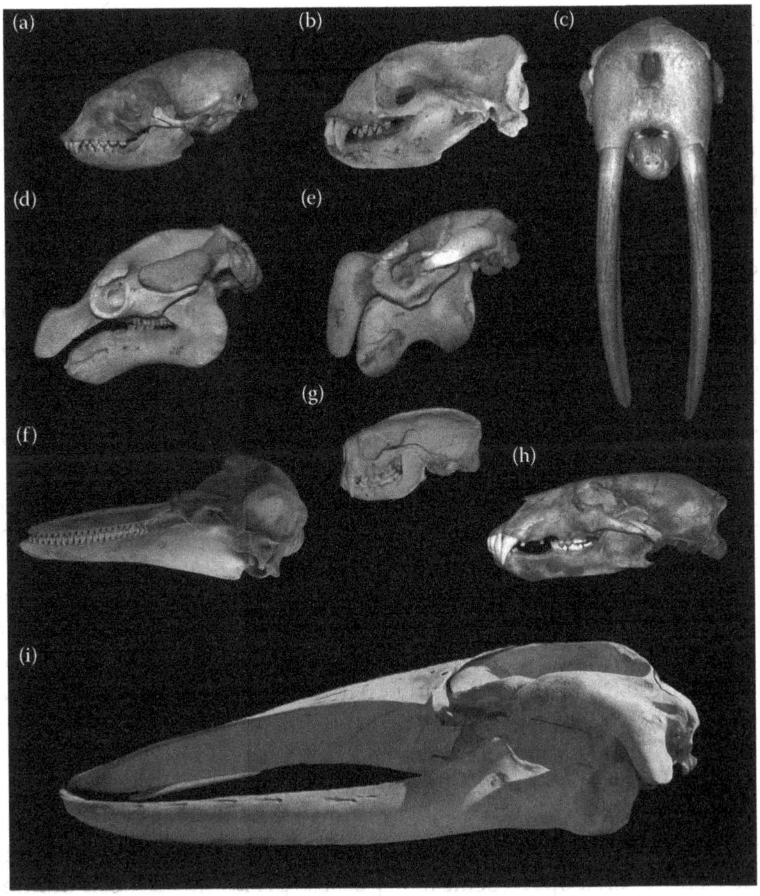

Figure 5.1 **(See color insert.)** Representative skulls of major marine mammal clades. (a) Phocidae (harbor seal, *Phoca vitulina*); (b) Otariidae (Steller's sea lion, *Eumetopias jubatus*); (c) Odobenidae (walrus, *Odobenus rosmarus*); (d) Trichechidae (West Indian manatee, *Trichechus manatus*); (e) Dugongidae (dugong, *Dugong dugon*); (f) Odontoceti (bottlenose dolphin, *Tursiops truncatus*); (g) Mustelidae (sea otter, *Enhydra lutris*); (h) Ursidae (polar bear, *Ursus maritimus*); and (i) Mysticeti (gray whale, *Eschrichtius robustus*). Skull images are not to scale.

The skulls and mandibles of pinnipeds are the least derived among marine mammals (with a few notable exceptions), and are more similar to those of canids. The modifications of craniodental morphology, particularly shape, have long been thought to be associated with feeding adaptations (King 1972). In general, pinniped crania are rounded and sharply delineation from the facial region of the skull; they do not possess sinuses with air chambers. The skulls of otariids are less variable than those of phocids, and many are sexually dimorphic. Compared to closely related terrestrial carnivores, pinniped dentition is reduced (~22–38 versus 44) (Berta et al. 2006). Although their dentition is heterodont, the premolars and molars are uniform in cusp number, size, and often shape, resulting in a virtually homodonty of these teeth. Hence, they are often collectively referred to as cheek teeth or post-canine teeth. As a consequence, pinnipeds have lost the shearing carnassial phenotype of terrestrial carnivores (Unger 2010).

Otariid skulls tend to be more dolichocephalic and with less variation of the post-canine teeth. Otariids usually possess deep transverse grooves on their incisors. The canines are large relative to phocids and are thought to be used for grip-and-tear feeding. The cheek teeth are uniformly homodont teeth but possess the addition of a large cingulum (shelf-like cusp) on the lingual surface, as well as a small cusp on the rostral surface of each cheek tooth. The degree of cusps varies greatly (King 1983; Hillson 2005; Unger 2010). Although distinctive phenotypes such as those found in crabeater and leopard seals are not found among otariids, feeding specializations still exist. Many otariids possess interesting feeding mechanisms for capturing prey. Antarctic fur seals feed heavily upon zooplankton, including krill. Their post-canine teeth are among the smallest of any fur seal, and hypothesized to be modified for straining krill (Bonner 1968; Repenning et al. 1971; Riedman 1990). The skulls of South American sea lions (*Otaria byronia*) are perhaps the most divergent. These skulls are large, robust, the upper palate is vaulted and elongated, the jaws are short and broad, and the orofacial muscles are well developed, broadening the snout further. This suite of characteristics is commonly associated with suction feeding specialization (King 1983) and indirect data strongly suggest that *Otaria* is capable of powerful subambient pressures. These traits are exemplified in the skulls of walruses, a well-known suction specialist.

The dentition of phocids is reduced relative to otariids (22–36 versus 34–38, respectively) (Berta et al. 2006). Among the phocids, there is a diversity of dental phenotypes most of which are likely adaptations for piscivory and teuthophagy, but outstanding exceptions for filter feeding exists. At least two species, the leopard seal (*Hydrurga leptonyx*) and the crabeater seal (*Lobodon carcinophaga*), possess intricate lophs of the cheek teeth that are hypothesized to assist in filter feeding. The lophs of the distinctive post-canine teeth of crabeater seals possess three long shearing cusps. Observations of captive crabeater seals suggest they have the capability to ingest krill using suction, then employing a filtering feeding mode that utilizes their elaborately lophed post-canine teeth (Ross et al. 1976; Klages and Cockcroft 1990). Leopard seals are known to feed upon large vertebrates, such as penguins and other seals, using a grip-and-tear feeding mode. The enlarged canines and strong jaw and neck muscles are useful traits for raptorial biting and grip-and-tear feeding modes. However, leopard seals also possess pronounced lophs on the post-canine teeth and are thought to be as effective for filtering as crabeater seals (Øritsland 1977; Hocking et al. 2013). Consumption of krill by leopard seals was once thought to be minimal but is now known to be important seasonally. The behavioral data demonstrate that they are capable of suction feeding when consuming krill (Hocking et al. 2013). Although the magnitude of this pressure remains unknown, a reinvestigation of the leopard teeth morphology in conjunction with a feeding study supports the hypothesis that they do use their post-canine teeth to filter feed (King 1983; Klages and Cockcroft 1990; Adam 2005; Hocking et al. 2013). Hence, many species likely modulate their feeding mode, and use multiple feeding depending upon the circumstances (Marshall et al. 2008, 2014, in press; Hocking et al. 2013, 2014). However, biting and suction feeding modes are likely still the most commonly used. Notable among phocids is the suction feeding specialty of bearded seals (*Erignathus barbatus*). Their dental and orofacial morphology, feeding performance (suction and hydraulic jetting), and trophic ecology converge with walruses. Bearded seals consume infaunal invertebrates such as bivalves and tubeworms, as well as fish. The upper palate of bearded seals is also vaulted, although not to the degree observed in South American sea lions and walruses. They are capable of generating up to 91.2 kPa of subambient pressure, very close to the capability of walruses (Marshall et al. 2008). Orofacial muscles close the upper and lower lateral lips to prevent negative pressure loss, whereas the broad lips at the anterior are pursed to create a circular aperture. These two features create a pipette-like structure that enhances suction generation and direct those subambient pressures in front of the animal to enhance subambient forces. Such behavior has also been reported for walruses (Fay 1982) but also for harbor seals (Marshall et al. 2014) and Steller sea lions (Marshall et al., in press). Although not as impressive as walruses, harbor seals and Steller sea lions can produce significant subambient pressures (45 kPa) and hydraulic jetting (Marshall et al. 2014, in press).

The skull of walruses is distinctive in its size, fusion, and generally derived condition. The tusks dominate its morphology. The maxillary bones in the walrus are enlarged to accommodate and anchor the tusks to the skull (Fay 1982; King 1983; Marshall 2002). The frontal orientation of the premaxillary broadens and shortens the rostrum, which is advantageous for benthic feeding. The tusks of walruses are not used for feeding. Instead, tusks are used for male–male interactions and for hauling out of the water onto ice. Walruses commonly use their tusks to pull and lift their bodies from the water, hence the derivation of their Latin name *Odobenus* (tooth walker). The tusks are ever-growing upper canines that can grow up to a meter in length in males (Fay 1982; King 1983; Unger 2010). The upper and lower deciduous incisors of walruses are present between the right and left canines (tusks); but these quickly fall out and are not replaced, leaving a wide space between the two tusks on the upper jaw and canines on the lower jaw. This loss creates a circular space that is part of the formation of a pipette and assists in maintaining subambient pressures. Walruses can generate subambient pressures as high as 108 kPa, which is just greater than 1 atmosphere of pressure (Kastelein et al. 1994). The movement of a piston-like tongue has been hypothesized to be responsible for the substantial intraoral subambient pressures measured (Gordon 1984). The vaulted palate enhances suction generation since there is a greater volume within the oral cavity to act on. Walruses are infaunal benthic specialists, consuming mostly bivalves and other infaunal invertebrates, but with occasional exceptions such as marine birds and other marine mammals. Functional studies have demonstrated that in addition to suction, walruses can perform the opposite behavior—hydraulic jetting. Large bivalves are excavated from the seafloor by alternating suction with hydraulic jetting to remove the sediment around their prey.

5.2.3.1 Suction and biting feeding modes

Unlike some odontocetes, it is noteworthy that extreme rostral elongation has not evolved among pinnipeds. This suggests that biting and suction feeding modes are not biomechanical trade-offs as in odontocetes but are perhaps synergistic in pinnipeds. Short, wide jaws with high mechanical advantage for biting feeding modes should also be advantageous for suction feeding if other morphological adaptations are also present. Indeed, new evidence from captive performance studies suggests that suction is widespread across most pinniped groups (Kastelein et al. 1994; Marshall et al. 2008; Hocking et al. 2013; Marshall et al. 2014, in press) and indirect evidence suggests many more species use suction (King 1983; Adam and Berta 2002; Hocking et al. 2013). Pinnipeds that are presumed, or known, to employ suction feeding tend to have short, wide rostra with jaws that have scoop-like anterior ends, and a long mandibular symphysis, or a mandible in which the ventral borders are angled toward the oral cavity. There is also a trend toward simplification of tooth morphology and reduced functionality, lost functionality, or loss of certain classes of teeth. The post-canine teeth of walruses, bearded seals, and elephant seals are simple oval peg-like teeth with simple cusps that are quickly worn flat with age, with little-to-no occlusal surface cusps or patterns (Fay 1982; King 1983; Unger 2010). These teeth may not be functional, or limited in function, in adults. Monk seals depart from this pattern. Although no specific functional feeding studies of monk seals have been conducted, animal-borne camera studies of feeding (Parrish et al. 2002), as well as anecdotal evidence strongly support a suction capability in these species (*Monachus* sp.). Their broad, robust teeth with heavy rugosities suggest these shallow water foragers crush many prey items. The biomechanics and shape of mandibles differ among pinnipeds (Jones et al. 2013). The shape of pinniped mandibles appears to separate due to phylogenetic relatedness and each family displays distinct morphologies. In addition, there are fundamental differences in jaw adductor muscular arrangement of otariids and the jaws of otariids are relatively longer than those of phocids. However, male–male combat and the influence of mating strategies appear to also act as a strong evolutionary driver in pinniped mandibular function (Jones et al. 2013) (Figure 5.2).

A feeding performance study of harbor seals (*Phoca vitulina*) (Marshall et al. 2014) suggests that suction may be widespread among phocids. Harbor seals are the ultimate generalists among pinnipeds (Figure 5.2). They are the most widely distributed phocid, and despite some disparity, there are up to five subspecies recognized (Rice 1998; Burns 2009). Harbor seals are opportunistic foragers that exhibit a generalized feeding ecology. For example, harbor seals feed upon a wide diversity of small- to medium-sized fish that include herring, anchovy, cod, hake, trout, smelt, shad, scorpionfish, rockfish, prickleback, greenling, sculpin, capelin, sandlance, salmon, flatfish, a variety cephalopods, and invertebrates that are mostly crab and shrimp species, but also include mollusks (Bigg 1981; Hoover 1988; Sharples et al. 2009; Kavanagh et al. 2010; Thomas et al. 2011; Brown et al. 2012; Bromaghin et al. 2013). Therefore, harbor seals are ideal candidates to test hypotheses regarding feeding performance and the ubiquity of suction feeding among phocids and all pinnipeds. In fact, animal-borne cameras attached to harbor seals provide convincing evidence that they likely use several feeding modes (suction and biting) (Bowen et al. 2002). The maximum hydraulic jetting force recorded was 53.9 kPa. As found with most marine mammal performance studies, suction and biting feeding modes were kinematically distinct. Suction was characterized by a significantly smaller gape and gape angle, pursing of the rostral lips to form a circular aperture, and pursing of the lateral lips to occlude lateral gape. Biting was characterized by a large gape and gape angle and lip curling to expose teeth. Harbor seals displayed a wide repertoire of behaviorally flexible feeding strategies, which likely forms the basis of their opportunistic, generalized feeding ecology and concomitant breadth of diet. It is suggested that most pinnipeds will fall into this behavioral pattern and suction capability.

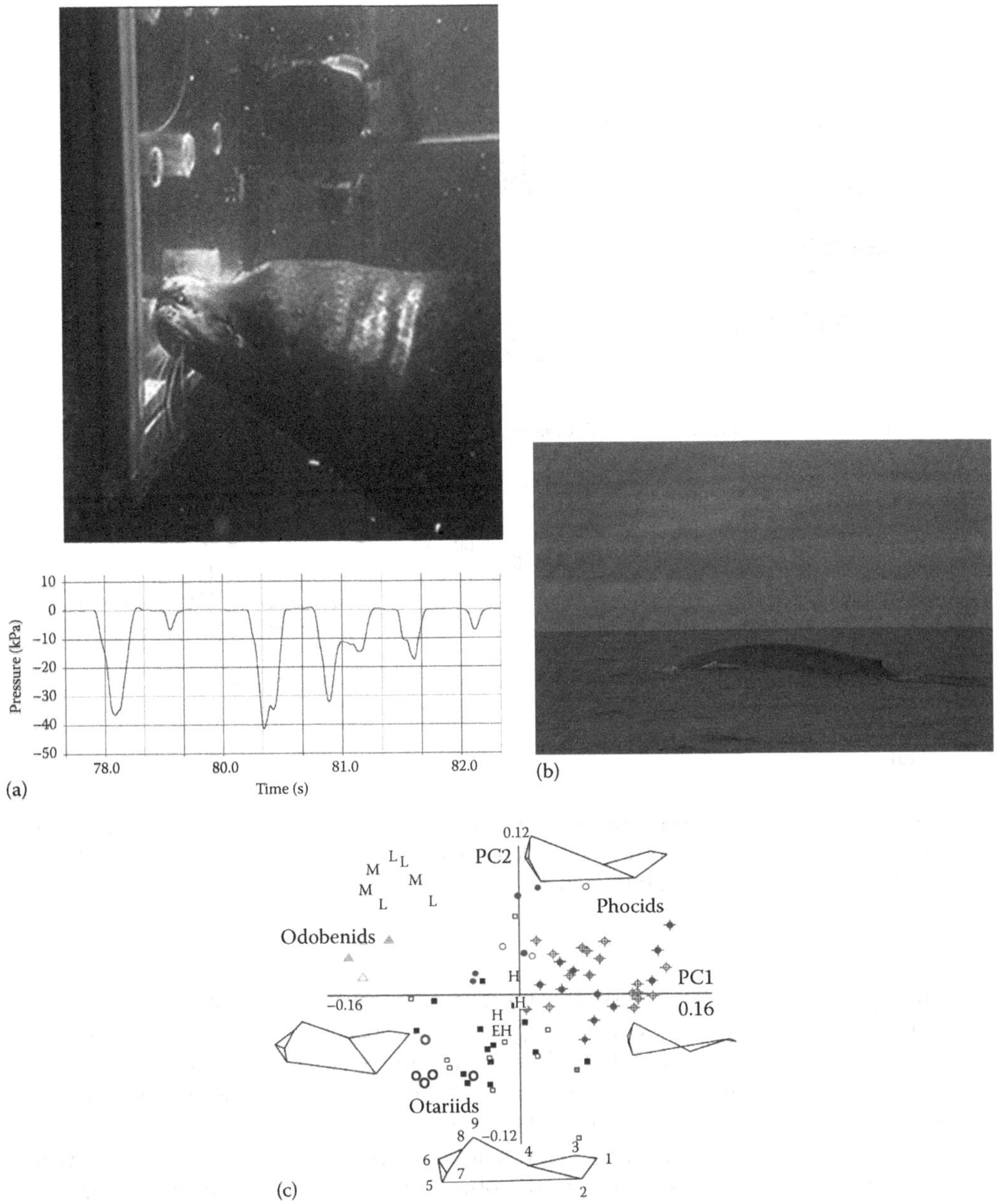

Figure 5.2 **(See color insert.)** Representative methods of marine mammal feeding studies. (a) Functional performance, kinematics, and biomechanics. (From Marshall, C.D. et al., *PLoS ONE*, 9, e86710, 2014, doi: 10.1371/journal.pone.0086710.) (b) Animal-borne tags on free-ranging animals. (Photo by Jeremy Goldbogen. Unpublished video tag from: Jeremy Goldbogen, Dave Cade, Ari Friedlaender, and John Calambokidis. NMFS Permits: #14534-2 and 14809.) (c) Comparative and integrative morphology, biomechanics, and geometric morphometrics. (From Jones, K.E. et al., *Anat. Rec.*, 296, 1049, 2013.)
(Continued)

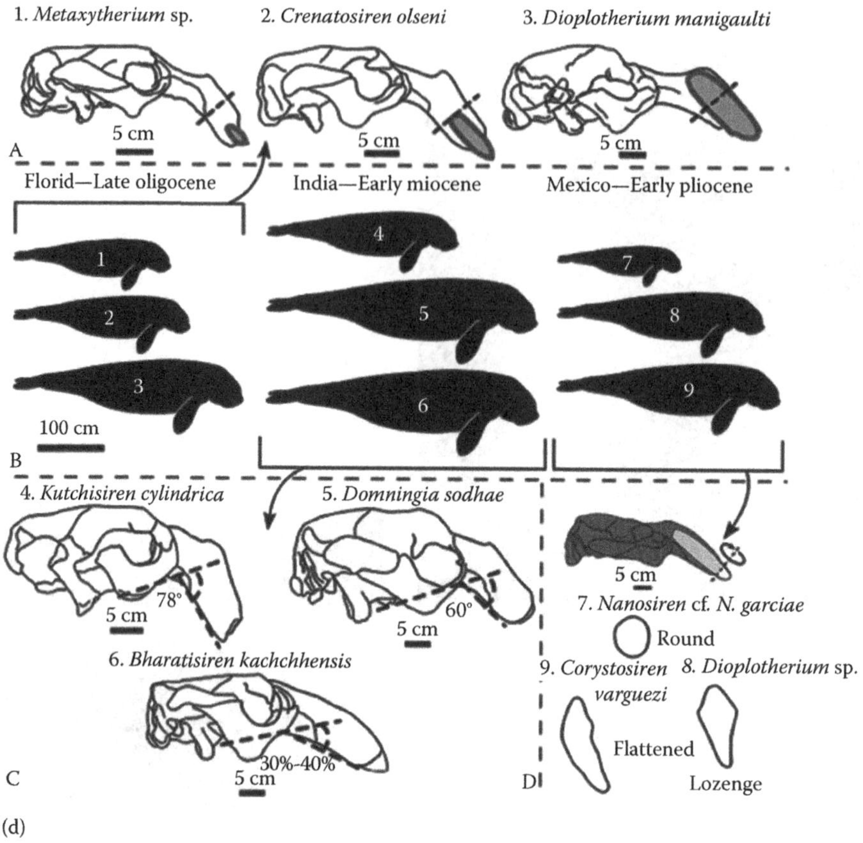

Figure 5.2 (Continued) **(See color insert.)** Representative methods of marine mammal feeding studies. (d) Integration of morphology and paleoecology. (From Velez-Juarbe, J. et al., *PLoS ONE*, 7, e31294, 2012, doi: 10.1371/journal.pone.0031294.)

5.2.4 Order Carnivora: Sea otters

Sea otters (*Enhydra lutris*) are the smallest and most recent group of marine mammals to return to marine habitats. They are thought to have arisen in the North Pacific during the Pleistocene (Leffler 1964; Mitchell 1966; Repenning 1976) and have only become fully aquatic in the last 1–3 million years (Berta and Sumich 1999). Taxonomically, they are in the order Carnivora, Family Mustelidae. However, two subfamilies are commonly recognized, the mustelinae (mainly terrestrial mustelids such as fishers, martens, and wolverines) and the Lutrinae (otters including sea otters). Their similarity to their aquatic otter relatives may be the reason why sea otter feeding has not received much attention. In fact, although some function studies of dietary adaptations have been conducted in terrestrial mustelids, there are few data for otters (Lutrinae) in general.

Although feeding adaptations have been examined to some extent in terrestrial mustelids (Riley 1985; Popowics 2003; Lee and Miller 2004; Abramov and Puzachenko 2005), few data exist for aquatic mustelids such as otters (Lutrinae). Morphological and behavioral diversity among otters is reflected in their diet and foraging behaviors (Radinsky 1981a,b; Kruuk et al. 1994; Hussain and Choudhury 1997; Lee and Miller 2004; Sacco and Van Valkenburgh 2004; Meiri et al. 2005; Goswami 2006; Van Valkenburg 2007; Wroe and

Milne 2007). For example, in river otters (*Lontra canadensis*), the digastric muscles are hypertrophied compared to terrestrial carnivores, which is thought to allow the rapid jaw closure necessary for catching elusive fish with their mouths underwater (Goswami 2006). This is also reflected in their cranial morphology; river otters possess broad mastoid processes where the enlarged digastric muscles originate (Goswami 2006). River otters also possess sharp carnassials necessary for piercing and shearing fish (Popowics 2003). In contrast, sea otters possess short, blunt skulls with bunodont dentition, used for crushing hard, benthic prey (Kenyon 1969).

However, recent functional data (Timm 2013) suggest that sea otters exhibit further specialized feeding methods. Sea otters forage on the bottom, in waters as deep as 40 meters (Kenyon 1969; Riedman and Estes 1990). Their diet is varied but shellfish and urchins comprise a large portion. Otters use their forepaws to excavate clams, and to pry shellfish and urchins from the rocky substrate, sometimes using tools. Food is usually consumed at the surface, and the behavioral observations suggest that otters do not use their teeth underwater, even when feeding on fish. Upon surfacing, fish are killed by a bite to the head. A rock or some other tool is usually carried in a flap of skin in the axilla region (under the arm) and is used to pound open shellfish. The spines of urchins are simply bitten off, and the test (shell) of the urchin is crushed with the cheek teeth. Their cheek teeth are broad, flat, and covered with thick enamel. The shearing cusps of the carnassial teeth have been lost; sea otters are adapted for crushing their food (Kenyon 1969).

It is commonly understood that sea otters are durophagous and forage primarily upon infaunal, but also epibenthic, invertebrates. However, Northern and Russian sea otters (*Enhydra lutris kenyoni* and *Enhydra lutris lutris*, respectively) also prey on epibenthic fish and occasionally other vertebrates (Riedman and Estes 1988). Sea otters differ from other otters in that the morphological traits are correlated with durophagy but also that durophagy is enhanced by an extreme blunt and wide skull and mandible (Timm 2013), and large occlusal surface area relatively to body size. Sea otters were found to possess a relatively high masseteric mechanical advantage compared to other otter species, which is thought to increase force at the most posterior molars as an adaptation for durophagy. This feature in other mammals is known to provide additional control over mastication (Radinsky 1985). A morphometric analysis of sea otters demonstrated that they exhibited taller and wider mandibular rami, and shorter, blunter skulls, than other otters (Timm 2013), which is consistent with increased bite force at the carnassials of other mammals (Sacco and Van Valkenburgh 2004; Figueirido et al. 2009; Nogueira et al. 2009). Furthermore, the blunt, crushing carnassials, and all molars in general, are known to possess relatively thicker enamel (Fisher 1941; Kenyon 1969) that is resilient in its structure (Chai et al. 2009) and is 2.5 times stronger than human enamel (Ziscovici et al. 2014). In addition, both morphological and kinematic investigations of sea otter biting demonstrates that they can display wide gapes (Timm 2013), while maintaining powerful bite forces, useful adaptations for feeding upon bivalves.

5.2.5 Order Carnivora: Polar bears

Polar bears (*Ursus maritimus*) are considered to be marine mammals since they spend a significant time in the marine environment and rely upon marine mammals to survive. This particular species arose only <1 million years ago from brown bears, which are omnivorous (Slater et al. 2010). Although they tend to be completely carnivorous (Slater et al. 2010), when required, polar bears will resort to omnivory and supplement its diet with fruits, vegetation, and various other food material (Iversen et al. 2014). However, recent data demonstrate that the cranial morphology of polar bears does diverge from brown bears and other ursids, reflecting the increased biomechanical pressures for consuming a primarily marine mammal diet (Slater et al. 2010). Polar bears tend to grasp their prey with their mouths and break the neck or skull of their prey with their large masticatory muscles and robust dentition. Interestingly, their cranial morphology and feeding biomechanics demonstrates a potential trade-off leaving polar bears less effective at processing tough food matter associated with omnivory (e.g., plants) (Slater et al. 2010). An ecomorphological study of craniodental morphology of all bears in relation to diet demonstrates significant morphological separation among bears that exhibit omnivory, herbivory, carnivory, and insectivory. Only polar and brown bears exhibit significant carnivory. Craniodental adaptations include a reduction in molar size, flexible mandibles, and relatively small carnassial blades. Polar bear feeding adaptations have more in common with omnivorous canids. The less than robust feeding apparatus that might be expected from polar bears can be explained by the fact that polar bears target ringed and bearded seal pups and seals that are under 2 years of age (King 1983). These seals are much smaller in body size and are much more vulnerable than larger prey. The size discrepancy suggests that polar bears can overpower their prey without the need for craniodental adaptations (Sacco and Van Valkenburgh 2004) that are typical in carnivorous canids that hunt large prey (Van Valkenburgh and Koepfli 1993). However, polar bears are capable of taking larger prey such as walruses and beluga whales (King 1983; Sacco and Van Valkenburgh 2004).

5.3 *Tools and methods for studying feeding strategies in marine mammals*

Marine mammals are notoriously difficult to study both in the wild and in captivity. As a consequence, much work on feeding mechanisms has been morphological. Morphology can be used to predict function, but with caution, and should be considered functional hypotheses until function can be verified by other methods. Regardless, craniodental morphology can be very instructive for investigating feeding mechanisms. Indeed, new methods such as computed tomography, 3D reconstruction, 3D printing, finite element analyses, physical and computation biomechanical modeling, materials science, and geometric morphometric are providing new and exciting frontiers in morphological research. Advances in camera, video, computing, and electrophysiology capabilities and the reduction in size and increased portability are now allowing pool-side performance experiments that could have only been conducted in the laboratory in the past. This translates into additional focused and experimental captive studies, which quantitatively measures kinematics, performance, physiology, and behavior. Such experiments interface well with field studies in which a cadre of animal-borne tags on free-ranging animals collects data from foraging bouts and events in the open ocean. Such tools include integrated systems that incorporate time–depth recorders with video systems, oceanographic data collection systems, and 3D movement sensors (accelerometers, magnetometers, and gyroscopes) that can record behavior, body movement, visual perspective, acoustics, and physiology. New unmanned aerial vehicles (UAVs) are providing new access and perspective to animal behavior and beyond. It is an exciting time to be a marine mammalogist.

5.4 *Future directions*

The future of functional studies of marine mammal feeding is bright. Advances in new technologies, as described earlier, will allow researchers to close the gap between experimental captive studies and field studies that are largely phenomological. As new fossils are discovered, new phylogenetic tools will allow researchers to investigate the evolution of feeding and begin to elucidate functional transitions from land to sea and the evolution of particular feeding modes and foraging behavior. In total, the impact of such new integration will advance our understanding of mammalian evolution and the functional pathways that aquatic mammals took during their transition to land as well as the ecological patterns of vertebrate natural history.

References

Abramov, A.V. and A.Y. Puzachenko. 2005. Sexual dimorphism of craniological characteristics in Eurasian badgers, *Meles* spp. (Carnviora: Mustelidae). *Zoologischer Anzeiger* 244: 11–29.

Adam, P.J. 2005. *Lobodon carcinophaga. Mammalian Species* 772: 1–4.

Adam, P.J. and A. Berta. 2002. Evolution of prey capture strategies and diet in the Pinnipedimorpha (Mammalia, Carnivora). *Oryctos* 4: 83–107.

Anderson, P.K. and A. Birtles. 1978. Behaviour and ecology of the dugong, *Dugong dugon* (Sirenia): Observations in Shoalwater and Cleveland Bays, Queensland. *Australian Wildlife Research* 5: 1–23.

Armfield, B.A., Z. Zheng, S. Bajpai, C. Vinyard, and J. Thewissen. 2013. Development and evolution of the unique cetacean dentition. *PeerJ* 1: e24. https://dx.doi.org/10.7717/peerj.24.

Arnold, S.J. 1983. Morphology, performance and fitness. *American Zoologist* 23: 347–361.

Berta, A. and J.L. Sumich. 1999. *Marine Mammals: Evolutionary Biology*. San Diego, CA: Academic Press.

Berta, A., J.L. Sumich, and K.M. Kovacs. 2006. *Marine Mammals: Evolutionary Biology*, 2nd edn. Boston, MA: Academic Press.

Best, R.C. 1981. Foods and feeding habits of wild and captive Sirenia. *Mammal Review* 11: 3–29. doi: 10.1111/j.1365-2907.1981.tb00243.x.

Bigg, M.A. 1981. Harbour seal, *Phoca vitulina* Linnaeus, 1758 and *Phoca largha* Pallas, 1811. In: *Handbook of Marine Mammals*. Seals, Vol. 2. S.H. Ridgway and R.J. Harrison, eds. London, UK: Academic Press, pp. 1–27.

Biknevicius, A.R. and B. Van Valkenburgh. 1996. Design for killing: Craniodental adaptations of predators. In: *Carnivore Behavior, Ecology, and Evolution*, Vol. 2. J.L. Gittleman, ed. Ithaca, NY: Cornell University Press, pp. 393–428.

Bloodworth, B. and C.D. Marshall. 2005. Feeding kinematics of *Kogia* and *Tursiops* (Odontoceti: Cetacea): Characterization of suction and ram feeding. *Journal of Experimental Biology* 208: 3721–3730.

Bloodworth, B. and C.D. Marshall. 2007. A functional comparison of the hyolingual complex in pygmy and dwarf sperm whales (*Kogia breviceps* & *K. sima*), and bottlenose dolphins (*Tursiops truncatus*). *Journal of Anatomy* 211: 78–91.

Bonner, W.N. 1968. The fur seal of South Georgia. *British Antarctic Survey Scientific Reports* 56: 1–81.

Bowen, W.D., D. Tully, D.J. Bones, B.M. Bulheier, and G.J. Marshall. 2002. Prey dependent foraging tactics and prey profitability in a marine mammal. *Marine Ecology Progress Series* 244: 235–245.

Bromaghin, J.F., M.M. Lance, E.W. Elliott, S.J. Jeffries, and A. Acevedo-Gutiérrez. 2013. New insights into the diets of harbor seals (*Phoca vitulina*) in the Salish Sea revealed by analysis of fatty acid signatures. *Fisheries Bulletin* 111: 13–26.

Brown, S.L., S. Bearhop, C. Harrod, and R.A. McDonald. 2012. A review of spatial and temporal variation in grey and common seal diet in the United Kingdom and Ireland. *Journal of the Marine Biological Association of the United Kingdom* 92: 1711–1722.

Burns, J.J. 2009. Harbor and spotted seal. In: *Encyclopedia of Marine Mammals*. W.F. Perrin, B. Würsig, and J.G.M. Thewissen, eds. San Diego, CA: Academic Press, pp. 533–542.

Carroll, A.M., P.C. Wainwright, S.H. Huskey, D.C. Collar, and R.G Turingan. 2004. Morphology predicts suction feeding performance in centrarchid fishes. *Journal of Experimental Biology* 207: 3873–3881.

Chai, H., J.J.-W. Lee, P.J. Constantino, P.W. Lucas, and B.R. Lawn. 2009. Remarkable resilience of teeth. *Proceedings of the National Academy of Sciences of the United States of America* 106: 7289–7293. doi: 10.1073pnas.0902466106.

Clarke, M.R. 1996. Cephalopods as prey. III. Cetaceans. *Philosophical Transactions of the Royal Society of London B* 351: 1053–1065. doi: 10.1098/rstb.1996.0093.
Dahl, T.M., C. Lydersen, K.M. Kovacs, S. Falk-Petersen, J. Sargent, I. Gjertz, and B. Gulliksen. 2000. Fatty acid composition of the blubber in white whales (*Delphinapterus leucas*). *Polar Biology* 23: 401–409.
Domning, D.P. 1980. Feeding position preference in manatees (*Trichechus*). *Journal of Mammalogy* 61: 544–547.
Domning, D.P. 1982. Evolution of manatees: A speculative history. *Journal of Paleontology* 56: 599–619.
Domning, D.P. 2001. Sirenians, seagrasses, and Cenozoic ecological change in the Caribbean. *Palaeogeography, Palaeoclimatology, Palaeoecology* 166: 27–50.
Domning, D.P. and L.A.C. Hayek. 1984. Horizontal tooth replacement in the Amazonian manatee (*Trichechus inunguis*). *Mammalia* 48: 105–127.
Domning, D.P. and L.A.C. Hayek. 1986. Interspecific and intraspecific morphological variation in manatees (Sirenia: *Trichechus*). *Marine Mammal Science* 2: 87–144. doi: 10.1111/j.1748-7692.1986.tb00034.x.
Emlin, J.M. 1966. The role of time and energy in food preference. *American Naturalist* 100: 611–617.
Fay, F.H. 1982. Ecology and biology of the Pacific walrus, *Odobenus rosmarus divergens* Illiger. North American Fauna No. 74. Washington, DC: United States Department of the Interior Fish and Wildlife Service, 279pp.
Figueirido, B., P. Palmqvist, and J.A. Perez-Claros. 2009. Ecomorphological correlates of craniodental variation in bears and paleobiological implications for extinct taxa: An approach based on geometric morphometrics. *Journal of Zoology (London)* 277: 70–80.
Fisher, E.M. 1941. Notes on the teeth of sea otters. *Journal of Mammalogy* 22: 428–433. doi: 10.2307/137493.
Fordyce, R.E. 1980. Whale evolution and Oligocene southern ocean environments. *Palaeogeography, Palaeoclimatology, Palaeoecology* 31: 319–336. doi: 10.1016/0031-0182(80)90024-3.
Fordyce, R.E. and L.G. Barnes. 1994. The evolutionary history of whales and dolphins. *Annual Review of Earth and Planetary Sciences* 22: 419–455. doi: 10.1146/annurev.ea.22.050194.002223.
Fordyce, R.E. and F.G. Marx. 2012. The pygmy right whale *Caperea marginata*: The last of the cetotheres. *Proceedings of the Royal Society of London B* 280: 20122645. http://dx.doi.org/10.1098/rspb.2012.2645.
Fudge, D.S., L.J. Szewciw, and A.N. Schwalb. 2009. Morphology and development of blue whale baleen: An annotated translation of Tycho Tullberg's classic 1883 paper. *Aquatic Mammals* 35: 226–252. doi: 10.1578/AM.35.2.2009.226.
Gaskin, D.E. 1982. *The Ecology of Whales and Dolphins*. New York: Heinemann.
Goldbogen, J.A., J. Calambokidis, D.A. Croll, M.F. McKenna, E. Oleson, J. Potvin, N.D. Pyenson, G. Schorr, R.E. Shadwick, and B. Tershy. 2012. Scaling of lunge feeding performance in rorqual whales: Mass-specific energy expenditure increases with body size and progressively limits diving capacity. *Functional Ecology* 26: 216–226.
Goldbogen, J.A., J. Calambokidis, R.E. Shadwick, E.M. Oleson, M.A. McDonald, and J.A. Hildebrand. 2006. Kinematics of diving and lunge-feeding in fin whales. *Journal of Experimental Biology* 209: 1231–1244.
Goldbogen, J.A., J. Potvin, and R.E. Shadwick. 2010. Skull and buccal cavity allometry increase mass-specific engulfment capacity in fin whales. *Proceedings of the Royal Society of London B* 277: 861–868.
Goldbogen, J.A., N.D. Pyenson, and R.E. Shadwick. 2007. Big gulps require high drag for fin whale lunge feeding. *Marine Ecology Progress Series* 349: 289–301. doi: 10.3354/ meps07066.
Gordon, K.R. 1984. Models of tongue movement in the walrus (*Odobenus rosmarus*). *Journal of Morphology* 182: 179–196. doi: 10.1002/jmor.1051820206.
Goswami, A. 2006. Morphological integration in the carnivoran skull. *Evolution* 60: 122–136.
Hartman, D.S. 1979. *Ecology and behavior of the manatee in Florida*. Special Publication No. 5. Pittsburgh, PA: American Society of Mammalogists, 153pp.
Heyning, J.E. and J.G. Mead. 1996. Suction feeding in beaked whales: Morphological and observational evidence. *Scientific Contributions Natural History Museum of Los Angles County* 464: 1–12.
Hillson, S. 2005. *Teeth*, 2nd edn. Cambridge, UK: Cambridge University Press.
Hocking, D.P., A.R. Evan, and E.M.G. Fitzgerald. 2013. Leopard seals (*Hydrurga leptonyx*) use suction and filter feeding when hunting small prey underwater. *Polar Biology* 36: 211–222. doi: 10.1007/ s00300-012-1253-9.

Hocking, D.P., M. Salverson, E.M.G. Fitzgerald, and A.R. Evans. 2014. Australian fur seals (*Arctocephalus pusillus doriferus*) use raptorial biting and suction feeding when targeting prey in different foraging scenarios. *PLoS ONE* 9: e112521. doi: 10.1371/journal.pone.0112521.

Hoover, A.A. 1988. Harbor seal. In: *Selected Marine Mammal of Alaska: Species Accounts with Research and Management Recommendations*. J.W. Lentfer, ed. Washington, DC: Marine Mammal Commission, pp. 125–157.

Hussain, S.A. and B.C. Choudhury. 1997. Distribution and status of the smooth-coated otter *Lutra perspicillata* in National Chambal Sanctuary, India. *Biological Conservation* 80: 199–206.

Iversen, M., J. Aars, T. Haug, I.G. Alsos, C. Lydersen, L. Bachmann, and K.M. Kovacs. 2014. The diet of polar bears (*Ursus maritimus*) from Svalbard, Norway, inferred from scat analysis. *Polar Biology* 36: 561–571.

Johnston, C. and A. Berta. 2011. Comparative anatomy and evolutionary history of suction feeding in cetaceans. *Marine Mammal Science* 27: 493–513. doi: 10.1111/j.1748-7692.2010.00420.x.

Jones, K.E. and A. Goswami. 2010. Quantitative analysis of the influences of phylogeny and ecology on phocid and otariid pinniped (Mammalia; Carnivora) cranial morphology. *Journal of Zoology (London)* 280: 297–308.

Jones, K.E., C.B. Ruff, and A. Goswami. 2013. Morphology and biomechanics of the pinniped jaw mandibular evolution without mastication. *Anatomical Record* 296: 1049–1063.

Kane, E.A. and C.D. Marshall. 2009. Comparative feeding kinematics and performance of odontocetes: Belugas, Pacific white-sided dolphins, and long-finned pilot whales. *Journal of Experimental Biology* 212: 3939–3950.

Kastelein, R.A., M. Muller, and A. Terlouw. 1994. Oral suction of a Pacific walrus (*Odobenus rosmarus divergens*) in air and under water. *Zeitschrift für Säugetierkunde* 59: 105–115.

Kastelein, R.A., C. Staal, A. Terlouw, and M. Muller. 1997. Pressure changes in the mouth of a feeding harbor porpoise. In: *The Biology of the Harbor Porpoise*. A.J. Read, P.R. Wiepkema, and P.E. Nachtigall, eds. Groningen, the Netherlands: DeSpil Publishers, pp. 279–291.

Kavanagh, A.S., M.A. Cronin, M. Walton, and E. Rogan. 2010. Diet of the harbour seal (*Phoca vitulina*) in the west and south-west of Ireland. *Journal of the Marine Biological Association of the United Kingdom* 90: 1517–1527.

Kenyon, K.W. 1969. The sea otter in the eastern Pacific Ocean. North American Fauna 68. Washington, DC: United States Department of the Interior Fish and Wildlife Service, 352pp.

Kier, W.M. and K.K. Smith. 1985. Tongues, tentacles and trunks: The biomechanics of movement in muscular-hydrostats. *Zoological Journal of the Linnean Society* 83: 307–324.

Kiltie, R.A. 1982. Bite force as a basis for niche differentiation between rain forest peccaries (*Tayassu tajacu* and *T. pecari*). *Biotropica* 14: 188–195.

King, J.E. 1972. Observations on phocid skulls. In: *Functional Anatomy of Marine Mammals*. J.E. Harrison, ed. New York, NY: Academic Press, pp. 81–115.

King, J.E. 1983. *Seals of the World*, 3rd edn. Ithaca, NY: Cornell University Press.

Klages, N.T.W. and V.G. Cockcroft. 1990. Feeding behavior of a captive crabeater seal. *Polar Biology* 10: 403–404. doi: 10.1007/BF00237828.

Kot, B.W., R. Sears, D. Zbinden, E. Borda, and M.S. Gordon. 2014. Rorqual whale (Balaenopteridae) surface lunge-feeding behaviors: Standardized classification, repertoire diversity, and evolutionary analyses. *Marine Mammal Science* 30: 1335–1357. doi: 10.1111/mms.12115.

Kruuk, H., B. Kanchanasaka, S. O'Sullivan, and S. Wanghongsa. 1994. Niche separation in three sympatric otters *Lutra perspicillata*, *L. lutra*, and *Aonyx cinerea* in Huai Kha Khaeng, Thailand. *Biological Conservation* 69: 115–120.

Lanyon, J.M. and G.D. Sanson. 2006a. Degenerate dentition of the dugong (*Dugong dugon*), or why a grazer does not need teeth: Morphology, occlusion and wear of mouthparts. *Journal of Zoology (London)* 268: 133–152.

Lanyon, J.M. and G.D. Sanson. 2006b. Mechanical disruption of seagrass in the digestive tract of the dugong. *Journal of Zoology (London)* 270: 277–289.

Lauder, G.V. 1985. Aquatic feeding in lower vertebrates. In: *Functional Vertebrate Morphology*. M. Hildebrand, D.M. Bramble, K.F. Liem, and D.B. Wake, eds. Cambridge, UK: Harvard University Press, pp. 210–229.

Lee, S. and P.J. Miller. 2004. Cranial variation in British mustelids. *Journal of Morphology* 260: 57–64.

Leffler, S.R. 1964. Fossil mammals from the Elk River formation, Cap Blanco, Oregon. *Journal of Mammalogy* 45: 53–61. doi: 10.2307/1377294.

Lindberg, D.R. and N.D. Pyenson. 2007. Things that go bump in the night: Evolutionary interactions between cephalopods and cetaceans in the tertiary. *Lethaia* 40: 335–343. doi: 10.1111/j.1502-3931.2007.00032.x.

Marsh, H., C.A. Beck, and T. Vargo. 1999. Comparison of the capabilities of dugongs and West Indian manatees to masticate seagrasses. *Marine Mammal Science* 15: 250–255.

Marshall, C.D. 2002. Marine mammal functional morphology. In: *Encyclopedia of Marine Mammals*. W.F. Perrin, B. Würsig, and H.G.M. Thewissen, eds. San Diego, CA: Academic Press, pp. 759–774.

Marshall, C.D. 2009. Feeding functional morphology. In: *Encyclopedia of Marine Mammals*. W.F. Perrin, B. Würsig, and H.G.M. Thewissen, eds. Boston, MA: Elsevier, pp. 406–414.

Marshall, C.D., L.A. Clark, and R.L. Reep. 1998a. The muscular hydrostat of the Florida manatee (*Trichechus manatus latirostris*) and its role in the use of perioral bristles. *Marine Mammal Science* 14: 290–303.

Marshall, C.D., G.D. Huth, V.M. Edmonds, D.L. Halin, and R.L. Reep. 1998b. Prehensile use of perioral bristles during feeding and associated behaviors of the Florida manatee (*Trichechus manatus latirostris*). *Marine Mammal Science* 14: 274–289.

Marshall, C.D., G.D. Huth, V.M. Edmonds, D.L. Halin, and R.L. Reep. 2000. Food-handling ability and feeding-cycle length of manatees feeding on several species of aquatic plants. *Journal of Mammalogy* 81: 649–658.

Marshall, C.D., K. Kovacs, and C. Lydersen. 2008. Feeding kinematics, suction, and hydraulic jetting capabilities in bearded seals (*Erignathus barbatus*). *Journal of Experimental Biology* 211: 699–708.

Marshall, C.D., H. Maeda, M. Iwata, M. Furuta, A. Asano, F. Rosas, and R.L. Reep. 2003. Orofacial morphology and feeding behaviour of the dugong, Amazonian, West African and Antillean manatees (Mammalia: Sirenia): Functional morphology of the muscular-vibrissal complex. *Journal of Zoology (London)* 259: 1–16.

Marshall, C.D., D. Rosen, and A.W. Trites. In press. Feeding kinematics and performance of basal otariid pinnipeds, Steller sea lions (*Eumetopias jubatus*), and northern fur seals (*Callorhinus ursinus*): Implications for the evolution of mammalian feeding. *Journal of Experimental Biology* 451: 91–97.

Marshall, C.D., S. Wieskotten, W. Hanke, F.D. Hanke, A. Marsh, B. Kot, and G. Dehnhardt. 2014. Feeding kinematics, suction, and hydraulic jetting performance of harbor seals (*Phoca vitulina*). *PLoS ONE* 9: e86710. doi: 10.1371/journal.pone.0086710.

Marx, F.G. and M.D. Uhen. 2010. Climate, critters, and cetaceans: Cenozoic drivers of the evolution of modern whales. *Science* 327: 993–996. doi: 10.1126/science.1185581.

Meiri, S., T. Dayan, and D. Simberloff. 2005. Variability and correlations in carnivore crania and dentition. *Functional Ecology* 19: 337–343.

Mitchell, E.D. 1966. Northeastern Pacific Pleistocene otters. *Journal of the Fisheries Research Board of Canada* 23: 1897–1911. doi: 10.1139/f66–177.

Nauwelaerts, S., C.D. Wilga, C.P. Sanford, and Lauder, G.V. 2007. Substrate passively improves suction feeding in benthic sharks. *Journal of the Royal Interface* 2: 341–345.

Nerini, M. 1984. A review of gray whale feeding ecology. In: *The Gray Whale Eschrichtius robustus*. M.L. Jones, S.L. Swartz, and S. Leatherwood, eds. Orlando, FL: Academic Press, Inc., pp. 423–449.

Nogueira, M.R., A.L. Peracchi, and L.R. Monteiro. 2009. Morphological correlates of bite force and diet in the skull and mandible of phyllostomid bats. *Functional Ecology* 23: 715–723.

Nowacek, D.P., A.S. Friedlaender, P.N. Halpin, E.L. Hazen, D.W. Johnston, A.J. Read, B. Espinasse, M. Zhou, and Y. Zhu. 2011. Super-aggregations of krill and humpback whales in Wilhelmina Bay, Antarctic Peninsula. *PLoS ONE* 6: e19173. doi: 10.1371/journal.pone.0019173.

Øritsland, T. 1977. Food consumption of seals in the Antarctic pack ice. In: *Adaptations within Antarctic Ecosystems*. G.A. Llano, ed., *Proceedings of the Third SCAR Symposium on Antarctic Biology*. Washington, DC: Smithsonian Institution, pp. 749–768.

Orton, L.S. and P.F. Brodie. 1987. Engulfing mechanisms of fin whales. *Canadian Journal of Zoology* 65: 2898–2907.

Parrish, F.A., K. Abernathy, G.J. Marshall, and B.M. Buhleier. 2002. Hawaiian monk seals (*Monachus schauinslandi*) foraging in deep-water coral beds. *Marine Mammal Science* 18: 244–258. doi: 10.1111/j.1748-7692.2002.tb01031.x.

Pauly, D., A.W. Trites, E. Capuli, and V. Christensen. 1998. Diet composition and trophic levels of marine mammals. *ICES Journal of Marine Science* 55: 467–481.

Perez-Barberia, F.J. and I.J. Gordon. 1999. The functional relationship between feeding type and jaw and cranial morphology in ungulates. *Oecologia* 118: 157–165.

Pinto, S.J.D. and R.E. Shadwick. 2013. Material and structural properties of fin whale (*Balaenoptera physalus*) Zwischensubstanz. *Journal of Morphology* 274: 947–955. doi: 10.1002/jmor.20154.

Popowics, T.E. 2003. Postcanine dental form in the Mustelidae and Viverridae (Carnivora: Mammalia). *Journal of Morphology* 256: 322–341.

Preen, A. 1995. Impacts of dugong foraging on seagrass habitats: Observational and experimental evidence for cultivation grazing. *Marine Ecology Progress Series* 124: 201–213.

Pyenson, N.D., J.A. Goldbogen, and R.E. Shadwick. 2013. Mandible allometry in extant and fossil Balaenopteridae: The largest vertebrate skeletal element and its role in rorqual lunge-feeding. *Biological Journal of the Linnean Society* 108: 586–599.

Pyenson, N.D., J.A. Goldbogen, A.W. Vogl, G. Szathmary, R.L. Drake, and R.E. Shadwick RE. 2012. Discovery of a sensory organ that coordinates lunge feeding in rorqual whales. *Nature* 485: 498–501.

Pyenson, N.D., N.P. Kelley, and J.F. Parham. 2014. Marine tetrapod macroevolution: Physical and biological drivers on 250 Ma of invasions and evolution in ocean ecosystems. *Palaeogeography, Palaeoclimatology, Palaeoecology* 400: 1–8. doi: 10.1016/j.palaeo.2014.02.018.

Pyenson, N.D. and D.R. Lindberg. 2011. What happened to gray whales during the Pleistocene? The ecological impact of sea-level change on benthic feeding areas in the North Pacific Ocean. *PLoS ONE* 6: e21295. doi: 10.1371/journal.pone.0021295.

Radinsky, L.B. 1981a. Evolution of skull shape in carnivores. 1. Representative modern carnivores. *Biological Journal of the Linnean Society* 15: 369–388.

Radinsky, L.B. 1981b. Evolution of skull shape in carnivores: 2. Additional modern carnivores. *Biological Journal of the Linnean Society* 16: 337–355.

Radinsky, L.B. 1985. Approaches in evolutionary morphology: A search for patterns. *Annual Review of Ecological Systems* 16: 1–14.

Ray, G.C. and W.E. Schevill. 1974. Feeding of a captive gray whale, *Eschrichtius robustus*. *Marine Fisheries Review* 36: 31–38.

Reep, R.L., C.D. Marshall, M.L. Stoll, and D.M. Whitaker. 1998. Distribution and innervation of facial bristles and hairs in the Florida *manatee (Trichechus manatus latirostris)*. *Marine Mammal Science* 14: 257–273.

Reep, R.L., M.L. Stoll, C.D. Marshall, B.L. Homer, and D.A. Samuelson. 2001. Microanatomy of facial vibrissae in the Florida manatee: The basis for specialized sensory function and oripulation. *Brain Behavior and Evolution* 58: 1–14.

Reidenberg, J.S. and J.T. Laitman. 1994. Anatomy of the hyoid apparatus in Odontoceti (toothed whales): Specializations of their skeleton and musculature compared with those of terrestrial mammals. *Anatomical Record* 240: 598–624.

Repenning, C.A. 1976. *Enhydra* and *Enhydriodon* from Pacific coast of North America. *Journal of Research of the U.S. Geological Survey* 4: 305–315.

Repenning, C.A., R.S. Peterson, and C.L. Hubbs. 1971. Contributions to the systematics of the southern fur seals, with particular reference to the Juan Fernández and Guadalupe species. In: *Antarctic Pinnipedia*. Antarctic Research Series, Vol. 18. W.H. Burt, ed. Washington, DC: American Geophysical Union, pp. 1–34.

Reynolds, J.E. and S.A. Rommel. 1996. Structure and function of the gastrointestinal tract of the Florida manatee, *Trichechus manatus latirostris*. *Anatomical Record* 245: 539–558.

Rice, D.W. 1998. *Marine Mammals of the World: Systematics and Distribution*. Special Publication No. 4. Lawrence, KS: The Society for Marine Mammalogy.

Riedman, M. 1990. *The Pinnipeds: Seals, Sea Lions, And Walruses*. Berkeley, CA: University of California Press.
Riedman, M.L. and J.A. Estes. 1988. Predation on seabirds by sea otters. *Canadian Journal of Zoology* 66: 1396–1402.
Riedman, M.L. and J.A. Estes. 1990. The sea otter (*Enhydra lutris*): Behavior, Ecology and Natural History. Biological Report 90(14). Washington, DC: U.S. Fish and Wildlife Service, pp. 126.
Riley, M.R. 1985. An analysis of masticatory form and function in three mustelids (*Martes americana, Lutra canadensis, Enhydra lutris*). *Journal of Mammalogy* 66: 519–528.
Rommel, S.A. 1990. Osteology of the bottlenose dolphin. In: *The Bottlenose Dolphin*. S. Leatherwood and R.R. Reeves, eds. San Diego, CA: Academic Press, pp. 29–49.
Rosas, F.C.W. 1994. Biology, conservation and status of the Amazonian manatee *Trichechus inunguis*. *Mammal Review* 24: 49–59.
Ross, G.J.B., F. Ryan, G.S. Saayman, and J. Skinner. 1976. Observations on two captive crabeater seals at the Port Elizabeth Oceanarium. *International Zoo Yearbook* 16: 160–164.
Sacco, T. and B. Van Valkenburgh. 2004. Ecomorphological indicators of feeding behaviour in bears (Carnivora: Ursidae). *Journal of Zoology (London)* 263: 41–54.
Savage, R.J.G., D.P. Domning, and J.G.M. Thewissen. 1994. Fossil sirenia of the West Atlantic and Caribbean region. V. The most primitive known sirenian, *Prorastomus sirenoides* Owen, 1855. *Journal of Vertebrate Paleontology* 14: 427–449.
Schoener, T.W. 1971. Theory of feeding strategies. *Annual Review of Ecological Systems* 2: 369–404. doi: 10.1146/annurev.es.02.110171.002101.
Shadwick, R.E., J.A. Goldbogen, J. Potvin, N.D. Pyenson, and A.W. Vogl. 2013. Novel muscle and connective tissue design enables high extensibility and controls engulfment volume in lunge-feeding rorqual whales. *Journal of Experimental Biology* 216: 2691–2701.
Sharples, R.J., B. Arrizabalaga, and P.S. Hammond. 2009. Seals, sandeels and salmon: Diet of harbour seals in St. Andrews Bay and the Bay Estuary, southeast Scotland. *Marine Ecology Progress Series* 390: 265–276.
Simon, M., M. Johnson, and P.T. Madsen. 2012. Keeping momentum with a mouthful of water: Behavior and kinematics of humpback whale lunge feeding. *Journal of Experimental Biology* 215: 3786–3798. doi: 10.1242/jeb.071092.
Simon, M., M. Johnson, P. Tyack, and Madsen, P.T. 2009. Behaviour and kinematics of continuous ram filtration in bowhead whales (*Balaena mysticetus*). *Proceedings of the Royal Society of London B* 276: 3819–3828. doi: 10.1098/rspb.2009.1135.
Slater, G.J., B. Figueirido, L. Louis, P. Yang, and B. Van Valkenburgh. 2010. Biomechanical consequences of rapid evolution in the polar bear lineage. *PLoS ONE* 5: e13870. doi: 10.1371/journal.pone.0013870.
Slijper, E.J. 1962. *Whales*. New York: Basic Books.
Steeman, M.E., M.B. Hebsgaard, R.E. Fordyce, S.Y.W. Ho, D.L. Rabosky, R. Nielsen, C. Rahbek, H. Glenner, M.V. Sørensen, and E. Willerslev. 2009. Radiation of extant cetaceans driven by restructuring of the oceans. *Systematic Biology* 58: 573–585. doi: 10.1093/sysbio/syp060.
Szewciw, L.J., D.G. de Kerckhove, G.W. Grime, and D.S. Fudge. 2010. Calcification provides mechanical reinforcement to whale baleen α-keratin. *Proceedings of the Royal Society of London* 277: 2597–2605. doi: 10.1098/rspb.2010.0399.
Timm, L.L. 2013. Feeding biomechanics & craniodental morphology in otters (Lutrinae). Doctoral dissertation, Texas A&M University, College Station, KS.
Thomas, A.C., M.M. Lance, S.J. Jeffries, B.G. Miner, and A. Acevedo-Gutiérrez. 2011. Harbor seal foraging response to a seasonal resource pulse, spawning Pacific herring. *Marine Ecology Progress Series* 441: 225–239.
Unger, P.S. 2010. *Mammal Teeth: Origin, Evolution and Diversity*. Baltimore, MD: Johns Hopkins Press, 304pp.
Van Valkenburgh, B. 1989. Carnivore dental adaptations and diet: A study of trophic diversity within guilds. In: *Carnivore, Behavior, Ecology, and Evolution*. J.L. Gittleman, ed. Ithaca, NY: Cornell University Press, pp. 410–436.
Van Valkenburgh, B. 2007. Déjà vu: The evolution of feeding morphologies in the Carnivora. *Integrative and Comparative Biology* 47: 147–163.

Van Valkenburgh, B. and K.P. Koepfli. 1993. Cranial and dental adaptations to predation in canids. *Symposia of the Zoological Society of London* 65: 15–37.
Van Wassenbergh, S. and P. Aerts. 2009. Aquatic suction feeding dynamics: Insights from computational modeling. *Journal of the Royal Society Interface* 6: 149–158. doi: 10.1098/rsif.2008.0311.
Velez-Juarbe, J., D.P. Domning, and N.D. Pyenson. 2012. Iterative evolution of sympatric seacow (Dugongidae, Sirenia) assemblages during the past ~26 million years. *PLoS ONE* 7: e31294. doi: 10.1371/journal.pone.0031294.
Wainwright, P.C. and D.R. Bellwood. 2002. Ecomorphology of feeding in coral reef fishes. In: *Coral Reef Fishes: Dynamics and Diversity in a Complex Ecosystem*. P.F. Sale, ed. Boston, MA: Academic Press, pp. 33–56.
Wainwright, P.C., A.M. Carroll, D.C. Collar, S.W. Day, T.E. Higham, and R.A. Holzman. 2007. Suction feeding mechanics, performance, and diversity in fishes. *Integrative and Comparative Biology* 47: 96–106.
Wainwright, P.C. and S.W Day. 2007. The forces exerted by aquatic suction feeders on their prey. *Proceedings of the Royal Society of London B* 4: 553–560.
Wainwright, P.C. and S.M. Reilly. 1994. Introduction. In: *Ecological Morphology, Integrative Organismal Biology*. P.C. Wainwright and S.M. Reilly, eds. Chicago, IL: Chicago Press, pp. 1–12.
Weihs, D. and P.W. Webb. 1984. Optimal avoidance and evasion tactics in predator-prey interactions. *Journal of Theoretical Biology* 106: 189–206.
Weijs, W.A. 1994. Evolutionary approach of masticatory motor patterns in mammals. In: *Advances in Comparative and Environmental Physiology*, Vol. 18. V.L. Bels, M. Chardon, and P. Vandewalle, eds. Berlin, Germany: Springer-Verlag, pp. 281–320.
Werth, A.J. 2000. A kinematic study of suction feeding and associated behavior in the long-finned pilot whale, *Globicephala melas* (Traill). *Marine Mammal Science* 16: 299–314.
Werth, A.J. 2001. How do mysticetes remove prey trapped in baleen? *Bulletin of the Museum of Comparative Zoology* 156: 189–203.
Werth, A.J. 2004. Models of hydrodynamic flow in the bowhead whale filter feeding apparatus. *Journal of Experimental Biology* 207: 3569–3580. doi: 10.1242/ jeb.01202.
Werth, A.J. 2006a. Odontocete suction feeding: Experimental analysis of water flow and head shape. *Journal of Morphology* 267: 1415–1428.
Werth, A.J. 2006b. Mandibular and dental variation and the evolution of suction feeding in Odontoceti. *Journal of Mammalogy* 87: 579–588.
Werth, A.J. 2007, Adaptations of the cetacean hyolingual apparatus for aquatic feeding and thermoregulation. *Anatomical Record* 290: 546–568. doi: 10.1002/ar.20538.
Werth, A.J. 2012. Hydrodynamic and sensory factors governing response of copepods to simulated predation by baleen whales. *International Journal of Ecology*. doi: 10.1155/2012/208913.
Woodward, B.L. and J.P. Winn. 2006. Apparent lateralized behavior in gray whales feeding off the central British Columbia coast. *Marine Mammal Science* 22: 64–73.
Wroe, S. and N. Milne. 2007. Convergence and remarkably consistent constraint in the evolution of carnivore skull shape. *Evolution* 61: 1251–1260.
Ziscovici, C., P.J. Constantino, T.G. Bromag, and A. van Casteren. 2014. Sea otter dental enamel is highly resistant to chipping due to its microstructure. *Biology Letters*. doi: 10.1098/rsbl.2014.0484.

chapter six

Diet and nutrition

Mark A. Hindell and Andrea Walters

Contents

6.1 Introduction

Diet and *nutrition* are fundamental in shaping the life histories and reproductive strategies of all animals but particularly so for marine mammals. This is because, despite their diverse evolutionary origins, all marine mammals returned to the sea from terrestrial antecedents to exploit the abundant *resources* available in the marine environment. In so doing, they have retained some of the traits of their land-based predecessors which dictate many aspects of their feeding biology. Primary among these is their need to breathe air, requiring them to spend time on the surface away from the prey fields. The tension between the need to obtain oxygen at the surface but to feed, hunt, and even digest food underwater in the absence of available oxygen is a fundamental challenge for marine mammals. Further, while cetaceans and sirenians can give birth and nurse young at sea, pinnipeds need to be on land (or ice) for reproduction, further influencing their relationship with marine-based food resources. During breeding and lactation pinnipeds either do not forage at all (many of the phocids) or they forage close to the breeding sites (many of the ottariids); both strategies have their own implications for diet.

The diverse phylogenetic origins of marine mammals also offer scientists an important opportunity to better understand the complex interplay between diet (or "energy in") and how marine mammals go about their aquatic lifestyles (or "energy out"). Biology has a long tradition of employing a comparative approach to understanding general underlying principles, and the marine mammals with their differing evolutionary histories, but convergent lifestyles are particularly valuable in this regard. By comparing and contrasting the diets and nutrition of marine mammals, we can learn a lot about how these influence life-history strategies.

As a group, marine mammals have a wide range of diets, from exclusive herbivory for the sireniens (manatees and dugongs), through grazers on lower *trophic levels* (mysticete whales), those that exploit mesopelagic fish and squid (most of the odontocete whales and pinnipeds), up to top predators that take other marine mammals (killer whales, *Orcinus orca*; polar bears, *Ursus maritimus*; and leopard seals, *Hydrurga leptonyx*). It is important to note that most of these dietary types are used by several orders of marine mammals. For example, both phocid seals and mysticete whales have evolved filter feeding as a mechanism for collecting small zooplankton. This convergence in diets and feeding techniques is a good example of how species with diverse phylogenetic origins are able to exploit the same abundant resources available in the marine environment. With the exception of the sirenians, all marine mammals are *carnivorous*, and indeed are likely to have had carnivorous terrestrial antecedents.

In this chapter, we review the diets of all the major families of marine mammals; this includes the sirenians, mysticete, and odontocete whales, as well as the three families of pinnipeds (ottarids, phocids, and odobenidae), but not sea otters (*Enhydra lutris*) or polar bears as these families are not exclusively marine. The overall scope of the chapter is to detail the range of diet types and, in so doing, highlight the diversity of taxa using each diet and their morphological convergence. We begin with an overview of *foraging* strategies as well as the morphological and behavioral adaptations and implications this has for their life history and broader ecology. As marine mammals are a particularly challenging group of organisms for dietary studies (due in large part to the difficulties in observing and accessing them in the ocean), we then include a brief review of the techniques that scientists use to quantify (with varying degrees of taxonomic and temporal resolution) what marine mammals eat.

6.2 *Life history and diet*

The biggest dichotomy in marine mammal life histories is the aquatic breeding of all the cetaceans and sirenians compared to the terrestrial (or ice) breeding of the pinnipeds. Both life-history strategies have implications for diet and nutrition. Many species of pinnipeds are central place foragers, meaning that they must regularly return to a specific location between foraging trips. This is most pronounced for those species that are income breeders. These are those species having too few *energy stores* to sustain the mother and her pup throughout the entire lactation period, requiring mothers to make regular trips to sea to replenish energy stores. A good example of this is the fur seals. This is in contrast to capital breeders, which have very large energy stores and do not need to feed during the lactation period. Capital breeders tend to be large and have very short lactation periods, ranging from less than a week in hooded seals (*Cystophora cristata*) to a month in Weddell seals (*Leptonychotes weddellii*). The requirement to breed and raise young onshore of the income breeders requires them to be central place foragers, at least during the pup-rearing period. In extreme cases, such as the Australian sea lion (*Neophoca cinerea*) where pups take

17 months or more to rear, diet opportunities are limited to regions very close to breeding sites, typically those areas mothers can reach in a few days. The limited foraging range of central place foragers restricts the diversity of foraging opportunities, and can possibly lead to local prey depletion around larger breeding colonies, ultimately limiting the population size. For central place foragers operating in temperate coastal waters, this can be particularly important because shallow, inshore waters can be relatively depauperate in prey compared to the open ocean, and this has been linked to the generally small population size and slow population growth rates of sea lions compared to fur seals (Arnould and Costa 2006). Outside the breeding season, individuals are less constrained and range widely allowing for a greater diversity of diets.

The pronounced sexual dimorphism (i.e., difference in body size between males and females within a species) of many species of land breeding marine mammals can also impose very different dietary requirements between males and females. Disparate body size among genders is typically related to mating strategies. Polygamous species, where males compete for access to females provide strong selection pressure for large males. Monogamous species, where there is less direct competition for mates, tend to have equivalent sized males and females. Pronounced differences in body size can have significant implications for diet and nutrition. The larger gender has greater energy requirements, in terms of both maintenance and energy stores for breeding, as well as different oxygen needs and different body forms. All of these factors can lead to size and gender-based differences in diet and foraging locations. A good example is southern elephant seals (*Mirounga leonina*) where males make more extensive use of continental shelf habitats than females and also feed on large *benthic* prey. Females tend to use oceanic waters and to feed on smaller *pelagic* fish and squid.

6.3 Feeding mechanisms

6.3.1 Filter feeding

Filter feeding, used by all the mysticete whales and two species of pinnipeds (crabeater seals, *Lobodon carcinophagus*; and leopard seals), is a foraging strategy that allows individuals to capture and process large quantities of very small prey in a single mouthful. All filter-feeding species need to feed on prey which occurs in dense aggregations. Two feeding adaptations have evolved to allow the exploitation of these prey: *baleen* (mysticete whales) and modified dentition (seals).

Filter feeding arose in both cetaceans and pinnipeds in response to the unique patterns of productivity and prey availability in marine ecosystems, particularly in the high latitudes during the spring and summer. Filter-feeding marine mammals primarily concentrate their foraging in polar and highly productive coastal upwelling regions. Prior to their exploitation by humans, the highest densities of mysticetes occurred in highly productive Southern Ocean waters. Crabeater seals, Antarctic fur seals (*Arctocephalus gazella*), and leopard seals are also found primarily in the Southern Ocean where seasonally dense aggregations of Antarctic *krill* (*Euphausia superba*) develop (Berta and Sumich 1999).

Studies of the diving behavior and daily movement patterns of right whales (*Eubalaena* spp.) have shown that they exploit dense aggregations of copepods that aggregate at oceanographic features such as fronts. Rorquals also follow seasonal and *diel* patterns in the abundance and behavior of their prey. In general, the distribution and movement patterns of most rorquals consist of a seasonal *migration* from high latitudes where foraging takes place to low latitudes where they mate and give birth. However, data from blue whales (*Balaenoptera musculus*) in the Pacific indicate that feeding also takes place at low latitude, upwelling-modified waters, and data from both the Pacific and the Indian Oceans indicate that some blue whales may remain at low latitudes year-round. Fin (*Balaenoptera physalus*) and blue whales foraging on krill off the coast of North America concentrate their foraging efforts on dense aggregations of krill that are deep (150–300 m) in the water column during the day and may cease feeding when the krill becomes more dispersed near the surface at night (Croll et al. 1998, 2005) (Figure 6.1).

Mysticetes lack teeth and, instead, have rows of baleen plates made of *keratin*, which hang from the upper jaw and are used to filter small prey items from the water. Similar to fingernails, the plates grow continuously from the base, but are worn by the movements of the tongue. The five species of right whales have very long baleen plates with finely frayed edges that allow very small plankton to be trapped. Right and bowhead (*Balaena mysticetus*) whales eat small crustaceans ranging from minute copepods less than 1 mm long, favored by the bowhead whale of the Northern seas and the pygmy right whale (*Caperea marginata*), to small euphausiids (krill) as much as 25 mm long, eaten by the southern right whale (*Eubalaena australis*). The bowhead whale is also known to eat small molluskan pteropods.

All of the eight species of rorquals have shorter baleen plates and, generally, favor larger prey than copepods (Berta and Sumich 1999). Blue whales feed almost exclusively upon krill in the Antarctic and other euphausiid species in the North Pacific and North Atlantic, although other rorquals have a more varied diet. Sei whales (*Balaenoptera borealis*) eat species of densely shoaling midwater crustaceans (krill and copepods), in addition to anchovy, cod, and assorted oceanic squid. The fin whale eats krill in the Antarctic, but broadens its diet to include schooling fish (clupeids), muscular squid, and copepods in the North Atlantic. Common minke whales (*Balaenoptera acutorostrata*) and Antarctic minke whales (*Balaenoptera bonaerensis*) feed on assorted crustaceans in the Arctic and Antarctic, respectively, but also include schooling fish in the North Pacific (anchovy) and North Atlantic (herring). They also take mid-water squid in southern

Figure 6.1 **(See color insert.)** An Antarctic krill swarm. (Photo by Steve Nicol.)

tropical waters and appear to rely more on fish than other baleen whales. Bryde's whales (*Balaenoptera brydei* and *Balaenoptera edeni*) eat crustaceans, including krill, in addition to various fish (mullet and anchovy in the Southern Hemisphere and anchovy in the North Atlantic). Humpback whales (*Megaptera novaeangliae*) mainly feed on krill in the Southern Hemisphere, but mainly schooling fish (anchovies and cod) in the Northern Hemisphere. Squid are also eaten by humpback whales. The diet of gray whales (*Eschrichtius robustus*) consists primarily of benthic gammarid amphipods, although they can forage on a wide variety of prey, including schooling mysids in some areas. Gray whales have very tough baleen plates that become worn, particularly on the right side, by rubbing on the seafloor from which it principally sucks, by a piston action with its tongue, bottom amphipod crustaceans but also mollusks and bristle worms.

In the Antarctic crabeater seals and leopard seals use heavily *cusped post-canine teeth* to sieve the highly abundant Antarctic krill (Figure 6.2). In captive feeding trials, leopard seals have been observed to use suction to draw small prey into the mouth followed by expulsion of ingested seawater through the sieve formed by post-canine teeth (Hocking et al. 2013). Little detailed information is available on the behavior used by crabeater seals, but diving behavior studies indicate that they track the diel migration of krill, with shallow dives performed during the night and deeper dives during the day (Burns et al. 2004, 2008; Wall et al. 2007). One other species, the Antarctic fur seal also includes Antarctic krill as an important diet item, but does this without modifying its dentition from the standard pinniped *homodont* structure. The feeding mechanism of this species is not well studied, but presumably they take individual krill in separate captures. This mechanism can only be energetically feasible in small predators, and as the body size increases the efficiency of the capture mechanism needs to increase.

The larger body size also provides an efficient energy store for wintering and long distance migration without feeding (Berta and Sumich 1999). Gray whales undergo the longest migration of any mammal, foraging during the summer and fall in the Bering Sea and Arctic Ocean when dense aggregations of benthic amphipods become available with the seasonal increase in productivity. Humpback whales seasonally migrate from breeding areas to higher latitude foraging areas where schooling fish and krill become

(a) (b)

Figure 6.2 **(See color insert.)** (a) Pygmy blue whale, feeding. Showing baleen, Portland, Victoria, Australia. (Photo by Paul Enser, Australian Antarctic Division, Kingston, South Australia, Australia.) (b) Crabeater seal head and shoulders illustrating the heavily cusped post-canine teeth used by this species to sieve Antarctic krill. (Photo by Mark A. Hindell.)

seasonally abundant (Berta and Sumich 1999). The timing of coastal migration patterns of the California blue whale appears to be linked to annual patterns in coastal upwelling and krill development patterns (Croll et al. 2005).

6.3.2 Pierce feeding

Another major feeding strategy among marine mammals is pierce feeding, and this is utilized by the odontocete whales (toothed whales), and most phocid and otariid seals. Pierce feeding involves catching squid and fish using sharp, homodont teeth and swallowing them whole, or mechanically breaking them into large pieces (Jones et al. 2013).

Species that employ pierce feeding display much greater diversity in size, habitat, and diet than *filter feeders*. This more generalized diet permits greater ecological and life-history diversity. The range of diet types extends from deep-diving squid specialists to top level predators. Additionally, whereas the diet of baleen whales is highly seasonal due to the seasonal nature of zooplankton *biomass* and production, that of odontocetes is generally more constant year-round.

There are 7 families and 52 species that comprise the suborder Odontocete, all of which utilize *echolocation* to locate their prey. Most species feed on fish and cephalopods, with species that live on the continental shelf eating muscular fish (herring, pilchards, whiting, and soles) and cephalopods (squid, cuttlefish, and octopods), and species that live in deeper oceanic waters eating mainly fish of the Myctophidae family, and soft-bodied, gelatinous squids. In contrast to mystecetes, most odontocete species generally chase, capture, and swallow single relatively large prey items (fish, squid, large crustaceans, and occasionally other marine mammals).

Broadly speaking, for dolphins (family Delphinidae) cephalopods comprise over 75% of the diet and fish make up the remainder. Depending on the species, cephalopod prey taken can be muscular squid for species living on the continental shelf, including the short-beaked common dolphin (*Delphinus delphis*) and the common bottlenose dolphin (*Tursiops truncatus*). The diet of other species, including the spinner dolphin (*Stenella longirostris*) and spotted dolphins (*Stenella attenuata* and *Stenella frontalis*), may include many soft-bodied oceanic species. Overall, dolphins, including species found in oceanic waters (including pilot whales, *Globicephala* spp.), favor muscular squid over soft-bodied species. Other species, such as Commerson's dolphins (*Cephalorhynchus commersonii*), eat some krill and killer whales also prey on seals and other crustaceans.

For the porpoises (family Phocoenidae), cephalopods are also the primary prey (75% of the diet) together with fish for half of the species in this family. Cephalopods comprise 50%–75% and fish comprise 25%–50% of the diet in the other half of the species. Porpoises are a family of inshore cetaceans, and their food consists of common muscular inshore squid, cuttlefish, and octopus, as well as fish such as herrings, whitings, and bottom-living soles.

Many aspects of the diet of beaked whales are poorly understood, but in general, more than half of the food of beaked whales (family Ziphidae) consists of cephalopods and the rest is fish. These whales are deep divers and at least one species favors soft-bodied squids. The number of teeth in this family is highly reduced and the two teeth in the lower jaw grow over the upper jaw and limit it to a narrow gape. In most species of beaked whales, females and juveniles lack erupted teeth and that of adult males are primarily used for intraspecific fighting (Heyning and Mead 1996). It has been hypothesized that some species may capture prey primarily by suction (Caldwell et al. 1966; Norris and Mohl 1983), and observations of live animals corroborate the anatomical findings that beaked whales use suction to acquire prey (Heyning and Mead 1996).

The diet of the narwhal (*Monodon monoceros*) includes fish (such as Greenland halibut and polar cod), muscular squid, and shrimp. Belugas (*Delphinapterus leucas*) feed on fish such as capelin and sand lance, as well as larger species (cod and flounder). Sand and bottom-living worms show that they probably feed on the bottom as well as in midwater.

Sperm whales (*Physeter macrocephalus*) are the largest odontocete species. They feed primarily on squid, although a few fish are taken as well (including sharks). Sperm whales dive deep in the water column to feed on deep-living oceanic squid, most of which are soft-bodied or gelatinous, luminous, and weak swimmers. Some sperm whales consume squid over 15 m in length, but most of the species they consume are remarkably small (0.5–7.0 kg). Pygmy and dwarf sperm whales (*Kogia breviceps* and *Kogia sima*) eat some of the same species as their larger counterparts, but their diet also includes muscular squids and octopods in continental shelf waters.

River dolphins (families Inidae, Pontoporiidae, Lipotidae, Platanistidae) inhabit freshwater systems of major rivers in South America, China, and the Indian subcontinent. The Franciscana dolphin (*Pontoporia blainvillei*) has a marine distribution and is endemic to the southwestern Atlantic where it is found in coastal and estuarine waters. Species found in river systems feed on a variety of freshwater fish (including sharks) and prawns, and occasionally freshwater turtles. Franciscanas eat mainly bottom-dwelling fish, coastal cephalopods, and several species of shrimp.

Pinnipeds are made up of 19 species of true, earless seals (*Phocidae*), 16 species of fur seals and sea lions (*Otariidae*) and walruses (*Odobenus rosmarus*), and the vast majority of these species use pierce feeding. Fish and cephalopods are the most common prey of pinnipeds. In the Southern Ocean, mesopelagic fish (e.g., myctophids, particularly *Gymnoscopelus* and *Electrona* spp.) and squid are important prey for many pinniped species, including southern elephant seals and Ross seals (*Ommatophoca rossi*). Although shelf and deep-sea squid are also an important part of the diet of adult southern elephant seals, which exhibit a range of feeding strategies from pelagic dives within the open waters of the Antarctic Circumpolar Current to *demersal* foraging over Antarctic continental shelves (Hindell et al. 1991; Biuw et al. 2007).

Mesopelgic fish and squid are an important part of the diet of several species of fur seals in the *Arctocephalus* genus, including Antarctic and subantarctic fur seals (*A. gazella* and *A. tropicalis*) in the southern Indian Ocean sector of the Southern Ocean (Lea et al. 2002; Makhado et al. 2008; Makhado et al. 2013). Other temperate species, such as New Zealand fur seals (*Arctocephalus forsteri*), feed on mesoplegic prey (myctophids) in oceanic (offshore) waters, in addition to arrow squid (*Nototodarus sloanii*) and other benthic fish species in continental shelf/slope waters (Harcourt et al. 2002). In the North Pacific, between California and Alaska, the northern fur seal (*Callorhinus ursinus*) eats primarily small shoaling fish (northern anchovy), Pacific herring, capelin, Pacific sandlance and whiting, salmon and muscular squids; while in the Bering Sea, the species feeds largely on walleye pollock, salmon, Pacific herring and sandlance, northern smooth tongue, and squid.

All species of sea lions include cephalopods in their diet, but only some eat fish. California, Galapagos, and Japanese sea lions (*Zalophus californianus*, *Zalophus wollebaeki*, and *Zalophus japonicas*) feed on shelf-living muscular squid and octopods, as well as oceanic, muscular squids; while Australian sea lions eat shelf octopods and cuttlefish. Steller sea lions (*Eumetopias jubatus*), which range throughout the North Pacific, consume a wide variety of fish such as walleye pollock, Atka mackerel, salmon, Pacific cod, hake, and herring, in addition to squid and octopod.

Other invertebrates, however, are important to several pinniped species. Euphausiids (krill) are an important part of the diet of crabeater seals, leopard seals, and Antarctic fur seals in the Antarctic (Doidge and Croxall 1985), and may also be an important part of the diet of southern elephant seal pups (Walters et al. 2014). In the Arctic, it is documented that harp seals (*Pagophilus groenlandicus*) to some extent feed on crustaceans (krill, pelagic amphipods), particularly pups and immature males (Sivertsen 1941; Sergeant 1973; Nilssen et al. 1992). Other phocid species found in the Arctic, such as bearded seals (*Erignathus barbatus*) are primarily benthic feeders, eating a variety of invertebrate prey (polychaetes, crustaceans, and mollusks) in addition to pelagic fish and a few species of octopods (Pauly et al. 1998; Hjelset et al. 1999; Hindell et al. 2012). The diet of walruses, which predominantly feed in shallow Arctic waters (typically less than 100 m deep), also consists largely of benthic invertebrate prey, including bivalves, gastropods, and polychaete worms (Sheffield and Grebmeier 2009), although larger prey (adult seals) are also taken.

Killer whales and leopard seals are two examples of top predator species which feed on a wide variety of prey. Killer whales feed on cetaceans, seals, seabirds, sea otters, dugongs, fish, and cephalopods. In the Northeast Pacific and Antarctica, populations of up to three sympatric ecotypes have been found to feed on different prey types. In the Northeast Pacific, "transient" killer whales feed mainly on marine mammals (especially harbor seals, *Phoca vitulina*), while "resident" forms feed primarily on fish, with salmon being particularly important in the diet of whales in British Columbia waters (Ford et al. 1998; Baird 2000; Pitman and Ensor 2003). The diet of a third ecotype ("offshores") that spends a limited time in coastal areas is less well-known but is thought to include fish (Jones 2006). In Antarctica, the leopard seal inhabits coastal areas of the Antarctic continent where it feeds on a range of prey types from small invertebrates to larger mammals. In the Antarctic Peninsula, the summer diet of the species includes Antarctic fur seals, gentoo (*Pygoscelis papua*), macaroni penguins (*Eudyptes chrysolophus*), fish, squid, and krill (e.g., Stone and Meier 1981), but mainly fish and krill during the winter (Lowry et al. 1988). While in Prydz Bay, east Antarctica, Adelie penguins (*Pygoscelis adeliae*) are an important prey item for seals during the summer, together with crabeater seals, benthic and pelagic fish, and to a lesser extent amphipods and krill (Hall-Aspland and Rogers 2004). Predation on southern elephant seal pups by this species has also been recorded at the Heard and Kerguelen Islands in the southern Indian Ocean sector (e.g., Borsa 1990).

6.3.3 *Suction feeding*

A number of species have evolved other novel feeding strategies (Klages and Cockcroft 1990; Adam and Berta 2002). In the Arctic, specialized *suction feeding* on bivalve mollusks is found in walrus and in bearded seals (Kastelein 1994; Adam and Berta 2002; Marshall et al. 2008). In the walrus and bearded seal, soft-bodied mollusks are removed from their shells by powerful suction forces created by the rapid retraction of the tongue (Kastelein 1994).

6.3.4 *Herbivory*

The only marine mammals that are exclusively herbivorous are the sirenians (manatees and dugongs), which have a very different phylogenetic origin to the cetaceans and pinnipeds.

Manatees (*Trichecus* spp.) and dugongs (*Dugong dugong*) feed on tropical grasses, roots, and rhizomes in nearshore areas in saline environments and on water hyacinths, water lilies, and other vegetation in rivers and lakes. The very high degree of dietary specialization in these species limits their distribution to shallow sea-grass habitats, which in turn often brings them into close contact with several conservation threats such as coastal pollution and boat strikes.

6.4 *Methods for studying diet in marine mammals*

Obtaining direct information on diet in often wide ranging and deep-diving marine species remains a major challenge (Biuw et al. 2007). The development of animal telemetry has greatly improved our understanding of how animals use the marine environment (Burns et al. 2004; Costa et al. 2010; Cotté et al. 2011; Arthur et al. 2015), but spatially explicit information on diet remains largely unavailable. By virtue of marine existence, the study of marine mammal diet is extremely difficult as many species live most of their lives submerged in open water, making them cryptic organisms when foraging. Assessing the diet of animals that have large foraging ranges is a particular problem (Field et al. 2007).

6.4.1 *Direct observation of feeding*

Direct observation of feeding relies on what can be observed from on or near the surface. Observations can be made from the air, a vessel, or from a vantage point on land. Much has been learned from these observations, especially when conducted in a systematic way. Direct observations of sub-surface feeding, however, are rare (Similä and Ugarte 1993) and hence, sub-surface feeding on bottom-dwelling (benthic) prey, such as flatfish, flounder, crabs and other invertebrates, and mid-water prey, such as myctophids and squid, are greatly underestimated in this method, as it requires the predator to bring its prey to the surface before consuming it (Figure 6.3).

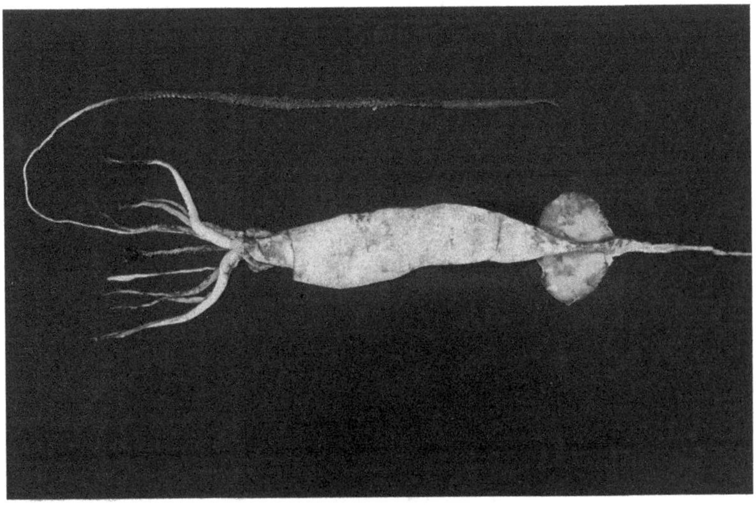

***Figure 6.3* (See color insert.)** Direct observation of sub-surface feeding events is rare. Sub-surface feeding on bottom-dwelling or mid-water prey, such as squid as shown here, are greatly underestimated in this method (*Chiroteuthis* sp.). (Photo by Glenn Jacobson, Australian Antarctic Division © Commonwealth of Australia.)

6.4.1.1 Traditional methods

Analyses of food remains in regurgitates or scats of living animals or in the stomachs and intestines of stranded animals have been the primary means for determining the diet and resource partitioning of marine mammal species (Hindell et al. 1995; Lea et al. 2002; Field et al. 2007). A particular disadvantage of stomach content and scat analysis is that they do not provide any indication of dietary variations over the annual cycle of animals, as they are restricted to the times of year animals are on-shore. For species which never come to shore, this technique is only helpful in the relatively rare occurrence of gaining access to specimens through misadventure (such as stranding) or through harvesting (such as whaling).

These methods rely on the presence and identification of structures representative of a typical meal, such as cephalopod jaws (commonly referred to as "beaks" due to their resemblance to the beaks of parrots), fish bones, or fish *otoliths. Cephalopod beaks* and fish otoliths are particularly useful diagnostic structures in the identification of prey because their size and shape vary considerably from species to species (Figure 6.4). Cephalopod beaks are composed of *chitin* and are not dissolved by digestive processes.

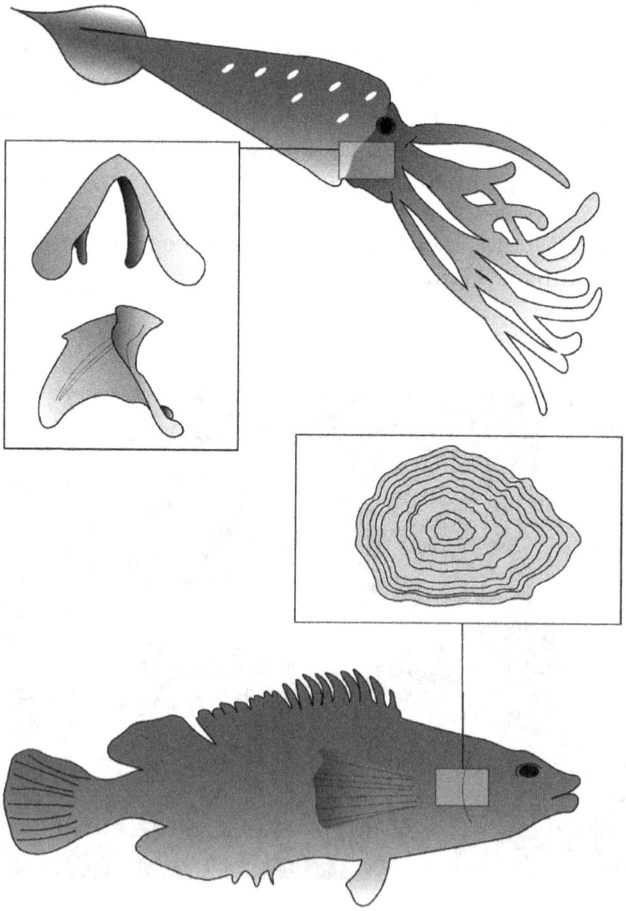

Figure 6.4 A schematic showing cephalopod jaws or "beaks" (left) and a fish ear stone (otolith; right) and where they are in the intact animal. (Schematic by Indi Hodgson-Johnston.)

Similarly, fish otoliths (ear stones) primarily composed of calcium carbonate are more resistant to *digestion* than bones.

These techniques provide detailed taxonomic data (Slip 1995) but are biased toward the most recent prey intake at the end of foraging trips, and hard part structures (or remnants of) prey that can be visually identified in stomach contents and feces (Staniland 2002). Soft-bodied species that have no hard parts are largely undetectable after digestion leading to complete under-representation of these groups in the diet. In the case of fecal studies, prey parts that do not survive digestion (or digest or erode at different rates) or whose hard parts are not consumed, are often missed. Controlled feeding trials of captive seals have revealed up to a 10-fold disparity with what has been recovered through fecal analysis (Tollit et al. 2003), leading to inappropriate conclusions about *niche* breadth utilization. Moreover, for a large proportion of cetacean species involved in single-stranding events (e.g., Gales et al. 1992), the cause of death is unknown but may be associated with old age, disease, or injury (Evans and Hindell 2004). In such instances, hard part identification from these animals may not be indicative of that of the larger population. However, dietary assessments of mass-stranded individuals are less biased by old, sick, or injured individuals and thus, more likely to be more representative of the larger, healthy population (Evans and Hindell 2004).

Due to limitations of conventional hard part diet analyses, researchers have increasingly turned to using biochemical markers contained in the tissues of predators to make inferences about what they have been eating. There are three types of markers, stable isotopes, fatty acids, and DNA, each of which has their own strengths and weaknesses. Reliable insights are most likely to be achieved by using a combination of techniques.

6.4.2 *Novel, new techniques to investigate the feeding ecology of marine mammals*

6.4.2.1 *Stable isotope ratios*

Over the past 20 years, stable isotope ratios of carbon ($^{13}C/^{12}C$; $\delta^{13}C$) and nitrogen ($^{15}N/^{14}N$; $\delta^{15}N$) have been increasingly used to study variation in resource and habitat use of elusive or highly migratory animals, such as marine top predators (Newsome et al. 2012). Their use in dietary studies is based on the fact that stable isotope ratios in the proteins of consumers reflect those of the proteins in their diet in a predictable manner (Hobson and Clark 1992). Although coarse in taxonomic resolution, the use of naturally occurring ratios of stable isotopes in animal tissues can be a powerful alternative method of dietary analysis. Stable isotope ratios can yield a data time-series from assimilated and not just ingested food (Tieszen et al. 1983).

Stable isotope ratios of carbon and nitrogen are the main elements used in dietary analyses (see reviews in Peterson and Fry 1987; Gannes et al. 1998; Kelly 2000; Dalerum and Angerbjörn 2005). This is because $\delta^{15}N$ exhibits a stepwise and predictable increase with trophic transfers. Consequently, the $\delta^{15}N$ values in the tissues of consumers tend to be relatively high compared to those of their diets and can, therefore, be used to estimate the diet and trophic position of consumers in a food web (McCutchan et al. 2003; Vanderklift and Ponsard 2003). Stable carbon isotope ratios also increase per trophic transfer but to a much lesser degree than $\delta^{15}N$. In the marine environment, carbon isotopes are mainly used to indicate the foraging habitats of predators (Kelly 2000; McCutchan et al. 2003).

Stable nitrogen and carbon isotope sources (e.g., $\delta^{15}N$ and $\delta^{13}C$) at the base of food chains vary spatially which is reflected in spatial variability in isotopic composition among food webs (Bearhop et al. 2004). Spatial variability in $\delta^{13}C$ can discriminate between inshore and offshore feeding at a range of spatial scales from oceanic (marine) habitats (Hobson et al. 1994) to lake (freshwater) food webs (Vander Zanden and Rasmussen 1999). Stable carbon values can also differentiate between pelagic and benthic contribution to food intake (Hobson et al. 1994; Cherel et al. 2011). Inshore/offshore and pelagic/benthic $\delta^{13}C$ gradients have been used as an effective way to investigate the habitats of coastal, neritic, and oceanic species of Antarctic fish, with inshore/benthic species having higher $\delta^{13}C$ values than offshore/pelagic species (Cherel et al. 2011).

Stable isotope composition turnover rates vary among tissues, with high rates in metabolically active tissues such as blood plasma and liver, somewhat lower in muscle, and lowest in long-lived tissue such as bone (Tieszen et al. 1983). Keratinous structures and whole blood are particularly suitable for studying temporal variation in diet. Since keratin is a highly stable structural protein, the $\delta^{13}C$ and $\delta^{15}N$ composition of keratin-based tissues, such as whiskers, remains unchanged after the completion of growth. Whiskers and other keratinous tissues, therefore, provide a temporal record of feeding dating back several months to years (Kernaléguen et al. 2012; Walters et al. 2014). By comparing the isotope ratios along the length of the whisker with those of putative prey items, changes in food sources and habitat can be surmised for the temporal span represented by the growth of the whisker. Whole blood, on the other hand, provides short- to medium-term dietary signals and can be used to examine diet in discrete temporal windows, including periods outside the limited sampling seasons of traditional dietary methods (Chaigne et al. 2013).

Interpretation of isotopic data from animals that move between areas of differing isotopic compositions (e.g., Schell et al. 1989) can be complicated (Hobson and Welch 1992), requiring knowledge of a migratory animal's breeding, wintering and stop over sites. Satellite and archival derived tracking information is increasingly used to provide this important contextual information. For example, analysis of stable isotope ratios in the whiskers of sub-yearling southern elephant seals in conjunction with satellite telemetry and environmental data has been successfully used to examine habitat use and diet during their first foraging migration (Walters et al. 2014; Box 6.1).

6.4.2.2 *Fatty acids*

Fatty acid signature analysis (FASA) (Iverson 1993) has emerged over the last few decades as another useful tool to investigate diet in marine mammals. As with other biochemical techniques used to reconstruct diet, FASA techniques do not rely on the recovery of prey hard parts and integrate the diet over ecologically significant periods of time (Newland et al. 2009). FASA techniques have been used qualitatively to infer trophic levels and spatial and temporal differences in diets both within and among species (Iverson et al. 1997; Smith et al. 1997; Beck et al. 2007). Unlike other nutrients, such as proteins that are readily broken down during digestion, fatty acids are released from ingested lipid molecules (e.g., triacylglycerols) during digestion but are not degraded (Iverson et al. 2004). Dietary fatty acids remain largely intact and those of carbon chain length >14 can be deposited in animal tissue with little or no modification (Smith et al. 1997). Owing to various restrictions and specifications in the biosynthesis and modification of fatty acids, only a relatively limited number of fatty acids can be biosynthesized by animals (Ackman 1980). This makes it possible to distinguish dietary and non-dietary components present in animal tissues, particularly in lipid-rich tissues such as blubber and milk (Iverson et al. 2004). Dietary fatty acid

BOX 6.1 THE USE OF STABLE ISOTOPES AND TRACKING INFORMATION TO IDENTIFY BROAD-SCALE FORAGING HABITAT USE AND DIETARY PREFERENCES

The trophic position of sub-yearling southern elephant seals from Macquarie Island (54°30′S, 158°57′E) in the Southern Ocean was estimated using stable carbon ($\delta^{13}C$) and nitrogen ($\delta^{15}N$) ratios along the length of the whisker, which provided a temporal record of prey intake (Figure 6.5). Satellite-relayed data loggers provided details on seal movement patterns, which were related to isotopic concentrations along the whisker. Animals fed in waters south of the Polar Front (>60°S) or within Commission for the Conservation of Antarctic Marine Living Resources (CCAMLR) Statistical Subareas 88.1 and 88.2, as indicated by both their depleted $\delta^{13}C$ (<–20%) values and tracking data (Figure 6.6). They predominantly exploited varying proportions of

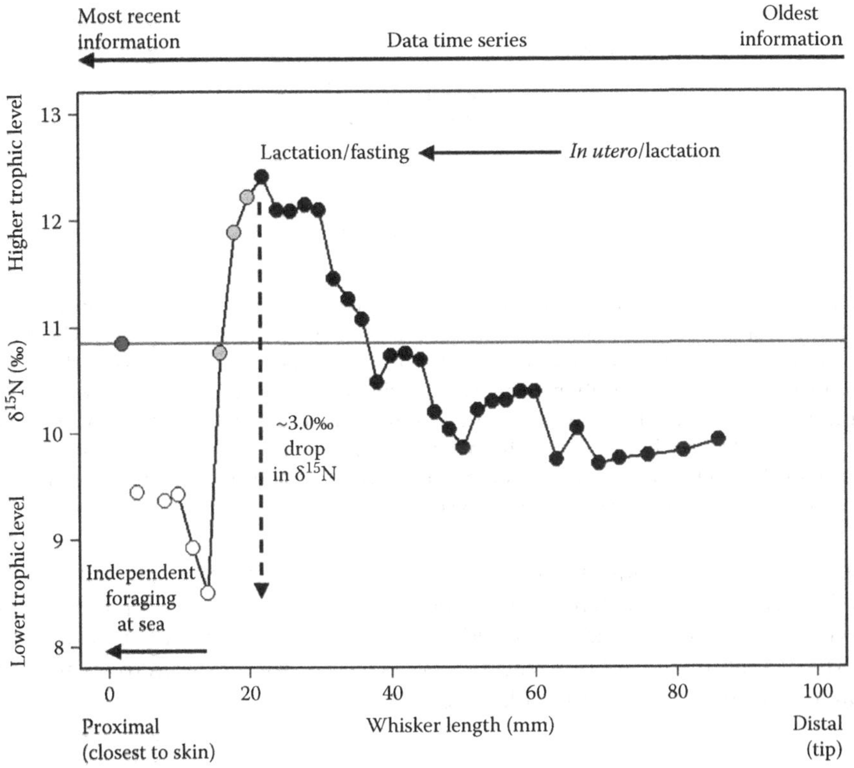

Figure 6.5 A schematic plot used to determine the shift to independent foraging along the post-trip whisker. We used stable nitrogen isotope ($\delta^{15}N$) values incorporated along the temporal span of the whisker as represented by the growth of the whisker from the distal (tip; oldest isotopic information) to proximal region (closest to the skin; most recent isotopic information). The solid line indicates where the pre-trip basal segment (dark gray symbol) intercepts $\delta^{15}N$ values along the length of the whisker. Solid arrows indicate the shift in food source along the temporal span from *in utero*/lactation to lactation/fasting (black symbols) to independent foraging at sea (open symbols). Dashed arrow indicates 3.9% drop in $\delta^{15}N$ (equivalent to one trophic level, 3.0%; grey symbols). The first 14 mm of the whisker represents independent foraging at sea. (From Walters, A. et al., *PLoS ONE*, 9(1), e86452, 2014.)

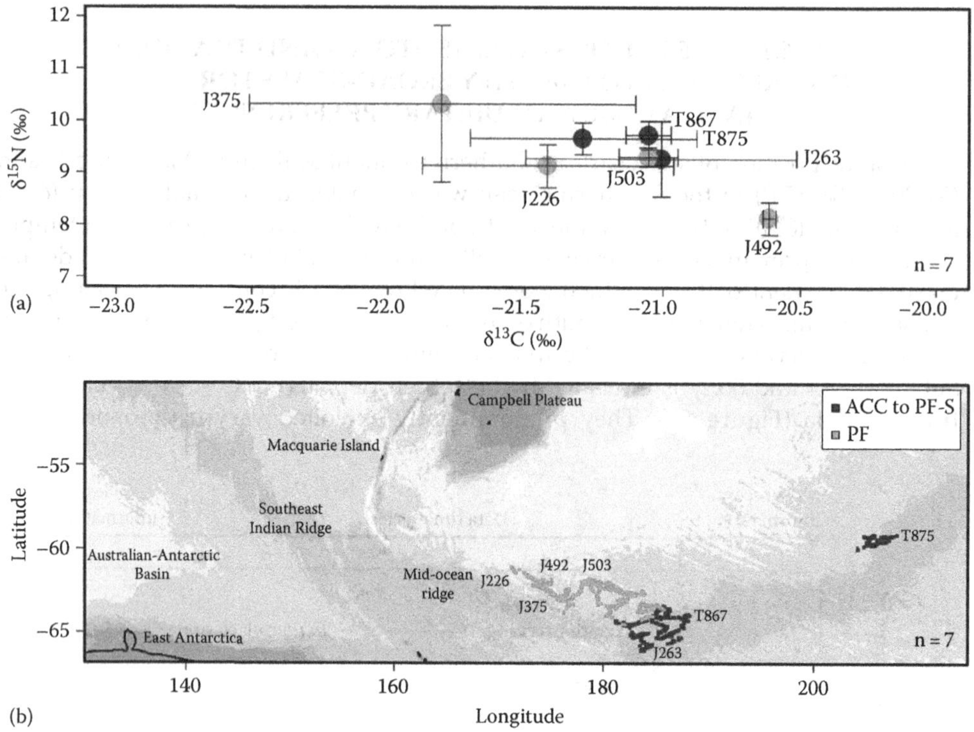

Figure 6.6 Area-restricted search locations and whisker stable isotope ratios of carbon ($\delta^{13}C$) and nitrogen ($\delta^{15}N$) values reflecting independent foraging at sea. (a) Mean whisker $\delta^{13}C$ and $\delta^{15}N$ values and (b) area-restricted search locations for seven subyearling elephant seals during their first migration from Macquarie Island are color-coded according to their foraging location. ACC to PF-S, Antarctic circumpolar current to polar front-south = dark gray; PF, polar front = light gray. Bathymetric features including the Southeast Indian ridge, Australian–Antarctic Basin and mid-ocean ridge are indicated in (b). Values are mean ± SD. (From Walters, A. et al., *PLoS ONE*, 9(1), e86452, 2014.)

mesopelagic fish and squid, and crustaceans, such as euphausiids, which have not been reported as a prey item for this species. Comparison of isotopic data between sub-yearlings, and 1-, 2- and 3-year olds indicated that sub-yearlings, limited by their size, dive capabilities, and prey capture skills to feeding higher in the water column, fed at a lower trophic level than older seals.

components (which appear in animal tissues) have a specific ecological origin that may be traced through a number of trophic levels (Iverson 1993). Thus, it is possible to consider the pattern of fatty acids in the adipose tissues of marine mammals as a signature that should reflect an integration of the fatty acid signatures of the major prey items in the diet (Iverson 1993).

6.4.2.3 *Molecular analysis of prey*

DNA-based techniques allow for the identification of prey species present in samples of stomach contents, vomit, and feces of predators. DNA metabarcoding has emerged as a powerful new tool with which to conduct dietary studies using fecal matter as a DNA source. Metabarcoding makes use of modern high-throughput DNA sequencing (HTS) technology which has simultaneously decreased the costs and increased the number of samples that can be analyzed concurrently as well as the efficiency by which samples can be processed (Deagle et al. 2009). DNA metabarcoding uses generic primers to amplify regions of the genome which are preserved across animal phyla. These regions are then sequenced using a HTS platform. The resulting sequences are then matched using bioinformatic techniques such as Basic Local Alignment Search Tools (BLAST) that interrogate databases for similar sequences from known taxa. As such, DNA metabarcoding offers the potential to generate genetic sequences of the complete diet of a predator and a relatively inexpensive means to conduct repeated sampling over time. However, despite the obvious benefits of DNA metabarcoding providing a much more accurate portrayal of the niche breadth of a predator, these techniques are also temporally restricted to dietary intake represented by the fecal sample obtained.

6.4.2.4 *Video and digital recording*

Video-data loggers ("crittercams") attached to free-ranging cetaceans and pinnipeds can be used to capture cryptic feeding events (prey–capture, predator–prey interactions) and characterize foraging strategies. Other devices such as synchronous motion and acoustic recording tags (DTAG) are also been increasingly used. A recent study used DTAGs to describe bottom side-roll feeding behavior in humpback whales along the seafloor (Ware et al. 2014).

Glossary

Baleen: Rows of plates made of dense, hair-like material (keratin), which hang from the upper jaw of whales in the Mysticeti order (the baleen whales). Baleen functions as a filter-feeding apparatus, allowing whales to filter small prey items (e.g., plankton) from surface waters.

Benthic: Living on the seafloor.

Biomass: A measure of the amount of biological (plant or animal) matter from living or recently living organisms in a given context or system.

Carnivorous: Feeding on other animals.

Cephalopod beaks: Chitinous mandibles or jaws of cephalopods (including squid, cuttlefish, and octopus); commonly referred to as "beaks" due to their resemblance to the beaks of parrots. Cephalopod beaks are hard, indigestible structures and consequently, they tend to be the only identifiable parts of cephalopod prey left in predator stomachs (pinnipeds and cetaceans).

Chitin: A highly insoluble nitrogen-containing polysaccharide, which forms a tough, semitransparent substance. Like cellulose, chitin functions as a structural polysaccharide and is the main component of arthropod exoskeletons and cephalopod beaks. Chitin is also found in the cell walls of some fungi and algae.

Cusped: Raised point or projection on the grinding surface of teeth.

Demersal: Living near the bottom of the sea.

Diel: Occurring on a 24 h cycle.

Digestion: The process whereby food and nutrients are rendered soluble and capable of being absorbed by an organism or by a cell.

Echolocation: Locating objects and investigating the surrounding environment by sensing the echoes returned by high-frequency sound waves emitted by an animal.

Energy store: The amount of energy stored in the body of an animal (usually as fat) that is surplus to current energy requirements.

Filter feeders: Marine mammals that use baleen (mysticete whales) or serrated teeth (some species of pinnipeds) to strain small particles of food (e.g., krill) from the water.

Foraging: The process of searching, capturing, and eating food (prey).

Herbivory: The eating of plants.

Homodont: Having teeth that are all of the same size and shape, not differentiated. Homodont dentition is found in the majority of vertebrates such as fish, amphibian, and reptiles, but only in some mammals (e.g., odontocetes: toothed whales).

Keratin: A fibrous protein rich in cystine, which forms the main structural component of hair (including mammalian whiskers otherwise known as vibrissae), nails and horn and the outer layer of human skin.

Krill: A general term used to describe small swarming crustaceans in the family Euphausiidae. Krill are a key food resource for filter feeding marine mammals.

Mesopelagic: Inhabiting or occurring in the ocean at depths between 200 and 1000 m.

Migration: The seasonal movement of animals between different geographic locations.

Niche: The ecological space or position occupied by a species within an ecological system.

Nutrition: The process by which an organism obtains the energy (nutrients) and other chemical elements it needs from its environment for its survival, growth, and reproduction.

Otoliths: Paired calcareous structures (earstones) used for balance and/or hearing in all teleost (i.e., bony) fish.

Pelagic: Relating to living or occurring in the open ocean.

Pierce feeding: Mode of feeding relating to the use of dentition to pierce and hold prey. Pierce feeding is used by most odontoceti (toothed) whales and pinnipeds that typically pursue, capture, and swallow whole prey (e.g., fish and/or squid).

Post-canine teeth: Teeth which are situated behind (posterior to) the canine teeth.

Resource: Any item, factor, or condition (e.g., food, mates, habitat) that contributes to a living.

Suction feeding: Mode of feeding using suction to capture prey; usually employs the use of the tongue as a piston to suck prey into a predator's mouth. This feeding adaptation is found in several species of toothed (odontoceti) whales and in some pinnipeds.

Trophic: Relating to the process of feeding and nutrition or feeding patterns.

Trophic level: The feeding position occupied by an organism in a food chain.

References

Ackman, R.G. 1980. Fish lipids. Part 1. In *Advances in Fish Science and Technology*, ed. J.J. Connell. Surrey, UK: Fishing News Books Ltd., 86–103.

Adam, P.J. and A. Berta. 2002. Evolution of prey capture strategies and diet in the Pinnipedimorpha (Mammalia, Carnivora). *Oryctos* 4:83–107.

Arnould, J. and D. Costa. 2006. Sea lions in drag, fur seals incognito: Insights from the Otariid Deviants. Paper read at *Sea Lions of the World: Proceedings of the Symposium Sea Lions of the World: Conservation and Research in the 21st Century*, University of Alaska Fairbanks, Fairbanks, AK.

Arthur, B., M. Hindell, M. Bester et al. 2015. Return customers: Foraging site fidelity and the effect of environmental variability in wide-ranging Antarctic fur seals. *PLoS ONE* 10(3):e0120888.

Baird, R.W. 2000. The killer whale: Foraging specializations and group hunting. In *Cetacean Societies*, eds. J. Mann, R.C. Connor, P.L. Tyack, and H. Whitehead. Chicago, IL: University of Chicago Press, 127–153.

Bearhop, S., C.E. Adams, S. Waldron, R.A. Fuller, and H. MacLeod. 2004. Determining trophic niche width: A novel approach using stable isotope analysis. *Journal of Animal Ecology* 73(5):1007–1012.

Beck, C.A., S.J. Iverson, W. Don Bowen, and W. Blanchard. 2007. Sex differences in grey seal diet reflect seasonal variation in foraging behaviour and reproductive expenditure: Evidence from quantitative fatty acid signature analysis. *Journal of Animal Ecology* 76(3):490–502.

Berta, A. and J.L. Sumich, eds. 1999. *Marine Mammals: Evolutionary Biology*. San Diego, CA: Academic Press.

Biuw, M., L. Boehme, C. Guinet et al. 2007. Variations in behavior and condition of a Southern Ocean top predator in relation to in situ oceanographic conditions. *Proceedings of the National Academy of Sciences* 104(34):13705–13710.

Borsa, P. 1990. Seasonal occurrence of the leopard seal, *Hydrurga leptonyx*, in the Kerguelen Islands. *Canadian Journal of Zoology* 68(2):405–408.

Burns, J.M., D.P. Costa, M.A. Fedak et al. 2004. Winter habitat use and foraging behavior of crabeater seals along the Western Antarctic Peninsula. *Deep Sea Research Part II: Topical Studies in Oceanography* 51(17–19):2279–2303.

Burns, J.M., M.A. Hindell, C.J.A. Bradshaw, and D.P. Costa. 2008. Fine-scale habitat selection of crabeater seals as determined by diving behavior. *Deep Sea Research Part II: Topical Studies in Oceanography* 55(3–4):500–514.

Caldwell, D.K., M.C. Caldwell, and D.W. Rice. 1966. Behavior of the sperm whale *Physeter catodon*. In *Whales, Dolphins, and Porpoises*, ed. K.S. Norris. Berkeley, CA: University of California Press, 678–718.

Chaigne, A., M. Authier, P. Richard, Y. Cherel, and C. Guinet. 2013. Shift in foraging grounds and diet broadening during ontogeny in southern elephant seals from Kerguelen Islands. *Marine Biology* 160(4):977–986.

Cherel, Y., P. Koubbi, C. Giraldo et al. 2011. Isotopic niches of fishes in coastal, neritic and oceanic waters off Adélie land, Antarctica. *Polar Science* 5(2):286–297.

Costa, D.P., L.A. Huckstadt, D.E. Crocker, B.I. McDonald, M.E. Goebel, and M.A. Fedak. 2010. Approaches to studying climatic change and its role on the habitat selection of Antarctic pinnipeds. *Integrative and Comparative Biology* 50(6):1018–1030.

Cotté, C., F. d'Ovidio, A. Chaigneau et al. 2011. Scale-dependent interactions of Mediterranean whales with marine dynamics *Limnology and Oceanography* 56(1):219–232.

Croll, D.A., B. Marinovic, S. Benson et al. 2005. From wind to whales: Trophic links in a coastal upwelling system. *Marine Ecology Progress Series* 289:117–130.

Croll, D.A., B.R. Tershy, R.P. Hewitt et al. 1998. An integrated approach to the foraging ecology of marine birds and mammals. *Deep Sea Research Part II: Topical Studies in Oceanography* 45(7):1353–1371.

Dalerum, F. and A. Angerbjörn. 2005. Resolving temporal variation in vertebrate diets using naturally occurring stable isotopes. *Oecologia* 144(4):647–658.

Deagle, B.E., R. Kirkwood, and S.N. Jarman. 2009. Analysis of Australian fur seal diet by pyrosequencing prey DNA in faeces. *Molecular Ecology* 18(9):2022–2038.

Doidge, D.W. and J.P. Croxall. 1985. Diet and energy budget of the Antarctic fur seal, *Arctocephalus gazella*, at South Georgia. In *Antarctic Nutrient Cycles and Food Webs*, eds. W.R. Siegfried, P.R. Condy, and R.M. Laws. Berlin, Germany: Springer-Verlag, 543–550.

Evans, K. and M.A. Hindell. 2004. The diet of sperm whales (*Physeter macrocephalus*) in southern Australian waters. *Ices Journal of Marine Science* 61(8):1313–1329.

Field, I.C., C.J.A. Bradshaw, J. van den Hoff, H.R. Burton, and M.A. Hindell. 2007. Age-related shifts in the diet composition of southern elephant seals expand overall foraging niche. *Marine Biology* 150(6):1441–1452.

Ford, J.K.B., G.M. Ellis, L.G. Barrett-Lennard, A.B. Morton, R.S. Palm, and K.C. Balcomb, III. 1998. Dietary specialization in two sympatric populations of killer whales (*Orcinus orca*) in coastal British Columbia and adjacent waters. *Canadian Journal of Zoology* 76(8):1456–1471.

Gales, G., D. Pemberton, and M. Clarke. 1992. Stomach contents of long-finned pilot whales (*Globicephala melas*) and Bottlenose dolphins (*Tursiops truncatus*) in Tasmania. *Marine Mammal Science* 8(4):405–413.

Gannes, L.Z., C. Martínez del Rio, and P. Koch. 1998. Natural abundance variations in stable isotopes and their potential uses in animal physiological ecology. *Comparative Biochemistry and Physiology—Part A: Molecular & Integrative Physiology* 119(3):725–737.

Hall-Aspland, S.A. and T.L. Rogers. 2004. Summer diet of leopard seals (*Hydrurga leptonyx*) in Prydz Bay, Eastern Antarctica. *Polar Biology* 27(12):729–734.

Harcourt, R.G., C.J.A. Bradshaw, K. Dickson, and L.S. Davis. 2002. Foraging ecology of a generalist predator, the female New Zealand fur seal. *Marine Ecology Progress Series* 227:11–24.

Heyning, J.E. and J.G. Mead. 1996. Suction feeding in beaked whales: Morphological and observational evidence. *Contributions in Science Natural History Museum of Los Angeles County* 464:1–12.

Hindell, M.A., C. Lydersen, H. Hop, and K.M. Kovacs. 2012. Pre-partum diet of adult female bearded seals in years of contrasting ice conditions. *PLoS ONE* 7(5):e38307.

Hindell, M.A., G.G. Robertson, and R. Williams. 1995. Resource partitioning in four species of sympatrically breeding penguins. In *The Penguins*, eds. P. Dann, I. Norman, and P. Reilly. Chipping Norton, New South Wales, Australia: Surrey Beatty & Sons, 196–215.

Hindell, M.A., D.J. Slip, and H.R. Burton. 1991. The diving behavior of adult male and female southern elephant seals, Mirounga-Leonina (*Pinnipedia, Phocidae*). *Australian Journal of Zoology* 39(5):595–619.

Hjelset, A.M., M. Andersen, I. Gjertz, C. Lydersen, and B. Gulliksen. 1999. Feeding habits of bearded seals (*Erignathus barbatus*) from the Svalbard area, Norway. *Polar Biology* 21(3):186–193.

Hobson, K.A. and R.G. Clark. 1992. Assessing avian diets using stable isotopes II: Factors influencing diet-tissue fractionation. *The Condor* 94(1):189–197.

Hobson, K.A., J.F. Piatt, and J. Pitocchelli. 1994. Using stable isotopes to determine seabird trophic relationships. *Journal of Animal Ecology* 63(4):786–798.

Hobson, K.A. and H.E. Welch. 1992. Determination of trophic relationships within a high Arctic marine food web using $\delta^{13}C$ and $\delta^{15}N$ analysis. *Marine Ecology Progress Series* 84:9–18.

Hocking, D.P., A.R. Evans, and E.M.G. Fitzgerald. 2013. Leopard seals (*Hydrurga leptonyx*) use suction and filter feeding when hunting small prey underwater. *Polar Biology* 36(2):211–222.

Iverson, S.J. 1993. Milk secretion in marine mammals in relation to foraging: Can milk fatty acids predict diet? *Symposia of the Zoological Society of London* 66:263–291.

Iverson, S.J., J.P.Y. Arnould, and I.L. Boyd. 1997. Milk fatty acid signatures indicate both major and minor shifts in the diet of lactating Antarctic fur seals. *Canadian Journal of Zoology* 75(2):188–197.

Iverson, S.J., C. Field, W.D. Bowen, and W. Blanchard. 2004. Quantitative fatty acid signature analysis: A new method of estimating predator diets. *Ecological Monographs* 74(2):211–235.

Jones, I.M. 2006. A northeast Pacific killer whale (*Orcinus orca*) feeding on a Pacific halibut (*Hippoglossus stenolepis*). *Marine Mammal Science* 22(1):198–200.

Jones, K.E., C.B. Ruff, and A. Goswami. 2013. Morphology and biomechanics of the pinniped jaw: Mandibular evolution without mastication. *The Anatomical Record* 296(7):1049–1063.

Kastelein, R.A. 1994. Oral suction of a Pacific walrus (*Odobenus rosmarus divergens*) in air and under water. *Mammalian Biology* 59:105.

Kelly, J.F. 2000. Stable isotopes of carbon and nitrogen in the study of avian and mammalian trophic ecology. *Canadian Journal of Zoology* 78(1):1–27.

Kernaléguen, L., B. Cazelles, J.P.Y. Arnould, P. Richard, C. Guinet, and Y. Cherel. 2012. Long-term species, sexual and individual variations in foraging strategies of fur seals revealed by stable isotopes in whiskers. *PLoS ONE* 7(3):e32916.

Klages, N.T.W. and V.G. Cockcroft. 1990. Feeding behaviour of a captive crabeater seal. *Polar Biology* 10(5):403–404.

Lea, M.A., Y. Cherel, C. Guinet, and P.D. Nichols. 2002. Antarctic fur seals foraging in the Polar Frontal Zone: Inter-annual shifts in diet as shown from fecal and fatty acid analyses. *Marine Ecology Progress Series* 245:281–297.

Lowry, L.F, J.W. Testa, and W. Calvert. 1988. Notes on winter feeding of crabeater and leopard seals near the Antarctic Peninsula. *Polar Biology* 8(6):475–478.

Makhado, A., M. Bester, S. Kirkman, P. Pistorius, J. Ferguson, and N. Klages. 2008. Prey of the Antarctic fur seal *Arctocephalus gazella* at Marion Island. *Polar Biology* 31(5):575–581.

Makhado, A.B., M.N. Bester, S. Somhlaba, and R.J.M. Crawford. 2013. The diet of the subantarctic fur seal *Arctocephalus tropicalis* at Marion Island. *Polar Biology* 36(11):1609–1617.

Marshall, C.D., K.M. Kovacs, and C. Lydersen. 2008. Feeding kinematics, suction and hydraulic jetting capabilities in bearded seals (*Erignathus barbatus*). *Journal of Experimental Biology* 211(5):699–708.

McCutchan, J.H., W.M. Lewis, C. Kendall, and C.C. McGrath. 2003. Variation in trophic shift for stable isotope ratios of carbon, nitrogen, and sulfur. *Oikos* 102(2):378–390.

Newland, C., I.C. Field, P.D. Nichols, C.J.A. Bradshaw, and M.A. Hindell. 2009. Blubber fatty acid profiles indicate dietary resource partitioning between adult and juvenile southern elephant seals. *Marine Ecology Progress Series* 384:303–312.

Newsome, S.D., J.D. Yeakel, P.V. Wheatley, and M.T. Tinker. 2012. Tools for quantifying isotopic niche space and dietary variation at the individual and population level. *Journal of Mammalogy* 93(2):329–341.

Nilssen, K.T., P.E. Grotnes, and T. Haug. 1992. The effect of invading harp seals (*Phoca groenlandica*) on local coastal fish stocks of North Norway. *Fisheries Research* 13(1):25–37.

Norris, K.S. and B. Mohl. 1983. Can odontocetes debilitate prey with sound? *The American Naturalist* 122(1):85–104.

Pauly, D., A.W. Trites, E. Capuli, and V. Christensen. 1998. Diet composition and trophic levels of marine mammals. *ICES Journal of Marine Science: Journal du Conseil* 55(3):467–481.

Peterson, B.J. and B. Fry. 1987. Stable isotopes in ecosystem studies. *Annual Review of Ecology and Systematics* 18:293–320.

Pitman, R.L. and P. Ensor. 2003. Three forms of killer whales in Antarctic waters. *Journal of Cetacean Research and Management* 5:131–139.

Schell, D.M., S.M. Saupe, and N. Haubenstock. 1989. Bowhead whale (*Balaena mysticetus*) growth and feeding as estimated by $\delta^{13}C$ techniques. *Marine Biology* 103(4):433–443.

Sergeant, D.E. 1973. Feeding, growth, and productivity of Northwest Atlantic Harp seals (*Pagophilus groenlandicus*). *Journal of the Fisheries Research Board of Canada* 30(1):17–29.

Sheffield, G. and J.M. Grebmeier. 2009. Pacific walrus (*Odobenus rosmarus divergens*): Differential prey digestion and diet. *Marine Mammal Science* 25(4):761–777.

Similä, T. and F. Ugarte. 1993. Surface and underwater observations of cooperatively feeding killer whales in northern Norway. *Canadian Journal of Zoology* 71(8):1494–1499.

Sivertsen, E. 1941. On the biology of the harp seal *Phoca groenlandica* Erxl. Investigations carried out in the White Sea 1925–1937. *Hoalrddets Skr* 26:1–166.

Slip, D. 1995. The diet of southern elephant seals (*Mirounga leonina*) from Heard Island *Canadian Journal of Zoology* 73(8):1519–1528.

Smith, S.J., S.J. Iverson, and W.D. Bowen. 1997. Fatty acid signatures and classification trees: New tools for investigating the foraging ecology of seals. *Canadian Journal of Fisheries and Aquatic Sciences* 54(6):1377–1386.

Staniland, I.J. 2002. Investigating the biases in the use of hard prey remains to identify diet composition using Antarctic fur seals (*Arctocephalus gazella*) in captive feeding trials. *Marine Mammal Science* 18(1):223–243.

Stone, S. and T. Meier. 1981. Summer leopard seal ecology along the Antarctic Peninsula. *Antarctic Journal of the United States* 16:151–152.

Tieszen, L.L., T.W. Boutton, K.G. Tesdahl, and N.A. Slade. 1983. Fractionation and turnover of stable carbon isotopes in animal tissues: Implications for $\delta^{13}C$ analysis of diet. *Oecologia* 57(1/2):32–37.

Tollit, D.J., M. Wong, A.J. Winship, D.A.S. Rosen, and A.W. Trites. 2003. Quantifying errors associated with using prey skeletal structures from fecal samples to determine the diet of Steller's sea lion (*Eumetopias jubatus*). *Marine Mammal Science* 19(4):724–744.

Vanderklift, M. and S. Ponsard. 2003. Sources of variation in consumer-diet $\delta^{15}N$ enrichment: A meta-analysis. *Oecologia* 136(2):169–182.

Vander Zanden, M.J. and J.B. Rasmussen. 1999. Primary consumer $\delta^{13}C$ and $\delta^{15}N$ and the trophic position of aquatic consumers. *Ecology* 80:1395–1404.

Wall, S.M., C.J.A. Bradshaw, C.J. Southwell, N.J. Gales, and M.A. Hindell. 2007. Crabeater seal diving behaviour in eastern Antarctica. *Marine Ecology Progress Series* 337:265–277.

Walters, A., M.A. Lea, J. van den Hoff et al. 2014. Spatially explicit estimates of prey consumption reveal a new krill predator in the Southern Ocean. *PLoS ONE* 9(1):e86452.

Ware, C., D.N. Wiley, A.S. Friedlaender et al. 2014. Bottom side-roll feeding by humpback whales (*Megaptera novaeangliae*) in the southern Gulf of Maine. *Marine Mammal Science* 30(2):494–511.

chapter seven

Water balance

Miwa Suzuki and Rudy M. Ortiz

Contents

7.1 Osmotic challenges for marine mammals

Constancy of the *milieu interieur* is critically important for Metazoan organisms because the cells can only live within a narrow range of physicochemical conditions necessary for proper cellular functions. Water and salt are fundamental for life and the main determinants of the osmolality of the extracellular fluid. In most vertebrates, water and salt balance is rigidly controlled to maintain the osmolality at about one-third of seawater by changing the transport of water and electrolytes to meet the demands induced by the external environments. When early vertebrates ascended upon dry land they evolved robust physiological mechanisms to conserve water and electrolytes. Subsequently, when these mammals radiated into a marine environment they had to evolve mechanisms to allow them to tolerate and thrive in an external milieu of high salinity with limited availability to freshwater. A number of morphological and physiological adaptations have allowed them to cope with these osmotic challenges. Additionally, other life-history traits such as mating, molting, and migration are associated with protracted periods of fasting, which place an additional osmoregulatory burden on marine mammals. The evolved physiological mechanisms have allowed marine mammals to conserve salts and water during prolonged fasting as well with virtually no consequences. In this chapter, the osmoregulatory mechanisms in marine mammals are reviewed.

7.2 Knowledge

7.2.1 Constancy of internal environment

When solutions with different solute concentrations are separated by a semi-permeable membrane, water molecules will move down their concentration gradient (from the compartment with the lower concentration of dissolved particles to the compartment with the greater concentration) through the membrane until the concentrations of the two compartments are equal, or reach an osmotic equilibrium. The process is defined as *osmosis*, and the power of the solution to draw water through the membrane is referred to as *osmotic pressure*. The osmolality of the plasma, therefore, is determined by the sum total of all the dissolved particles in a solution (water) such as electrolytes, glucose, and urea (blood urea nitrogen, BUN) and can be approximated by the following formula:

$$\text{Plasma osmolality} = \text{Na (mEq/l)} * 2 + \frac{\text{glucose (mg/dl)}}{18} + \frac{\text{BUN (mg/dl)}}{2.8}$$

Because the concentrations of electrolytes, especially sodium, potassiums, and chloride, in plasma are rigidly controlled so is the osmolality. The maintenance of plasma constituents within defined ranges suitable for cellular activity represents the homeostatic control afforded by adaptable osmoregulatory mechanisms. Aside from the potential impacts of the environment on plasma osmolality, diet and prolonged fasting can also contribute to alterations in osmolality. For example, protein loading from protein-rich diets can influence plasma urea concentration, which would be reflected by an increase in plasma osmolality. Plasma osmolality in humans ranges from 275 to 290 mOsm/kg; however, those in marine mammals are comparatively higher (Table 7.1). If the osmotic gap between the internal fluid and the surrounding medium (seawater) is reduced, the energetic costs for maintaining plasma osmolality in seawater may be comparatively reduced. Therefore, the presence of higher plasma osmolality might be advantageous for marine mammals that expend more energy for thermoregulation and locomotion in the water (Williams et al. 2001; Williams and Worthy 2002). The higher concentrations of plasma electrolytes and urea in marine mammals may contribute to their relatively higher plasma osmolality (Table 7.1) (reviewed in Ortiz 2001). The comparatively higher plasma urea concentrations in marine mammals, especially in cetaceans and sea otter, may confer an additional beneficial in the conservation of water (see Section 7.2.2).

The regulation of water and salt balance in the internal milieu is dependent on the physiological responses to the changes in input and losses from the kidney (Eaton and Pooler 2009). Under the condition of excessive salt-loading and limited water, it is necessary

Table 7.1 Osmolality and concentrations in electrolytes and urea in plasma of marine mammals

Species	Sex	Osmolality (mOsm/l)	Sodium (mEq/l)	Chloride (mEq/l)	Glucose (mg/dl)	Urea (mg/dl)	References
Seawater		1000	470	548	—	—	
Human		275–290	135–145	98–108	80–110	8–20	
Cetacea							
Bottlenose dolphin	M	331 ± 11	153 ± 7	106 ± 9	—	—	Ridgway et al. (1970)
Bottlenose dolphin	F	335 ± 14	155 ± 7	110 ± 8	—	—	Ridgway et al. (1970)
Bottlenose dolphin		341 ± 9	154 ± 3	—	—	—	Ortiz and Worthy (2000)
Pacific white-sided dolphin	M	332 ± 9	157 ± 9	107 ± 9	—	—	Ridgway et al. (1970)
Pacific white-sided dolphin	F	335 ± 13	153 ± 7	108 ± 8	—	—	Ridgway et al. (1970)
Killer whale	M	337 ± 17	155 ± 8	112 ± 7	—	—	Ridgway et al. (1970)
Short-finned pilot whale	M	333 ± 15	149 ± 4	109 ± 9	—	—	Ridgway et al. (1970)
Dall's porpoise	M	—	155 ± 9	107 ± 6	—	—	Ridgway et al. (1970)
White whale		325 ± 12	163 ± 8	114 ± 5	112 ± 20	115 ± 22	St. Aubin et al. (2001)
Amazon River dolphin	M	315 ± 11	144 ± 5	102 ± 11	—	—	Ridgway et al. (1970)
Amazon River dolphin	F	315 ± 15	142 ± 5	98 ± 6	—	—	Ridgway et al. (1970)
Finless porpoise (Yangtze)		329 ± 7	153 ± 2	108 ± 3	139 ± 21	101 ± 19	Guo et al. (2014)
Finless porpoise (Bohai)		362 ± 15	156 ± 8	117 ± 8	149 ± 40	86 ± 19	Guo et al. (2014)
Fin whale		364 ± 29	164 ± 9	121 ± 14	—	146 ± 22	Kjeld (2001)
Sei whale		—	155 ± 1	117 ± 2	—	—	Kjeld (2003)
Pinnipeds							
Harbor seal (mother)	F	318 ± 2	149 ± 1	107 ± 1	—	25 ± 2	Schweigert (1993)
Harbor seal (pup)		304 ± 2	144 ± 1	99 ± 2	—	20 ± 2	Schweigert (1993)
Harp seal		338 ± 2	162 ± 1	114 ± 1	—	—	Storepheier and Nordøy (2001)
Harp seal		331 ± 1	150 ± 1	111 ± 1	—	—	How and Nordøy (2007)
Northern elephant seal (pup)		304	147	107	—	—	Ortiz et al. (2003)
South American fur seal		325 ± 3	169 ± 4	77 ± 7	—	31 ± 16	Schweigert (1993)
South American fur seal		336 ± 14	149 ± 6	119 ± 5	—	68 ± 14	Le Bas (2003)
South American sea lion		329 ± 9	173 ± 4	90 ± 2	—	10 ± 1	Schweigert (1993)
Steller sea lion		—	149 ± 2	111 ± 2	—	—	Rosen et al. (2004)
Sirenia							
West Indian manatee (FW)		298 ± 2	147 ± 2	89 ± 2	—	—	Ortiz et al. (1998)
West Indian manatee (SW)		314 ± 3	159 ± 4	102 ± 3	—	—	Ortiz et al. (1998)
West Indian manatee		311 ± 6	151 ± 4	89 ± 4	83 ± 29	14 ± 4	Medway et al. (1982)
Sea otter		—	152	105	—	126	Costa (1982)

Note: Seawater and human values are provided for comparison and reference.

to excrete more salt with less water to maintain homeostasis. This process is facilitated by the fact that some marine mammals can concentrate urine above that of seawater (approximately 1000 mOsm/kg), which affords a net gain in free water. The *maximum urine osmolality* represents their maximal urine concentrating ability, which reaches more than 2000 mOsm/kg in pinnipeds (Tarasoff and Toews 1972; Hong et al. 1982; Skog and Folkow 1994), dolphins (Ridgway 1972), and sea otters (*Enhydra lutris*) (Costa 1982) (Table 7.2). The *urine-to-plasma ratio in osmolality* is used as another index of concentrating ability, and this ratio is greater in dolphins and seals than in humans (Table 7.2). However, both of these

Table 7.2 Maximum urine osmolalities reported and urine-to-plasma ratios calculated for various marine mammals

Species	Maximum urine osmolality (mOsm/kg)	Maximum urine to plasma ratio	Body weight (kg)	Reference
Terrestrial				
Elephant shrew	3118	—	0.05	Leon et al. (1983)
Human	1330	4.6	70	Miles et al. (1954)
Cattle (heifers)	1289	—	232	Weeth and Lesperance (1965)
Terrestrial (hot arid areas)				
Domestic cat	3100	10	3–5	Willmer et al. (2005)
Vampire bat	4650	14	0.05	Willmer et al. (2005)
Kangaroo rat	5500	16	0.07–0.17	Willmer et al. (2005)
Hopping mouse	9374	—	0.029	MacMillen and Lee (1969)
	9400	25	—	Schmidt-Nielsen (1990)
Sheep	3200	—	50	Macfarlane et al. (1961)
East African Antelope	4700	—	79	Schoen (1972)
Marine				
Bottlenose dolphin	2658	7.3	(300)[a]	Ridgway (1972)
Fin whale	1300	3.6	(75,000)[a]	Kjeld (2001)
South America fur seals	2344	6.9	(♂200, ♀50)[a]	La Bas (2003)
Gray seal	2161	6.0	(♂310, ♀186)[a]	Skog and Folkow (1994)
Gray seal (pups)	2741	—	(11–20)[a]	Reilly (1991)
Baikal seal	2374	6.9	(90)[a]	Hong et al. (1982)
Hooded seal	1472	4.4	(♂150, ♀110)[a]	Skalstad and Nordøy (2000)
Hooded seal	1502	—	(♂150, ♀110)[a]	Verlo (2012)
Harp seal	1739	5.4	(♂135, ♀120)[a]	Skalstad and Nordøy (2000)
West Indian manatee	1200	3.7	(1,590)[a]	Irvine et al. (1980)
Sea otter	2130	6.7	(45)[a]	Costa (1982)

Note: Values of some terrestrial mammals are provided for reference.

[a] Data of body weight in parenthesis are from the *FAO Species Identification Guide: Marine Mammals of the World* (1993).

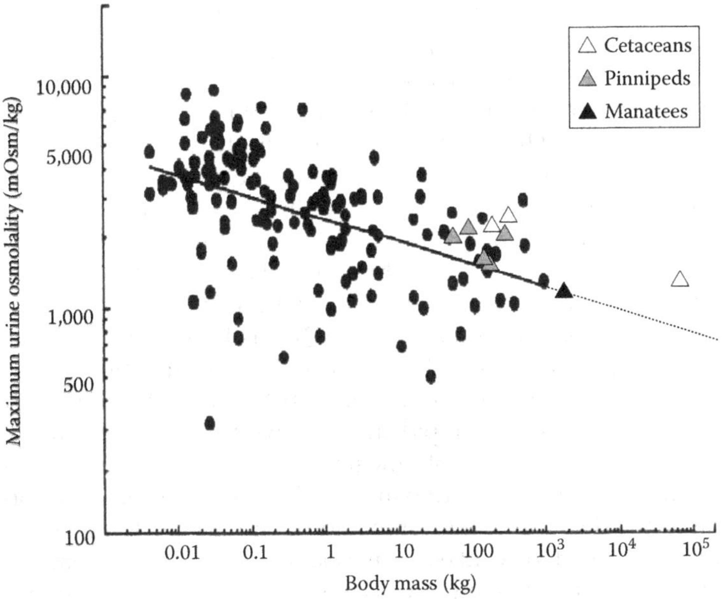

Figure 7.1 Interspecific allometry between maximum urine osmolality and body mass for 146 species of mammals including animals inhabiting arid regions. Maximum urine osmolalities recorded for various marine mammals are represented by triangles. (Modified from Beuchat, C.A., *Am. J. Physiol.*, 258(2), R298, 1990a. With permission.)

indices of urine concentrating ability in marine mammals are lower than those reported for many small terrestrial animals, such as the hopping mouse, the kangaroo rat, and the domestic house cat (Table 7.2). While the maximal urine concentrating ability in marine mammals is not exceptional or superlative, it is sufficient to have allowed this group of mammals to adapt to and tolerate their potentially challenging and demanding environment. So, why aren't marine mammals the champions of urine concentrating ability?

Let us consider the issue of body mass scaling in our answer. Marine mammals are among the largest mammals on earth regardless of habitat. Small mammals, generally, demonstrate a much greater urine concentrating ability than larger mammals. Because small mammals have much greater surface-to-volume ratios, they also have higher rates of evaporative and respiratory water loss (Beuchat 1990b). These higher rates of water loss may represent a greater need to conserve water. Even in the fin whale (*Balaenoptera physalus*), which has a body mass one thousand times larger than a human, the maximum urine osmolality recorded (1300 mOsm/kg) is only slightly better than that in a human. Therefore, maximal urine concentrating ability does not appear to scale with body mass across mammals. However, maximal urine concentrating ability in marine mammals may be greater than in terrestrial mammals of similar body size as shown in Figure 7.1.

7.2.2 *Morphological adaptations*

The gastrointestinal tract and the kidneys are primary organs which are responsible for maintaining water and electrolyte homeostasis. In addition, marine mammals have developed specialized structures in the skin and the respiratory system to minimize evaporative water loss. Modifications of these organs and structures contribute to the conservation of body water.

7.2.2.1 Gastrointestinal tract—The entrance for water

The intestines are critical for promoting net water gain in marine mammals because they likely meet a significant portion of their water requirements from pre-formed water in their diets when not fasting (Irving et al. 1935; Smith 1936; Fetcher 1939; Pilson 1970; Depocas et al. 1971; Ortiz et al. 1978; Hui 1981). Water is, generally, absorbed in the small and large intestines. The intestine, especially the small intestine, is longer in cetaceans and pinnipeds than those in terrestrial carnivores (Helm 1983; Goodman-Lowe et al. 2001; Williams et al. 2001). Therefore, elongation of the small intestine may be advantageous for water absorption. Generally, water is absorbed through cell junction of the absorptive enterocyte (the paracellular pathway) in terrestrial mammals (Barrett 2005). However, it was reported recently that the water channel aquaporin-1 (AQP1) is located at the apical membrane of the absorptive enterocyte in the small intestine of bottlenose dolphin (*Tursiops truncatus*) unlike terrestrial mammals (Suzuki 2010). The distribution of AQP1 along the apical membrane potentially contributes to the rapid and effective absorption of water from the lumen into the body across the cellular membrane (the transcellular pathway). These anatomical and histological features of the intestine may facilitate the efficient absorption of water.

In addition to the demand to conserve and obtain free water, marine mammals in a hyperosmotic habitats need to avoid excessive Na^+ ingestion, either from their prey directly or indirectly during the act of eating. Cetaceans possess a very muscular sphincter at the entrance of the esophagus that is well developed and forms a very narrow (3–5 mm) opening leading to the stomach that may function to only allow prey to pass without much seawater (Harrison et al. 1970). Reidenberg (2007) suggested that the tongue of cetaceans may function in squeezing water out of the mouth, minimizing ingestion of saltwater.

7.2.2.2 Kidney—The principal regulator

The kidneys are responsible, among other functions, for the regulation of water and salt balance. Electrolytes, water, and certain metabolic by-products in the plasma are filtered by the kidneys through a series of processes that constitute selective filtration, secretion, and excretion. While the processes are consistent throughout mammals, the anatomy of the kidneys in marine mammals is distinctive from terrestrial mammals. The kidneys of cetaceans and pinnipeds are reniculated and comprised of hundreds to thousands of reniculi, which contain divided cortical tissue and a single medullary pyramid, generally inserted in a single calyx and bundled within the kidney capsule (Ommanney 1932; Bester 1975; Hedges et al. 1979; Vardy and Bryden 1981; Stoskopf and Herbert 1990) (Figure 7.2). In manatees, the kidney is superficially lobulated (Hill and Reynolds 1989) and is elongated and unlobulated in the dugong (*Dugong dugon*) (Batrawi 1953, 1957; Marsh et al. 1978). Similar to cetaceans and pinnipeds, sea otters possess a lobulated kidney and possess slightly greater urine concentrating ability than river otters (Hoover and Tyler 1986). The kidneys in marine mammals are relatively larger than in terrestrial mammals (Beuchat 1996) and may be an adaptation for a deep-diving lifestyle and/or a direct consequence of an enlargement of their body mass, independent of any osmoregulatory burden (Pfeiffer 1997; Maluf and Gassman 1998; Ortiz 2001).

The kidneys consist of a cortex and medulla. Plasma is filtered in the glomeruli in the cortex, and the original urine flows to the proximal tube in the cortex, where electrolytes and organic nutrients including urea are reabsorbed. Subsequently, the urine enters the descending thin loop of Henle in the medulla, where necessary molecules are absorbed and then goes up in the loop of Henle connecting to the collecting tubules in the cortex. Several collecting tubules converge into the collecting duct where water is finally absorbed,

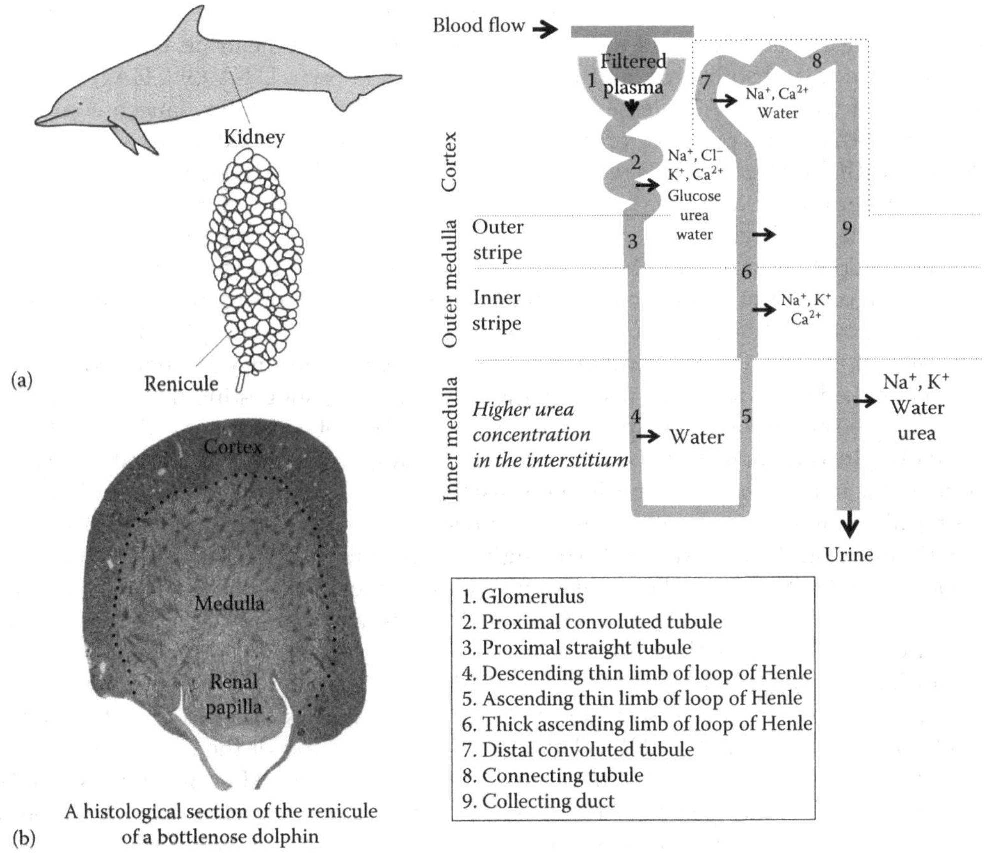

Figure 7.2 **(See color insert.)** Kidney structure. The kidneys of marine mammals are lobulated and composed of hundreds to thousands of renicules (a). Each renicule possesses its independent cortex and medulla that empty into a common renal pelvis (b).

and urine flows from the duct to the bladder to be excreted (Figure 7.2). The renal medulla contains the loop of Henle and the collecting duct that contribute to urine concentrating ability. The *relative medullary thickness* (RMT: the thickness of the medulla relative to the size of the kidney) increases in proportion with urine concentrating ability (Beuchat 1990a) and is quite variable in marine mammals. In cetaceans, the RMT is comparatively high (bottlenose dolphin = 6.0, Baird's beaked whale [*Berardius bairdii*] = 4.9) (Suzuki et al. 2008), whereas in pinnipeds that is much more variable (1.1 in the cape fur seal [*Arctocephalus pusillus*] and bearded seal [*Erignathus barbatus*]) (Sperber 1944; Bester 1975) to 7.5 in the Weddell seal (*Leptonychotes weddellii*) (Vardy and Bryden 1981), indicating the relationship between the RMT and urine concentrating ability may vary among species. The fact that the medulla of the kidney is six times thicker than the cortex in the West Indian manatee (*Trichechus manatus*) suggests that this species possesses an enhanced ability to concentrate urine (Hill and Reynolds 1989; Maluf 1989).

Urea is the end-product of proteins, and simultaneously it is a key molecule for the production of concentrated urine. Accumulation of urea in the renal medulla is a significant contributing factor to urinary concentration. Urea is reabsorbed from urine across the epithelium of the inner-medullary collecting duct and recycled among circulation and the renal ducts. Accumulated urea contributes to increased osmolality in the medullary interstitium resulting in reabsorption of water and the concentration of the urine. Urea transporters (UT) function to reabsorb urea, and two subtypes are known, UT-A (SLC14A2) and UT-B (SLC14A1), of which UT-A2 is essential for urine concentrating ability. Janech et al. (2002) first demonstrated the presence of two UT-A2 transcripts in the kidney of a short-finned pilot whale (*Globicephala macrorhynchus*). The transporters had only one protein kinase C site, different from that of terrestrial mammals, which have two sites, suggesting that unique cellular mechanisms regulate this transporter. Birukawa et al. (2008) reported UT-A2 structures with variation in the number of protein kinase C sites in baleen whales. In addition, plasma urea concentrations are comparatively higher in cetaceans than those in terrestrial mammals (Artiodactyla), which may reflect the carnivorous feeding habits of cetaceans (Birukawa et al. 2005). When compared to the mouse, plasma urea concentration in cetaceans is markedly elevated as a result of lower urea cycling suggesting that the regulation of urea may be an important adaption for the conservation of body water (Miyaji et al. 2010).

Acidic glucolipids include gangliosides and sulfoglycolipids, and the latter may contribute to the maintenance of ionic homeostasis because the expression of sulfoglycolipids is upregulated in the renal-collecting duct in response to increased environmental osmolality (Niimura and Ishizuka 1990). The higher total concentrations of sulfoglycolipids in the kidneys of the Steller sea lion (*Eumetopias jubatus*), melon-headed whale (*Peponocephala elecrta*), and rough-toothed dolphin (*Steno bredanensis*) are thought to contribute to their adaptation to a marine habitat (Nagai et al. 2008). Sulfoglycolipids may also serve as a counter-ion for interstitial ammonium accumulated in the medulla and papilla and may be necessary for ammonium secretion into the collecting duct, resulting in the urinary excretion of H^+ (Stettner et al. 2013). This process is important for the regulation of blood pH and provides evidence for the importance of the kidney in pH in marine mammals.

Immersion in water generally induces *diuresis* (increased excretion of urine) (Šrámek et al. 2000). Water immersion increases cardiac output and arterial pressure (Epstein 1992; Wilcock et al. 2006), and ultimately, the acute distention of arteries induces a baro-reflex that suppresses the anti-diuretic hormone, vasopressin, resulting in a diuresis. This reflex arc is known as the Henry-Gauer reflex (Gauer and Henry 1976). However, this reflex may be permanently suppressed and/or non-existent in marine mammals as an adaptation to conserve water during chronic and/or frequent immersion (Ortiz 2001). During immersion of harbor seal (*Phoca vitulina*), urine flow ceased (Murdaugh et al. 1961b). This dramatic reduction in urine flow is likely the result of greatly reduced *glomerular filtration rate* (GFR: the rate at which a volume of blood passes through the glomeruli) because GFR was reported to decrease with induction of the diving reflex in seals (Ladd et al. 1951; Davis et al. 1983). The immersion- and/or diving-induced decrease in GFR corresponds well with a decrease in blood flow (ml/min) to the kidneys (Zapol et al. 1979). Therefore, the shunting of regional blood flow to the kidneys appears to be the principal hemodynamic alteration that evolved to allow marine mammals to conserve water during diving/immersion.

7.2.2.3 Other organs—Synergies to conserve body water

The epidermis (skin) serves as a boundary between the internal body and the external environment. The cetacean epidermis is composed of lipokeratinocytes that contain not only keratin filaments but also lipid droplets, likely serving as a barrier to a hypertonic environment (Menon et al. 1986). Semi-aquatic and aquatic mammals have high numbers of lipid vesicles in the cells of the *strata granulosum* and the *strata corneum*, which may function as a barrier and possess a rich distribution of Na^+/K^+ exchanger-1 (Meyer et al. 2011). In addition, sweat glands are absent in marine mammals, and this is likely because there is no need for evaporative cooling of the body in aquatic environments (Irving et al. 1935; Ridgway 1972). Water loss through cutaneous evaporation is considered negligible in phocids due to the lack of sweat glands (Lester and Costa 2006). Exceptionally, active sweating from the bare skin of the flippers was observed in the California sea lion (*Zalophus californianus*) under hot condition (Matsuura and Whittow 1974). Regardless, the characteristics of the skin such as low moisture permeability and high lipid content may contribute to the conservation of body water. Indeed, Kjeld (2003) calculated that in the sei whale (*Balaenoptera borealis*) and fin whale that they have no water loss from skin. On the other hand, the skin may serve as an avenue of water flux in Cetacea, although Telfer et al. (1970) concluded the idea was unlikely. Hui (1981) calculated that water flux across skin may account for as much as 70% of total flux in fasting dolphins. Furthermore, delphinids in freshwater may obtain water from the water flux across the skin dependent on the osmotic gradient of the environment (Andersen and Nielsen 1983). However, further studies of a more comprehensive nature are needed to confirm that water flux across the skin is viable in marine mammals, especially in cetaceans.

Respiratory evaporation is another avenue for water loss. In phocids, body water is conserved by cooling the expired air by nasal a *countercurrent heat exchanger* (parallel pipes of flowing fluids at different temperatures in opposite directions to exchange their heat content) resulting in the reduction of evaporative water loss (Huntley et al. 1984; Folkow and Blix 1987). *Respiratory water loss* through evaporation was calculated to be lower in dolphins than in terrestrial mammals of similar body mass (Coulombe et al. 1965). This reduction in respiratory water loss was likely the result of controlling temperature and pressure within the respiratory system and lowering ventilation rate (Coulombe et al. 1965). Aside from conserving water, the respiratory system may also serve as a site of water gain as the air passing through the nasal passage across the nasal mucosa condenses because of the nasal heat exchange mechanism (Depocas 1971). While metabolic water and mariposia accounted for 93% of total daily water influx in experimentally dehydrated hooded seals, respiratory water influx was calculated to contribute as much as 7% of this influx (Alvira-Iraizoz 2014). Skalstad and Nordøy (2000) calculated that the respiratory water intake accounted for approximately 5% of total water intake in the hooded seal (*Cysophora cristata*) and harp seal (*Pagophilus groenlandicus*); however, they noted that the net water flux through respiration is negative, so more water is lost than gained. Nonetheless, the nasal countercurrent heat exchanger abates the conservation of water loss during respiration in marine mammals.

7.2.3 *Source of water*

The issue of freshwater availability that confronts marine mammals is often likened to that of stranded drifters at sea because they cannot access freshwater. However, unlike drifters at sea, the availability of freshwater for marine mammals is much less of a life crisis. Their ability to obtain the requisite amount of freshwater to maintain their life at sea remains a fascinating question.

7.2.3.1 *Diet is the main source of free water*

In general, the available data, with few exceptions, suggest that the vast majority of marine mammals that live in hypertonic habitats do not actively consume water. For the few studies of mammals in freshwater environments, the data would suggest that active drinking is common but not exceptional (Ortiz 2001), and most likely because these species must guard against the potential for overconsumption inducing hyponatremia or hemodilution. Regardless, *pre-formed water* in the diet of marine mammals is a principal source of free water.

Table 7.3 Content of water and electrolytes in various diets of marine mammals

Species	Water (%)	Lipid (%)	Protein (%)	Carbohydrate (%)	Sodium (mg)
Fish					
Pollack	80	0.2	18	0.1	130
Flounder	78	1.3	20	0.1	110
Lizardfish	78	0.8	20	0.8	120
Barracudas	73	7.2	19	0.1	120
Bonito	72	0.5	26	0.1	330
Herring	66	15	17	0.1	110
Mackerel	65–78	3–12	17–21	0.1–0.3	81–140
Sardine	65	14	20	0.7	120
Saury	56	25	19	0.1	130
Invertebrates					
Clam	89	0.5	6	1.8	780
Squid	79–84	0.3–1.5	15–18	0.1–0.4	170–280
Sea urchin	74	4.8	16	3.3	220

Source: Data are from the Food Composition Database of the Ministry of Education, Culture, Sports, Science and Technology in Japan.

The prey of marine mammals such as fish, squid, and a variety of mollusks contain a wide range of water (70%–90% of the body mass) (Table 7.3). Many studies have demonstrated that marine mammals can meet the water requirement to maintain homeostasis primarily from pre-formed water in their diet (Irving et al. 1935; Smith 1936; Fetcher 1939; Pilson 1970; Depocas et al. 1971; Hui 1981). Thus, these studies reinforce the importance of effective and efficient absorption of water from the GI tract as described previously (see Section 7.2.2.1).

In addition, *metabolic water* (the water produced during substrate-level metabolism), from primarily lipid oxidation, is robustly conserved by both the kidney and reduced respiratory loss. For example, for every 100 g of carbohydrate, lipid, and protein metabolized, 55 ml, 107 ml, and 41 ml of metabolic water, respectively, are produced. Baleen whales, some pinnipeds, and dugong consume prey with higher salt content such as squid, krill, and sea-grasses. While these marine mammals may also gain free water contained in their prey, they must also rely on water derived from respiratory metabolism to maintain water balance (Kjeld 2003), which likely reflects the advantage of the relatively high metabolic rates of most marine mammals (Hinga 1979; Williams et al. 2001). In dugong, a strictly marine inhabitant, 69% of their water influx is derived from pre-formed water and the rest from metabolic water since mariposia is not thought to be a significant component of their intake (Lanion et al. 2006).

7.2.3.2 Water balance during fasting

The absolute deprivation of food and water, or fasting, is a natural component of the life history of marine mammals, although the duration varies among species and taxonomic groups. Therefore, the fasting induces an additional burden on their capabilities to maintain water and ion balance during this period. Given that the duration of fasting in many of these mammals is on the order of many months (3–4), most incredible is that dehydration is not a consequence of these fasts such as those in the harp seal and northern elephant seal (*Mirounga angustirostris*) (Ortiz et al. 1978, 2000; Worthy and Lavigne 1982; Castellini et al. 1990; Houser et al. 2001). However, plasma osmolality and Na^+ concentration increased 35% and 7.5%, respectively, at the end of fasting in the gray seal (*Halichoerus grypus*) suggesting some degree of dehydration (Nordøy et al. 1992). In bottlenose dolphin, fasting (days) did not significantly increase plasma osmolality with levels remaining comparable to fed dolphins suggesting that acute periods of food deprivation are inconsequential (Ortiz et al. 2010; Ridgway and Venn-Watson 2010; Venn-Watson et al. 2011). Other than plasma osmolality, the total protein and electrolyte concentrations, *hematocrit* (Hct: the volume percentage of red blood cells in blood) is another clinical indicator of hydration state (Shirreffs 2000). However, its use in marine mammals is not as predictable and indicative. In postweaning northern elephant seal pups, Hct increased 35% over the course of 8 weeks of fasting (Ortiz et al. 2000); however, the other indices of dehydration (plasma osmolality, $[Na^+]$, and total protein) were not altered suggesting that the increase in Hct was likely the result of increased red blood cell volume and not loss of water from the blood (classical dehydration). This is further supported by the fact that the variability in Hct among marine mammals may be attributed to an increase in red blood cell volume associated with their deep-diving ability (Kohin 1998). Therefore, changes in Hct may vary independent of hydration state in deep-diving mammals and may not be a reliable indicator of dehydration (Castellini et al. 1996).

Fasting marine mammals conserve body water by a combination of mechanisms. Fasting also alters renal hemodynamics (blood flow), which in turn may impart an effect on the reabsorption of filtered electrolytes and free water to maintain fluid and electrolyte homeostasis. In fasting pinnipeds, urine osmolality is transiently increased before returning to baseline suggesting that renal mechanisms are dynamically changing in response to fasting duration (Skog and Folkow 1994; Ortiz et al. 1996). Prolonged fasting (2.5 months) in northern elephant seal pups decreases protein catabolism, which reduces the nitrogen (primarily urea) load on the kidneys, GFR, and consequently urine volume (Pernia et al. 1989; Adams and Costa 1993). Lactating elephant seals can prevent urinary water loss during fasting by increasing fractional reabsorption of urea, despite an increase in GFR in the presence of elevated protein catabolism necessary for supporting lactation (Crocker et al. 1998). Furthermore, respiratory evaporative water loss is reduced, resulting in water conservation in fasting northern elephant seals (Lester and Costa 2006). Collectively, these mechanisms comprise a suite of physiological alterations that contribute to the conservation of body water.

Contrary to concentrating urine in fasting pinnipeds, cetaceans excrete diluted urine in response to fasting as an acute phase response (Bentley 1963; Telfer et al. 1970; Hui 1981; Ridgway and Venn-Watson 2010). The oxidation of fat can produce more metabolic water than the catabolism of the other substrates, and fasting marine mammals rely primarily on the metabolic water from the oxidation of their vast fat stores (Worthy and Lavigne 1987). Acute food deprivation in the bottlenose dolphin quickly increased plasma non-esterified fatty acids (NEFAs) suggesting that lipid metabolism is increased to help alleviate the potential for nutritional and osmotic stress (Ortiz et al. 2010). Furthermore, food deprivation also induced metabolic water production in the West Indian manatee as in pinnipeds and cetaceans (Ortiz et al. 1999) suggesting that metabolic water is a significant source of free water across all marine mammals. In addition to the conservation of body water, electrolyte homeostasis is also maintained during fasting. Remarkably, the nearly 3-month long fasts of the northern elephant seal are characterized by the maintenance of plasma electrolyte concentrations, which is likely achieved by increased renal reabsorption of Na^+ and K^+ at the expense of H^+ (Ortiz et al. 2000).

7.2.3.3 *Drinking of seawater and freshwater*

Mariposia, or the voluntary and deliberate consumption of seawater, is not common in marine mammals, despite the fact that the vast majority of these animals can produce more concentrated urine than seawater (Table 7.2). The consumption of seawater in cetaceans is thought to occur incidentally as a function of capturing and eating prey while submerged. Fasting dolphins consumed seawater at rates ranging between 4.5 and 13 ml/kg/day (Telfer et al. 1970; Hui 1981), which are not exceptional and more indicative of volumes more closely related to incidental ingestion than active consumption. In fin and sei whales, which primarily consume crustaceans that have a high salt content, mariposia only contributes to 1%–2% of estimated total water ingestion (Kjeld 2003) and only 2%–3% in bottlenose dolphins (Ridgway 1972) suggesting that mariposia is a very small fraction of water intake across cetaceans. In the harbor seal, seawater ingestion accounted for 9.2% and 7.3% of total water flux in fed and fasted animals, respectively, suggesting that the small volume of seawater intake is the consequence of incidental ingestion and that mariposia is not essential to maintain water balance (Depocas et al. 1971).

Because most marine mammals can concentrate their urine greater than seawater, and thus, potentially gain free water from active consumption, mariposia may not only serve in an osmoregulatory capacity. Mariposia has been reported in pinnipeds inhabiting temperate regions, and thought to counter thermal stress. Seawater drinking has been reported in the Galápagos fur seal (*Arctocephalus galapagoensis*) that inhabit a tropical region, but it is not observed in the Antarctic fur seal (*Arctocephalus gazella*) living in colder regions (Costa and Gentry 1986; Costa and Trillmich 1988). Fasting or dehydration may also induce mariposia. Mariposia was reported in dehydrated harp seals (How and Nordøy 2007) and in a number of fasting otariids (Gentry 1981). Hooded seals with no access to seawater exhibited an increase in blood osmolality and plasma urea, and the values returned to normal when the seals were allowed access to seawater (Storeheier and Nordøy 2001; How and Nordøy 2007). Additionally, dehydrated hooded seals drank on average 1900 ml of seawater a day (Verlo 2012). Skalstad and Nordøy (2000) calculated that hooded and harp seals may drink 300 ± 55 and 900 ± 12 ml/day, respectively, amounting to 14% and 27% of total water turnover, respectively, at low ambient temperature.

The herbivorous dugong, whose habitat is strictly marine, may require more water than carnivorous marine mammals to facilitate the efficient fermentation of sea grasses. Additionally, the anatomy of the dugong kidney suggests that they can drink seawater and gain free water (Lanyon et al. 2006). In contrast, the West Indian manatee that does not drink appreciable volumes of seawater (Ortiz et al. 1999) is found in both hypo- and hyperosmotic environments. The relatively high water turnover rate recorded in dugong suggests the presence of mariposia or sufficiently high metabolic rate to maintain water balance (Lanyon et al. 2006). Although manatees have the high ability to concentrate their urine above seawater (Irvine et al. 1980), water turnover rates are not indicative of active seawater consumption (Ortiz et al. 1999), and it is likely that they need to excrete excessive salt that is the consequence of the ingestion of plants in marine and estuarine environment (Reich and Worthy 2006).

The sea otter may be the only marine mammal that actively consumes seawater to eliminate urea–nitrogen load (Costa 1982). As sea otters can excrete Na^+ and Cl^- in much greater concentrations than that in seawater, they can obtain a net gain in free water by consuming seawater (Costa 1982).

Although the number of marine mammals that inhabit strictly freshwater or have ready access to freshwater environments is small, freshwater drinking is common among those species that have been studied, including pinnipeds that inhabit on ice packs. The harp seal consumes freshwater in the form of ice cubes while in captivity (Renouf et al. 1990). Under captive/experimental conditions, harbor seals maintained at high ambient temperature drank freshwater (Irving et al. 1935) or after ingestion of 1 l of seawater (Albrecht 1950). In the West Indian manatee, water turnover rates were greater in animals maintained in freshwater compared to those in seawater suggesting that freshwater consumption was a significant component of their water flux (Ortiz et al. 1999).

7.2.4 Infusion experiments: Water, electrolytes, and organic molecules

The infusions of water, electrolytes, or organic molecules provide important experimental protocols to help elucidate mechanisms that regulate body water and salts. With respect to osmoregulatory mechanisms, the intravenous infusion of hyper- or hypo-osmotic solutions and the subsequent monitoring of the humoral, excretory, and/or renal responses provide insight to the physiological mechanisms and adjustments to recover homeostasis. Infusion studies with freshwater, hypertonic saline, mannitol, and isotonic gelatin among others have been performed in a variety of marine mammals to help reveal their osmoregulatory capabilities.

7.2.4.1 Loading of hypertonic/hyperosmotic saline

In terrestrial mammals, the ingestion of seawater typically induces excretion of excess salts and water resulting in dehydration without the ability of producing urine that is more concentrated than the ingested seawater. Excessive salt intake (sodium chloride; NaCl) has consequences in the form of diarrhea, vomiting, edema, and elevated arterial blood pressure (Boyd et al. 1966; Meneely and Battarbee 1976; Thompson 2011). Generally, marine mammals are thought to be resistant to seawater ingestion because they have adapted to living in the marine habitats.

Hypertonic saline infusion of various concentrations induces urine production with excretion of Na^+ and Cl^- in bottlenose dolphin (Fetcher and Fetcher 1942) and seals (Albrecht 1950; Bradley et al. 1954; Tarasoff and Toews 1972; Hong et al. 1982; Skog and Folkow 1994; Storeheier and Nordøy 2001; Ortiz et al. 2002; How and Nordøy 2007). Although these marine mammals can produce concentrated urine with a relatively high osmolality (Bester 1975; Costa 1982; Maluf 1989; Ortiz 2001), they do not usually concentrate Na^+ and Cl^- above that of seawater. At least in pinnipeds, the tolerance to salt loading seems to vary among species. For example, in the harbor seal, vomiting and diarrhea were induced by the loading of 2.6 ml seawater/kg of body weight and the continuous intravenous infusion of hyperosmotic saline resulted in death in two harbor seals (Albrecht 1950). In the California sea lion, similar consequences were reported with the loading of 1.5 ml/kg (Ridgway 1972). However, the acute intravenous bolus infusion of iso- and hypertonic saline with equivalent amounts of Na^+ (310 mEq) had no apparent consequences on northern elephant seal pups suggesting that some pinnipeds are resistant to excessive salt loading or that the differences in the experimental protocols accounted for the disparate physiological responses (Ortiz et al. 2002). Similarly, in the bottlenose dolphin, given 4 ml/kg (Ridgway 1972) or 4 l (Ridgway and Venn-Watson 2010) of seawater by gavage did not demonstrate any clinical signs of seawater toxicosis. In the dolphins, seawater ingestion initially induced a diuresis with high urine concentrations of Na^+, Cl^-, and K^+, with a relatively hyperosmotic urine continuously produced (Ridgway and Venn-Watson 2010). Collectively, these observations suggest that the tolerance to a hyperosmotic load is species-specific among marine mammals and depends on the extent of the Na^+ load.

Infusion of mannitol, which can increase osmolality of a solution independent of changes in electrolyte concentrations, has also been performed in some marine mammals. Mannitol infusion in gray seals decreased plasma electrolyte and urea concentrations without changing plasma osmolality (Skog and Folkow 1994) indicative of a hemodilution effect.

In addition, the excretion of osmotic solutes following infusion of mannitol required 2.5% of total body water in these seals suggesting that they regulate plasma osmolality by rapid excretion of diluted urine (Skog and Folkow 1994).

7.2.4.2 *Loading of isotonic water*

Gelatin is proteinaceous matrix void of electrolytes that makes colloid particles in a solution, thus, its infusion can be used to expand extracellular volume. Excessive colloid particles in the blood will increase plasma osmolality resulting in an osmotic gradient that favors the movement of water from the intracellular into the extracellular space. The infusion of gelatin in harbor seal increased urine volume and osmolality, excretion, resulting from an increase in GFR without changing plasma osmolality (Murdaugh et al. 1961a). This observation is significant because it suggests that plasma volume is regulated, at least in part, by volume receptors (likely vasopressinergic) that lead to a diuresis (as would be expected by suppression of vasopressin).

7.2.4.3 *Loading of freshwater*

All mammals require access to freshwater in some form to survive; however, this can be especially challenging for animals living in a strictly marine habitat or arid environment. Paradoxically, the excessive loading of solute-free water can also be intoxicating, resulting in hemodilution, hyponatremia, pulmonary edema, coma, and ultimately death (Gardner 2002). Thus, the intake of freshwater needs to be properly regulated as with the ingestion of electrolytes. From an experimental perspective, freshwater loading causes diuresis to maintain the appropriate plasma osmolality.

In bottlenose dolphin, the effects of freshwater ingestion seem to be incongruent. Malvin and Rayner (1968) reported no change in urine osmolality and urine flow following the loading of 4 L of freshwater. Whereas Ridgway and Venn-Watson (2010) described that 2–4 L of freshwater ingestion lead to a diuresis, associated with decreases in plasma and urine osmolalities and electrolyte concentrations (sodium and chloride), indicative of a state of hypocholoremia, hyponatremia, and hemodilution. In pinnipeds, freshwater ingestion also induces diuresis (Albrecht 1950; Ladd et al. 1951; Bradley et al. 1954; Tarasoff and Toews 1972; Hong et al. 1982; Skog and Folkow 1994). The infusion of freshwater in pinnipeds induced: (1) an increase in urine flow and/or volume (Bradley et al. 1954; Tarasoff and Toews 1972; Ortiz et al. 2002); (2) a decrease in urinary osmolality (Ladd et al. 1951; Tarasoff and Toews 1972; Hong et al. 1982; Ortiz et al. 2002); (3) a decrease in excreted Na^+ and K^+ (Bradley et al. 1954); and (4) an increase in the excretion of Na^+, K^+, Cl^-, and/or urea (Hong et al. 1982; Ortiz et al. 2002). Changes in GFR in response to freshwater loading are also inconsistent among studies in seals. For example, in the harbor seal, no changes were reported (Bradley et al. 1954; Murdaugh et al. 1961a), whereas in the Baikal seal (*Pusa sibirica*), harbor seal, and ringed seal (*Pusa hispida*) (Ladd et al. 1951; Hong et al. 1982) and fasting northern elephant seal pups (Ortiz et al. 2002), GFR increased with freshwater loading. These data indicate that freshwater loading results in increase in urine volume and fractional clearance of water and electrolytes, and decrease in water reabsorption in the renal tubule. As the consequences of excessive freshwater ingestion (infusion) were previously mentioned, it is of note that harbor seals died from water intoxication induced by excessive loading (Ladd et al. 1951).

Similarly, in manatees, which inhabit both freshwater and marine habitats, freshwater exposure is associated with reduced plasma and urine Na^+, Cl^-, and osmolality (Ortiz et al. 1998) while water flux is nearly twice that for animals in fresh water versus saltwater (Ortiz et al. 1999).

Collectively, freshwater loading causes diuresis in marine mammals. However, as they cannot excrete urine with a lower osmolality than that of plasma, this diuresis leads to a loss of electrolytes, resulting in hyponatremia, hypocholoremia, and hemodilution. Also, the few freshwater studies that exist suggest that marine mammals may not have especially well-developed mechanisms to excrete excessive water efficiently while conserving electrolytes. If so, this provides some insight to the evolution of physiological mechanisms necessary to tolerate life without access to freshwater.

7.2.4.4 *Loading of organic molecules*

The mechanisms necessary to maintain water balance are changing continuously in response to the status and content of food consumption because the constitution of diets (organic nutrients, electrolytes, pre-formed water, etc.) varies. High protein diets promote urine concentrating ability and GFR in human and other mammals to excrete metabolites (primarily nitrogenous) arising from the excessive protein (Epstein et al. 1957; Bouby et al. 1988; Brändle et al. 1996). As pinnipeds and cetaceans are carnivorous, the consequence is the need to excrete the end-product of protein (urea/ammonia) and excessive electrolytes efficiently. Generally, in marine mammals, feeding leads to an increase in GFR, renal plasma flow, and urine flow rate (Hiatt and Hiatt 1942; Schmidt-Nielsen et al. 1959; Malvin and Rayner 1968; Ridgway 1972). These changes increase urine production, which can lead to an increase in the excretion of Na^+, K^+, and urea/ammonia (Schmidt-Nielsen et al. 1959; Malvin and Rayner 1968; Ridgway 1972). Ridgway and Venn-Watson (2010) demonstrated the combined loading of seawater and protein to bottlenose dolphin increased plasma osmolality, urine flow, and urine concentrations of Na^+, K^+, and urea. These changes were enhanced and lasted longer than those in the group infused with seawater alone. These data suggest that the ability to concentrate urine above seawater in marine mammals is an important mechanism adapted to tolerate a carnivorous feeding behavior in a marine habitat.

7.2.5 *Hormonal control*

Hormones are important extrinsic factors that regulate signaling pathways at the cellular level to help maintain homeostasis of water and electrolyte balance. Generally, in terrestrial mammals, a suite of vasoactive hormones such as arginine vasopressin, angiotensin II, atrial natriuretic peptide, vasoactive intestinal peptide, and urotensin II contribute to the acute (minutes–hours) perturbations in water and electrolyte balance, whereas steroids (aldosterone and cortisol), prolactin, and growth hormones are more involved in longer-term responses (McCormick and Bradshaw 2006). In marine mammals, studies have focused more on the acute responses of the hormones to hyperosmotic water ingestion or fasting than on the chronic effects.

7.2.5.1 *Arginine vasopressin*

Thirst, primarily in response to the low availability of freshwater, stimulates a cascade of neural and humoral responses that culminates in the prevention of water loss by concentrating the urine. In mammals, the volume of excreted water is regulated tightly by the octapeptide, *arginine vasopressin* (AVP; previously, antidiuretic hormone [ADH]). The secretion of AVP from the posterior pituitary is primarily in response to an increase in plasma osmolality or plasma volume following the stimulation of osmo- or baroreceptors, respectively (Dunn et al. 1973; Robertson et al. 1976). The binding of AVP to its receptor (V2R) in the principal cells in the collecting duct initiates the insertion of water channels (aquaporins) into the apical side of the cell. The aquaporins facilitate the absorption of water down its concentration gradient that is achieved primarily by the concentrations of urea and sodium (Coleman et al. 2000; Frøkiær et al. 2003) (Figure 7.3). However, the traditional function of AVP in marine mammals remains controversial and not well-defined.

In pinnipeds, AVP potentially functions as an anti-diuretic as in terrestrial mammals. Administration of synthetic AVP (pitressin) in harbor seals decreased urine flow and increased urinary electrolyte concentration (Bradley et al. 1954) and increased urine osmolality and osmotic clearance (Page et al. 1954) suggesting that AVP induced a typical

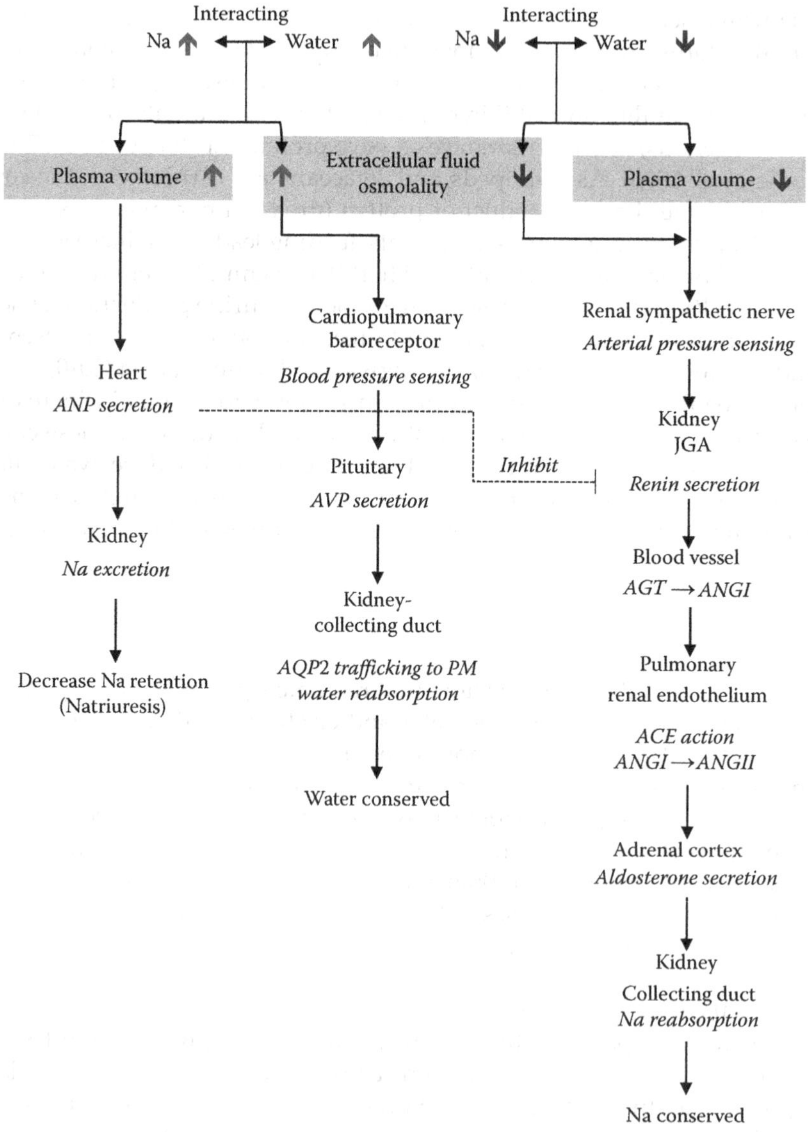

Figure 7.3 **(See color insert.)** General schematics of the functions of osmoregulatory hormones. (ACE, angiotensin I-converting enzyme; AGT, angiotensinogen; ANG, angiotensin; ANP, atrial natriuretic peptide; AVP, arginine vasopressin; juxta-glomerular apparatus cells in the kidney; PM, plasma membrane.)

functional response resulting in increased water reabsorption. Fasting induced an increase in plasma AVP concentration and urine osmolality that were positively correlated, and decreased urine flow in Baikal and ringed seals (Hong et al. 1982), consistent with AVP function. Furthermore, in the gray seal, fasting increased urine and plasma osmolality along with an increase in plasma AVP (Skog and Folkow 1994). Collectively these data suggest that the observed increases in urine osmolality are the result of AVP-mediated increase in tubular water reabsorption. In the West Indian manatee, a significant correlation between plasma vasopressin and osmolality was detected further suggesting that the typical regulation of AVP secretion is conserved among these marine mammals (Ortiz et al. 1998).

Conversely, in fasting northern elephant seal pups increasing urine osmolality was associated with decreasing plasma AVP in the presence of constant plasma osmolality suggesting that increased urine concentration was achieved independent of AVP (Ortiz et al. 1996). Furthermore, in fasting elephant seal pups, an acute bolus infusion of AVP induced an osmotic diuresis with no change in free water clearance suggesting that acutely elevated AVP does not have a typical mammalian function (Ortiz et al. 2003). However, in response to an iso- and hyperosmotic saline challenge in fasting elephant seal pups, urinary AVP excretion and free water clearance were inversely correlated (i.e., increased AVP was correlated with an increase in free water reabsorption) suggesting that under these stimulatory conditions AVP possesses a typical mammalian function (Ortiz et al. 2002). Collectively, these data in fasting elephant seal pups suggest that static, chronic measures of plasma AVP and osmolality are not sufficient indicators of AVP function and that acute challenges may better elucidate the functionality of AVP in marine mammals.

Additionally, circulating concentrations of AVP often fail to be directly related to plasma osmolality in cetaceans. AVP is synthesized and secreted into circulation in dolphins as would be expected for any other mammal; however, plasma AVP concentrations in dolphins are relatively low for mammals (Malvin and Rayner 1968; Malvin et al. 1971; Ortiz et al. 2000, 2010). Suzuki et al. have also detected a AVP-like molecule by highly sensitive matrix-assisted laser desorption/ionization-time of flight mass spectrometry from the pituitary of a bottlenose dolphin; however, plasma AVP levels were not correlated with either plasma or urine osmolality (personal observation). Plasma AVP concentrations were also not correlated with urine flow in fasting bottlenose dolphins suggesting that AVP does not significantly contribute to the regulation of water retention (Malvin et al. 1971). While a postprandial increase in plasma NaCl levels were associated with an increase in urinary AVP levels suggesting that AVP secretion at least is regulated in a typical manner (Ballarin et al. 2011), the anti-diuretic function of AVP is still inconclusive in cetaceans.

7.2.5.2 Renin–angiotensin aldosterone system

Renin–angiotensin aldosterone system (RAAS) is important for the control of sodium excretion and blood pressure. As shown in Figure 7.3, *renin* secretion from the juxta-glomerular (JG) cells in the kidney is stimulated by renal sympathetic nerve activity in response to changes in renal arterial pressure or reduced delivery of tubular Na^+. Conversely, renin is suppressed by an increase in tubular Na^+ delivery to the JG cells. Secreted renin acts to convert circulating *angiotensinogen* to *angiotensin I*, which is subsequently converted to *angiotensin II* (Ang II) by the *angiotensin-converting enzyme* (ACE) (Ichikawa and Harris 1991). In turn, Ang II promotes the secretion of *aldosterone* from adrenal gland, and aldosterone then stimulates the renal reabsorption of Na^+ distal tubule and collecting duct resulting in a decrease in urinary Na^+ excretion (Funder 1993; Eaton and Pooler 2009).

In marine mammals, RAAS is present and appears to possess a typical mammalian function. Eichelberger et al. (1940) confirmed the vasoconstrictive action of renin extracted from the bottlenose dolphin in dogs. A positive correlation between plasma renin activity (PRA; an indicator of Ang I generation) and plasma aldosterone concentration was observed in the California sea lion (Malvin et al. 1978), fasting elephant seal pups (Ortiz et al. 2000), the bottlenose dolphin (Malvin et al. 1978), and manatees (Ortiz et al. 1998). When sodium availability is reduced, the sensitivity of RAAS appears to be increased in the case of elephant seals (Ortiz et al. 2000). Aldosterone concentration correlated negatively to water efflux rate in fasting elephant seals suggesting contribution of RAAS to water conservation during fasting (Ortiz et al. 2006). Because manatees that inhabit freshwater have reduced availability to Na^+ (Best 1981; Ortiz et al. 1999), manatees are more sensitive to RAAS than in pinnipeds and cetaceans (Ortiz 2001). Collectively, RAAS is a significant endocrine mechanism in the regulation of water and electrolytes in those marine mammals studied.

7.2.5.3 Natriuretic peptides

The family of *natriuertic peptides* consists of *atrial natriuertic peptide* (ANP), *brain natriuertic peptide* (BNP), and *C-type natriuretic peptide* (CNP), and the peptides defend against excessive salt and water retention, and also influence renal hemodynamic and direct tubular actions resulting in diuretic and natriuretic actions (Levin et al. 1998). ANP is secreted mainly from the atrial wall in response to increased cardiac pressure and distention (Stanton 1991). ANP promotes natriuresis by increasing renal filtration, inhibiting Na^+ absorption by suppressing renin and aldosterone release and antagonizing the actions of Ang II (Light et al. 1989). While ANP may be thought to be a highly functional and relevant hormone in marine mammals because of the abundance of Na^+ in their environment, only a few reports of changes in natriuretic peptides and their potential osmoregulatory function exist.

Precursors of ANP and BNP were cloned in the Pacific white-sided dolphin (*Lagenorhynchus obliquidens*), bottlenose dolphin, and Dall's porpoise (*Phocoenoides dalli*), and their presence in the plasma of bottlenose dolphins were also detected (Naka et al. 2007). Zenteno-Savin and Castellini (1998a) reported plasma concentrations of an *ANP-like* substance in six pinniped species. Sleep apnea, consistent with how animals hold their breath during diving, increased plasma ANP concentration in the northern elephant and Weddell seals suggesting that the bradycardia associated with breath-holding decreased cardiac output and increased intra-cardiac pressure (Zenteno-Savin and Castellini 1998b). Urinary ANP excretion was more than doubled in response to a hyperosmotic saline infusion in fasting elephant seal pups, but levels were not significantly increased with an iso-osmotic infusion (Ortiz et al. 2002). However, only the hyperosmotic infusion was associated with a sustained elevation in plasma Na^+ suggesting that ANP is responsive to the typical mammalian stimulus in pinnipeds (Ortiz et al. 2002). Further studies to ascertain the functional relevance of ANP and the other natriuretic peptides in marine mammals are warranted.

7.3 Toolbox

7.3.1 General tools

The large body masses and strictly aquatic life styles in many groups (cetaceans and sirenians) of marine mammals make investigations into the osmoregulatory mechanisms in these animals extremely challenging. Early studies were limited to comparisons of hormones and electrolytes among different species as well as descriptions of electrolyte responses and changes in water flux rates to exogenous manipulations. Descriptions of the osmoregulatory mechanisms of marine mammals were based on inferences from these comparative studies and our existing knowledge of these mechanisms gleaned from detailed studies in humans and common laboratory models. These studies that measured primarily plasma and urine concentrations of variables used commonly employed techniques and methodologies such as antibody-based (radioimmunoassays followed by enzyme-linked immunosolvent assay) and photometric methods. Furthermore, early histological analyses of the kidneys of marine mammals employed traditional methods commonly used in other vertebrates. The advent of technological advancements has provided the use of time-resolved fluorescent immunoassay (TR-FIA), molecular techniques such as qRT-PCR and Western blotting, kinetic enzyme activity assays, and immunohistochemistry. While none of these techniques and methods is considered especially advanced in today's laboratory, obtaining fresh or *in vivo/in situ* renal tissue or gaining access to the kidney to make real-time renal functional studies is extremely challenging. Therefore, assessing the renal responses as measured by changes in urinary variables to exogenous perturbations of water and electrolytes in addition to the quantification of expressions of renal genes and proteins (most likely from postmortem samples) remain principal methodologies to further elucidate the renal mechanisms. However, inferences from basic renal studies in more traditional mammalian models will continue to be important to help interpret discoveries made in marine mammals.

7.3.2 *Isotopic dilution*

The use of *isotopic dilution* to estimate total body water (TBW) pool size, *water turnover rates* (r_{H_2O}), and energetics has been applied across a broad range of animals and has provided a significant tool for the study of water and energy metabolism (kinetics) (Hevesy and Hofer 1934; Lifson and McClintock 1966; Lifson et al. 1997; Nagy and Costa 1980; Schoeller et al. 1986; Speakman 1997; Ortiz et al. 1999). While this technique requires the acceptance of a number of assumptions and the validation for its specific application, and has some degree of error of estimation (Lifson and McClintock 1966; Culebras and Moore 1977; Nagy and Costa 1980; Wong et al. 1987; Speakman et al. 1993; Schoeller and Hnilicka 1996), it continues to be well received and widely used technique. The measurement of TBW is based on the fact that both of the hydrogen atoms of H_2O are completely exchangeable with isotopic H_2O labeled with either *deuterium* (D or ^2H) or *tritium* (^3H) (Culebras and Moore 1977). While deuterium is a stable, non-radioactive isotope of hydrogen (^1H), ^3H is radioactive; regardless, both behave exactly as the hydrogens of water, and thus, serve as ideal tracers for studying the kinetics of water within the body. The excellent work of the earlier investigators (H. Hevesy, E. Hofer, N. Lifson, R. McClintock, F.D. Moore, K.A. Nagy, D.A. Schoeller, J.R. Speakman, W.W. Wong, and others) helped establish the framework for applying this technique in marine mammals. The first application of isotopic dilution to measure TBW and r_{H_2O} was performed in fasting northern elephant seal pups by Ortiz et al. (1978). While the use of ^{24}Na in delphinids was reported in 1970, this technique was not performed in the animals directly but rather added to their tank water to track the appearance of Na^+ in the animals after 24 h (Telfer et al. 1970). Nonetheless, this work provided a useful and alternative application of isotopic dilution for studying water flux in a group of marine mammals. Soon after the work of Ortiz et al. (1978), a number of papers employing the use of isotopic dilution to measure water kinetics and metabolism emanated from the team at the University of California, Santa Cruz (Costa et al. 1986, 1989;

Worthy et al. 1992; Crocker et al. 1998; Houser et al. 2001) that helped establish the foundation for its use in other labs studying similar interests in marine mammals (i.e., Ortiz et al. 1999; Champagne et al. 2005; Houser et al. 2007).

In theory, the application of the technique is relatively easy now that it has been well established. The biggest challenges are the cost of the isotopes, especially deuterium and ^{18}O (if studying energetics by doubly labeled water) and, if the animals are very large, handling the animals to obtain blood or tissue for subsequent analyses. In practicum, the dilution, or equilibration, of the isotope with the body's total water pool, or compartment, provides a relatively accurate estimate of the actual TBW pool. In general, an initial blood sample is taken prior to administration of the isotope to determine background levels of the isotope being used since many of the isotopes are naturally present in the animal's environment. If background levels are detected, they should be subtracted from the subsequent levels measured post-dosing. Animals then receive a mass-specific dose of the isotope, thus weighing the animals prior to dosing is necessary and can provide an additional challenge depending on size and temperament of the animal. For these reasons, sedation of the animal, especially with most marine mammals is necessary, and thus, can induce additional expenses. To ensure accuracy of the method, the vehicle used for dosing should be pre-weighed to the greatest degree of accuracy of the scale and again after dosing to accurately calculate the actual amount of isotope infused. Furthermore, route of dosing (i.e., gavage or intravenous) should also be flushed with sterile saline to further ensure complete administration of the isotope. After dosing of the isotope, subsequent blood (or urine) samples are obtained at specific and well-defined intervals to develop a dilution (or disappearance) curve. If using an isotope of hydrogen in the form of water, the labeled water must then be extracted from the blood samples (i.e., as demonstrated in Ortiz et al. 1978) for later quantification of the isotope. The labeled water is lyophilized from the whole blood samples and a freeze trap method is common (Ortiz et al. 1978; Byers 1979). The labeled water is now ready for measurement (i.e., D_2O can be measured by infrared analyzer or ^3H by scintillation). Once the values from each sample at each time point have been obtained, *the instantaneous dilution space* (IDS) of the isotope is calculated as the y-intercept (T_0) from a multipoint (depending on the number of post-dosing samples) regression of natural log of the isotope concentration versus time (Ortiz et al. 1978; Byers and Schelling 1986; Ortiz et al. 1999). Isotopic dilution space (usually in liters) is calculated as: IDS = $D/\mathrm{CF}*[T_0]$, where D is the administered dose of isotope, CF is the correction factor for the difference in the mass of unlabeled water versus that of the labeled water (i.e., 1.105 is the correction for the density of D_2O versus unlabeled H_2O), and $[T_0]$ is the concentration of the isotope at time zero, or the IDS (Byers and Schelling 1986; Ortiz et al. 1999). The IDS provides an estimate of TBW. Water turnover rate (r_{H_2O}) can then be estimated from the product of TBW and the slope (K) of the multi-point regression constructed from the dilution curve. In cases where animals lose body mass during the study period, r_{H_2O} can be estimated using the equations presented by Nagy and Costa (1980) that account for changes in the size of the water pool since TBW scales with body mass. The isotopic half time ($t_{1/2}$) of the isotope in the body pool is calculated as $0.693/K$, where 0.693 is a constant.

While a vast majority of tracer studies in marine mammals assess the dilution of a single- or dual-label isotope technique, isotopic analysis by nuclear magnetic resonance (NMR) provides a very important and useful alternative approach. The cost and availability of an NMR may be problematic and challenging its use has been successfully deployed in the study of glucose metabolism in elephant seal pups (Champagne et al. 2012). Spectral isotopic analysis by NMR allows for the simultaneous determination of the enrichment of isotopes at multiple atomic positions. The advantage of this technique is that the positional tracer enrichment can be used to determine the isotopic flux within multiple metabolic pathways simultaneously during a single tracer study (i.e., Champagne et al. 2012).

7.3.3 Genome analysis

While inferences of functional relevance based on genetic analyses such as comparing genomes among species requires a precautionary approach, the extensive information obtained from genome-scale analyses should be appreciated. The molecular analyses that allow for comparisons of gene sequences is indispensable and can provide insight to the nature and extent of selective pressures that contributed to the evolution of certain osmoregulatory mechanisms. Such studies have been performed in cetaceans focusing on genes related to proteins that contribute to osmoregulation. In 11 cetacean species, genes coding for proteins involved in osmoregulation demonstrated positive selection with functional changes in their amino acid sequences (Xu et al. 2013; Yim et al. 2014). Positive selection of AQP2 and urea transporter-A (SLC14A2) in various cetaceans suggests that the evolution of these genes was an important event in the development of enhanced capacity for water and urea transport in the renal tubules (Xu et al. 2013) that can contribute to increased urine concentrating ability. A series of positively selected amino acid residues identified in angiotensinogen and ACE (two principal proteins of RAAS) of cetaceans was described suggesting that the evolution of RAAS was a significant adaptation for the maintenance of water and electrolyte balance in response to a hyperosmotic environment (Xu et al. 2013). Furthermore, functional changes in the sequences of five genes encoding for RAAS proteins were reported in cetaceans (Yim et al. 2014). Nery et al. (2013) demonstrated that genes related to proteins facilitating renal development (SMAD1, NPNT, LEF1, SERPINF1, AQP2) are positively selected in bottlenose dolphins when compared to the cow lineage. Collectively, these data suggest that the radiation of mammals to a marine environment was associated with the evolution of genes encoding for proteins related to osmoregulation. While the comprehensive, panoramic investigations of genome-wide analyses provide a wealth of information, inferences of physiological mechanisms based on such discoveries need to be made with the appropriate precautions and confirmed by functional studies.

7.4 Unresolved questions

Plasma osmolality of marine mammals is maintained at relatively higher levels, attributed to higher concentrations of plasma electrolytes and urea. The difference in average plasma osmolality between some terrestrial and marine mammals can be as much as 75 mOsm/L, and the average concentration of plasma Na^+ (155 mEq/L) for marine mammals would be indicative of dehydration in humans. What osmoregulatory mechanisms have marine mammals evolved to facilitate resistance to the detriments of such elevations in the osmotic and electrolyte content of their extracellular fluid? While the effects of ion channels and organic molecule transporters, organic anions and cations, and the kinetics of intra- and extracellular water all likely contribute, research in these areas in marine mammals under varying conditions are scarce or non-existent. As Ridgway and Venn-Watson (2010) have suggested, the end-product of protein metabolism (i.e., urea and maybe ammonia) to regulate plasma osmolality via the concentration of urine is an relatively unexplored area of interest in marine mammals, and further investigation along these lines has the potential to reveal very interesting and unique discoveries. The specific contribution of urea to osmoregulation remains unclear in marine mammals. Regarding the hormonal control of osmoregulation, their longer-term responses to guard against excess salt environments are not well-defined, although the many systems must have evolved to allow for their tolerance to an aquatic lifestyle of varying osmotic habitats. While many challenges exist that limit the capacity to fully elucidate many unresolved questions of the osmoregulatory abilities of marine mammals, further investigation in this area is fully warranted and encouraged.

Glossary

Aldosterone: A steroid hormone secreted by the adrenal cortex that primarily regulates the reabsorption of Na^+ in the kidney and colon.

Angiotensin II: An eight amino acid peptide cleaved from angiotensin I by angiotensin-converting enzyme that primarily regulates blood pressure and the adrenal secretion of aldosterone.

Angiotensinogen: An alpha-2 globulin produced by the liver and serves as the precursor protein for the formation of angiotensin I.

Arginine vasopressin: A nine amino acid peptide secreted by neurohypophysis (posterior pituitary) that primarily regulates the reabsorption of free water in the collecting duct of the renal tubule in mammals.

Atrial natriuretic peptide: A 28 amino acid peptide produced by the atrium that functions primarily to induce urinary Na^+ excretion (natriuresis).

Concentration gradient: From the compartment with the lower concentration of dissolved particles to the compartment with the greater concentration.

Countercurrent heat exchanger: A design in which fluids at sufficiently different temperatures flow closely in opposite directions to facilitate heat exchange down the temperature gradient.

Diuresis: Increased urine output.

Glomerular filtration rate: The rate at which a volume of blood passes through the glomeruli.

Hemodilution: The dilution or decreased concentration of the plasma and its constituents.

Hyperosmotic: Increased osmotic content of a solution.

Instantaneous dilution space: The calculated estimate of the space or volume into which a particle would be diluted at the moment the particle is introduced into that space. Defined by the *y*-intercept of a multi-point regression during isotopic dilution studies.

Isotopic dilution: A technique to study the kinetics of an isotopically labeled molecule that emulates the endogenous, unlabeled molecule. The rate at which an isotope is diluted in a compartment estimates the turnover of the endogenous, unlabeled molecule.

Mariposia: The voluntary and deliberate consumption of seawater.

Metabolic water: The water produced during substrate-level metabolism.

***Milieu interieur*:** The internal environment, in reference to Claude Bernard's description of the physiological control of homeostasis.

Osmotic pressure: The pressure produced by the concentration of dissolved particles in solution.

Pre-formed water: The water that occurs naturally in food, for example.

Relative medullary thickness: The thickness of the medulla relative to the size of the kidney.

Reniculi: Refers to a specific design of the mammalian kidney in which the entire kidney is compartmentalized into discrete, independent functional units (cortex and medulla) that empty into a common renal pelvis. Externally the kidney appears *lobulated* where each lobe in cetaceans, pinnipeds, otters, and polar bears is a reniculus and collectively the kidney is referred as multireniculate.

References

Adams, S.H. and D.P. Costa. 1993. Water conservation and protein metabolism in northern elephant seal pups during the postweaning fast. *Journal of Comparative Physiology B* 163(5): 367–373.

Albrecht, C.B. 1950. Toxicity of seawater in mammals. *American Journal of Physiology* 163(2): 370–385.

Alvira-Iraizoz, F. 2014. Sampling site and potential errors in estimating total body water and water turnover rate in fasting hooded seals (*Cystophora cristata*). Master thesis, University of Tromsø, Tromsø, Norway.

Andersen, S.H. and E. Nielsen. 1983. Exchange of water between the harbor porpoise, *Phocoena phocoena* and the environment. *Experientia* 39(1): 52–53.

Ballarin, C., L. Corain, A. Peruffo, and B. Cozzi. 2011. Correlation between urinary vasopressin and water content of food in the bottlenose dolphin (*Tursiops truncatus*). *Young* 500(17): 9–14.

Barrett, K.E. 2005. *Water and Electrolyte Absorption and Secretion. Gastrointestinal Physiology*. New York: McGraw-Hill, pp 79–100.

Batrawi, A. 1953. The external features of the dugong kidney. *Bulletin of the Zoological Society of Egypt* 11: 12–13.

Batrawi, A. 1957. The structure of the dugong kidney. *Publications of the Marine Biological Station Al-Ghardaga* 9: 51–68.

Bentley, P.J. 1963. Composition of the urine of the fasting humpback whale (*Megaptera nodosa*). *Comparative Biochemistry and Physiology* 10(3): 257–259.

Best, R.C. 1981. Foods and feeding habits of wild and captive sirenia. *Mammal Review* 11(1): 3–29.

Bester, M.N. 1975. The functional morphology of the kidney of the cape fur seal, *Arctocephalus pusillus (Schreber)*. *Modogua Series* 2(4): 69–92.

Beuchat, C.A. 1990a. Body size, medullary thickness, and urine concentrating ability in mammals. *American Journal of Physiology* 258(2): R298–R308.

Beuchat, C.A. 1990b. Metabolism and the scaling of urine concentrating ability in mammals: Resolution of a paradox? *Journal of Theoretical Biology* 143(1): 113–122.

Beuchat, C.A. 1996. Structure and concentrating ability of the mammalian kidney: Correlations with habitat. *American Journal of Physiology—Regulatory Integrative and Comparative Physiology* 40(1): R157–R179.

Birukawa, N., H. Ando, M. Goto, N. Kanda, L.A. Pastene, H. Nakatsuji, H. Hata, and A. Urano. 2005. Plasma and urine levels of electrolytes, urea and steroid hormones involved in osmoregulation of cetaceans. *Zoological Science* 22(11): 1245–1257.

Birukawa, N., H. Ando, M. Goto, N. Kanda, L.A. Pastene, and A. Urano. 2008. Molecular cloning of urea transporters from the kidneys of baleen and toothed whales. *Comparative Biochemistry and Physiology Part B: Biochemistry and Molecular Biology* 149(2): 227–235.

Boyd, E.M., M.M. Abel, and L.M. Knight. 1966. The chronic oral toxicity of sodium chloride at the range of the LD50 (0.1 L). *Canadian Journal of Physiology and Pharmacology* 44(1): 157–172.

Bouby, N., M.M. Trinh-Trang-Tan, D. Laouari, C. Kleinknecht, J.P. Grünfeld, W. Kriz, and L. Bankir. 1988. Role of the urinary concentrating process in the renal effects of high protein intake. *Kidney International* 34(1): 4–12.

Bradley, S.E., G.H. Mudge, and W.D. Blake. 1954. The renal excretion of sodium, potassium and water by the harbor seal (*Phoca vitulina* L.): Effect of apnea; sodium, potassium and water loading; pitressin; and mercurial diuresis. *Journal of Cellular and Comparative Physiology* 43(1): 1–22.

Brändle, E., H.G. Sieberth, and R.E. Hautmann. 1996. Effect of chronic dietary protein intake on the renal function in healthy subjects. *European Journal of Clinical Nutrition* 50(11): 734–740.

Byers, F.M. and G.T. Schelling (1986). Evaluation of deuterium oxide dilution systems for estimating body composition of beef cattle. PR—Texas Agricultural Experiment Station.

Castellini, J.M., M.A. Castellini, and M.B. Kretzmann. 1990. Circulatory water concentration in suckling and fasting northern elephant seal pups. *Journal of Comparative Physiology B* 160(5): 537–542.

Castellini, J.M., H.J. Meiselman, and M.A. Castellini. 1996. Understanding and interpreting hematocrit measurements in pinnipeds. *Marine Mamma. Science* 12(2): 251–264.

Champagne, C.D., D.S. Houser, and D.E. Crocker. 2005. Glucose production and substrate cycle activity in a fasting adapted animal, the northern elephant seal. *Journal of Experimental Biology* 208(5): 859–868.

Champagne, C.D., D.S. Houser, M.A. Fowler, D.P. Costa, and D.E. Crocker. 2012. Gluconeogenesis is associated with high rates of tricarboxylic acid and pyruvate cycling in fasting northern elephant seals. *American Journal of Physiology—Regulatory, Integrative and Comparative Physiology* 303(3): R340–R352.

Coleman, R.A., D.C. We, J. Liu, and J.B. Wade. 2000. Expression of aquaporins in the renal collecting tubule. *American Journal of Physiology—Renal Physiology* 279(5): 874–883.

Costa, D.P. 1982. Energy, nitrogen, electrolyte flux and sea water drinking in the sea otter *Enhydra lutris*. *Physiological Zoology* 55(1): 35–44.

Costa, D.P., J.P. Croxall, and C.D. Duck. 1989. Foraging energetics of Antarctic fur seals in relation to changes in prey availability. *Ecology* 70(3): 596–606.

Costa, D.P. and R.L. Gentry. 1986. *Reproductive Energetics of the Northern Fur Seal. Fur Seals: Maternal Strategies at Land and Sea*. Princeton, NJ: Princeton University Press, pp. 79–101.

Costa, D.P., B.J. Le Boeuf, A.C. Huntley, and C.L. Ortiz. 1986. The energetics of lactation in the northern elephant seal, *Mirounga angustirostris*. *Journal of Zoology* 209(1): 21–33.

Costa, D.P. and F. Trillmich. 1988. Mass changes and metabolism during the perinatal fast: A comparison between Antarctic (*Arctocephalus gazella*) and Galápagos fur seals (*Arctocephalus galapagoensis*). *Physiological Zoology* 61(2): 160–169.

Coulombe, H.N., S.H. Ridgway, and W.E. Evans. 1965. Respiratory exchange in two species of porpoise. *Science* 149(3679): 86–88.

Crocker, D.E., P.M. Webb, D.P. Costa, and B.J. Le Boeuf. 1998. Protein catabolism and renal function in lactating northern elephant seals. *Physiological Zoology* 71(5): 485–491.

Culebras, J.M. and F.D. Moore. 1977. Total body water and the exchangeable hydrogen. I. Theoretical calculation of nonaqueous exchangeable hydrogen in man. *American Journal of Physiology* 232(1): R54–R59.

Davis, R.W., M.A. Castellini, G.L. Kooyman, and R. Maue. 1983. Renal glomerular filtration rate and hepatic blood flow during voluntary diving in Weddell seals. *American Journal of Physiology—Regulatory, Integrative and Comparative Physiology* 245: R743–R748.

Depocas, F., J. Hart, and H.D Fisher. 1971. Seawater drinking and water flux in starved and fed harbor seals, *Phoca vitulina*. *Canadian Journal of Physiology and Pharmacology* 49(1): 53–62.

Dunn, F.L., T.J. Brennan, A.E. Nelson, and G.L. Robertson. 1973. The role of blood osmolality and volume in regulating vasopressin secretion in the rat. *Journal of Clinical Investigation* 52(12): 3212–3219.

Eaton, D.C. and J.P. Pooler. 2009. *Control of Sodium and Water Excretion: Regulation of Plasma Volume and Plasma Osmolality and Renal Control of Systemic Blood Pressure. Vander's Renal Physiology*, 7th edn. New York: McGraw-Hill.

Eichelberger, L., L. Leiter, and E.M.K. Geiling. 1940. Water and electrolyte content of dolphin kidney and extraction of pressor substance (renin). *Experimental Biology and Medicine* 44(2): 356–359.

Epstein, F.H., C.R. Kleeman, S. Pursel, and A. Hendrikx. 1957. The effect of feeding protein and urea on the renal concentrating process. *Journal of Clinical Investigation* 36(5): 635–641.

Epstein, M. 1992. Renal effects of head-out water immersion in humans: A 15-year update. *Physiological Reviews* 72(3): 563–621.

Fetcher, E.S. 1939. The water balance in marine mammals. *The Quarterly Review of Biology* 14(4): 451–459.

Fetcher, E.S. and G.W. Fetcher. 1942. Experiments on the osmotic regulation of dolphins. *Journal of Cellular and Comparative Physiology* 19(1): 123–130.

Folkow, L.P. and A.S. Blix. 1987. Nasal heat and water exchange in gray seals. *American Journal of Physiology* 253(2): R883–R889.

Frøkiær, J., C. Li, Y. Shi, A. Jensen, H. Prætorius, H. Hansen, O. Topcu, C. Sardeli, W. Wang, T.H. Kwon, and S. Nielsen. 2003. Renal aquaporins and sodium transporters with special focus on urinary tract obstruction. *Acta Pathologica, Microbiologica et Immunologica Scandinavica, Supplementum* 109: 71–79.

Funder, J.W. 1993. Aldosterone action. *Annual Review of physiology* 55(1): 115–130.

Gardner, J.W. 2002. Death by water intoxication. *Military Medicine* 167(5): 432–434.

Gauer, O.H. and J.P. Henry. 1976. Neurohormonal control of plasma volume. *International Review of Physiology* 9: 145–190.

Gentry, R.L. 1981. Seawater drinking in eared seals. *Comparative Biochemistry and Physiology Part A: Physiology* 68(1): 81–86.

Goodman-Lowe, G.D., S. Atkinson, and J.R. Carpenter. 2001. Gross anatomy of the digestive tract of the Hawaiian monk seal, *Monachus schauinslandi*. *Pacific Science* 55(4): 399–407.

Guo, A., Y. Hao, J. Wang, Q. Zhao, and D. Wang. 2014. Concentrations of osmotically related constituents in plasma and urine of finless porpoise (*Neophocaena asiaeorientalis*): Implications for osmoregulatory strategies for marine mammals living in freshwater. *Zoological Studies* 53: 25–30.

Harrison, R.J., F.R. Johnson, and B.A. Young. 1970. The oesophagus and stomach of dolphins (Tursiops, Delphinus, Stenella). *Journal of Zoology* 160(3): 377–390.

Hedges, N.A., D.E. Gaskin, and G.J.D. Smith. 1979. Renicular morphology and renal vascular system of the harbour porpoise *Phocoena phocoena* (L.). *Canadian Journal of Zoology* 57(4): 868–875.

Helm, R.C. 1983. Intestinal length of three California pinniped species. *Journal of Zoology* 199(3): 297–304.

Hevesy, G.V. and E. Hofer. 1934. Elimination of water from the human body. *Nature* 134(87): 9.

Hiatt, E.P. and R.B. Hiatt. 1942. The effect of food on the glomerular filtration rate and renal blood flow in the harbor seal (*Phoca vitulina* L.). *Journal of Cellular and Comparative Physiology* 19(2): 221–227.

Hill, D.A. and J.E. Reynolds III. 1989. Gross and microscopic anatomy of the kidney of the West Indian manatee, *Trichechus manatus* (Mammalia: Sirenia). *Cells Tissues Organs* 135(1): 53–56.

Hinga, K.R. 1979. The food requirement of whales in the Southern Hemisphere. *Deep Sea Research Part A: Oceanographic Research Papers* 26(5): 569–577.

Hong, S.K., R. Elsner, J.R. Claybaugh, and K. Ronald. 1982. Renal functions of the Baikal seal *Pusa sibirica* and ringed seal *Pusa hispida*. *Physiological Zoology* 55(3): 289–299.

Hoover, J.P. and R.D. Tyler. 1986. Renal function and fractional clearances of American river otters (*Lutra canadensis*). *Journal of Wildlife Diseases* 22(4): 547–556.

Houser, D.S., C.D. Champagne, and D.E. Crocker. 2007. Lipolysis and glycerol gluconeogenesis in simultaneously fasting and lactating northern elephant seals. *American Journal of Physiology—Regulatory, Integrative and Comparative Physiology* 293(6): R2376–R2381.

Houser, D.S., D.E. Crocker, P.M. Webb, and D.P. Costa. 2001. Renal function in suckling and fasting pups of the northern elephant seal. *Comparative Biochemistry and Physiology Part A: Molecular & Integrative Physiology* 129(2): 405–415.

How, O.J. and E.S Nordøy. 2007. Seawater drinking restores water balance in dehydrated harp seals. *Journal of Comparative Physiology B* 177(5): 535–542.

Hui, C. 1981. Seawater consumption and water flux in the common dolphin *Delphinus delphis*. *Physiological Zoology* 54(4): 430–440.

Huntley, A.C., D.P. Costa, and R.D. Rubin. 1984. The contribution of nasal countercurrent heat exchange to water balance in the northern elephant seal, *Mirounga angustirostris*. *Journal of Experimental Biology* 113(1): 447–454.

Ichikawa, I. and R.C. Harris. 1991. Angiotensin actions in the kidney: Renewed insight into the old hormone. *Kidney International* 40(4): 583–596.

Irvine, A.B., R.C. Neal, R.T. Cardeilhac, J.A. Popp, F.H. Whiter, and R.C. Jenkins. 1980. Clinical observations on captive and free-ranging West Indian manatees, *Trichechus manatus*. *Aquatic Mammals* 8: 2–10.

Irving, L., K.C. Fisher, and F.C. McIntosh. 1935. The water balance of a marine mammal, the seal. *Journal of Cellular and Comparative Physiology* 6(3): 387–391.

Janech, M.G., R. Chen, L. Klein, M.W. Nowak, W. McFee, R.V. Paul, W.R. Fitzgibbon, and D.W. Ploth. 2002. Molecular and functional characterization of a urea transporter from the kidney of a short-finned pilot whale. *American Journal of Physiology—Regulatory, Integrative and Comparative Physiology* 282(5): R1490–R1500.

Kjeld, M. 2001. Concentrations of electrolytes, hormones, and other constituents in fresh postmortem blood and urine of fin whales (*Balaenoptera physalus*). *Canadian Journal of Zoology* 79(3): 438–446.

Kjeld, M. 2003. Salt and water balance of modern baleen whales: Rate of urine production and food intake. *Canadian Journal of Zoology* 81(4): 606–616.

Kohin, S. 1998. Respiratory physiology of northern elephant seal pups: Adaptations for hypoxia, hypercapnia and hypometabolism. PhD dissertation, University of California, Santa Cruz, CA.
Ladd, M., L.G. Raisz, C.H. Crowder, and L.B. Page. 1951. Filtration rate and water diuresis in the seal, *Phoca vitulina*. *Journal of Cellular and Comparative Physiology* 38(2): 157–164.
Lanyon, J.M., K. Newgrain, and T.S.S. Alli. 2006. Estimation of water turnover rate in captive dugongs (*Dugong dugon*). *Aquatic Mammals* 32(1): 103–108.
Le Bas, A.E. 2003. Renal handling of water, urea and electrolytes in wild South America fur seal (*Arctocephalus australis*). *Latin American Journal of Aquatic Mammals* 2(1): 13–20.
Leon, B., A. Shkolnik, and T. Shkolnik. 1983. Temperature regulation and water metabolism in the elephant shrew *Elephantulus edwardi*. *Comparative Biochemistry and Physiology Part A: Physiology* 74(2): 399–407.
Lester, C.W. and D.P. Costa. 2006. Water conservation in fasting northern elephant seals (*Mirounga angustirostris*). *Journal of Experimental Biology* 209(21): 4283–4294.
Levin, E.R., D.G. Gardner, and W.K. Samson. 1998. Natriuretic peptides. *New England Journal of Medicine* 339 (5): 321–328.
Lifson, N., G.B. Gordon, and R. McClintock. 1997. Measurement of total carbon dioxide production by means of D2O181. *Obesity Research* 5(1): 78–84.
Lifson, N. and R. McClintock. 1966. Theory of use of the turnover rates of body water for measuring energy and material balance. *Journal of Theoretical Biology* 12(1): 46–74.
Light, D.B., E.M. Schwiebert, K.H. Karlson, and B.A. Stanton. 1989. Atrial natriuretic peptide inhibits a cation channel in renal inner medullary collecting duct cells. *Science* 243(4889): 383–385.
Macfarlane, W.V., R.J.H. Morris, B. Howard, J. McDonald, and O.E. Budtz-Olsen. 1961. Water and electrolyte changes in tropical merino sheep exposed to dehydration during summer. *Crop and Pasture Science* 12(5): 889–912.
MacMillen, R.E. and A.K. Lee. 1969. Water metabolism of Australian hopping mice. *Comparative Biochemistry and Physiology* 28(2): 493–514.
Maluf, N.S.R. 1989. Renal anatomy of the manatee, *Trichechus manatus (Linnaeus)*. *American Journal of Anatomy* 184(4): 269–286.
Maluf, N.S.R. and J.J. Gassman. 1998. Kidneys of the killer whale and significance of reniculism. *The Anatomical Record* 250(1): 34–44.
Malvin, R.L., J.P. Bonjour, and S. Ridgway. 1971. Antidiuretic hormone levels in some cetaceans. *Experimental Biology and Medicine* 136(4): 1203–1205.
Malvin, R.L. and M. Rayner. 1968. Renal function and blood chemistry in Cetacea. *American Journal of Physiology* 214(1): 187–191.
Malvin, R.L., S. Ridgway, and L. Cornell. 1978. Renin and aldosterone levels in dolphins and sea lions. *Experimental Biology and Medicine*, 157(4): 665–668.
Marsh, H., A.V. Spain, and G.E. Heinsohn. 1978. Physiology of the dugong. *Comparative Biochemistry and Physiology Part A: Physiology* 61(2): 159–168.
Matsuura, D.T. and G.C. Whittow. 1974. Evaporative heat loss in the California sea lion and harbor seal. *Comparative Biochemistry and Physiology Part A: Physiology* 48(1): 9–20.
McCormick, S.D. and D. Bradshaw. 2006. Hormonal control of salt and water balance in vertebrates. *General and Comparative Endocrinology* 147(1): 3–8.
Medway, W., M.L. Bruss, J.L. Bengtson, and D.J. Black. 1982. Blood chemistry of the West Indian manatee (*Trichechus manatus*). *Journal of Wildlife Diseases* 18(2): 229–234.
Meneely, G.R. and H.D. Battarbee. 1976. Sodium and potassium. *Nutrition Reviews* 34(8): 225–235.
Menon, G.K., S. Grayson, B.E. Brown, and P.M. Elias. 1986. Lipokeratinocytes of the epidermis of a cetacean (*Phocena phocena*). *Cell and Tissue Research* 244(2): 385–394.
Meyer, W., J. Schmidt, J. Kacza, R. Busche, H.Y. Naim, and R. Jacob. 2011. Basic structural and functional characteristics of the epidermal barrier in wild mammals living in different habitats and climates. *European Journal of Wildlife Research* 57(4): 873–885.
Miles, B.E., A. Paton, and H.E. De Wardener. 1954. Maximum urine concentration. *British Medical Journal* 2(4893): 901–905.
Miyaji, K., K. Nagao, M. Bannai, H. Asakawa, K. Kohyama, D. Ohtsu, F. Terasawa, S. Ito, H. Iwao, N. Ohtani, and M. Ohta. 2010. Characteristic metabolism of free amino acids in cetacean plasma: Cluster analysis and comparison with mice. *PLoS ONE* 5(11): e13808.

Murdaugh, H.V., W.L. Mitchell, J.C. Seabury, and H.O. Sieker. 1961a. Volume receptors and post-prandial diuresis in the seal (*Phoca vitulina* L.). *Experimental Biology and Medicine* 108(1): 16–18.

Murdaugh, H.V., B. Schmidt-Nielsen, J.W. Wood, and W.L. Mitchell. 1961b. Cessation of renal function during diving in the trained seal (*Phoca vitulina*). *Journal of Cellular and Comparative Physiology* 58(3): 261–265.

Nagai, K.I., K. Tadano-Aritomi, Y. Niimura, and I. Ishizuka. 2008. Higher expression of renal sulfoglycolipids in marine mammals. *Glycoconjugate Journal* 25(8): 723–726.

Nagy, K.A. and D.P. Costa. 1980. Water flux in animals: Analysis of potential errors in the tritiated water method. *American Journal of Physiology* 238(5): R454–R465.

Naka, T., E. Katsumata, K. Sasaki, N. Minamino, M. Yoshioka, and Y. Takei. 2007. Natriuretic peptides in cetaceans: Identification, molecular characterization and changes in plasma concentration after landing. *Zoological Science* 24(6): 577–587.

Nery, M.F., D.J. González, and J.C. Opazo. 2013. How to make a dolphin: Molecular signature of positive selection in cetacean genome. *PLoS ONE* 8(6): e65491.

Niimura, Y. and I. Ishizuka. 1990. Adaptive changes in sulfoglycolipids of kidney cell lines by culture in anisosmotic media. *Biochimica et Biophysica Acta—Molecular Cell Research* 1052(2): 248–254.

Nordøy, E.S., D.E. Stijfhoorn, A. Råheim, and A.S. Blix. 1992. Water flux and early signs of entrance into phase III of fasting in grey seal pups. *Acta Physiologica Scandinavica* 144(4): 477–482.

Ommanney, F.D. 1932. The urogenital system of the fin whale (*Balaenoptera physalus*) with appendix: The dimensions and growth of the kidneys of blue and fin whales. *Discovery Reports* 5: 363–466.

Ortiz, C.L., D. Costa, and B.J. Le Boeuf. 1978. Water and energy flux in elephant seal pups fasting under natural conditions. *Physiological Zoology* 51(2): 166–178.

Ortiz, R.M. 2001. Osmoregulation in marine mammals. *Journal of Experimental Biology* 204(11): 1831–1844.

Ortiz, R.M., S.H. Adams, D.P. Costa, and C.L. Ortiz. 1996. Plasma vasopressin levels and water conservation in fasting, postweaned northern elephant seal pups (*Mirounga angustirostris*). *Marine Mammal Science* 12(1): 99–106.

Ortiz, R.M., D.E. Crocker, D.S. Houser, and P.M. Webb. 2006. Angiotensin II and aldosterone increase with fasting in breeding adult male northern elephant seals (*Mirounga angustirostris*). *Physiological and Biochemical Zoology* 79(6): 1106–1112.

Ortiz, R.M., B. Long, D. Casper, C.L. Ortiz, and T.M. Williams. 2010. Biochemical and hormonal changes during acute fasting and re-feeding in bottlenose dolphins (*Tursiops truncatus*). *Marine Mammal Science* 26(2): 409–419.

Ortiz, R.M., C.E. Wade, D.P. Costa, and C.L. Ortiz. 2002. Renal responses to plasma volume expansion and hyperosmolality in fasting seal pups. *American Journal of Physiology—Regulatory, Integrative and Comparative Physiology* 282(3): R805–R817.

Ortiz, R.M., C.E. Wade, and C.L. Ortiz. 2000. Prolonged fasting increases the response of the renin–angiotensin–aldosterone system, but not vasopressin levels, in postweaned northern elephant seal pups. *General and Comparative Endocrinology* 119(2): 217–223.

Ortiz, R.M., C.E. Wade, C.L. Ortiz, and F. Talamantes. 2003. Acutely elevated vasopressin increases circulating concentrations of cortisol and aldosterone in fasting northern elephant seal (*Mirounga angustirostris*) pups. *Journal of Experimental Biology* 206(16): 2795–2802.

Ortiz, R.M., G.A.J. Worthy, and F.M. Byers. 1999. Estimation of water turnover rates of captive West Indian manatees (*Trichechus manatus*) held in fresh and salt water. *Journal of Experimental Biology* 202(1): 33–38.

Ortiz, R.M., G.A.J. Worthy, and D.S. MacKenzie. 1998. Osmoregulation in wild and captive West Indian manatees (*Trichechus manatus*). *Physiological Zoology* 71: 449–457.

Page, L.B., J.C. Scott-Baker, G.A. Zak, E.L. Becker, and C.F. Baxter. 1954. The effect of variation in filtration rate on the urinary concentrating mechanism in the seal, *Phoca vitulina L. Journal of Cellular and Comparative Physiology* 43(3): 257–269.

Pernia, S.D., D.P. Costa, and C.L. Ortiz. 1989. Glomerular filtration rate in weaned elephant seal pups during natural, long term fasts. *Canadian Journal of Zoology* 67(7): 1752–1756.

Pfeiffer, C.J. 1997. Renal cellular and tissue specializations in the bottlenose dolphin (*Tursiops truncatus*) and beluga whale (*Delphinapterus leucas*). *Aquatic Mammals* 23(2): 75–84.

Pilson, M.E. 1970. Water balance in California sea lions. *Physiological Zoology* 43(4): 257–269.

Reich, K.J. and G.A. Worthy. 2006. An isotopic assessment of the feeding habits of free-ranging manatees. *Marine Ecology Progress Series* 322: 303–309.

Reidenberg, J.S. 2007. Anatomical adaptations of aquatic mammals. *The Anatomical Record* 290(6): 507–513.

Reilly, J.J. 1991. Adaptations to prolonged fasting in free-living weaned gray seal pups. *American Journal of Physiology—Regulatory, Integrative and Comparative Physiology* 260(2): R267–R272.

Renouf, D., E. Noseworthy, and M.C. Scott. 1990. Daily fresh water consumption by captive harp seals (*Phoca groenlandica*). *Marine Mammal Science* 6(3): 253–257.

Ridgway, S. and S. Venn-Watson. 2010. Effects of fresh and seawater ingestion on osmoregulation in Atlantic bottlenose dolphins (*Tursiops truncatus*). *Journal of Comparative Physiology B* 180(4): 563–576.

Ridgway, S.H. 1972. *Homeostasis in the Aquatic Environment. Mammals of the Sea: Biology and Medicine.* Springfield, IL: Charles C Thomas, pp. 590–747.

Ridgway, S.H., J.G. Simpson, G.S. Patton, and W.G. Gilmartin. 1970. Hematologic findings in certain small cetaceans. *Journal of the American Veterinary Medical Association* 157 (5): 566–575.

Robertson, G.L., R.L. Shelton, and S. Athar. 1976. The osmoregulation of vasopressin. *Kidney International* 10(1): 25–37.

Rosen, D.A., G.D. Hastie, and A.W. Trites. 2004. Searching for stress: Hematologic indicators of nutritional inadequacies in steller sea lions. *Proceedings of the Fifth Comparative Nutrition Society Symposium*, pp. 145–149.

Schmidt-Nielsen, B., H.V. Murdaugh, R. O'Dell, and J. Bacsanyi. 1959. Urea excretion and diving in the seal (*Phoca vitulina L.*). *Journal of Cellular and Comparative Physiology* 53(3): 393–411.

Schmidt-Nielsen, K. 1990. *Animal Physiology: Adaptation and Environment,* 4th edn. New York: Cambridge University Press.

Schoeller, D.A. and J.M. Hnilicka. 1996. Reliability of the doubly labeled water method for the measurement of total daily energy expenditure in free-living subjects. *Journal of Nutrition* 126(1): 348S–354S.

Schoeller, D.A., E. Ravussin, Y. Schutz, K.J. Acheson, P. Baertschi, and E. Jequier. 1986. Energy expenditure by doubly labeled water: Validation in humans and proposed calculation. *American Journal of Physiology* 250(2): R823–R830.

Schoen, A. 1972. Studies on the environmental physiology of a semi-desert antelope, the dik-dik. *East African Agricultural and Forestry Journal* 37 (4): 325–330.

Schweigert, F.J. 1993. Effects of fasting and lactation on blood chemistry and urine composition in the grey seal (*Halichoerus grypus*). *Comparative Biochemistry and Physiology Part A: Physiology* 105(2): 353–357.

Shirreffs, S.M. 2000. Markers of hydration status. *Journal of Sports Medicine and Physical Fitness* 40(1): 80–84.

Skalstad, I. and E.S. Nordøy. 2000. Experimental evidence of seawater drinking in juvenile hooded (*Cystophora cristata*) and harp seals (*Phoca groenlandica*). *Journal of Comparative Physiology B* 170(5–6): 395–401.

Skog, E.B. and L.P. Folkow. 1994. Nasal heat and water exchange is not an effector mechanism for water balance regulation in grey seals. *Acta Physiologica Scandinavica* 151(2): 233–240.

Smith, H.W. 1936. The composition of urine in the seal. *Journal of Cellular and Comparative Physiology* 7(3): 465–474.

Speakman, J. 1997. *Doubly Labelled Water: Theory and Practice.* New York, Springer.

Speakman, J.R., K.S. Nair, and M.I. Goran. 1993. Revised equations for calculating CO_2 production from doubly labeled water in humans. *American Journal of Physiology—Endocrinology and Metabolism* 264(6): E912–E912.

Sperber, I. 1944. Studies on the mammalian kidney. *Zoologiska bidrag från Uppsala* 22: 249–432.

Šrámek, P., M. Šimečková, L. Janský, J. Šavlíková, and S. Vybiral. 2000. Human physiological responses to immersion into water of different temperatures. *European Journal of Applied Physiology* 81(5): 436–442.

Stanton, B.A. 1991. Molecular mechanisms of ANP inhibition of renal sodium transport. *Canadian Journal of Physiology and Pharmacology* 69(10): 1546–1552.

St. Aubin, D.J., S. Deguise, P.R. Richard, T.G. Smith, and J.R. Geraci. 2001. Hematology and plasma chemistry as indicators of health and ecological status in beluga whales, *Delphinapterus leucas. Arctic* 54(3): 317–331.

Stettner, P., S. Bourgeois, C. Marsching, M. Traykova-Brauch, S. Porubsky, V. Nordström, C. Hopf et al. 2013. Sulfatides are required for renal adaptation to chronic metabolic acidosis. *Proceedings of the National Academy of Sciences* 110(24): 9998–10003.

Storeheier, P.V. and E.S. Nordøy. 2001. Physiological effects of seawater intake in adult harp seals during phase I of fasting. *Comparative Biochemistry and Physiology Part A: Molecular & Integrative Physiology* 128(2): 307–315.

Stoskopf, M.K. and D. Herbert. 1990. Selected anatomical features of the sea otter (*Enhydra lutris*). *Journal of Zoo and Wildlife Medicine* 21(1): 36–47.

Suzuki, M. 2010. Expression and localization of aquaporin-1 on the apical membrane of enterocyte in small intestine of bottlenose dolphin. *Journal of Comparative Physiology B* 180(2): 229–238.

Suzuki, M., N. Endo, Y. Nakano, H. Kato, T. Kishiro, and K. Asahina. 2008. Localization of aquaporin-2, renal morphology and urine composition in the bottlenose dolphin and the baird's beaked whale. *Journal of Comparative Physiology B* 178(2): 149–156.

Tarasoff, F. and D. Toews. 1972. The osmotic and ionic regulatory capacities of the kidney of the harbor seal, *Phoca vitulina. Journal of Comparative Physiology* 81(2): 121–132.

Telfer, N., L.H. Cornell, and J.H. Prescott. 1970. Do dolphins drink water? *Journal of the American Veterinary Medical Association* 157(5): 555–558.

Thompson, L. 2011. Sodium chloride (salt). In: *Veterinary Toxicology: Basic and Clinical Principles*, 2nd edn, Gupta, R.C., ed. Waltham, MA: Academic Press, pp. 461–464.

Vardy, P.H. and M.M. Bryden. 1981. The kidney of *Leptonychotes weddelli* (Pinnipedia: Phocidae) with some observations on the kidneys of two other southern phocid seals. *Journal of Morphology* 167(1): 13–34.

Venn-Watson, S., K. Carlin, and S. Ridgway. 2011. Dolphins as animal models for type 2 diabetes: Sustained, post-prandial hyperglycemia and hyperinsulinemia. *General and Comparative Endocrinology* 170(1): 193–199.

Verlo, A. 2012. Seawater consumption in dehydrated hooded seals (*Cystophora cristata*). Master thesis, University of Tromsø, Tromsø, Norway.

Weeth, H.J. and A.L. Lesperance. 1965. Renal function of cattle under various water and salt loads. *Journal of Animal Science* 24(2): 441–447.

Wilcock, I.M., J.B. Cronin, and W.A. Hing. 2006. Physiological response to water immersion. *Sports Medicine* 36(9): 747–765.

Williams, T.M., J. Haun, R.W. Davis, L.A. Fuiman, and S. Kohin. 2001. A killer appetite: Metabolic consequences of carnivory in marine mammals. *Comparative Biochemistry and Physiology Part A: Molecular & Integrative Physiology* 129(4): 785–796.

Williams, T.M. and G.A. Worthy. 2002. Anatomy and physiology: The challenge of aquatic living. In: *Marine Mammal Biology: An Evolutionary Approach*, Hoelzel, A.R., ed. Hoboken, NJ: Wiley-Blackwell, pp. 73–97.

Willmer, P., G. Stone, and I. Johnston. 2005. *Environmental Physiology of Animals*, 2nd edn. Hoboken, NJ: Wiley-Blackwell, pp. 393–443.

Worthy, G.A.J. and D.M. Lavigne. 1982. Changes in blood properties of fasting and feeding harp seal pups, *Phoca groenlandica*, after weaning. *Canadian Journal of Zoology* 60(4): 586–592.

Worthy, G.A.J. and D.M. Lavigne. 1987. Mass loss, metabolic rate and energy utilization by harp and gray seal pups during the postweaning fast. *Physiological Zoology* 60(3): 352–364.

Worthy, G.A.J., P.A. Morris, D.F. Costa, and B.J. Le Boeuf. 1992. Moult energetics of the northern elephant seal (*Mirounga angustirostris*). *Journal of Zoology* 227(2): 257–265.

Wong, W.W., L.S. Lee, and P.D. Klein, 1987. Deuterium and oxygen-18 measurements on microliter samples of urine, plasma, saliva, and human milk. *American Journal of Clinical Nutrition* 45(5): 905–913.

Xu, S., Y. Yang, X. Zhou, J. Xu, K. Zhou, and G. Yang. 2013. Adaptive evolution of the osmoregulation-related genes in cetaceans during secondary aquatic adaptation. *BMC Evolutionary Biology* 13(1): 189.

Yim, H.S., Y.S. Cho, X. Guang, S.G. Kang, J.Y. Jeong, S.S. Cha, H.M. Oh et al. 2014. Minke whale genome and aquatic adaptation in cetaceans. *Nature Genetics* 46: 88–92.

Zapol, W.M., G.C. Liggins, R.C. Schneider, J. Qvist, M.T. Snider, R.K. Creasy, and P.W. Hochachka. 1979. Regional blood flow during simulated diving in the conscious Weddell seal. *Journal of Applied Physiology* 47: 968–973.

Zenteno-Savin, T. and M.A. Castellini. 1998a. Plasma angiotensin II, arginine vasopressin and atrial natriuretic peptide in free ranging and captive seals and sea lions. *Comparative Biochemistry and Physiology Part C: Pharmacology, Toxicology and Endocrinology* 119(1): 1–6.

Zenteno-Savin, T. and M.A. Castellini. 1998b. Changes in the plasma levels of vasoactive hormones during apnea in seals. *Comparative Biochemistry and Physiology Part C: Pharmacology, Toxicology and Endocrinology* 119(1): 7–12.

chapter eight

Fasting

David Rosen and Allyson Hindle

Contents

8.1 Snapshot: Living without food

Many marine mammals undergo regular, prolonged periods of forgoing food. Surprisingly, these fasting episodes often coincide with life stages when additional energy is needed for important tasks: long-distance migrations, growth in offspring, lactation in females, and territory defense in males. How can these critical activities be supported during fasting without animals starving to death?

The answer lies in several specific, but equally important questions:

- What is the difference between fasting and starvation?
- How does an animal's physiology change over the course of the fast?
- What types of on-board energy reserves provide fuel for the fasting animal?
- How can an animal minimize energy requirements during a fast?
- What hormones control fasting physiology in marine mammals?

Predictable fasting periods generally occur as part of an animal's natural life history. Suites of behavioral and physiological adaptations permit survival across these periods with minimal long-term consequences.

Although a general model describes fasting physiology in vertebrates, marine mammal studies suggest that this group does not always follow expectations—in fact, even closely related species seem to cope with fasting in very different ways. Many different marine mammals withstand extended fasts; however the ways that they do so are constantly surprising scientists. Not only is the physiology of marine mammals specialized to cope with episodes of fasting, but these mechanisms can be altered over the course of the fast to adjust to ongoing changes. As is typical in other fasting-adapted vertebrates, lipid stores provide the primary support for this physiology. However, marine mammals uniquely tap into lipids stored in their expansive hypodermal blubber layer. While fasting often occurs during energetically demanding life-history stages (e.g., periods of reproduction and growth), marine mammals extend fasting tolerance by saving energy through reducing other energetic costs. The physiological and energetic changes that support and regulate fasting are hormonally controlled. A diversity of responses across species, and divergence between marine and terrestrial mammals in the suite of key hormones that regulate fasting provide an important example of the unique physiology of marine mammals.

8.1.1 *What is fasting and when does it occur?*

Fasting refers to the physiological state of living with a complete lack of food intake. In the strictest sense, short-term fasting occurs constantly in most mammals, since individuals are not continuously consuming prey—or more accurately, processing previously acquired prey. In this chapter, we will not be discussing such short-term events (what is referred to as *dynamic equilibrium*) (Kleiber 1975). Rather, we will address "fasting" as a complete lack of food intake over longer intervals of days or months.

It might be surprising to learn that scientists have not agreed upon clear definitions for the terms "fasting state" versus "starvation state." Both imply a lack of food intake, and while they are sometimes used interchangeably, they should more appropriately differentiate two different physiological states. More specifically, the terms fasting and starvation distinguish differences in the severity of the effects of a lack of food intake. Whereas fasting can be sustained over a period of time, a "starvation state" is characterized by life-threatening or potentially irreversible physiological and anatomical changes.

Fasting and starvation can also be distinguished by cause or motivation (McCue 2010). A fasting animal is unwilling to eat (or forage) even though food is available, due to a non-compatible requirement, such as reproduction (breeding, nursing), or thermoregulation, etc. While fasting also occurs during hibernation or estivation, no pinnipeds or cetaceans exhibit these strategies (although pregnant female polar bears, *Ursus maritimus*, which are sometimes discussed with marine mammals, hibernate in the winter). In contrast, starvation would be said to occur when an animal is willing to consume food but some extrinsic factor (e.g., lack of prey, inclement weather, and predator avoidance) is limiting this ability.

This difference between these states is tied to predictability; fasting is a natural part of life history and may, therefore, occur very predictably. Since fasting is "natural," and timed to life history, an animal will be more physiologically able to withstand this lack of food intake at certain times of the year or life stage. We, therefore, expect that the impacts of a natural fasting episode be less detrimental than those incurred during "unpredicted" episodes of starvation.

8.1.1.1 Pinnipeds

Almost all pinniped species fast, which is partly why most of our knowledge about marine mammal fasting physiology comes from this group. Their fasting episodes also coincide with periods when they are available on land for intensive study.

Importantly, most pinnipeds experience their first fasts during their initial year of life, a time when physical growth is critical to survival, and their ability to forage is physically and behaviorally limited. After a period of near-constant energy intake via maternal milk during the nursing period, the pups of many phocid species are abruptly weaned and undergo a postweaning fast. This period often corresponds to their annual molt; in northern elephant seals (*Mirounga angustirostris*) this postweaning fast lasts for 9–12 weeks (Reiter et al. 1978). For the most part, pups remain on land during this period assumedly to minimize the costs and tenure of the molt. Fasting in these pups is, therefore, a consequence of the environmental boundary imposed by the molt. Extended postweaning fasts may also provide a window for the physiological development of diving capacity (Bennett et al. 2010; Burns et al. 2014).

Some otariid pups also undergo postweaning fasts, also often associated with molts, although many otariids wean later in life compared to phocids. Otariid pups may also avoid complete fasting in the postweaning period due to greater experience and physiological capacity for independent foraging (Fowler et al. 2007). The more fundamental difference between phocid and otariids pups is that otariid pups characteristically contend with regular fasts throughout their nursing period when their mothers regularly return to sea to feed. The complex manner in which females devote time and energy to lactation while managing their own fast across the nursing period is covered in Chapter 10. However, the nursing/foraging cycle of otariid females can be viewed as a balancing act of maternal and offspring fasting episodes. The length of maternal foraging trips and the resulting fasting episodes of pups can last from days to weeks, depending on species and stage of lactation. These episodes can make up a significant portion of the total nursing period for pups. For example, episodes of 4–5-day foraging bouts by Antarctic fur seal mothers (*Arctocephalus gazelle*) mean that their pups are potentially fasting for 71% of the total nursing period (Boyd 1991; Lunn et al. 1993). Conversely, the time that the mother is attending her pup on land to provide adequate milk necessitates that the female is fasting while expending tremendous levels of energy through lactation.

Pinniped males also fast during the breeding season. For many species, breeding strategies rely upon continuously defending territories on beaches or breathing holes underwater, often for months at a time. This requires them to expend large amounts of energy fighting and mating while being precluded from feeding over the entire reproductive season. Male northern elephant seals have one of the highest sustained rates of energy expenditure of any male pinniped (Crocker et al. 2012), losing about a third of their starting body mass over the 3-month breeding period. As in many pinniped species with territorial breeding systems, it has been observed that males are in very poor physical shape by the end of the breeding season, and many are often not observed in subsequent years (LeBoeuf 1974). These individuals are presumed to have died due to the high rates of protein catabolism over the extended, expensive fast, although increased oxidative stress may play a role (Sharick et al. 2014).

8.1.1.2 Cetaceans

Many large baleen whales travel great distances across oceans, alternating between summer feeding grounds in high-latitude cold waters, with warm-water, low-latitude winter reproductive areas. Many whale species are observed to fast on these calving grounds.

This fasting is likely due to limited food availability in these low-productivity waters (Horwood 1990). This fasting is a trade-off for the warm waters that may be necessary to rear young in a more hospitable thermal environment, or even to avoid predators (Corkeron and Connor 1999; Mehta et al. 2007). Similar patterns are not generally observed in smaller bodied odontocetes, but there is evidence of fasting during periodic "maintenance" migrations undertaken by Orcas (*Orca orcinus*) (Durban and Pitman 2011).

8.1.2 Energy substrates during a fast

Energy is constantly required to maintain cellular and biochemical homeostasis. As we shall discuss later, although fasting can be associated with strategies to limit and streamline energy expenditure requirements, energy use will always exceed external energy intake, which is zero. Energy requirements during fasting episodes are, therefore, met solely by substantial contributions from *endogenous* (internal) *body reserves*. *Catabolism* is the process by which energy-containing molecules, or substrates, are broken down to release energy. Fuel stores are unequally accumulated within body tissues in three forms: fat, protein, and carbohydrate.

The most common energy source for a feeding organism is glucose, a carbohydrate that is polymerized and stored as glycogen in the liver and muscles. Despite the importance of glucose in meeting the energy requirements of body tissues, including the brain and red blood cells, glycogen stores are relatively small in most marine mammals. By far, lipids and proteins provide the largest energy reserves. Lipids, stored as fat, are the preferred fuel to sustain fasting for several reasons. They have the highest caloric density, meaning that more energy is liberated per gram of fat catabolized (approx. 40 kJ g^{-1} depending on the composition of lipid stores) compared to other substrates (proteins and carbohydrate provide approx. 18 kJ g^{-1}) (Schmidt-Nielsen 1997). Lipids also provide more metabolic water on breakdown than do proteins (107 g versus 40 g 100 g^{-1} tissue), an important resource that is also limited or lacking during fasting (Figure 8.1).

Unlike many terrestrial mammals that store fat in visceral deposits, the majority of lipids used as an energy reserve for marine mammals are situated in the hypodermal (subcutaneous) blubber layer. This blubber layer can account for 50% of the total body mass,

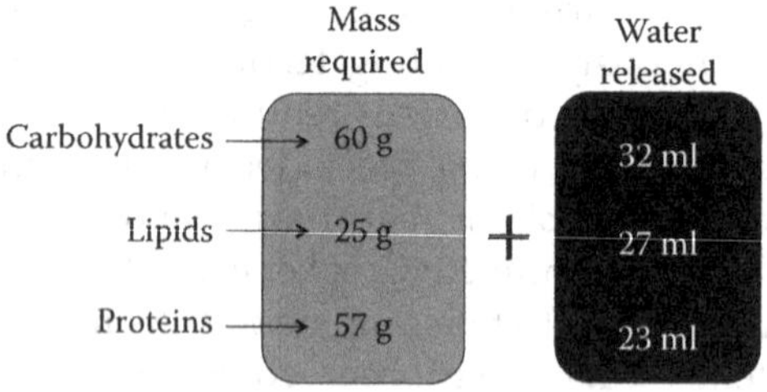

Figure 8.1 Lipids are the most energy dense substrate; less lipid mass is required to generate 1000 kJ of energy compared to carbohydrates or proteins. These estimates are respectively based on glucose as the carbohydrate source, the complete oxidation of an "average" fat, and the catabolism of a protein to urea. Different sources or configurations of carbohydrate, lipid, and protein energy stores will yield slightly different values. Although catabolism of carbohydrates to generate 1000 kJ of energy produces the most metabolic water, lipids yield the most water per gram. (Data from Schmidt-Nielsen, K., *Animal Physiology: Adaptation and Environment*, 5th edn., Cambridge University Press, Cambridge, UK, 1997; Edney, E.B., Metabolic water, in: *Water Balance in Land Arthropods*, Springer, 1977.)

and while providing critical insulation, its extent is generally far greater than required for thermoregulation (see Chapter 9). During *lipolysis*, lipases hydrolyze *triacylglycerols* in the blubber layer into *glycerol* and *free fatty acids* (FFA). Glycerol is available as an immediate energy source. FFA must be activated for transport into the mitochondria, where they are broken down to acetyl-coA subunits (via beta-oxidation), which enter the citric acid cycle. Fatty acids are also partially oxidized in the liver into four-carbon *ketone bodies*, which are subsequently oxidized as a fuel by other tissues (Figure 8.2).

In addition to their lower energy density, it is least desirable to catabolize proteins because they are energetically expensive to make and are a critical element of tissues. All enzymes, molecular chaperones, receptors, and many building blocks of the cell itself are proteins. Protein catabolism, therefore, diminishes lean body tissue. In fact,

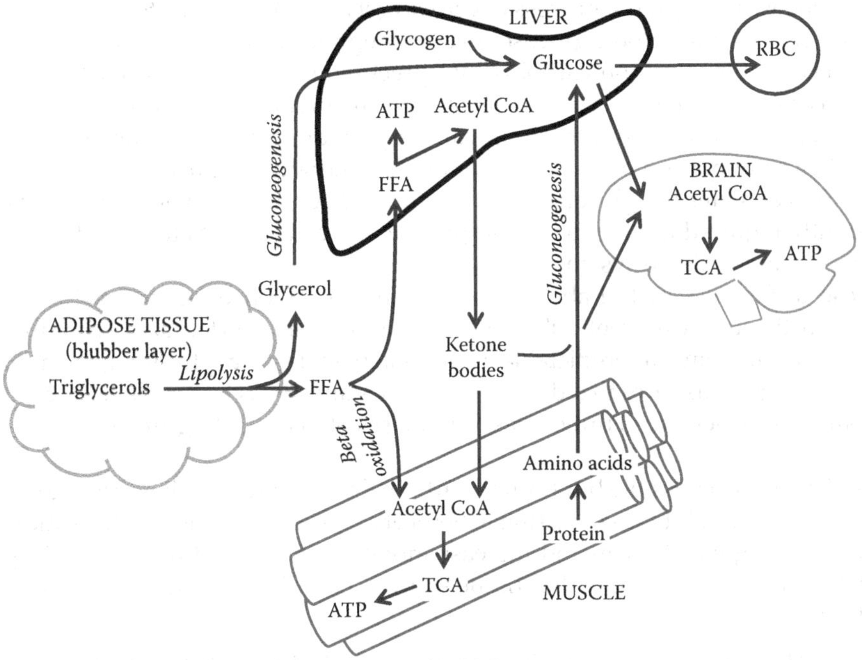

Figure 8.2 **(See color insert.)** Summary of some of the major metabolic pathways related to energy production during a fast. Major pathways are illustrated for catabolism of lipids from adipose tissue and protein from muscles, as well as production of glucose in the liver from either stored glycogen or from gluconeogenesis from other precursors. Triacylglycerols catabolized via lipolysis from adipose tissues form the major source of energy during fasting. The resulting free fatty acids (FFA) are transformed via beta-oxidation into acetyl coenzyme A (acetyl CoA), which produces energy (ATP) at target tissues via the tricarboxylic acid (TCA) cycle. Some of the FFA that enter the liver are oxidized directly, but most are converted to ketone bodies (acetoacetate and β-hydroxybutyrate), which are released into the blood (the liver cannot use ketone bodies itself). The ketone bodies are oxidized in the mitochondria of target cells into acetyl-CoA, which enters the TCA cycle. Certain tissues require glucose to function. This can be originally met by liver glycogen stores, which are rapidly depleted. New glucose can be formed in the liver (gluconeogenesis) from the smaller glycerol segment of triacylglycerols and from amino acids freed via protein catabolism (proteolysis). The central nervous system can derive some of its energy from oxidation of ketone bodies, but red blood cells are obligate glucose consumers. In the later stages of fasting, protein becomes a more dominant metabolic substrate. (Adapted from Lieberman, M. and Marks, A.D., *Marks' Basic Medical Biochemistry: A Clinical Approach*, Lippincott Williams & Wilkins, Baltimore, MD, 2009.)

death by starvation is linked to depletion of lean body mass; therefore, limiting protein catabolism during the fast has a direct effect on survival (Øritsland 1990; Øritsland and Markussen 1990). Protein catabolism also generates nitrogen end-products that must be recycled or removed in urine. This is at odds with the strategy to minimize urinary water loss observed in many animals without access to external water stores (see Chapter 7).

However, protein stores do play an important role in providing glucose during a fast. A constant supply of glucose is required by neurons and aspects of the central nervous system (which lack enzymes to oxidize FFA, although they can use ketone bodies for a portion of their energy requirements) and red blood cells (which lack mitochondria and therefore, the enzymes for both FFA and ketone oxidation). Glycogen stores in the liver are usually depleted within a few days of fasting. Fortunately, the liver and kidneys can then produce glucose through the process of *gluconeogenesis*, which helps to maintain a reduced but relatively constant blood glucose level throughout extended fasts. Gluconeogenesis forms glucose from the non-carbohydrate precursors *glycerol* (from *lipolysis* of adipose triacylglycerols), and *amino acids* (via *proteolysis*, the breakdown of proteins to individual amino acids). Because most fatty acids cannot provide carbon for gluconeogenesis, only the small glycerol portion of the vast store of food energy contained in adipose tissue triacylglycerols can enter the gluconeogenic pathway. Thus, some level of protein catabolism is constantly required during fasting to provide the components to manufacture glucose for these glucose-obligate systems.

These differences in the abundance, benefits, and consequences of fueling fasting metabolism through different substrates results in many vertebrates, including marine mammals, undergoing a predictable progression in fasting physiology. This series of adaptive changes was formalized in the late 1980s and 1990s by scientists such as George Cahill and Yves Cherel into three "classical phases of fasting" (Figure 3).

Phase I: Readily accessible but limited carbohydrates stores (such as glycogen) are used to fuel the initial stages of fasting (Cherel et al. 1988). Animals also transition to a protein-sparing metabolism and increase mobilization of lipid resources. Despite energetic contributions from other tissue sources, carbohydrates are depleted in hours to days.

Phase II: This period represents the most physiologically stable fasting state that can last for weeks to months, where fasting marine mammals rely almost entirely on lipid stores. Despite this strong metabolic preference, not all biochemical processes can be met by lipid catabolism alone. Protein sources play a limited but vital role by supplying amino acids to replenish circulating glucose via gluconeogenesis. Animals able to accumulate greater body fat reserves prior to fasting will sustain this physiology longer; however, lipid stores are obviously finite.

Phase III: The blubber layer cannot be completely depleted without infringing on thermoregulatory capabilities, causing further energetic imbalances (Rosen et al. 2007). At this point, marine mammals must end their protein-sparing strategy; however, increased protein catabolism represents the "end-game" physiological state associated with impending exhaustion of fuel reserves. Although Phase III physiology can be reversible upon re-feeding, at this point animals must quickly end their fast or suffer severe physical consequences, including death. For this reason, reliance on protein catabolism only lasts days to weeks, and fasting in most marine mammals does not typically extend into Phase III.

Body mass dynamics often reflect these changes in fasting physiology (in combination with changes in energy expenditure discussed in Section 8.2.3). Rates of mass loss tend to decline early in the fasting period, reaching stable and relatively low rates for the majority of the episode. A terminal switch to protein catabolism is often accompanied by rapidly increasing rates of mass loss.

The biochemical signatures of shifts in substrate catabolism include levels of enzymes necessary to breakdown different fuels, as well as metabolite by-products and end-products of those reactions. Circulating FFA derived from lipid (triacylglycerol) breakdown can undergo beta-oxidation and be used directly as a fuel source by tissues, or they can be partially oxidized in the liver to produce the ketone bodies β-hydroxybutyrate (β-OHB) and acetoacetate (Figure 8.2). Water-soluble ketones can then be metabolized by cells instead of glucose. Ketone bodies are able to pass the blood–brain barrier, providing a portion of the vital fuel for the brain and central nervous system. As a result of these shifts in metabolic substrate, plasma levels of FFA and β-OHB increase rapidly during the initial phases of fasting and remain high as indicators of an almost complete reliance on lipid catabolism (Figure 8.3). Subsequent increases in protein breakdown are mirrored by increases in rates of nitrogen excretion and blood urea nitrogen (BUN) levels.

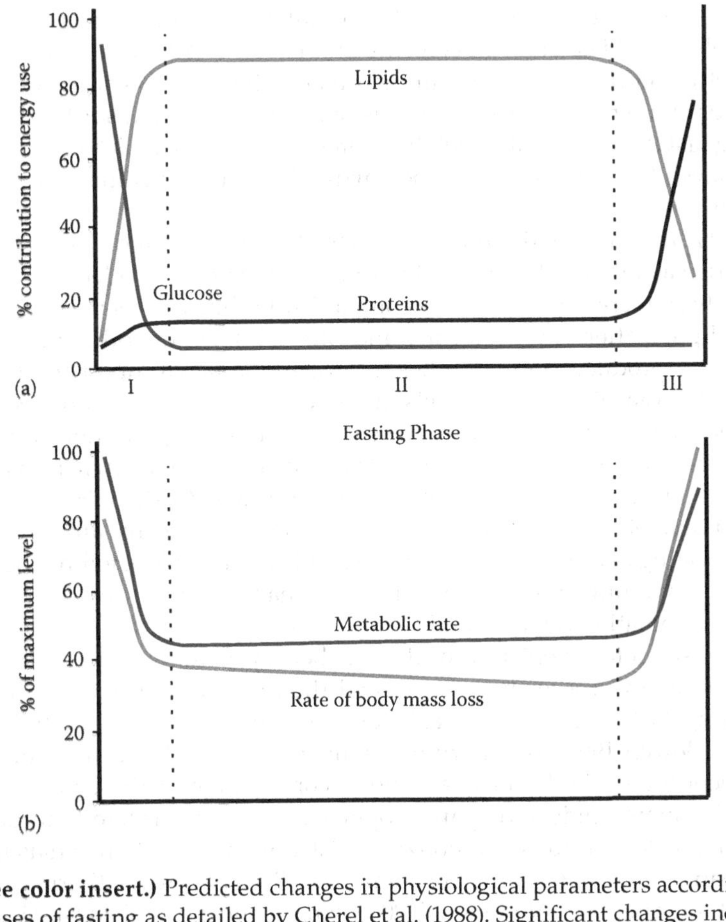

Figure 8.3 **(See color insert.)** Predicted changes in physiological parameters according to the classical three phases of fasting as detailed by Cherel et al. (1988). Significant changes include a shift in metabolic fuel use (a) and rates of body mass loss and mass-specific metabolic rates (b).

8.1.2.1 Pinnipeds

8.1.2.1.1 Phocids One of the best-studied fasting scenarios is the postweaning fast of phocid seals, which can last up to 10 weeks. For harp seals (*Phoca groenlandica*), 86%–97% of energy is derived from lipids in the first 8 weeks of their fast (Worthy and Lavigne 1983; Nordøy et al. 1990, 1993), and 72%–80% of this energy comes from blubber lipids (Siverston 1941; Stewart and Lavigne 1980; Worthy and Lavigne 1983; Kovacs and Lavigne 1985). Extended fasting episodes (>8 weeks) are associated with increased protein catabolism (but still only up to 15.7%) (Worthy and Lavigne 1983). In postweaning gray seal pups (*Halichoerus grypus*), the energetic contribution from lipids can reach 94%–97% (Nordøy and Blix 1985; Øritsland et al. 1985; Reilly 1991). Similarly, northern elephant seal pups may derive up to 98% of their energetic requirements from lipids over their 10-week fast (Pernia et al. 1980; Houser and Costa 2001).

While phocid seals rely extensively on lipid metabolism during the bulk of their postweaning fast, the underlying metabolic shift is not instantaneous. Many appear to rely initially on protein stores (up to 1–3 weeks) (Worthy and Lavigne 1983, 1987; Rea 1995). For example, during the initial 3 days of the very short hooded seal (*Cystophora cristata*) pup fasts, 16% of the total mass loss was from fat while 28% was from protein (the remainder was water) (Lydersen et al. 1997). The pups also have high initial rates of mass loss, which rapidly decrease to lower, stable rates by day 5 (Bowen et al. 1987). Harp seal pups similarly utilize greater levels of protein in the initial 2 weeks of their fast before switching almost entirely to lipid catabolism (Worthy and Lavigne 1983, 1987). This protein to lipid switch is partly reflected in changes in their blood chemistry (Nordøy et al. 1993). While gray seal pups exhibit uniformly low levels of protein oxidation starting at day 4 of their postweaning fast (Nordøy et al. 1990) their rates of mass loss, which continue to drop over the first 10 days of fasting, appear to be partly disconnected from changes in metabolic substrate use.

Plasma FFA increase continuously across the fasts of harp and gray seal pups (Nordøy and Blix 1991), reaching levels even higher than seen in northern elephant seals (Castellini et al. 1987). These increased FFA are accompanied by linearly increasing β-OHB in harp and gray seal pups after day 8 (Nordøy and Blix 1991). On the other hand, northern elephant seal pups produce ketone bodies almost immediately upon the onset of fasting (Castellini and Costa 1990). These levels increase until ~day 55, before declining sharply, soon followed by the pup's departure to sea. It is interesting that northern elephant seal pups likely have the longest postweaning fast and seem able to switch to a total reliance on lipid metabolism almost immediately (Castellini and Costa 1990; Adams and Costa 1993). Protein catabolism indicative of "end-stage" fasting is scarce among phocid seals, as most begin foraging before lipid reserves are overly depleted (Castellini and Rea 1992). An exception are "starveling" post-molt northern elephant seal pups, which are not indicative of the general population (Houser and Costa 2003).

As previously noted, depletion of the blubber layer can have thermoregulatory consequences, and many studies have examined the thermal effects of changes in blubber reserves during fasting in phocid seal pups (e.g., Irving and Hart 1957; Worthy 1991; Muelbert and Bowen 1993). Northern elephant seal pups exhibit differing energy usage patterns depending on body mass and body composition at the end of nursing. Fatter pups catabolize more lipids and spare proportionally more protein than do leaner pups. As a consequence, leaner pups, catabolizing relatively more protein versus lipid, conclude their postweaning fast with sufficient remaining lipid stores for thermoregulation during their first foraging trip (Noren et al. 2003; Noren and Mangel 2004). There is evidence that species which normally fast in water are more adapted to preserving their lipid layer for thermoregulation than species which fast on land (Worthy and Lavigne 1987). Experimentally, gray seal pups (which normally fast on land) had thinner blubber layers after a 10-week fast in water compared those fasted on land, whereas harp seals (which normally fast partly in water) showed no differences in body composition whether fasting on land or in water (Worthy and Lavigne 1987).

Maintaining stable circulating glucose levels can be particularly challenging for pups, and initial drops are sometimes observed early in the fast, before production catches up to utilization. Harp seal pups maintain high, constant plasma levels of glucose over 32-day weaning fasts, except for an early, insignificant decline (Worthy and Lavigne 1982; Nordøy et al. 1993). In contrast, gray seal pups exhibit a constant 20% decline in glucose levels that reflects decreasing replacement rates over their 52-day fasts (Nordøy et al. 1990). Plasma glucose concentrations in northern elephant seal pups are quite variable. While they may sometimes decline slightly during the fast, they generally remain at consistent high levels, with no obvious changes related to either the start of weaning or the decrease in circulating ketones that is thought to signal the end of weaning (Costa and Ortiz 1982; Castellini and Costa 1990; Ortiz et al. 2001).

Breeding male phocids also rely on lipid metabolism during their extensive fasts. Male northern elephant seals meet only 7% of energy expenditure through protein breakdown (Crocker et al. 2012). Protein use declines with initial proportion of body fat (so thinner males use more protein catabolism), emphasizing the importance of adequate pre-fast lipid stores. Circulating metabolites in these fasting males differ from expectations for fasting mammals relying on lipolysis and also differ from fasting females and pups (Castellini and Costa 1990; Houser et al. 2007). β-OHB levels are very low and only increase slightly during fasting in contrast to females and pups, suggesting ketone regulation differs with life history. Further, BUN levels (and serum FFA) are consistent over the fasting and are unrelated to protein catabolism (Crocker et al. 2012).

Fasting may also occur outside of the breeding season, and some experiments have attempted to mimic these "unpredicted" but natural fasting episodes. Juvenile harbor seals (*Phoca vitulina*) experimentally fasted for 2 weeks derived most (~75%) of their energy from fat, showed a marked (20%) metabolic decrease (Markussen 1995; Markussen et al. 1992), and a linear decline in body mass typical of lipid-based metabolism.

8.1.2.1.2 Otariids Fasts endured by otariid pups during maternal foraging trips can last from days to weeks. Antarctic fur seal pups transition to a lipid-based metabolism within 2–3 days of their 5-day nursing fasts (Arnould et al. 2001). Protein turnover accounts for only 5.4% of total energy expenditure over the course of their fast, during which they exhibit decreases in plasma BUN, triglyceride concentrations, and circulating glucose levels and increases in β-OHB.

Subantarctic fur seal pups (*Arctocephalus tropicalis*) undergo relatively prolonged nursing fasts compared to most other otariids, which initially last 10 days but extend up to 3–4 weeks just prior to weaning (Georges and Guinet 2000). On average, mass loss during these fasts comprised 56% lipids and 10% protein (the remainder was water), suggesting that 93% of their energy requirements derived from lipids (Beauplet et al. 2003). Curiously, protein catabolism was twice as high in female pups, despite their greater lipid reserves.

Based on blood chemistry, 6-week-old Steller sea lion pups (*Eumetopias jubatus*) make a rapid metabolic transition to fasting within 16 h, with rapidly decreased plasma BUN concentrations and increased β-OHB (Rea et al. 2000). Subsequent increases in plasma BUN implies that the pups reverted to protein catabolism after only 2.5 days of fasting. Older Steller sea lions undergoing 7–14 days of experimental fasts (Rea et al. 2009) also demonstrated both age- and season-specific responses. Similar to fasting pups, BUN decreased rapidly in both juveniles and subadults. This decline was more rapid for juveniles during the non-breeding season, when animals had slightly higher pre-fasting lipid reserves compared to during the breeding season. Significantly increased BUN concentrations observed at the end of the non-breeding season fasts suggest that subadult Steller sea lions are unable to maintain a protein-sparing metabolism for a full 14 days during this season. Subadult and juvenile sea lions also exhibited lower circulating ketone body concentrations compared to pups, suggesting age-related differences in substrate use. Breeding male otariids also undergo substantial fasts, but almost nothing is known about their physiology during this period.

8.1.2.2 Other marine mammals

We know virtually nothing about the metabolic basis of fasting physiology in any other marine mammal species. Sea otters are very intolerant to fasting, becoming rapidly hypoglycemic after even overnight fasts. Presumably, this is due to their relative small lipid stores coupled with a very high metabolic rate. Cetaceans presumably support their fasting from massive lipid reserves contained in their blubber. Working backward from the sodium (eliminating saltwater ingestion) and urea concentrations (eliminating protein catabolism) measured in the urine of migrating humpback whales, the metabolic water derived from lipids was determined after a 3-month fast (Bentley 1963). This single study estimated that 97% of catabolized on-board energy reserves (by tissue weight) came from fat and 3% came from protein. Polar bears are also extremely tolerant to fasting; summer land-based bears display extremely elevated serum FFA levels, for example (Nelson et al. 1983). As Arctic seal predators, they rely heavily on seasonal sea ice to access productive feeding grounds, and can experience extended periods where marine food is inaccessible. They are also known to recycle nitrogen from urea during fasting to limit protein loss.

8.1.3 Energetic conservation during fasting

The impact of fasting depends on the accrued energetic deficit, which is a direct consequence of an animal's energy expenditure. Therefore, it is not surprising that animals may make behavioral and physiological adjustments to minimize their energy requirements during a fast. What is surprising is that energetic conservation efforts often co-occur with energetically expensive, critical life-history requirements that constrain the degree and potential avenues of energy conservation. The most energetically costly concurrent demands are likely aspects of reproduction (in females: lactation and mating; in males: defense of breeding territories).

The molting period is another inflexible life-history requirement that imposes additional energy requirements but is associated with fasts for many marine mammals. Molting has required direct costs (skin and hair replacement) and potential indirect costs (thermoregulation). The latter is reduced in pinnipeds by hauling out on land, given the decreased thermal conductivity of air versus water. Some species, such as elephant seals, have a more condensed and drastic molt than others; the shorter molting period allows them to haul out and fast during the entire process. Killer whales migrate to warmer waters to molt, and there is evidence that they do not forage during these trips (Durban and Pitman 2011).

The assumption is that regenerating skin in warmer waters provides a net energy savings despite the additional locomotory costs and extended fast that it requires.

Potential energetic savings can be realized through decreases in activity levels, thermoregulatory costs, and Resting Metabolic Rate (Cherel et al. 1988). Reduced physical activity will obviously lower energy use, however, the option to do so is highly dependent on concurrent behavioral requirements. For example, baleen whales that undertake seasonal migrations between calving and foraging grounds may experience extended fasting periods while coping with substantial locomotory costs. These migrations can be unbelievably extensive; humpback whales (including mothers and calves) migrating from Antarctic feeding grounds to Pacific wintering areas off the coast of Central America travel approximately 8300 km.

In some mammals, thermoregulatory costs may be curtailed by either a controlled decrease in core body temperature, or a decrease in the defended (effective) core body mass. Such thermoregulatory adaptations may manifest rapidly at the onset of the fast, but there is little evidence for this among marine mammals due to their expansive insulation.

Metabolic depression is a term for the decrease in *Resting Metabolic Rate* (RMR)—the total energy requirements of an inactive animal within its thermoneutral zone. It is often cited as a common physiological adaptation to periodic food shortages among many species (Keys et al. 1950) that serves to limit rates of mass loss despite insufficient energy intake (Guppy and Withers 1999). This strategy may be key to maximizing protein sparing during extended fasts (Henry et al. 1988), by limiting glucose consumption rates and, therefore, catabolism of protein for gluconeogenesis. Among vertebrates, rapid depression of RMR is often observed early in the fast, leading to a new steady state fasting level (i.e., the amount of energy expended per kg body mass). As mass-specific RMR remains constant, subsequent decreases in overall (or absolute) metabolic rate throughout the fast are due to decreasing body mass. There are also indications of increasing metabolic rate toward the end of the natural fasting period, as cell and tissues are prepared for the resumption of foraging and digestion.

Depressed RMR can occur by at least three avenues: selective loss of metabolically active tissue, downregulation of energetically expensive processes, and reduction of cellular metabolism. While protein catabolism and loss of lean tissue will reduce total metabolism (versus loss of metabolically inert lipids), mass-specific decreases in metabolism can most effectively be produced by targeted catabolism of an animal's most energy-demanding tissues. One common strategy is to shrink portions of the digestive tract, which represents a large portion of total body mass and requires continuous turnover of the cell lining, which is relatively expensive to maintain. Fortunately, during periods of fasting, the gut is not required and can be "restructured" when food again becomes available. The kidney is another metabolically expensive organ. However, unlike the gut, there is no evidence that the kidney becomes physically smaller during fasts. This is partly due to its complexity, and partly due to the fact that it is still required to process metabolic waste, and to help maintain fluid and electrolyte homeostasis through resorption of endogenous electrolytes and body water (see Chapter 7). However, energy is saved if the level of processing through the kidney is significantly reduced during fasts (usually measured as changes in glomerular filtration rate), due to decreased protein turnover and increased water conservation. Metabolic depression can also occur through downregulation of cellular processes. This hypometabolic state is not the result of major biochemical reorganization (Guppy and Withers 1999) but results from molecular controls operating at a level "above" that of allosteric regulation of enzymes and "below" that of gene expression. The net result is a "coordinated inactivation of many cellular processes" (Storey and Storey 2004).

8.1.3.1 *Pinnipeds*

8.1.3.1.1 Phocids Several studies have examined changes in RMR during the postweaning fast of phocid pups. Collectively, these illustrate apparent species-specific differences in patterns of changes in metabolism during the fast. Fasting gray seal pups demonstrate the expected rapid decrease in mass-specific RMR, which reaches a stable level 45% below initial levels by day 10 and continues largely unchanged through day 47 of the fast (Nordøy et al. 1990). A subsequent study confirmed the constancy of post-weaning mass-specific metabolism in both gray seals and harp seals fasting for 8 weeks in air, although they also found an increase in RMR over the fast for individuals fasting (but not tested) in water, potentially due to differences in blubber and mass loss incurred due to higher thermal costs (Worthy and Lavigne 1987). Such changes in RMR seem to conform to the standard fasting model.

However, a study on northern elephant seals revealed a different pattern. During the course of their 10-week molting fast, metabolism of pups gradually declined by 35% (Rea and Costa 1992). This may explain why fasting pups lose mass at a faster rate during the first 4 weeks of the fast than during the following 4-week period. Metabolism also notably increased during the final week of the fast, just prior to when they set off to sea to forage.

A few studies have been conducted on fasting in older phocids. Both experimentally fasted juvenile and adult harbor seals and gray seals display rapid (within 24 h) decreases in mass-specific metabolism (Markussen et al. 1992; Boily and Lavigne 1995; Markussen 1995), which is indicative of a rapid entrance into an adaptive "fasting state."

8.1.3.1.2 Otariids Many otariids pups are minimally active during the fasting period. For example, subantarctic fur seal pups spend more time sleeping than other otariid pups, contributing to lower daily rate of energy expenditure and lower rates of mass loss during their nursing fasts (Beauplet et al. 2003).

Several studies have examined changes in the RMR of otariid pups during the nursing period. Mass-specific RMR decreased significantly while Antarctic fur seal pups were fasting over 5 days (normal for maternal foraging trips) (Arnould et al. 2001). However, it is unclear whether part of the large drop might have been due to a large increase in metabolism attributable to digestion of the milk (i.e., the heat increment of feeding) in their stomachs at the start of the fasting period, pointing out a difficulty in measuring and interpreting resting metabolism in intermittent feeders. Subantarctic fur seal pups show increasing fasting abilities, including a decrease in mass-specific metabolism during their onshore fasts, which themselves increase from 5 to 30 days over the course of their extended period of maternal investment (Verrier et al. 2011).

However, the metabolic response of mammals during life-history stages or seasons when food shortages normally occur may be different from the response during periods when food shortages are unexpected. Young Steller sea lion pups (6–14 weeks old) significantly decrease metabolism in response to 48 h fasts. However, 6–24-month-old northern fur seals (*Callorhinus ursinus*) showed no changes in RMR when subject to a similar fast (Rosen et al. 2014). The difference might be attributable to the fact that, at this age, Steller sea lions would still be undergoing natural fasts during the nursing stage, while northern fur seals would already be weaned. By contrast, juvenile Steller sea lions outside their natural fasting period showed a 31% decrease in RMR over a 14-day experimental fast (Rosen and Trites 2002).

8.1.3.2 Other marine mammals

Polar bears likely exhibit episodes of metabolic depression throughout the year, and perhaps episodes of hibernation during denning (Nelson et al. 1983). Curiously, there are no published studies of energy-saving strategies among any other types of marine mammals. Hence, our vision of what makes up a "marine mammal" response is likely highly skewed. This is obviously an important area for future comparative research into the bioenergetic strategies of fasting among marine mammals.

8.1.4 Hormonal controls

Changes in an animal's physiology during a fast—including metabolic rate, appetite, growth, and choice of metabolic substrate—are all part of a controlled shift that is mediated by a suite of hormones. In this section, we will focus on those hormones thought to be most important in fasting physiology of marine mammals, but the list is by no means exhaustive.

Glucocorticoids are steroid hormones, known most commonly as a biochemical marker of physiological stress ("stress hormone"). However, they have multiple specific purposes (such as within the immune system) including serving an important regulatory role during fasting. In marine mammals, the primary glucocorticoid is *cortisol.* Plasma cortisol levels increase during prolonged fasting. Among its functions, cortisol helps to provide energy by increasing lipolysis and the associated mobilization of FFA. However, it also assists in maintaining circulating glucose concentrations via increased gluconeogenesis from protein stores. It is believed that during fasts the impacts of increased cortisol on lipid mobilization predominate over protein wasting effects (Ortiz et al. 2001).

The increase in lipolysis and lipid oxidation during extended fasting is thought to be regulated by increased levels of *growth hormone* (GH) and decreased levels of *leptin.* GH is a peptide hormone that, true to its name, stimulates cellular growth. However, it also functions as a "stress hormone," increasing in response to fasting in most species. It is important in the conservation of protein during fasting by raising the concentration of free fatty acids (via increased lipolysis). In non-fasting animals it also stimulates production of the hormone IGF-1 (*insulin-like growth factor 1*), which is important in protein anabolism. However, during fasting there is both an elevation of GH and suppression of IGF-1 (possibly due to reduced hepatic GH receptors) (Crocker et al. 2012). Inhibition of IGF-1 secretion allows hypersecretion of GH during fasting without diverting energy to tissue growth (Crocker et al. 2012).

Leptin is a relatively recently discovered hormone that is often referred to as the "satiety hormone." It is made by fat cells, and in feeding animals it regulates the amount of fat stored in the body through adjustment of the hunger response. In fed animals, as fat deposition surpasses a critical point, the fat cells release increasing levels of leptin, decreasing the sensation of hunger (increasing satiety), and increasing energy expenditures, promoting lipid oxidation and reducing triacylglycerol synthesis. However, episodes of fasting are usually associated with lowered leptin concentrations as part of an animal's strategy for limiting energy expenditures (Crocker et al. 2012).

This is partly because decreases in leptin are also associated with decreases in the *thyroid hormones* T3 (triiodothyronine) and T4 (thyroxine) in fasted animals. These hormones, produced by the thyroid gland, are primarily responsible for regulation of metabolism, and therefore decreased levels facilitate the decreases in metabolic expenditures associated with metabolic depression. This is despite the fact that the energy-demanding process of fat mobilization might require higher thyroid hormone levels (Ortiz et al. 2001). During extended fasting, reductions occur in levels of both T4 and T3, the former assisted by the conversion of T4 to the biologically inactive reverse T3 (rT3).

Ghrelin antagonizes the action of leptin and is known as the *hunger hormone*. Ghrelin is produced by specialized cells that line the stomach and the pancreas when the stomach is empty, while secretion stops when the stomach is stretched. It acts on hypothalamic brain cells, and its neural receptors are found on the same cells in the brain as the receptors for leptin. It serves to both increase hunger and to increase gastric acid secretion and gastrointestinal motility to prepare the body for food intake. It increases both appetite and fat mass by triggering receptors that stimulate production of neuropeptide Y. Ghrelin also functions as a growth hormone-releasing peptide. While leptin and ghrelin usually work in opposite directions in feeding animals, ghrelin levels should be lower in naturally fasting animals as a means of suppressing appetite (including suppression of neuropeptide Y).

However, as will become apparent, the results of experimental studies so far indicate that we have only begun to understand the hormonal control of physiological processes during fasting in marine mammals and, in fact, a new theoretical framework may be required given how poorly the experimental data to date match our current expectations.

8.1.4.1 Pinnipeds

As with many aspects of fasting physiology, most studies of hormonal changes have been conducted on northern elephant seal pups. In this species, many of the hormonal changes that support fasting occur earlier than expected, indicating that preparatory actions occur prior to weaning. Cortisol levels have been observed to increase between early and late nursing, and then again during early fasting, with the largest increase (more than doubling) seen between early and late fasting (Ortiz et al. 2003). All thyroid hormones decrease from early to late nursing, and then do not increase during fasting (except for a small increase in total T4). No significant changes have been observed in leptin levels during the fast (Ortiz et al. 2001), suggesting that leptin does not have an expected role in regulating body fat in fasting (or nursing) northern elephant seal pups (Ortiz et al. 2003).

Studies on the pups of other phocid species also yield unexpected results. Plasma cortisol levels in harp seal pups remained stable throughout their fast, and at levels equal to older, feeding pups (Nordøy et al. 1993). Neither was there any observed decrease in thyroid hormones. Similarly, plasma cortisol levels remained relatively low and stable in gray seals, except for a rise in some animals toward the end of their 52-day fasts (Nordøy et al. 1990).

In juvenile elephant seals, cortisol concentrations do not change during their seasonal fast, and neither do T3. However, both T4 and growth hormone concentrations decrease dramatically during the fast (Kelso et al. 2012).

Some of the hormone changes observed in adult male northern elephant seals during the breeding fast are closer to those predicted by fasting theory (Crocker et al. 2012). For example, there was a 43% decrease in GH, with no matching decrease in IGF-1. This reduction in GH should function to reduce lipolysis and increase hepatic glucose production, which would seem maladaptive in a fasting animal. However, it has been proposed that reductions in GH may be required to suppress more serious anabolic actions given that some level of protein catabolism is required for gluconeogenesis. While leptin concentrations decreased 11% (although they did not follow changes in fat mass), ghrelin concentrations did not change. This suggests that the high levels of ghrelin and dropping levels of leptin may function to suppress the drive to forage while optimizing rates of lipid oxidation (Crocker et al. 2012). The high levels of ghrelin also run counter to the observed reductions in GH, as increased ghrelin is usually associated with increased GH production in fasting animals. This suggests a loss of the ability of ghrelin to stimulate GH secretion in fasted adult seals. Further, the observed decreases in leptin were not associated with any significant changes in thyroid hormones. An exception was that changes in total T3 were directly related to changes in daily energy expenditure. Cortisol levels also did not change during the fast.

8.1.4.2 Cetaceans

Only a single study has examined multiple hormone changes in a fasting cetacean. This study was conducted on two fasting adult bottlenose dolphins, a species that likely rarely experiences prolonged fasts (Ortiz et al. 2010). The results confirmed the expected switch to lipid metabolism, but the swiftness of the response was more typical of mammalian species not adapted to regular fasting episodes. Plasma fatty acids doubled by 24 h and increased 2.5-fold by 38 h of fasting. Conversely, BUN decreased 17% by 24 h of fasting and 22% by 38 h. Plasma glucose decreased 25% between 14 and 24 h and levels returning to baseline by 38 h of fasting.

Neither plasma total T3 nor free T4 were changed. Mean total T4 increased 19% by 38 h of fasting, while mean rT3 showed an initial 30% decrease by 24 h of fasting, but returned to baseline levels by 38 h. The increase in total T4 might be due to decreased clearance rates (versus increased production) while the eventual recovery of rT3 by 38 h might reflect preferential deiodination of T4 to decrease cellular metabolism. While measured plasma cortisol levels were undetectable, these results may not be typical given values reported in other studies (Thomson and Geraci 1986; Ortiz and Worthy 2000).

8.2 Toolbox

8.2.1 Methods of measuring mass loss/body condition

Most pinniped studies use serial measures of body composition and body mass to calculate changes in the mass of specific tissues over time (usually differentiated into lipid and fat-free or lean mass). While a variety of methods can be used to determine body composition (whole-body dissections, direct imaging, ultrasonic measurements of lipid depth, and bioelectrical impedance), the most common method currently employed uses dilution of a chemical marker to indirectly estimate body composition through measures of whole body water content.

This method is based on the knowledge that different tissue types have different water content. The differential water content of tissues is the basis for how your home scale takes body fat measurements. While the technique has been detailed elsewhere (Reilly and Fedak 1990; Iverson et al. 2010), briefly, a small dose of isotopically distinct water (either deuterium oxide or tritiated water) is injected into the animal, and allowed time (usually several hours) to equilibrate with the rest of the animal's body water. A blood sample is analyzed for the resulting concentration of the chemical marker in the serum which, combined with the known amount injected, yields an estimate of the total body water. This value is then converted to estimates of lean and lipid mass through published mathematical equations previously generated from empirical studies (often involving carcass analysis). The accuracy of the technique is dependent upon the applicability of the mathematical models converting body water to body composition (these equations are often species- and even age-specific) and the hydration state of the animal (which may be an issue in fasting animals).

For many marine mammals, including most cetaceans, it is impractical to consider any techniques that require repeated capture and handling, and so new, innovative measures of physiological condition have to be developed. Photogrammetry—using photographs to make morphological measurements—have been investigated as a possible tool (De Bruyn et al. 2009), but is usually limited to detecting large changes in "body condition" (Pettis et al. 2004). The animal's diving behavior, which can be more readily monitored via attached dive recorders, may also provide an indirect measure of its relative body stores. For many marine mammals, a portion of the natural dive sequence is made up of an unpowered glide. The rate at which an animal ascends or descends through the water during a glide is dependent upon the hydrodynamic drag (a factor of body shape), water depth, and the buoyancy of the individual. An individual's buoyancy is altered by changes in the relative proportion of lipid mass in their body (as well as factors as the amount of air they dive with). Hence, rates of ascent/descent while gliding and changes in stroke rate have been used to detect differences in body condition in a number of species of marine mammals (Watanabe et al. 2006; Aoki et al. 2011; Schick et al. 2013). There have even been attempts to determine the metabolic status of cetaceans by chemically analyzing their captured exhalations (Aksenov et al. 2014).

8.2.2 Measuring metabolism and calculating metabolic depression

Metabolic rate is technically the amount of energy liberated or expended in a given unit of time by an animal. While the earliest studies of metabolism measured the amount of heat an animal produced (often by measuring the change in temperature of a surrounding water bath or ice mixture), later studies realized that aerobic metabolism consumed a set amount of oxygen (and produced a set amount of carbon dioxide) depending on the exact fuel source. Respirometry, the science of measuring the rate of oxygen consumption, has become the standard method of measuring rates of metabolism. The method is quite simple; usually ambient air is drawn at a known rate through a sealed chamber containing the organism (or at least within which it must breath). The excurrent airflow is sampled to determine the concentration of oxygen (and often carbon dioxide). Knowing the rate of airflow, and the difference in gas concentrations between the sampled and the ambient air (which is essentially constant) allows one to calculate the rate of oxygen consumption. Often, metabolic rate is presented as a rate of oxygen consumption; while this can be converted to a rate of energy use, this step also involves several assumptions. One of these assumptions is the nature of the metabolic substrate, which can be elucidated by examining the ratio of carbon dioxide produced to oxygen consumed (known as the respiratory quotient).

Unfortunately, respirometry has limited applications in the field. Scientists have developed a number of proxies to indirectly estimate rates of energy expenditure from wild marine mammals, including heart rate, flipper strokes, and body acceleration (Iverson et al. 2010). By far, the most common method involves the differential turnover of two isotopically labeled waters, known as the doubly labeled water method (Lifson and McClintock 1966). Still, most of these methods cannot provide estimates of Resting Metabolic Rate, but only estimate an animal's average metabolic rate over a period of time, known as its field metabolic rate (FMR), which may be affected by parallel changes in activity, thermoregulation, and other factors.

Determining changes in resting metabolism over the course of a fast is complicated by the fact that the animal is simultaneously losing body mass, which will inherently decrease total energy expenditures. In addition, body mass loss during different phases of the fast may be due to loss of tissues that are metabolically active (protein) or relatively metabolically inert (lipid stores). For metabolic depression to occur, the decrease in metabolic rate during the fast must not be a mere consequence of body mass loss. Some authors have suggested that true metabolic depression can be said to occur if there is a clear decrease in this rate when it is expressed per unit body mass (mass-specific metabolism) (Cherel et al. 1988). While some might argue that this incorrectly assumes a specific relationship between body mass and metabolic rate, it is certainly a conservative way of calculating changes in metabolism.

8.2.3 *Plasma metabolites as biomarkers of fasting phase*

We can analyze the metabolites present in a blood sample for clues about the fasting state of an animal. Metabolites are the end-products and the intermediaries of all chemical reactions in the body. They can, therefore, provide evidence about the rate and amount of specific substrate types that are broken down across time during fasting. Examples of these metabolites are mentioned earlier in this chapter and include glucose, FFA, and BUN.

Metabolites can be analyzed in two ways to study fasting. Specific known biomarkers associated with metabolic processes can be analyzed in a plasma or serum sample. Glucose and BUN, for example, are routinely measured in human and veterinary clinical applications and are available on standard blood chemistry panels. FFA vary by chain length and number of double bonds, meaning that researchers must decide whether to simply measure all circulating lipid, or to measure the specific composition of the lipid pool. The latter requires more sophisticated analytical methods, including gas chromatography. Metabolites can also be analyzed by large-scale screens of the metabolome (the entire metabolite population), taking advantage of rapidly advancing technology and analytical methods to find patterns and signals in large data sets. The benefit of evaluating the entire metabolome is that more data are retrieved from the single sample, which allow further investigation into the details of substrate turnover and the interaction between fasting and other physiological processes (e.g., lactation, stress, molt).

Metabolites are small molecular weight chemicals that are identified by metrics such as their mass, charge, pH, and hydrophobicity. Metabolomics platforms at research facilities and in industry combine liquid or gas chromatography to separate metabolites by their chemical characteristics from a biosample, and then the mass of separated metabolites is determined by mass spectrometry. Knowing the specifics of mass along with charge, etc., identifies metabolites from catalogs of hundreds of known (and synthetic) chemicals; these platforms are also able to quantify each metabolite within a biosample. Using metabolites as the target of large-scale screening is especially useful for studying marine mammals because they can be consistently identified by their composition and does not rely on comprehensive genomic information (needed to identify proteins and transcripts based on their sequences), which is not yet available for many marine mammal species.

A limitation to using metabolite signatures to study fasting is that they require tissue or blood sample collection. Samples must also be quickly processed (e.g., to separate the plasma or serum from red blood cells) and stabilized by freezing to limit the degradation of metabolites. Hence, we know the most about the biochemistry of fasting in elephant seals, and it is not hard to imagine that this is partly because they fast on accessible beaches. Collecting and carefully preserving a blood sample from a migrating, fasting whale in pelagic waters is obviously more daunting.

8.2.4 Turnover of metabolic tracers

Similar to the doubly labeled water method, the dilution and breakdown of tracer chemicals can be used to track substrate use during fasting. This requires an even more complicated field sampling strategy than a single blood draw. An initial blood sample is collected for baseline information, and subsequent samples are collected to determine the rate of tracer chemical turnover. Potentially, several samples must be collected to confirm this rate. Often, the animal's blood volume (typically measured by dilution of IV-injected Evans blue dye) is also determined. Depending on the timeframe and the study species, it is necessary to sedate the animals to collect blood or give injections. The length of these procedures can complicate sampling strategy. However, this type of experiment provides very useful, specific data about the metabolism of substances and has been successfully accomplished in fasting elephant seals for labeled urea, fatty acids, and glucose, to name a few (Pernia et al. 1980; Castellini et al. 1987; Houser and Costa 2001).

8.3 Lingering mysteries

There are many specific details of fasting physiology that we have only begun to explore. While our depth of understanding of basic fasting physiology is rapidly increasing, its breadth is still very narrow. The bulk of our knowledge on hormonal control mechanisms, for example, comes from a single species, the northern elephant seal. Studies with other pinnipeds have supported a basic conclusion that marine mammals are well-adapted to their natural fasts, but the manner in which they accomplish this feat differs significantly between species and from terrestrial mammals. But these scientific glimpses are exceedingly limited. Perhaps the most important focus for future research is to explore fasting physiology in a wider range of species, with particular emphasis on cetaceans. This increased knowledge will also allow us to determine what evolutionary processes have honed species-specific fasting strategies, even among closely related animals.

The question of the fasting capacity of marine mammals is of more than mere academic interest but also has important conservation implications. For example, in 2014–2015, unprecedented numbers of stranded, starving California sea lions (*Zalophus californianus*) began arriving on beaches of the U.S. west coast. Scientists believe that anomalously warm coastal waters shifted the food base, requiring lactating mothers to prolong their trips to sea. The result is an extension beyond natural fasting durations that the pups cannot endure, driving them into the water to begin their own foraging too early.

With concern increasing over the impacts to wild populations of both natural and anthropogenic disturbance, it is timely that we start addressing how predictable periods of fasting interact with unpredicted perturbations. How much of an additional physiological burden are imposed by threats such as disease, pollution, and human harassment? How might survival and reproductive capacity be impacted by environmental changes resulting in slightly longer or more expensive fasts, slightly smaller energy reserves, or altered seasonal timing of fasts? Future research will help us understand how changing oceans could affect marine mammal populations.

Glossary

Endogenous body reserves: Internal tissues that are broken down (*catabolized*) by an animal as fuel when energetic expenditures are greater than energy intake.

Fasting: Voluntary cessation of food intake.

Glucocorticoids: A group of steroid hormones that are important in carbohydrate, lipid, and protein metabolism, as well as immune and stress responses. In marine mammals, the prominent glucocorticoid is *cortisol.*

Gluconeogenesis: The process of manufacturing the required fuel *glucose* from the non-carbohydrate precursors *glycerol* (from *triacylglycerols*) and *amino acids* (derived from proteins). Gluconeogenesis occurs mainly in the liver and kidneys.

Hormones: Chemical messengers that are produced in the body that control and regulate the activity of certain cells or organs. This includes hormones that affect appetite (leptin, ghrelin), growth (growth hormone [GH], *insulin-like growth factor 1* [IGF-1]), and metabolism (*thyroid hormones,* T3 and T4).

Ketone bodies: Carbon structures that derive from oxidized fatty acids and are manufactured in the liver, which can be used as a fuel source during fasting.

Resting Metabolic Rate (RMR): A measure of energy expenditure reflecting the total energy requirements of an inactive animal within its thermoneutral zone.

Triacylglycerols: A high-energy lipid source derived from *glycerol* and three fatty acids. It is the primary lipid in the blubber layer of marine mammals and can be broken down via *lipolysis* into *glycerol* and *free fatty acids* (FFA).

References

Adams, S.H. and D.P. Costa. 1993. Water conservation and protein metabolism in northern elephant seal pups during the postweaning fast. *Journal of Comparative Physiology B* 163:367–373.

Aksenov, A.A., L. Yeates, A. Pasamontes et al. 2014. Metabolite content profiling of bottlenose dolphin exhaled breath. *Analytical Chemistry* 86:10616–10624.

Aoki, K., Y.Y. Watanabe, D.E. Crocker et al. 2011. Northern elephant seals adjust gliding and stroking patterns with changes in buoyancy: Validation of at-sea metrics of body density. *Journal of Experimental Biology* 214(17):2973–2987.

Arnould, J.P.Y, J.A. Green, and D.R. Rawlins. 2001. Fasting metabolism in Antarctic fur seal (*Arctocephalus gazella*) pups. *Comparative Biochemistry and Physiology A* 129:829–841.

Beauplet, G., C. Guinet, and J.P.Y. Arnould. 2003. Body composition changes, metabolic fuel use, and energy expenditure during extended fasting in subantarctic fur seal (*Arctocephalus tropicalis*) pups at Amsterdam Island. *Physiological and Biochemical Zoology* 76(2):262–270.

Bennett, K.A., B.J. McConnell, S.E. Moss, J.R. Speakman, P.P. Pomeroy, and M.A. Fedak. 2010. Effects of age and body mass on development of diving capabilities of gray seal pups: Costs and benefits of the postweaning fast. *Physiological and Biochemical Zoology* 83(6):911–923.

Bentley, P.J. 1963. Composition of the urine of the fasting humpback whale (*Megaptera nodosa*). *Comparative Biochemistry and Physiology* 10(3):257–259.

Boily, P. and D.M. Lavigne. 1995. Resting metabolic rates and respiratory quotients of gray seals (*Halichoerus grypus*) in relation to time of day and duration of food deprivation. *Physiological Zoology* 68(6):1181–1193.

Bowen, W.D., D.J. Boness, and O.T. Oftedal. 1987. Mass transfer from mother to pup and subsequent mass loss by the weaned pup in the hooded seal, *Cystophora cristata*. *Canadian Journal of Zoology* 65(1):1–8.

Boyd, I.L. 1991. Environmental and physiological factors controlling the reproductive cycles of pinnipeds. *Canadian Journal of Zoology* 69(5):1135–1148.

Burns, J.M., K. Lestyk, D. Freistroffer, and M.O. Hammill. 2014. Preparing muscles for diving: Age-related changes in muscle metabolic profiles in harp (*Pagophilus groenlandicus*) and hooded (*Cystophora cristata*) seals. *Physiological and Biochemical Zoology* 88(2):167–182.

Castellini, M.A. and D.P. Costa. 1990. Relationship between plasma ketones and fasting duration in neonatal elephant seals. *American Journal of Physiology* 259(5):R1086–R1089.

Castellini, M.A., D.P. Costa, and A.C. Huntley. 1987. Fatty acid metabolism in fasting elephant seal pups. *Journal of Comparative Physiology B* 157(4):445–459.

Castellini, M.A. and L.D. Rea. 1992. The biochemistry of natural fasting at its limits. *Experientia* 48(6):575–582.

Cherel, Y., J.-P. Robin, and Y. Le Maho. 1988. Physiology and biochemistry of long-term fasting in birds. *Canadian Journal of Zoology* 66(1):159–166.

Corkeron, P.J. and R.C. Connor. 1999. Why do baleen whales migrate? *Marine Mammal Science* 15(4):1228–1245.

Costa, D.P. and C.L. Ortiz. 1982. Blood chemistry homeostasis during prolonged fasts in the northern elephant seal. *American Journal of Physiology* 242(5):R591–R595.

Crocker, D.E., D.S. Houser, and P.M. Webb. 2012. Impact of body reserves on energy expenditure, water flux, and mating success in breeding male northern elephant seals. *Physiological and Biochemical Zoology* 85(1):11–20.

Crocker, D.E., R.M. Ortiz, D.S. Houser, P.M. Webb, and D.P. Costa. 2012. Hormone and metabolite changes associated with extended breeding fasts in male northern elephant seals (*Mirounga angustirostris*). *Comparative Biochemistry and Physiology A* 161(4):388–394.

De Bruyn, P.J., M.N. Bester, A.R. Carlini, and W.C. Oosthuizen. 2009. How to weigh an elephant seal with one finger: A simple three-dimensional photogrammetric application. *Aquatic Biology* 5:31–39.

Durban, J.W. and R.L. Pitman. 2011. Antarctic killer whales make rapid, round-trip movements to subtropical waters: Evidence for physiological maintenance migrations? *Biology Letters* 8:274–277. doi: rsbl20110875.

Edney, E.B. 1977. Metabolic water. In *Water Balance in Land Arthropods.*, Vol. 9: Zoophysiology and Ecology, eds. W.S. Hoar, B. Hoelldobler, H. Langer, and M. Lindauer. Berlin, Germany: Springer-Verlag.

Fowler, S.L., D.P. Costa, J.P.Y. Arnould, N.J. Gales, and J.M. Burns. 2007. Ontogeny of oxygen stores and physiological diving capacity in Australian sea lions. *Functional Ecology* 21(5):922–935.

Georges, J.-Y. and C. Guinet. 2000. Maternal care in the subantarctic fur seals on Amsterdam Island. *Ecology* 81(2):295–308.

Guppy, M. and P. Withers. 1999. Metabolic depression in animals: Physiological perspectives and biochemical generalizations. *Biological Review* 74(1):1–40.

Henry, C.J., J.P. Rivers, and P.R. Payne. 1988. Protein and energy metabolism in starvation reconsidered. *European Journal of Clinical Nutrition* 42(7):543–549.

Horwood, J. 1990. *Biology and Exploitation of the Minke Whale*. Boca Raton, FL: CRC Press, Inc.

Houser, D. and D. Costa. 2001. Protein catabolism in suckling and fasting northern elephant seal pups (*Mirounga angustirostris*). *Journal of Comparative Physiology B* 171(8):635–642.

Houser, D.S. and D. Costa. 2003. Entrance into stage III fasting by starveling northern elephant seal pups. *Marine Mammal Science* 19(1):186–197.

Houser, D.S., C.D. Champagne, and D.E. Crocker. 2007. Lipolysis and glycerol gluconeogenesis in simultaneously fasting and lactating northern elephant seals. *American Journal of Physiology—Regulatory, Integrative and Comparative Physiology* 293(6):R2376–R2381.

Irving, L. and J.S. Hart. 1957. The metabolism and insulation of seals as bare-skinned mammals in cold water. *Canadian Journal of Zoology* 35(4):497–511.

Iverson, S.J., C.E. Sparling, T.M. Williams, S.L.C. Lang, and W.D. Bowen. 2010. Measurement of individual and population energetics of marine mammals. In *Marine Mammal Ecology and Conservation: A Handbook of Techniques*, eds. I.L. Boyd, W.D. Bowen, and S.J. Iverson. Oxford, UK: Oxford University Press.

Kelso, E.J., C.D. Champagne, M.S. Tift, D.S. Houser, and D.E. Crocker. 2012. Sex differences in fuel use and metabolism during development in fasting juvenile northern elephant seals. *Journal of Experimental Biology* 215(15):2637–2645.

Keys, A., A. Brozek, A. Henschel, O. Micckelsen, and H.L. Taylor. 1950. *The Biology of Human Starvation*. Minneapolis, MN: University of Minnesota Press.

Kleiber, M. 1975. *The Fire of Life: An Introduction to Animal Energetics*. New York: Robert E. Krieger Publ. Co.

Kovacs, K.M. and D.M. Lavigne. 1985. Neonatal growth and organ allometry of northwest Atlantic harp seals (*Phoca groenlandica*). *Canadian Journal of Zoology* 63(12):2793–2799.
LeBoeuf, B.J. 1974. The hectic life of the alpha bull: Elephant seal as fighter and lover. *Psychology Today* 8(5):104–108.
Lieberman, M. and A.D. Marks. 2009. *Marks' Basic Medical Biochemistry: A Clinical Approach*. Baltimore, MD: Lippincott Williams & Wilkins.
Lifson, N. and R. McClintock. 1966. Theory and use of the turnover rates of body water for measuring energy and material balance. *Journal of Theoretical Biology* 12(1):46–74.
Lunn, N.J., I.L. Boyd, T. Barton, and J.P. Croxall. 1993. Factors affecting the growth rate and mass at weaning of Antarctic fur seals at Bird Island, South Georgia. *Journal of Mammalogy* 74(4):908–919.
Lydersen, C., K.M. Kovacs, and M.O. Hammill. 1997. Energetics during nursing and early postweaning fasting in hooded seal (*Cystophora cristata*) pups from the Gulf of St Lawrence, Canada. *Journal of Comparative Physiology B* 167(2):81–88.
Markussen, N.H. 1995. Changes in metabolic rate and body composition during starvation and semistarvation in harbour seals. In *Developments in Marine Biology 4: Whales, Seals, Fish and Man*, eds. A.S. Blix, L. Walløe, and Ø. Ulltang. Amsterdam, the Netherlands: Elsevier.
Markussen, N.H., M. Ryg, and N.A. Øritsland. 1992. Metabolic rate and body composition of harbour seals, *Phoca vitulina*, during starvation and refeeding. *Canadian Journal of Zoology* 70(2):220–224.
McCue, M.D. 2010. Starvation physiology: Reviewing the different strategies animals use to survive a common challenge. *Comparative Biochemistry and Physiology A* 156(1):1–18.
Mehta, A.V., J.M. Allen, R. Constantine et al. 2007. Baleen whales are not important as prey for killer whales *Orcinus orca* in high-latitude regions. *Marine Ecology Progress Series* 348:297–307.
Muelbert, M.M.C. and W.D. Bowen. 1993. Duration of lactation and postweaning changes in mass and body composition of harbour seal, *Phoca vitulina*, pups. *Canadian Journal of Zoology* 71(7):1405–1414.
Nelson, R.A., G.E. Folk Jr., E.W. Pfeiffer, J.J. Craighead, C.J. Jonkel, and D.L. Steiger. 1983. Behavior, biochemistry, and hibernation in black, grizzly, and polar bears. *Bears: Their Biology and Management* 5:284–290.
Nordøy, E.S., A. Aakvaag, and T.S. Larsen. 1993. Metabolic adaptations to fasting in harp seal pups. *Physiological Zoology* 66(6):926–945.
Nordøy, E.S. and A.S. Blix. 1985. Energy sources in fasting gray seal pups evaluated with computed tomography. *American Journal of Physiology* 249(18):R471–R476.
Nordøy, E.S. and A.S. Blix. 1991. Glucose and ketone body turnover in fasting gray seal pups. *Acta Physiologica Scandinavica* 141(4):565–571.
Nordøy, E.S., O.C. Ingebretsen, and A.S. Blix. 1990. Depressed metabolism and low protein catabolism in fasting gray seal pups. *Acta Physiologica Scandinavica* 139(2):361–369.
Noren, D.P., D.E. Crocker, T.M. Williams, and D.P. Costa. 2003. Energy reserve utilization in northern elephant seal (*Mirounga angustirostris*) pups during the postweaning fast: Size does matter. *Journal of Comparative Physiology B* 173(5):443–454.
Noren, D.P. and M. Mangel. 2004. Energy reserve allocation in fasting northern elephant seal pups: Inter-relationships between body condition and fasting duration. *Functional Ecology* 18(2):233–242.
Øritsland, N.A. 1990. Starvation survival and body composition in mammals with particular reference to *Homo sapiens*. *Bulletin of Mathematical Biology* 52(5):643–655.
Øritsland, N.A. and N.H. Markussen. 1990. Outline of a physiologically-based model for population energetics. *Ecological Modelling* 52:267–286.
Øritsland, N.A., A.J. Pasche, N.H. Markussen, and K. Ronald. 1985. Weight loss and catabolic adaptations to starvation in gray seal pups. *Comparative Biochemistry and Physiology A* 82(4):931–933.
Ortiz, R.M., D.S. Houser, C.E. Wade, and C.L. Ortiz. 2003. Hormonal changes associated with the transition between nursing and natural fasting in northern elephant seals (*Mirounga angustirostris*). *General and Comparative Endocrinology* 130(1):78–83.
Ortiz, R.M., B. Long, D. Casper, C.L. Ortiz, and T.M. Williams. 2010. Biochemical and hormonal changes during acute fasting and re-feeding in bottlenose dolphins (*Tursiops truncatus*). *Marine Mammal Science* 26(2):409–419.

Ortiz, R.M., D.P. Noren, B. Litz, and C.L. Ortiz. 2001. A new perspective on adiposity in a naturally obese mammal. *American Journal of Physiology Endocrinology and Metabolism* 281:E1347–E1351.

Ortiz, R.M., C.E. Wade, and C.L. Ortiz. 2001. Effects of prolonged fasting on plasma cortisol and TH in postweaned northern elephant seal pups. *American Journal of Physiology* 280(3):R790–R795.

Ortiz, R.M. and G.A.J. Worthy. 2000. Effects of capture on adrenal steroid and vasopressin concentrations in free-ranging bottlenose dolphins (*Tursiops truncatus*). *Comparative Biochemistry and Physiology A* 125(3):317–324.

Pernia, S.D., A. Hill, and C.L. Ortiz. 1980. Urea turnover during prolonged fasting in the northern elephant seal. *Comparative Biochemistry and Physiology B* 65(4):731–734.

Pettis, H.M., R.M. Rolland, P.K. Hamilton, S. Brault, A.R. Knowlton, and S.D. Kraus. 2004. Visual health assessment of North Atlantic right whales (*Eubalaena glacialis*) using photographs. *Canadian Journal of Zoology* 82(1):8–19.

Rea, L.D. 1995. Prolonged fasting in pinnipeds. PhD Thesis, University of Alaska, Fairbanks, AK.

Rea, L.D., M. Berman-Kowalewski, D.A.S. Rosen, and A.W. Trites. 2009. Seasonal differences in biochemical adaptation to fasting in juvenile and subadult Steller sea lions (*Eumetopias jubatus*). *Physiological and Biochemical Zoology* 82(3):236–247.

Rea, L.D. and D.P. Costa. 1992. Changes in standard metabolism during long-term fasting in northern elephant seal pups (*Mirounga angustirostris*). *Physiological Zoology* 65(1):97–111.

Rea, L.D., D.A.S. Rosen, and A.W. Trites. 2000. Metabolic response to fasting in 6-week-old Steller sea lion pups (*Eumetopias jubatus*). *Canadian Journal of Zoology* 78(5):890–894.

Reilly, J.J. 1991. Adaptations to prolonged fasting in free-living weaned gray seal pups. *American Journal of Physiology* 260(29):R267–R272.

Reilly, J.J. and M.A. Fedak. 1990. Measurement of the body composition of living gray seals by hydrogen isotope dilution. *Journal of Applied Physiology* 69(3):885–891.

Reiter, J., N.L. Stinson, and B.J. LeBoeuf. 1978. Northern elephant seal development; transition from weaning to nutritional independence. *Behavioral Ecology and Sociobiology* 3(4):337–367.

Rosen, D.A.S. and A.W. Trites. 2002. Changes in metabolism in response to fasting and food restriction in the Steller sea lion. *Comparative Biochemistry and Physiology B* 132(2):389–399.

Rosen, D.A.S., B.L. Volpov, and A.W. Trites. 2014. Short-term episodes of imposed fasting have a greater effect on young northern fur seals (*Callorhinus ursinus*) in summer than in winter. *Conservation Physiology* 2(1):1–9.

Rosen, D.A.S., A.J. Winship, and L.A. Hoopes. 2007. Thermal and digestive constraints to foraging behaviour in marine mammals. *Philosophical Transactions, Royal Society of London B* 362(1487):2151–2168.

Schick, R.S., L.F. New, L. Thomas et al. 2013. Estimating resource acquisition and at-sea body condition of a marine predator. *Journal of Animal Ecology* 82(6):1300–1315.

Schmidt-Nielsen, K. 1997. *Animal Physiology: Adaptation and Environment*, 5th edn. Cambridge, UK: Cambridge University Press.

Sharick, J.T., J.P. Vazquez-Medina, R.M. Ortiz, and D.E. Crocker. 2014. Oxidative stress is a potential cost of breeding in male and female northern elephant seals. *Functional Ecology* 29(3):367–376.

Siverston, E. 1941. *On the Biology of Harp Seal, Phoca groenlandica, Erxl., Investigations Carried Out in the White Sea 1925–1937*, Vol. 26. Oslo, Norway: Hvalradets Skrifter.

Stewart, R.E.A. and D.M. Lavigne. 1980. Neonatal growth of northwest Atlantic harp seals, *Pagophilus groenlandicus*. *Journal of Mammalogy* 61(4):670–680.

Storey, K.B. and J.M. Storey. 2004. Metabolic rate depression in animals: Transcriptional and translational controls. *Biological Reviews* 79(1):207–233.

Thomson, C.A. and J.R. Geraci. 1986. Cortisol, aldosterone, and leucocytes in the stress response of bottlenose dolphins, *Tursiops truncatus*. *Canadian Journal of Fisheries and Aquatic Sciences* 43(5):1010–1016.

Verrier, D., R. Groscolas, C. Guinet, and J.P.Y. Arnould. 2011. Development of fasting abilities in subantarctic fur seal pups: Balancing the demands of growth under extreme nutritional restrictions. *Functional Ecology* 25(3):704–717.

Watanabe, Y., E.A. Baranov, K. Sato, Y. Naito, and N. Miyazaki. 2006. Body density affects stroke patterns in Baikal seals. *Journal of Experimental Biology* 209(17):3269–3280.

Worthy, G.A.J. 1991. Insulation and thermal balance of fasting harp and gray seal pups. *Comparative Biochemistry and Physiology* 100A(4):845–851.
Worthy, G.A.J. and D.M. Lavigne. 1982. Changes in blood properties of fasting and feeding harp seal pups, *Phoca groenlandica,* after weaning. *Canadian Journal of Zoology* 60(4):586–592.
Worthy, G.A.J. and D.M. Lavigne. 1983. Energetics of fasting and subsequent growth in weaned harp seal pups, *Phoca groenlandica. Canadian Journal of Zoology* 61(2):447–456.
Worthy, G.A.J. and D.M. Lavigne. 1987. Mass loss, metabolic rate, and energy utilization by harp and gray seal pups during the postweaning fast. *Physiological Zoology* 60(3):352–364.

chapter nine

Thermoregulation

Michael A. Castellini and Jo-Ann Mellish

Contents

9.1 The big picture challenge

One of the first questions when seeing a seal or a whale in the ocean is to ask "How do they stay warm in that cold water?" As humans, we cannot survive for long in the ocean because it is much cooler in absolute terms than our body temperature, and it also carries heat away from our bodies faster than air. Yet, many marine mammals spend their entire lives in this environment. The obvious answer to this dilemma is that they often have very thick fur or blubber that insulates them from the colder water. While this may be true in many cases, it creates another challenge that is usually not considered: How do they cool off if they are exercising very hard, come into very warm water, or haul out on a hot sandy beach? If the fur and blubber are so effective that they can protect them from water as much as 40°C colder than their core body temperature, then it must be a problem for them to cool off or dump heat.

The answer to this apparent conflict is that marine mammals have developed highly evolved *thermoregulatory* systems that not only allow them to conserve heat under cold conditions, but to release heat under warm conditions. Of course, a species' ability to conserve versus dissipate heat varies widely given the global distribution of this large group of mammals. For instance, some cetaceans spend their entire lives in ice-laden polar waters, and therefore would only need to disperse excess heat when exercising very hard. Other cetaceans migrate regularly between polar and tropical waters. Many pinnipeds haul out onto beaches where it is very hot, but spend considerable amounts of time in very cold waters. The polar pinnipeds encounter the reverse, where temperatures in air are dramatically lower than the cold water. A suite of physiological, anatomical, and biochemical tools, unique to each species in some detail, enables them to maintain their body temperatures in the face of these many environmental changes. In this section, we will look at the basic principles of heat flow and discuss how they apply to marine mammal thermoregulation in general, as well as some species-specific details.

Like almost all of their terrestrial relatives, marine mammals are *endothermic homeotherms*. That is, they generate their own heat and hold a constant body temperature well above that of the environment. These groups were called *warm blooded* in older literature, to distinguish them from the *cold blooded* reptiles, amphibians, and so on. In modern terminology, lizards, snakes, fish, and other groups are considered *ectothermic heterotherms*. This means that their body temperature mostly fluctuates with their surrounding environment. A few groups cross these lines, such as some of the large, warm-bodied tunas and sharks. They are considered *ectothermic homeotherms*, in that they generate and conserve a large amount of their own heat in some areas of their bodies. The true mammalian hibernators (e.g., Arctic ground squirrels; *Spermophilus parryii*) are *endothermic heterotherms* that let their body temperature fluctuate widely under specific conditions, but generate their own heat to warm up.

9.2 Basics of heat flow

The basic equation of heat balance involves moving heat energy through *conduction*, *convection*, *radiation*, and *evaporation*. In conduction, heat flows from a warm body that is in contact with a colder body following the principles of the *Fick equation*. The rate of heat transfer is determined by the area of contact, the thickness of the surface that the heat must travel through, the nature of the material in contact with the object, and the temperature difference between the warm and colder bodies as below:

$$\text{Rate} = \frac{\text{kAT}_2 - \text{T}_1}{\text{thickness}}$$

where the rate of heat flow is directly proportional to the thermal conductivity of the material (k; cal s^{-1} cm^{-1} $°C^{-1}$), area of contact (A; cm^2) normal to the heat flow direction, the temperature difference ($T_1 - T_2$; °C) between the two bodies, and inversely proportional to the thickness of the material (cm). One can visualize this by thinking about being outside on a cold day with a warm, down jacket. We know that our down jacket with its great insulation capacity works better than a thin, cotton jacket. Similarly, if we can curl up into a smaller size, we will stay warmer. All of these are practical aspects of the Fick equation describing heat flow.

Consider each of the Fick variables: Conduction is the direct contact transfer of heat energy from a warm body to a cooler body. A common misperception is that cold flows from the cold to the warm body. Energy can only move down a gradient, not up. For marine mammals, this is heat that will be lost to any contact material that is cooler than their external body temperature, be it land, sea ice, water, or air. *Convection* is the loss of heat through the combined effect of conduction and the movement of air or water molecules. We commonly know this in air as wind chill. Wind chill cannot be measured simply on a thermometer. When the air temperature is –25°F, the absolute temperature will be the same whether the wind is blowing or not, despite how much colder it may feel. This function is important not only for marine mammals that haul out on land or ice, but also when animals encounter strong water currents or are swimming rapidly through the water, which can drive the cooling power up to almost 100 times that of air (Schmidt-Nielsen 1997). *Radiation* is the release of thermal energy from all matter, which does not require a medium for transfer. The sun heats the earth through radiation, as there is no air in space to facilitate conductive transfer of energy. A seal warming on the ice surface by orienting to the sun is an excellent example of radiative heat gain. *Evaporation* is the application of body heat to disperse surface water. Convection can speed up the process by pulling away the moisture on the skin surface. Humans recognize this phenomenon as sweating. Evaporation can play a significant role in the thermoregulation of those marine mammal species that spend large amounts of time in air.

Body size and the ratio of surface area to volume (SA:V) are also important to the Fick equation. Consider how the relative surface area of a simple sphere increases as the volume goes down (volume = $4/3\Pi r^3$; surface area $4\Pi r^2$). A sphere with a diameter of 10 cm has surface area of 314 cm^2, a volume of 524 cm^3, and a SA:V 0.60. If we reduce the diameter to 5 cm, the surface area is 78.5 cm^2, the volume is 65.5 cm^3, and the SA:V is now increased to 1.2. A quick shortcut to this calculation is that the SA:V of a sphere is given by the value of 3/r. This SA:V ratio means that an otter will have a proportionately larger surface area than a small dolphin. However, that same dolphin has a much higher SA:V than a whale. All else being equal, the Fick equation dictates that a smaller animal will experience greater heat loss than a larger animal because its surface area is proportionately larger.

Finally, when comparing heat loss in air or water, an essential point to emphasize is that the *heat capacity* of water is about 25× that of air. Applied to the Fick principle, this means that all other factors being held constant (temperature differential, surface area, etc.), heat will flow out of a warm body 20–25 times faster in water than in air. For a marine mammal, being in the water implies a heat challenge significantly greater than to a comparable terrestrial mammal, or even to itself when on land. Cetaceans are always confronted with the greater heat loss capacity of water.

9.2.1 *Heat flow applied to marine mammals*

Marine mammals are endothermic homeotherms, subject to all of the components of the Fick equation (conduction, convection, radiation, and evaporation), albeit to varying degrees. They are primarily susceptible to heat loss through conduction (in all circumstances) and convection when in the water. Even though there is limited radiative heat exchange in water, this component can come into play to a much larger degree for the pinnipeds when hauled out on beaches or ice (Mellish et al. 2015). There is some evidence that seals will haul out on beaches for pupping or during the annual molt when radiative heat gain is most likely to offset the losses of conduction and convection (Boily 1995; Hind and Gurney 1998). While there is no direct observation of sweating in marine mammals as a method of heat loss, there is histological evidence of sweat glands in at least some species (Ling 1965; Bryden and Button 1977; Rotherham et al. 2005; Khamas et al. 2012). This makes evolutionary sense in that these species spend most of their time in the water, where evaporation as a means to offset excess body heat is impossible. Instead, many species will employ behavioral tactics to compensate when cooling is necessary, such as

Table 9.1 Thermal conductance of various materials

Silver	1.0181
Water	0.0013
Organic tissue	0.0011
Leather	0.0004
Crude oil	0.00025
Blubber (as pelt)	0.0001
Fur	0.00009
Air	0.00005
Eider down	0.00001

Sources: Data from the *CRC Handbook of Chemistry and Physics* and Bryden (1964); Elam et al. (1989); Schmidt-Nielsen (1997); Folk et al. (1998); Bagge et al. (2012).

Note: Units are cal s^{-1} cm^{-1} $°C^{-1}$.

moving to cooler waters, or entering the water from the land. Northern elephant seals (*Mirounga angustirostris*) will flip sand over their backs to reduce radiative heat gain on hot and sunny days when they are hauled out along the California beaches (Heath and Schusterman 1975).

9.2.2 *What is insulation?*

Insulation acts by changing the thickness measurement of Fick equation (5 cm of an insulator is better than 2 cm for reducing heat flow) and/or by changing the *thermal conductivity* (k) of the material (Table 9.1). Thermal conductivity and insulation are the inverse of each other. Returning to our winter jacket analogy, feather down traps large amounts of air and therefore is far superior for insulation when compared to silver. However, this means that down has a very low thermal conductivity, whereas our poor insulating silver is an excellent conductor. Several classic textbooks deal with the many intricacies and equations of thermoregulation, including *Animal Physiology* (Schmidt-Nielsen 1997) and *Principles of Integrative Physiology* (Folk et al. 1998). MacArthur, Kooyman, and Castellini provide reviews of thermoregulation with marine mammal perspectives (Kooyman 1981; MacArthur 1989; Castellini 2009).

9.3 *Insulation in marine mammals*

Given what we now know about the basic principles of heat transfer in general, how can we apply this to marine mammals? As we mentioned earlier, insulation via fur or blubber is a well-known mechanism for effective thermoregulation in the cold ocean. However, beyond the initial identification of these two characteristics, the details of how fur and blubber work are quite complex.

9.3.1 *Fur*

The best fur is an effective insulator, and therefore poor heat conductor, by trapping air between hairs and creating a barrier to keep the skin dry. The thermal gradient defined by the Fick principle (Figure 9.1a) follows a path from the warm skin surface, through the fur, and finally to the external environment. Consider that the common rat has a hair density of 95 hairs/mm^2, compared to the sea otter with 1188 hairs/mm^2, and fur seal at about 400 hairs/mm^2 (Fish et al. 2002; Liwanag et al. 2012a). Sea otter baby fur is so dense, full

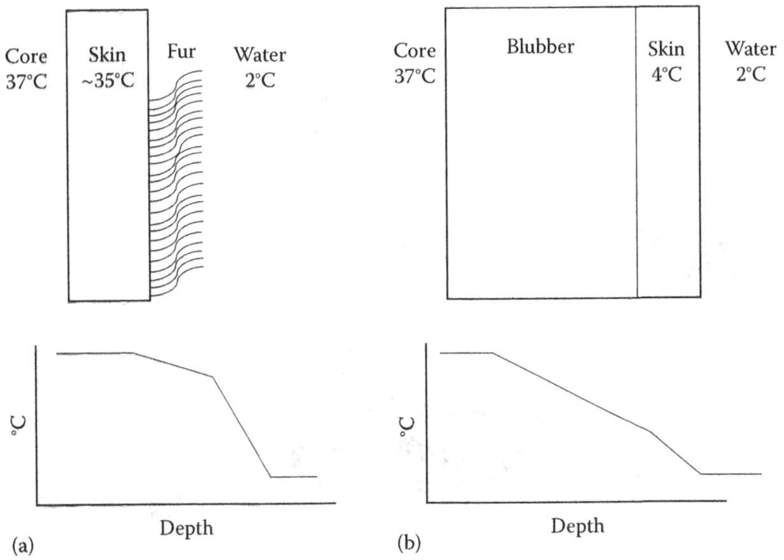

Figure 9.1 Thermal gradients from core body temperature to water. (a) For a fur-bearing marine mammal with water-impermeable fur with entrapped air. Most of the thermal gradient is within the fur and the skin temperature is close to core temperature. (b) For a blubber-bearing marine mammal, with little or no fur. In this case, most of thermal gradient is within the blubber and the skin temperature is kept only a few degrees above water temperature.

of air and buoyant, that the pup cannot dive and floats around on the surface waiting for its mother to come back from foraging trips. The thick, luxurious fur of these two species inspired the Russian exploration of Alaska, and fur trading fueled the regional economy for many years. By contrast, the hair densities of harbor seals and sea lions are less than 25 hairs/mm^2 and of course, cetaceans do not have body hair (Liwanag et al. 2012a). Fish et al. (2002) provide an excellent review of the buoyancy and hydrodynamic characteristics of fur and hair. Thermal budgets for survival of newborn pups have been built for fur seals and models how heat transfer, fur condition, weather, and other thermal factors can influence population survival (Trites 1990).

The temperature gradient for marine mammals with water-resistant fur (e.g., otters, fur seals) is entirely external to the skin surface. The energetic trade-off is that these animals must maintain their fur in prime condition to keep the protective air barrier intact. Sea otters can spend up to 20% of their *Daily Metabolic Rate* (DMR) solely on fur grooming (Walker et al. 2008). A fur coat fouled with oil loses its protective air barrier and rapidly becomes a very poor insulator (Costa and Kooyman 1982; Table 9.1), which was the unfortunate situation for sea otters in the vicinity of the 1989 Exxon Valdez oil spill. A large rescue effort included the cleaning of oiled otters followed by extended rehabilitation to allow time for grooming and lipid replacement for the natural water-repellent oil that was unavoidably stripped by the process (Williams and Davis 1990).

Some pinniped species are born with a very thick *lanugo* that is excellent at keeping the pup warm in air (Figure 9.2). However, lanugo is not water-repellent, and is mostly useless as a thermal barrier in the water as described by Burns (1970). In the *pagophilic* seals (seals that breed on and are associated with sea-ice), pups stay mostly on the sea-ice surface until they accumulate sufficient blubber reserves from the lipid-rich milk provided by their mothers to provide insulation, for example, harp seals (Lavigne and Kovacs 1988; Gmuca et al. 2015).

Figure 9.2 **(See color insert.)** A young Weddell seal pup and nursing mother. Note the lanugo coat of the pup. The fur of the mother is kept cool and snow can sit on the surface without melting, except where the pup has been nursing. (Photo by M. Castellini, MMPA Permit #801.)

They shed their lanugo as they approach weaning and begin to take on the coloration and appearance of the adults with thinner, coarser hair. The extreme example of this is the hooded seal, which is weaned in a mere 4 days due to a phenomenal efficiency of fat transfer (Bowen et al. 1985; Mellish et al. 1999). Hooded seals, however, have a pre-loaded fat layer of about 15% body composition and, therefore, do not have as thick a lanugo, nor need as much time as most other ice seals to deposit an insulatory layer of blubber.

Not all marine mammal fur is comparable to the water-impermeable coat on the sea otter. Seals and sea lions have a very rough, short, and relatively low density fur that affords little insulation because it allows the skin to become wet. The evolutionary pressure to maintain the fur likely comes from the protection it provides to the skin from the rough surfaces of the ground, ice, and so on rather than thermal contributions. In this case, the Fick equation dictates that the thermal gradient is internal to the animal, as the skin surface may be only a few degrees warmer than the environment, if at all (Figure 9.1b). In these pinniped species, and all cetaceans, the insulating layer is usually found in the form of blubber. An interesting variant on this pattern is the heavily furred polar bear, in that the fur allows water to permeate to the skin. For these bears, the fur acts mainly to keep them warm in the air, but fat layers beneath the skin are the primary thermal barrier when in water (Frisch et al. 1974).

Morphological differences in marine mammal hair are found on two levels: in general, not only is marine mammal hair shorter, denser, and flatter than other carnivores, but those species that rely heavily on fur for heat retention have longer cuticle scales than those that do not (Liwanag et al. 2012a).

9.3.2 Blubber

The first commonly held concept to dismiss when considering blubber is that it is simply a collection of homogenous fat cells, similar to body fat in most terrestrial mammals. By contrast, blubber is a highly specialized organ comprised of a loose collagen protein matrix, interspersed with lipid depots known as *adipocytes*. Therefore, while a solid mass of beef fat can be melted into an amorphous liquid, blubber can be heated and still maintain its shape due to the protein matrix. Raw or heated, blubber can be sliced, very unlike the material left over from heating regular animal fat.

Blubber is the product of evolutionary and adaptation pressures that go beyond thermal insulation capacity. It is a source of fuel and energy for fasting marine mammals (see Chapters 6 and 10), and is laid on in very large amounts for animals preparing for the low-food time of year. By the very process of metabolizing the lipid stored in the blubber, marine mammals produce significant amounts of metabolically derived water, which is then used for water balance during fasting or migration (see Chapters 7 and 10). It is also buoyant, because of the high lipid content (over 90% lipid), and therefore influences hydrodynamic lift (see Chapter 1). The protein matrix provides shape, which comes into play for hydrodynamic design and water flow. Finally, it also provides a significant and thick shield for injury and damage caused by fighting (examples include male Steller sea lions, elephant seals). All of these factors will influence the deposition and utilization of blubber in addition to its needs for thermal insulation (Rosen et al. 2007).

The lipid composition and protein matrix of blubber can vary by species and within a single animal. For the most part, cetacean blubber varies with depth, so that the chemical and structural properties next to the skin are not the same as those next to the deep muscle (Lowenstine and Osborn 1990; Haldiman and Tarpley 1993). The extremely thick blubber of the Arctic bowhead whale can reach depths as great as 50 cm (Haldiman and Tarpley 1993). At the skin surface, the blubber is relatively cold, very structural, and firm. The cold water tends to solidify the lipids and there is a high protein content to provide skin surface structure (Haldiman and Tarpley 1993; Koopman 2007). Moving toward the warmer core of the body, the blubber becomes more flexible and has a reduced protein matrix. By the time the blubber contacts the muscle, it is extremely flexible and oily, with little or no discernable structure remaining (Lowenstine and Osborn 1990; Haldiman and Tarpley 1993). Consequently, blubber samples taken from whales must be carefully calibrated to the depth at which they were collected. Further, the blubber near the surface tends to be metabolically stable and structural, while the blubber nearer to the muscle is more heavily used for metabolic fuel and water production. This is an extremely important and highly debated point for field methods that use dart biopsies on cetaceans to assess genetics, contaminant loads, and stable isotope status (Krahn et al. 2004). The outer skin covering of the blubber is easily collected via dart for genetics analyses. However, the tissue near the skin collected by the dart tends to be more structural blubber. Because of the high lipid content of blubber and because many *organo-chlorine* (OC) contaminants (e.g., DDT and PCBs) are highly *lipophilic* (dissolve easily in lipids), blubber should be a great sample tissue for contaminant load (see Chapter 14). However, if the cetacean sequesters ingested OCs in the outer blubber layer where they are not metabolized and relatively harmless, then samples from the surface blubber may be both relatively high in OC content, and yet not reflective of OCs that could damage the animal if metabolized. Both the chapters on health and disease cover this issue in greater detail (see Chapters 13 and 14).

To add more complications, blubber depth, distribution and lipid content varies dramatically between otariids and phocids, and by species within each order. In general, seals have thicker blubber with a relatively homogenous distribution across the body, and high lipid content. The depth of the blubber layer in otariids can vary by a factor of 3 or more depending on where on the body it is measured (e.g., Steller sea lions; Mellish et al. 2007). Blubber of both pinnipeds and cetaceans is also allocated for the various needs of energy versus thermoregulation according to species-specific maps (Koopman 1998;

Koopman et al. 2002). A seasonality effect leads to thicker blubber tending to be present in winter months for phocids (Rosen and Renouf 1997; Nilssen et al. 2001), but this effect does not come into play until adulthood (Pitcher 1986).

Given all of these caveats that can influence blubber content and structure, we can turn now to its thermal characteristics. As seen in Table 9.1, lipid has a high thermal insulation (low conductivity) capacity and forms an effective insulator against the cold for marine mammals. Therefore, the entire Fick thermal gradient occurs through the blubber of the marine mammal, from the warm muscle to the cold skin. The thicker the blubber, the greater insulation for the cetacean or the pinniped. Blubber can vary from a few mm in some body regions of the fur seals and seal lions, to over 35 cm in the bowhead. Cold water pinnipeds in the polar regions, such as the Weddell seal, can carry blubber of 8 cm thickness or more (Mellish et al. 2011). Table 9.2 contains blubber measurements for many species of marine mammals. How we measure blubber is discussed in Section 9.6.

Recent research by the authors has been focused on the role of blubber relative to thermal conditions for Antarctic ice seals. Castellini et al. looked at blubber depths via ultrasound in all four Antarctic ice seals (Weddell, leopard, crabeater, and Ross) taken at the same time of year in the same general area of the Ross Sea. They concluded that Weddell seals might have far more blubber than would be required solely for insulation needs, and that this extra blubber might be necessary for the energetic costs of breeding and nursing (Castellini et al. 2009). Mellish et al. (2015) examined multiple life stages of Weddell seals during the pupping season in great detail using both blubber ultrasound and *thermal imaging*. Adult females who do not have a pup in a given year but have returned to the colony to breed are 25% heavier overall, with 60% greater fat reserves than those females who have weaned a pup. While in general, the average body surface temperature of these seals was best modeled by body mass, air temperature, and wind speed, the two groups of females differed further in the ways they lost heat. The Fick equation once again comes into play, where the females who had depleted much of their blubber energy to support a pup experienced higher heat loss due to conduction compared to their non-breeding counterparts.

At the beginning of this section, we noted that using insulation alone as a means to reduce heat loss did not comprise the whole story of thermal balance in marine mammals. We also need to consider that these animals will encounter situations where they must deal with an excess of heat. Typical terrestrial outlets for heat are sweating and panting, but marine mammals do neither. The answer lies in the fact that blubber is an organ with

Table 9.2 Blubber depths in marine mammals

Bowhead whale	Average ~35
Sperm whale	15–20
Pygmy sperm whale	3
Orca	7–10
Common dolphin	1.5
Weddell seal	3–4.5
Harbor seal	2–2.5
California sea lion	<1
Manatee	3.5

Sources: Data from Haldiman and Tarpley (1993); Fadely (1997); Pabst et al. (1999); Luque and Aurioles-Gamboa (2001); Smith and Worthy (2006); Castellini et al. (2009); George (2009); Bagge et al. (2012); Ashley (2014).

Note: Depth is in cm.

a suite of blood vessels that run through it to the skin surface. These veins and arterioles can be opened or shut and blood flow to the skin surface can be controlled. When the animal is too hot, it opens up the vessels, essentially blushes, and hot blood flows to the skin surface to effectively dump the heat. Both authors have had the unusual fortune to see seals steaming on the ice surface, even in subzero temperatures. This effect is often described as a window in thermal imaging literature (Mauck et al. 2003; Nienaber et al. 2010). By contrast, if the animal is cold and needs to conserve heat, it will shut down the blood flow through the blubber and reduce heat loss. Therefore, the very same animal that can be on the surface giving off enough heat to steam and melt into the ice surface, can under other conditions have skin and fur cool enough to not even melt snow that has fallen on it during a storm. In both pinnipeds and cetaceans, this is taken even further by controlling blood flow to the relatively free of blubber, flippers, and flukes. These appendages have almost no insulation through blubber or fur, allowing for large amounts of heat to be released when blood is shunted through them (MacArthur 1989). Conversely, when these appendages are shut down, there is almost no blood flow to those sites. This phenomenon is routinely experienced by field biologists attempting blood sample collection by venipuncture in a flipper or fluke. If the animal is warm, blood flow will allow for relatively easy sample collection if the needle is placed properly, however, if the animal is cold, it is essentially impossible to collect a sample regardless of the skill of the collector. The very large surface areas of the flippers of sea lions are thought to play a major role in dumping of heat (Beenijes 2006), especially the hind limbs that are not used for propulsion.

Cooling mechanisms are not limited to the extremities. For example, the internal testes of male cetaceans are completely enclosed inside the blubber layer of the animal. There is anatomical evidence that male cetaceans route cool blood coming from the dorsal fins to the testes, presumably to keep these reproductive organs from overheating (Rommel et al. 1992). Direct measurement of surface temperatures in sea lions has shown that the shoulders and hips are warmer than the trunk in active and inactive animals alike, and this was not a function of blubber depth (Willis and Horning 2005). The fact that blubber alone does not dictate heat loss but merely plays a role in the Fick regime has been reinforced in several species (Mellish et al. 2013).

Before we leave the discussion on the biochemical nature of blubber, it is important to discuss *brown adipose fat*. Some terrestrial newborn mammals contain an amount of a specialized fat termed brown fat at key locations in their bodies for the first days after birth. This fat is extremely metabolically active through massive lipid oxidation and futile biochemical cycling that generates a large amount of heat as a waste product. The brown fat does not appear to have a role in caloric energy balance, instead it is used as a heat-generating organ. Many decades ago, there were studies that investigated whether marine mammals had brown fat to help newborn pups survive being born in extremely cold conditions. The only evidence of brown fat was its discovery in harp seals in 1979 (Blix et al. 1979), along with a recent description of the uncoupling mechanisms that allow the futile cycling (Pearson et al. 2014). The use of brown fat to generate heat is generally not thought to be of significance for most marine species. Pabst et al. (1999) provide a summary of many of the functional morphology aspects of heat flow and blubber characteristics in marine mammals.

9.4 Heat generation in marine mammals

We now turn to the other side of the equation and discuss the mechanisms of heat generation by marine species. As noted above, they are endothermic homeotherms and create most of their heat (barring external radiative or conductive heat gain) through metabolic processes.

Like all mammals, the normal body temperature of most marine species is around 37°C. There is nothing particularly special about 37°C, but rather it is the constancy of that temperature that is more important (Hochachka and Somero 1984). Homeotherms aim to hold body temperature constant because it is biochemically more efficient than having a body temperature that fluctuates widely. There is a chemical principle called the Q_{10} concept that has described the change in metabolic rate driven by changes in temperature. For mammals, a 10°C tissue temperature change induces a biochemical change in reaction of about 2x, but this varies through the many thousands of biochemical reactions in the body. Because biochemical *homeostatis* (constancy of biochemical reactions) is important for effective metabolic control, it is advantageous to not allow body temperature to fluctuate too widely. Therefore, it is not the 37°C that is so important, but rather that it is mostly constant. *Hyperthermia,* or when the body temperature rises to dangerous levels, is a problem because it causes significant disruption of the delicate biochemical reaction balances in the body. For example, some Antarctic fish that live in water at −1.8°C will die of hyperthermia at only 6°C–7°C (Hochachka and Somero 1984). They do not die because tissues are breaking down from the temperature, but because of Q_{10} imbalances in the many biochemical reactions in their bodies. Similarly, hyperthermia in a marine mammal would be dangerous because of loss of biochemical balance. Therefore, the need to thermoregulate in a marine mammal is to provide a relatively constant body temperature and get neither too cold nor too hot.

Surprisingly, the generation of heat by all mammals, including the marine species, is actually a highly inefficient waste product of basic cellular metabolism. Almost all energy in cellular metabolism is needed to maintain ion balance across cell walls. Mammals have relatively leaky cell walls compared to ectotherms, and the cellular ion pumps must constantly work to keep the ion balance between the inside of outside of the cells (Stevens 1973; Else and Hulbert 1981; Hochachka and Guppy 1987). This means that they consume vast amounts of ATP to maintain those ion pumps, and the consumption of ATP in general, is only about 25% efficient. That is, only 25% of the ATP is used to drive the ion pumps, and the rest is lost as waste heat. The ion leakiness of mammalian cells and the vast amount of waste heat is thought to have been important in the evolution of the group, because it allowed the expansion of mammals to colder regions on the planet (Hammel 1976).

The *Kleiber principle,* covered in-depth in Chapter 3, is also involved in endothermic heat production and energy consumption. It describes how the relationship between metabolic rate and body mass decreases in a logarithmic function, so that small animals have a much higher metabolic rate per body mass than larger animals. The sea otter has a very high metabolic rate/body mass compared to a larger pinniped or cetacean (Morrison et al. 1974). Recall that the smaller sea otter also has a higher SA:V ratio. It is tempting to relate the loss of heat due to the higher SA:V ratio, but that ratio has a log exponent of −0.66, while the Kleiber line has a log exponent of −0.75. Further, the Kleiber line also has been seen in ectothermic species, meaning the loss of heat by a mammal as the surface area goes up relative to body mass is not the driver for the Kleiber-derived increase in metabolic rate.

Knowing these relationships, we can return to an aspect of body mass and metabolic rate that was discussed by Williams and Maresh (Figure 3.2, Chapter 3). They showed that the metabolic rate of a marine mammal is roughly 2× that of a terrestrial mammal of the same mass. For many years it was thought that this was to offset the heat loss of being in water, which is 25× more a thermal challenge than being on land. However, as pointed out by Williams and Maresh, "thermoregulation may not be a problem for all but the smallest of marine mammals and that other traits in marine mammals set metabolism."

Sea otters have a very high metabolic rate and probably even use the heat from food digestion for thermal balance (Morrison et al. 1974; Costa and Kooyman 1984; Yeates et al. 2007). Remember our point about the SA:V ratio and that dumping of heat may be more important for larger species. This leads to an essential point about the thermal biology of marine mammals: they are not warmer than terrestrial mammals, yet they appear to generate more heat. Therefore, the fundamental difference between marine and terrestrial mammals lies almost entirely in their thermoregulatory control of heat loss, rather than specialized systems to generate heat.

9.5 *Heat balance versus other metabolic demands*

Exercise and digestion (specific dynamic action, SDA) are two activities that increase Resting Metabolic Rate and provide additional heat sources in all mammals, and this is no different for marine mammals (Costa and Kooyman 1984; Rosen et al. 2007). Under most circumstances, this extra heat can be conserved to offset heat in colder water, or they can release heat in a warmer environment. However, this balance becomes more complex when the animals are diving. Williams and Maresh (Chapter 3) discussed the need to reduce oxygen consumption while still meeting the increased metabolic demands of diving. Previously, we noted that ion pumping through cell wall channels accounted for most of the daily metabolic demands of mammalian cells. The Hochachka laboratory argued that if marine mammals could channel arrest and reduce ion leakiness, and therefore the need for ion pumping, they could lower their metabolic rate and dive for longer times (Hochachka and Guppy 1987). This would also reduce their heat production during diving with the potential to create a thermal deficit (Elsner 2015). In the 1970s–1980s, Kooyman et al. (1980) and Hill et al. (1987) were able to place central arterial temperature devices into freely long-diving Weddell seals and found that deep circulating body temperature did decrease during diving, but only by 1°C–2°C. Since that time, a suite of studies have examined body temperature in diving seals (and penguins) and found that temperature does fluctuate throughout the body during diving, and that even a few degrees drop can provide some reduction in metabolic demand as defined by the Q_{10} relationship (Ponganis et al. 2003; Meir and Ponganis 2010). However, for species such as elephant seals that remain at sea for months and dive over 90% of their time at sea, they would not be able to tolerate a continued drop in body temperature. This returns us to the discussions in this book in Chapters 1 through 3, and others about swimming patterns, muscle use and the energetics of diving. There is a clear relationship between diving physiology, metabolic rate, and thermal biology for these species. Therefore, while measuring thermal characteristics in the laboratory or on the beach is important, you also have to consider how those relationships may change during diving. For example, we know that blood flow to the liver and kidney (important for digestion) is reduced during free diving in Weddell seals (Davis et al. 1983). This will reduce not only the body temperature, but delay any resulting SDA from their nutrient clearance and blood filtering reactions.

Another consideration about the differences between studying thermal biology on land or in a small pool is that a swimming marine mammal will be moving quickly (usually ~2 m/s) and the convective heat loss will be significant. Of particular interest is the bowhead whale with its extremely thick blubber swimming through ice-laden waters. It is possible that these whales have some unusual thermal properties. Recall that their blubber layers can be up to 50 cm thick, which is at the high end of all marine mammals. Some believe this thick blubber is necessary as a fuel source for their long migrations and is selected more for energy than for thermal needs. Others propose that the thick blubber acts as a shield against potential body damage as they work through the heavy Arctic ice. George working with Alaska native hunters was able to obtain deep body temperature profiles from bowheads and found that they were cooler than expected (33°C–34°C) (George 2009). They are so heavily insulated (perhaps over insulated) that one intriguing model suggested that they could swim through liquid oxygen at −183°C and still remain at 33°C–34°C in their core (Hokkanen 1990), and others estimated that the animals were "100× over insulated" (Kanwisher and Ridgway 1983). Elsner et al. (2004) has looked at the circulatory system in the flukes of bowheads and found a suite of blood vessels that could be utilized to conserve or dump heat as discussed previously. How these interesting animals balance their thermal needs will be questions for many investigators yet to come.

9.6 Toolbox

Because of the many different types of studies involved in the field of thermoregulation, there are a wide range of technical and scientific tools that are used by scientists to investigate questions about body temperature, heat loss, insulation, and more. From relatively simple remote thermometers to complex heat flow devices, the field is one where technical advancement draws heavily on medical and engineering approaches to make a rapidly changing toolbox for data collection and theoretical models.

9.6.1 Measuring blubber

Much of the early knowledge about blubber was limited in scope and was collected from harvested animals. Because blubber depth on the body can vary from site to site on the body, it was standard to measure the blubber thickness on a seal at the midline on the ventral side (xiphosternal), if only one measurement could be taken (Pitcher 1986; Fadely 1997).

However, advances in non-imaging *ultrasound* technology in the 1980s changed how we could measure blubber. Ultrasound functions on the principle that sound travels at different speeds through different types of tissues. In some cases, such as bone, the sound does not travel through at all, but instead is refracted. Ultrasound allowed for the estimation of blubber depth in phocids, in particular, given their relatively uniform blubber layer and clear signal at the blubber–muscle interface (Gales and Burton 1987; Slip et al. 1992). This led to the ability to perform time-series studies in individuals, and the important finding that not only does blubber depth vary by body site, it can vary seasonally and is quite species-specific, for example, Rosen and Renouf (1997) and Mellish et al. (2007).

These first portable ultrasound devices provided a simple LED scale of blubber depth, but not long after, "advanced graphic display imaging" ultrasound revolutionized the ability to detect blubber depth both rapidly and effectively (Mellish et al. 2004). As these units have shrunk from the size of a wheeled cart to the size of a backpack, and now to the size of a smartphone, they have become a part of regular health examinations for cetaceans and pinnipeds at aquaria around the world, and are found in the gear complement of most field physiologists (Noren et al. 2008; Mellish et al. 2011; Hoopes et al. 2014).

In its simplest form, one can model the body shape of a marine mammal as a core of muscle, surrounded by a ring of blubber and build a series of volume cones to estimate body mass, blubber content, and so on (Castellini et al. 2009). However, blubber is almost non-existent in the flippers and tails, usually thin around the head, and so forth. As described in Chapter 10, it is most accurate to obtain a suite of blubber thickness measurements around the body and then build a model of total blubber content, as defined for your particular species (Shero et al. 2014; Shuert et al. 2015).

Ultrasound methods work well when the animal in question can be handled. However, there are many cases where the animals cannot be touched (almost all wild cetaceans) or where they are not restrained easily (many wild pinnipeds). In those cases, there have been great advances made in photographic methods to produce images where measurements can be made of width, girth, and so on. Using controls, studies of live and dead specimens and geometry, values of blubber depth and total body mass can be modeled using photographic techniques. However, a basic assumption in these methods is that a marine mammal with large amounts of blubber appears different than the same animal with less blubber. The concern here is the structural nature of blubber from the protein matrix. It is possible to remove a great deal of the lipid from blubber without changing its shape, because of the protein matrix. Therefore, the same whale under low blubber fat and high blubber fat conditions may appear to have the same width and girth.

9.6.2 Infrared thermography

Similar to the leaps in ultrasound technology, the rapid development of microelectronics has provided research scientists with accurate and inexpensive laser-guided infrared remote thermometers that can be used to measure the skin surface temperature of all kinds of marine mammals. This is clearly a phenomenal advancement over physically placing a thermometer or thermode on the skin surface of a whale or seal, which has, by the way, been done (Boyd 2000). However, an even more impressive advancement has been easy-to-use thermal imaging cameras that measure the temperature profile of the entire animal in a single frame. One of the pioneering applications of this technology to wildlife research was by one of the authors of Chapter 3, where the differences in heat transfer in elephant ears was defined (Williams 1990). The method has since been applied to numerous thermoregulatory studies in terrestrial species (McCafferty et al. 1998, 2013; Tattersall and Cadena 2010). There have been several applications of thermal imaging specific to pinnipeds, ranging from tracking the energy budgets of newly weaned gray seals in comparison to their environment (McCafferty et al. 2005), to the variation in heat loss by territorial male elephant seals during battle (Norris et al. 2010). The thermal windows mentioned previously provide evidence beyond the visual that marine mammals can and will use increased blood flow to the skin to enhance evaporative heat loss (Mauck et al. 2003). With some care, thermal images of individuals can be used in biophysical modeling to estimate metabolic heat loss (McCafferty et al. 2011).

The Weddell seal is mentioned many times in this chapter, due to their large size, accessibility during the breeding season, and deep southern environment that provides an excellent tool for modeling. In a recent study, Mellish et al. (2015) combined both thermal imaging and heat flux sensor deployments in Weddell seals as a thermoregulatory model for phocid seals in a wide range of mass and condition states. The overall body surface temperature of the seals was a surprising 14°C on average, despite the below freezing air temperatures. However, on particularly cold days, there could be as little as a 2°C–4°C difference between the outer temperature of the skin and the environment, which according to the Fick principle, would greatly limit the amount of heat lost to the environment.

The overall temperatures of the seals, in a five-fold range of body mass and three-fold range in blubber depth, were largely influenced by a combination of the mass of the animal, the ambient air temperature, and the wind speed. Unlike more temperate pinniped species, there was no regional difference in heat loss across the body. Using heat transfer modeling, it was possible to estimate that radiation contributed more than half of the total heat loss in dry animals resting on the ice, regardless of the size of the individual. Conduction to the ice surface accounted for about one-third, and convection came in last place with about 15% of total heat loss. Further, the studies were able to estimate how environmental conditions could impact heat loss in these animals, providing an example of how drastic changes in energy budgets can occur with changes in an animal's environment. For example, a small 120 kg Weddell seal (or other similarly sized seal) in low wind conditions of 4 m/s will lose 52 W, however, if you increase the wind to a moderate (for Antarctica) wind speed of 17 m/s, the heat loss jumps to 124 W.

This is just a small sample of the species-specific information that we can now model due to the improvement of two types of imaging tools—infrared (thermal) and ultrasound. These technologies have a multitude of diverse applications that we only expect to grow, given their no or low-contact requirements, high portability, and increasingly affordable prices. However, caution must be exercised in that these images are providing the combined effect of the anatomy (insulation), physiology (metabolic state), and environment. A major limitation of thermal imaging cameras is that they can only be used in air.

9.6.3 *Heat flow underwater*

Throughout this section, we have brought up the issue that because the heat capacity of water is so high, heat will flow out from the warm marine mammal to the cold water at elevated rates. We have also noted that diving will reduce metabolic rate, and therefore reduce internal heat generation. However, we have mostly discussed the characteristics and measurements of heat flow in air. How can one measure heat flow in marine mammals while underwater?

The Fick equation provides some background for this problem. In general, the surface area, inside to outside temperature differential, and the thickness of a material are usually the easiest values to obtain. However, the thermal conductivity is usually very difficult, especially in living tissue (such as blubber) given the dynamic nature of blood flow. Early in the study of potential impacts of oil development in Alaska, research teams investigated the impact of oil on the thermal properties of sea otter, walrus, and seal pelts. They used a Fick device where a pelt was put in a water bath with the skin side of the pelt sealed directly on a hot plate set to 37°C, and water flowing over the pelt above held at 1°C–2°C. They used a device that measured how much electrical energy was needed to keep the hot plate at 37°C. Conduction and convection were accounted for by maintaining the water at a constant temperature of 1°C–2°C, and the hot plate was stabilized at 37°C. The SA was measured as was the depth of the pelt. All the Fick variables were known, except for the thermal conductivity of the pelt, which could then be solved using simple algebra (Kooyman et al. 1977). More modern versions use computer-aided analyses of pelts using suites of thermocouples and standard reference materials with known heat conductance (Liwanag et al. 2012b). While this method works well in the laboratory for pelts, it was not useful for studies of live animals, either in air or in water.

The invention of electronic heat flow disks was a breakthrough in the field of thermoregulation for both cetaceans and pinnipeds. These small disks can measure the temperature difference between the skin surface where they are attached, the water temperature surrounding them, they are of a precise thickness and their thermal conductivity is accurately known. Therefore, using Fick theory, they can be used to calculate the heat flow that is moving through them. Noren et al. (1999) were able to adapt this technology by putting the disks on a tool that could be placed against the skin of dolphins accustomed to divers. In this way, they were able to measure heat flow from the surface of the dolphin skin at multiple sites around the body, after the animals had been resting, exercising, and so on. They found that most of their *extra heat* from exercise was actually dissipated when they returned to the surface and that the diving responses over-rode the need to dump heat to the periphery. Evidence for this included that heat flow from the fins and flukes decreased when the animals were underwater, suggesting a reduction in blood flow which would be consistent with blood flow redistribution during diving.

A pair of highly trained sea lions at the Vancouver Aquarium provided the stage for application of this method to otariid seals (Willis et al. 2005). Custom-made housings allowed for a temporary attachment of heat flux sensors for up to 7 days during routine swim bouts. This allowed the determination of correction factors required, as they discovered that the attachment itself could greatly impact the heat flux readings. Pilot deployments of this method led to a recent full-scale effort with Weddell seals in McMurdo Sound, Antarctica.

Hindle et al. (in press) deployed skin surface heat flux sensors on free-ranging Weddell seals and recorded heat loss in both air and water over days to weeks (Figures 9.3 and 9.4). By analyzing additional baseline information about each individual (body size and condition, insulation, and surface temperature patterns from infrared thermograms, mentioned above), they were able to suggest a method to integrate point measurements of heat flux across the body into a total measurement of whole-animal heat loss.

While adult female Weddell seals in very good condition with high blubber insulation showed little need for additional thermoregulatory heat production in air or in cold

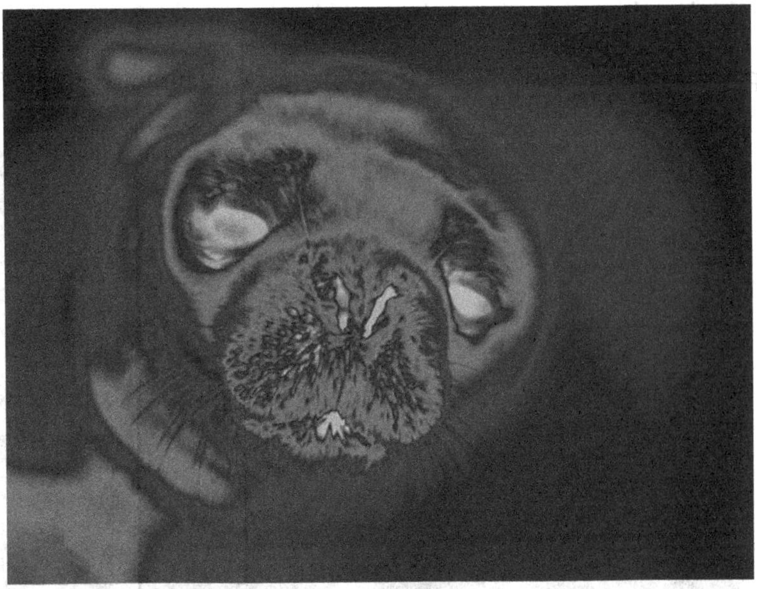

Figure 9.3 **(See color insert.)** An infrared image reveals the hot (red) and cold (blue) surface temperatures on the face of a Weddell seal pup. Most of the pup surface remains cool, with the exception of the un-insulated eyes, lightly insulated head, and highly vascularized muzzle. (Photo by J. Mellish, NMFS 15748.)

Figure 9.4 **(See color insert.)** A group of resting Weddell seals on the summer sea ice of McMurdo Sound, Antarctica. These animals can face winter air temperatures in this region as low as −50°C, before factoring in the effects of convection, or *wind chill*. (Photo by D. Uhlmann, NMFS 1034–1854.)

Antarctic water (−1.9°C), the smallest bodied pups consistently showed obligate thermoregulatory heat costs. Juvenile Weddell seals (yearlings), having the least blubber insulation, lost heat at the fastest rate as their swim speed (and therefore convective water flow over their bodies) increased. This type of information will help scientists to understand when higher thermoregulatory heat production becomes necessary to maintain stable core body temperature, and when changes in the environment or in animal conditions impose thermoregulatory costs for wild marine mammals (Figure 9.5).

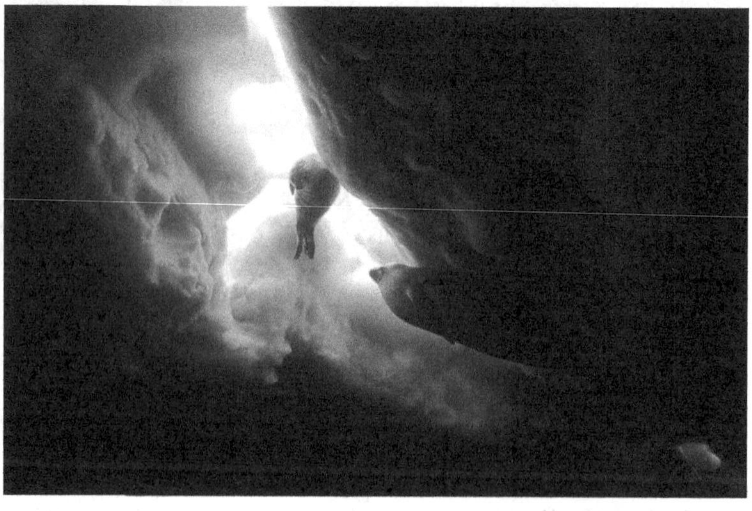

Figure 9.5 **(See color insert.)** A Weddell seal mother watches her young pup explore the comparatively stable −2°C water of McMurdo Sound, Antarctica. (Credit to H. Kaiser, NMFS 15748.)

9.6.4 Designer temperature telemetry

The development of ever more sophisticated dive recorders is discussed in the "ToolBox" sections of many chapters in this book. They are important for the study of thermoregulation, because they can collect skin surface temperatures, stomach, deep muscle, and any other temperature from wherever the investigator is able to place a temperature probe. Temperature data from these recorders have been integral in the models of diving blood flow, identifying tissues that are metabolically depressed during diving, and in understanding the overall thermal balance of animals underwater (Andrews 1998; Ponganis et al. 2004; Kuhn and Costa 2006; Meir and Ponganis 2010). They can even be used to measure food ingestion by showing the temperature change in the stomach after a marine mammals has consumed a cold fish meal. In a very recent development, specialized units have been designed to be placed permanently in the body cavity of wild Steller sea lions, transmitting archived data only when the animal dies and the tags are released from the decomposing or consumed body. These life-history transmitters (LHX) use temperature as one of a suite of measured parameters being used to provide not only individual known fate mortality, but data on the diving behavior, birth rate (for females only), and inferences on the cause of death. Upcoming modifications to this technology will also allow for interim uplinks of core body temperature in live animals to nearby receivers placed at strategic haul outs (Horning and Mellish 2009, 2012, 2014).

9.7 Unsolved and future questions

All of the marine mammals we have discussed so far have been carnivores that are active swimmers and strong divers, with a high metabolic rate. The sirenia (dugongs: *Dugong dugon*, and manatees: *Trichechus* spp.) are warm water vegetarians with a low metabolic rate. They are not phylogenetically related to any of the pinnipeds or cetaceans and are most closely related to the elephants. They are marine mammals, but in a group by themselves. However, because they live in a marine environment, they must be able to dive, hold their breath, swim and have many of the same evolutionary adaptation pressures that face other marine mammals. While sirenia are discussed in several of the other chapters of this book from those perspectives, we look here at their thermal biology.

Returning to Fick for our discussion, sirenia are relatively large (small SA:V ratio), tend to inhabit warm water (reduced convective heat loss) and will even move into warm water springs during the colder winter months (behavioral thermoregulation). They have a thick blubber layer (insulation), but no fur or hair. Their body temperature is lower and, therefore, the temperature gradient from body core to water is reduced (Gallivan et al. 1983). However, despite these heat-conserving characteristics, they must still compensate for a low metabolic rate. Clearly, this balance works well and they are a successful group, but they have a very low environmental thermal window. That is, they do not have much ability to move away from warm water. The endangered status of many sirenian populations is a result of human interference and loss of habitat that allows them to stay in warm waters (Fertl et al. 2005).

9.7.1 Steller's sea cow

In terms of thermal biology, the extinct Steller's sea cow (*Hydrodamalis gigas*) is a fascinating case study. It was a vegetarian like dugongs and manatees, however, unlike extant sirenia, the Steller's sea cow lived in the shallow intertidal areas from southeast

Alaska to Japan. It was hunted to extinction by the same fur traders in search of otter and seal pelts, and sadly the last one was seen in 1768.

Just how did the sea cow, a low-metabolic rate vegetarian, survive in subarctic waters? While we do not have living representatives to apply our tools and models to, we do know from historical records and recovered bones that the sea cow was the largest of all the non-cetacean marine mammals. It approached 8 m in length, reached 10 metric tons in mass, had very thick skin and may have had such a significant fat layer that it was unable to dive efficiently because of its high buoyancy (Scheffer 1972; Anderson and Domning 2002). The only reasonable answer to the Fick requirements is that their overwhelming mass and blubber combined with a relatively small surface area to volume ratio afforded them a much smaller heat loss than their much smaller current relatives. How interesting a situation would it be to have the large, low metabolic rate sea cow still swimming alongside the sea otter: the smallest of all marine mammals with the highest of all metabolic rates? Anderson (1995) wrote about the co-evolutionary aspects of sea otters and sea cows in the same nearshore niche, with a discussion of the curious potential impact of overlapping diets of sea cows and sea urchin prey.

9.7.2 Climate change

Of interest is the question about how climate change and ocean warming would impact marine mammals. From the basic and fundamental aspects of heat balance discussed here, the answer is that it would probably have no impact (Huntington and Moore 2008; Castellini 2012). As you have learned, marine mammals maintain a constant internal body temperature in the face of massive temperature swings seasonally, on migrations, off and onto beaches, and in and out of freezing seawater. A 1°C–2°C water temperature change due to climate shift is insignificant and easily accommodated by their physiology. However, these oceanic temperature changes will noticeably impact the distribution and abundance of their prey. Oceanic temperatures might impact the migration routes of the large whales as ice, prey, or other oceanic patterns shift. Warming temperatures can and will continue to alter the ice pack upon which the polar seals, walrus, and bears depend for rest and breeding. It is on these bases of predicted changes in ice platform that the recent listings of Endangered under the U.S. Endangered Species Act for polar bears and some of the Arctic seals have been enacted (Kuhn 2010).

9.8 Conclusions

There are several concepts that are essential in a summary on the thermal biology of marine mammals. First, they not only have to stay warm when in cold water but must also be able to remove excess heat when needed. Second, they do not have any specialized heat organs and must be able to conserve or dump heat from normal metabolic activity, exercise, and food digestion. They do this with very fine control of blood and heat flow through blubber or through the use of selective blood flow to skin areas without fur or blubber. Because they span such a large size range (from sea otters to blue whales), the relative heat flow issues differ based on mass, volume, and surface area. Water impermeable fur is used for insulation by a few species, but most use blubber for insulation. However, blubber is also used as a fuel, a water source, and for buoyancy control, and therefore cannot be selected solely for thermal needs. It is very likely that only the smallest individuals of a given species are ever in a cold stress situation, but rather the most common thermoregulatory challenge for marine mammals is how to prevent overheating.

Glossary

Adipocytes: Cells that contain fat.
ATP: Adenosine triphosphate. Energy compound for metabolism.
Conduction: Transfer of heat by direct contact between two bodies.
Convection: *Wind chill.* Heat loss by moving air or water over surface.
Daily Metabolic Rate: Energy cost of an animal over a 24-h period.
Ectothermic: *Cold blooded* animal where body temperature depends on environment.
Endothermic: *Warm blooded* animal where body heat is generated internally.
Fick equation: Defines variables that determine heat transfer between two bodies.
Heat capacity: Heat energy required to raise 1 g of material by 1°C.
Heterothermic: Animals with a variable body temperature.
Homeostasis: Constancy of metabolism.
Homeothermic: Animals with a constant body temperature.
Hyperthermia: Body temperatures high above normal.
Lanugo: *Pup fur.* Very fine and good insulation in air. Poor in water.
Lipophilic: Compounds that dissolve easily in lipids.
Organo-chlorines: Organic compounds containing chlorine. Common contaminants.
Pagophilic: *Ice loving.* Animals that live closely associated with ice.
Radiation: Heat transfer through electromagnetic radiation.
Specific dynamic action: Heat release through the process of food digestion.
Thermal conductivity: Description of how well a material will conduct heat energy.
Thermal imaging: Method to visualize surface temperatures.
Thermoregulatory: Referring to ability to control heat loss and gain.
Ultrasound: Use of sound waves to visualize tissue type.

References

Anderson, P.K. 1995. Competition, predation, and the evolution and extinction of Steller's sea cow, *Hydrodamalis gigas. Marine Mammal Science* 11(3):391–394.

Anderson, P.K. and D.P. Domning. 2002. Steller's sea cow. In: *Encyclopedia of Marine Mammals*, ed. W.F. Perrin, B. Wursig, and J.G.M. Thewissen. San Diego, CA: Academic Press, pp. 1178–1181.

Andrews, R.D. 1998. Remotely releasable instruments for monitoring the foraging behaviour of pinnipeds. *Marine Ecology-Progress Series* 175:289–294.

Ashley, C. 2014. *The Yankee Whaler.* Newburyport, MA: Dover Publications.

Bagge, L.E., H.N. Koopman, S.A. Rommel, W.A. McLellan, and D.A. Pabst. 2012. Lipid class and depth-specific thermal properties in the blubber of the short-finned pilot whale and the pygmy sperm whale. *Journal of Experimental Biology* 215(24):4330–4339.

Beenijes, M.P. 2006. Behavioral thermoregulation of the New Zealand sea lion (*Phocarctos hookeri*). *Marine Mammal Science* 22(2):311–325.

Blix, A.S., H.J. Grav, and K. Ronald. 1979. Some aspects of temperature regulation in newborn harp seal pups. *American Journal of Physiology* 236:R188–R197.

Boily, P. 1995. Theoretical heat flux in water and habitat selection of phocid seals and beluga whales during the annual molt. *Journal of Theoretical Biology* 172(3):235–244.

Bowen, W.D., O.T. Oftedal, and D.J. Boness. 1985. Birth to weaning in four days: Remarkable growth in the hooded seal, *Cystophora cristata. Canadian Journal of Zoology* 63:2841–2846.

Boyd, I.L. 2000. Skin temperatures during free-ranging swimming and diving in Antarctic fur seals. *Journal of Experimental Biology* 203(12):1907–1914.

Bryden, M.M. 1964. Insulating capacity of the subcutaneous fat of the southern elephant seal. *Nature* 203:1299–1300.

Bryden, M.M. and M.A. Button. 1977. The structure of sweat glands in the Weddell seal, *Leptonychotes weddelli. Journal of Anatomy* 124: 514.

Burns, J.J. 1970. Remarks on the distribution and natural history of pagophilic pinnipeds in the Bering and Chukchi Seas. *Journal of Mammalogy* 28:445–454.
Castellini, M. 2009. Thermoregulation. In: *Encyclopedia of Marine Mammals*, 2nd edn, ed. W.F. Perrin, B. Wursig, and J.G.M. Thewissen. Academic Press, New York, pp. 1166–1171.
Castellini, M. 2012. Life under water: Physiological adaptations to diving and living at sea. *Comprehensive Physiology* 2:1889–1919.
Castellini, M.A., S.J. Trumble, T.L. Mau, P.K. Yochem, B.S. Stewart, and M.A. Koski. 2009. Body and blubber relationships in Antarctic pack ice seals: Implications for blubber depth patterns. *Physiological and Biochemical Zoology* 82(2):113–120.
Costa, D.P. and G.L. Kooyman. 1982. Oxygen consumption, thermoregulation, and the effect of fur oiling and washing on the sea otter, *Enhydra lutris*. *Canadian Journal of Zoology* 60(11):2761–2767.
Costa, D.P. and G.L. Kooyman. 1984. Contribution of specific dynamic action to heat balance and thermoregulation in the sea otter, *Enhydra lutris*. *Physiological Zoology* 57:199–202.
Davis, R.W., M.A. Castellini, G.L. Kooyman, and R.A. Maue. 1983. Renal glomerular filtration rate and hepatic blood flow during voluntary diving in Weddell seals. *American Journal of Physiology* 245:R743–R748.
Elam, S.K., I. Tokura, K. Saito, and R.A. Altenkirch. 1989. Thermal conductivity of crude oils. *Experimental Thermal and Fluid Science* 2(1):1–6.
Else, P.L. and A.J. Hulbert. 1981. Comparison of the "mammal machine" and the "reptile machine": Energy production. *American Journal of Physiology* 240:R3–R9.
Elsner, R. 2015. *Diving Seals and Meditating Yogis: Strategic Metabolic Retreats*. Chicago, IL: Universtiy of Chicago Press.
Elsner, R., J.C. George, and T. O'hara. 2004. Vasomotor responses of isolated peripheral blood vessels from bowhead whales: Thermoregulatory implications. *Marine Mammal Science* 20(3):546–553.
Fadely, B.S. 1997. Investigations of harbor seal (*Phoca vitulina*) health status and body condition in the Gulf of Alaska. University of Alaska Fairbanks, Fairbanks, AK.
Fertl, D., A.J. Schiro, G.T. Regan et al. 2005. Manatee occurrence in the northern Gulf of Mexico, west of Florida. *Gulf and Caribbean Research* 17:69–94.
Fish, F.E., J. Smelstoys, R.V. Baudinette, and P.S. Reynolds. 2002. Fur doesn't fly, it floats: Buoyancy of pelage in semi-aquatic mammals. *Aquatic Mammals* 28(2):103–112.
Folk, G.E., M.L. Riedesel, and D.L. Thrift. 1998. *Principles of Integrative Environmental Physiology*. San Francisco, CA: Austin and Winfield.
Frisch, J., N.A. Øritsland, and J. Krog. 1974. Insulation of furs in water. *Comparative Biochemistry and Physiology Part A: Physiology* 47(2):403–410.
Gales, N.J. and H.R. Burton. 1987. Prolonged and multiple immobilizations of the southern elephant seal using ketamine hydrochloride-xylazine hydrochloride or ketamine hydrochloride-diazepam combinations. *Journal of Wildlife Diseases* 23(4):614–618.
Gallivan, G.J., R.C. Best, and J.W. Kanwisher. 1983. Temperature regulation in the Amazonian manatee *Trichechus inunguis*. *Physiological Zoology* 56:255–262.
George, J.C. 2009. Growth, morphology and energetics of bowhead whales (*Balaena mysticetus*): University of Alaska Fairbanks, Fairbanks, AK.
Gmuca, N.V., L.E. Pearson, J.M. Burns, and E.M. Liwanag. 2015. The fat and the furriest: Morphological changes in harp seal fur with ontogeny. *Physiological and Biochemical Zoology* 88(2):158.
Haldiman, J.T. and R.J. Tarpley. 1993. Anatomy and physiology. In *The Bowhead Whale*, eds. J.J. Burns, J.J. Montague, and C.J. Cowles. Lawrence, KS: Allen PR.
Hammel, H.T. 1976. On the origin of endothermy in mammals. *Israel Journal of Medical Sciences* 12(9):905–915.
Heath, M.E. and R.J. Schusterman. 1975. "Displacement" sand flipping in the northern elephant seal (*Mirounga angustirostris*). *Behavioral Biology* 14(3):379–385.
Hill, R.D., R.C. Schneider, G.C. Liggins et al. 1987. Heart rate and body temperature during free diving of Weddell seals. *American Journal of Physiology* 253:R344–R351.
Hind, A.T. and W.S.C. Gurney. 1998. Are there thermoregulatory constraints on the timing of pupping for harbour seals? *Canadian Journal of Zoology* 76(12):2245–2254.

Hindle, A.G., M. Horning, and J. Mellish. in press. Estimating total body heat dissipation in air and water from skin surface heat flux telemetry in Weddell seals. *Animal Biotelemetry.*
Hochachka, P.W. and M. Guppy. 1987. *Metabolic Arrest and the Control of Biological Time.* Cambridge, UK: Harvard University Press.
Hochachka, P.W. and G.N. Somero. 1984. *Biochemical Adaptation.* Princeton, NJ: Princeton University Press.
Hokkanen, J.E. 1990. Temperature regulation of marine mammals. *Journal of Theoretical Biology* 145(4):465–485.
Hoopes, L.A., L.D. Rea, A. Christ, and G.A.J. Worthy. 2014. No evidence of metabolic depression in Western Alaskan juvenile Steller sea lions (*Eumetopias jubatus*). *PLoS ONE* 9(1):e85339.
Horning, M. and J. Mellish. 2009. Spatially explicit detection of predation on individual pinnipeds from implanted post-mortem satellite data transmitters. *Endangered Species Research* 10:135–143.
Horning, M. and J.A.E. Mellish. 2012. Predation on an upper trophic marine predator, the Steller sea lion: Evaluating high juvenile mortality in a density dependent conceptual framework. *PLoS ONE* 7(1):e30173.
Horning, M. and J.A.E. Mellish. 2014. In cold blood: Evidence of Pacific sleeper shark (*Somniosus pacificus*) predation on Steller sea lions (*Eumetopias jubatus*) in the Gulf of Alaska. *Fishery Bulletin* 112(4):297–310.
Huntington, H.P. and S.E. Moore. 2008. Assessing the impacts of climate change on Arctic marine mammals. *Ecological Applications* 18(sp2):S1–S2.
Kanwisher, J.W. and S.H. Ridgway. 1983. The physiological ecology of whales and porpoises. *Scientific American* 248:111–119.
Khamas, W.A., H. Smodlaka, J. Leach-Robinson, and L. Palmer. 2012. Skin histology and its role in heat dissipation in three pinniped species. *Acta Veterinaria Scandinavica* 54:46.
Koopman, H.N. 1998. Topographical distribution of the blubber of harbor porpoises (*Phocoena phocoena*). *Journal of Mammalogy* 79(1):260–270.
Koopman, H.N. 2007. Phylogenetic, ecological, and ontogenetic factors influencing the biochemical structure of the blubber of odontocetes. *Marine Biology* 151(1):277–291.
Koopman, H.N., D.A. Pabst, W.A. McLellan, R.M. Dillaman, and A.J. Read. 2002. Changes in blubber distribution and morphology associated with starvation in the Harbor porpoise (*Phocoena phocoena*): Evidence for regional differences in blubber structure and function. *Physiological and Biochemical Zoology* 75(5):498–512.
Kooyman, G.L. 1981. *Weddell Seal: Consummate Diver.* Cambridge, UK: Cambridge University Press.
Kooyman, G.L., R.W. Davis, and M.A. Castellini. 1977. Thermal conductance of immersed pinniped and sea otter pelts before and after oiling with Prudhoe Bay crude. In: *Fate and Effects of Petroleum Hydrocarbons in Marine Organisms and Ecosystems*, ed. D.A. Wolfe. Oxford: Pergamon Press, pp. 151–157.
Kooyman, G.L., E.A. Wahrenbrock, M.A. Castellini, R.W. Davis, and E.E. Sinnett. 1980. Aerobic and anaerobic metabolism during voluntary diving in Weddell seals: Evidence of preferred pathways from blood chemistry and behavior. *Journal of Comparative Physiology* 138:335–346.
Krahn, M.M., D.P. Herman, G.M. Ylitalo et al. 2004. Stratification of lipids, fatty acids and organochlorine contaminants in blubber of white whales and killer whales. *Journal of Cetacean Research and Management* 6:175–189.
Kuhn, C.E. and D.P. Costa. 2006. Identifying and quantifying prey consumption using stomach temperature change in pinnipeds. *Journal of Experimental Biology* 209(22):4524–4532.
Kuhn, M. 2010. Climate change and the polar bear: Is the endangered species act up to the task. *Alaska Law Review* 27:125.
Lavigne, D.M. and K.M. Kovacs. 1988. *Harps & Hoods: Ice-Breeding Seals of the Northwest Atlantic.* Waterloo, Ontario, Canada: University of Waterloo Press.
Ling, J.K. 1965. Functional significance of sweat glands and sebaceous glands in seals. *Nature* 208(5010):560–562.
Liwanag, H.E.M., A. Berta, D.P. Costa, M. Abney, and T.M. Williams. 2012a. Morphological and thermal properties of mammalian insulation: The evolution of fur for aquatic living. *Biological Journal of the Linnean Society* 106(4):926–939.

Liwanag, H.E.M., A. Berta, D.P. Costa, S.M. Budge, and T.M. Williams. 2012b. Morphological and thermal properties of mammalian insulation: The evolutionary transition to blubber in pinnipeds. *Biological Journal of the Linnean Society* 107(4):774–787.

Lowenstine, L.J. and K.G. Osborn. 1990. Practical marine mammal microanatomy for pathologists. In: *CRC Handbook of Marine Mammal Medicine: Health, Disease and Rehabilitation*, ed. L. Dierauf, pp. 287–290. Boca Raton, FL: CRC Press.

Luque, S.P. and D. Aurioles-Gamboa. 2001. Sex differences in body size and body condition of California sea lion (*Zalophus californianus*) pups from the Gulf of California. *Marine Mammal Science* 17(1):147–160.

MacArthur, R.A. 1989. Aquatic mammals in cold. In *Animal Adaptation to Cold*. Vol. 4, ed. L.C.H. Wang, 289-325, Berlin, Germany: Springer-Verlag.

Mauck, B., K. Bilgmann, D.D. Jones, U. Eysel, and G. Dehnhardt. 2003. Thermal windows on the trunk of hauled-out seals: Hot spots for thermoregulatory evaporation? *Journal of Experimental Biology* 206(10):1727–1738.

McCafferty, D.J., C. Gilbert, W. Paterson et al. 2011. Estimating metabolic heat loss in birds and mammals by combining infrared thermography with biophysical modelling. *Comparative Biochemistry and Physiology A—Molecular & Integrative Physiology* 158(3):337–345.

McCafferty, D.J., C. Gilbert, A.M. Thierry, J. Currie, Y. Le Maho, and A. Ancel. 2013. Emperor penguin body surfaces cool below air temperature. *Biology Letters* 9(3):20121192.

McCafferty, D.J., J.B. Moncrieff, I.R. Taylor, and G.F. Boddie. 1998. The use of IR thermography to measure the radiative temperature and heat loss of a barn owl (*Tyto alba*). *Journal of Thermal Biology* 23(5):311–318.

McCafferty, D.J., S. Moss, K. Bennett, and P.P. Pomeroy. 2005. Factors influencing the radiative surface temperature of grey seal (*Halichoerus grypus*) pups during early and late lactation. *Journal of Comparative Physiology B—Biochemical Systemic and Environmental Physiology* 175(6):423–431.

Meir, J.U. and P.J. Ponganis. 2010. Blood temperature profiles of diving elephant seals. *Physiological and Biochemical Zoology* 83(3):531–540.

Mellish, J., J. Nienaber, L. Polasek, and M. Horning. 2013. Beneath the surface: Profiling blubber depth in pinnipeds with infrared imaging. *Journal of Thermal Biology* 38(1):10–13.

Mellish, J.A.E., A.G. Hindle, and M. Horning. 2011. Health and condition in the adult Weddell seal of McMurdo Sound, Antarctica. *Zoology* 114(3):177–183.

Mellish, J.A.E., M. Horning, and A.E. York. 2007. Seasonal and spatial blubber depth changes in captive harbor seals (*Phoca vitulina*) and Steller's sea lions (*Eumetopias jubatus*). *Journal of Mammalogy* 88(2):408–414.

Mellish, J.A., P.A. Tuomi, and M. Horning. 2004. Assessment of ultrasound imaging as a noninvasive measure of blubber thickness in pinnipeds. *Journal of Zoo and Wildlife Medicine* 35(1):116–118.

Mellish, J.E., A.G. Hindle, J.D. Skinner, and M. Horning. 2015. Heat loss in air of an Antarctic marine mammal, the Weddell seal. *Journal of Comparative Physiology B* 185(1):143–152.

Mellish, J.E., S.J. Iverson, W.D. Bowen, and M.O. Hammill. 1999. Fat transfer and energetics during lactation in the hooded seal: The roles of tissue lipoprotein lipase in milk fat secretion and pup blubber deposition. *Journal of Comparative Physiology B—Biochemical Systemic and Environmental Physiology* 169(6):377–390.

Morrison, P., M. Rosenmann, and J.A. Estes. 1974. Metabolism and thermoregulation in the sea otter. *Physiological Zoology* 47:218–229.

Nienaber, J., J. Thomton, M. Horning, L. Polasek, and J. Mellish. 2010. Thermal windows in seals and sea lions. *Journal of Thermal Biology* 35:435–440.

Nilssen, K.T., T. Haug, and C. Lindblom. 2001. Diet of weaned pups and seasonal variations in body condition of juvenile Barents Sea harp seals *Phoca groenlandica*. *Marine Mammal Science* 17(4):926–936.

Noren, D.P., T.M. Williams, P. Berry, and E. Butler. 1999. Thermoregulation during swimming and diving in bottlenose dolphins, *Tursiops truncatus*. *Journal of Comparative Physiology B—Biochemical Systemic and Environmental Physiology* 169(2):93–99.

Noren, S.R., L.E. Pearson, J. Davis, S.J. Trumble, and S.B. Kanatous. 2008. Different thermoregulatory strategies in nearly weaned pup, yearling, and adult Weddell seals (*Leptonychotes weddelli*). *Physiological and Biochemical Zoology* 81(6):868–879.

Norris, A.L., D.S. Houser, and D.E. Crocker. 2010. Environment and activity affect skin temperature in breeding adult male elephant seals (*Mirounga angustirostris*). *Journal of Experimental Biology* 213(24):4205–4212.

Pabst, D.A., S.A. Rommel, and W.A. McLellan. 1999. The functional morphology of marine mammals. In: *Biology of Marine Mammals*, ed. J.E. Reynolds and S.A. Rommel, pp. 15–72. Washington DC: Smithsonian Institution Press.

Pearson, L.E., H.E.M. Liwanag, M.O. Hammill, and J.M. Burns. 2014. Shifts in thermoregulatory strategy during ontogeny in harp seals (*Pagophilus groenlandicus*). *Journal of Thermal Biology* 44:93–102.

Pitcher, K.W. 1986. Variation in blubber thickness of harbor seals in southern Alaska. *Journal of Wildlife Management* 50:463–466.

Ponganis, E.P., R.P. van Dam, T. Knower, D.H. Levenson, and K.V. Ponganis. 2004. Deep dives and aortic temperatures of emperor penguins: New directions for bio-logging at the isolated dive hole. *Memoirs of National Institute of Polar Research Special Issue* 58:155–161.

Ponganis, P.J., R.P. Van Dam, D.H. Levenson, T. Knower, K.V. Ponganis, and G. Marshall. 2003. Regional heterothermy and conservation of core temperature in emperor penguins diving under sea ice. *Comparative Biochemistry and Physiology Part A: Molecular & Integrative Physiology* 135(3):477–487.

Rommel, S.A., D.A. Pabst, W.A. Mclellan, J.G. Mead, and C.W. Potter. 1992. Anatomical evidence for a countercurrent heat-exchanger associated with dolphin testes. *Anatomical Record* 232(1):150–156.

Rosen, D.A.S. and D. Renouf. 1997. Seasonal changes in blubber distribution in Atlantic harbor seals: Indications of thermodynamic considerations. *Marine Mammal Science* 13(2):229–240.

Rosen, D.A.S., A.J. Winship, and L.A. Hoopes. 2007. Thermal and digestive constraints to foraging behaviour in marine mammals. *Philosophical Transactions of the Royal Society B—Biological Sciences* 362(1487):2151–2168.

Rotherham, L.S., M. van der Merwe, M.N. Bester, and W.H. Oosthuizen. 2005. Morphology and distribution of sweat glands in the Cape fur seal, *Arctocephalus pusillus pusillus* (Carnivora: Otariidae). *Australian Journal of Zoology* 53(5):295–300.

Scheffer, V.B. 1972. The weight of the Steller sea cow. *Journal of Mammalogy* 53:912–914.

Schmidt-Nielsen, K. 1997. *Animal Physiology: Adaptation and Environment*. Cambridge, UK: Cambridge University Press.

Shero, M.R., L.E. Pearson, D.P. Costa, and J.M. Burns. 2014. Improving the precision of our ecosystem calipers: A modified morphometric technique for estimating marine mammal mass and body composition. *PLoS ONE* 9(3).

Shuert, C.R., J.P. Skinner, and J.E. Mellish. 2015. Weighing our measures: Approach-appropriate modeling of body composition in juvenile Steller sea lions (*Eumetopias jubatus*). *Canadian Journal of Zoology* 93:177–180.

Slip, D.J., H.R. Burton, and N.J. Gales. 1992. Determining blubber mass in the southern elephant seal, Mirounga-Leonina, by ultrasonic and isotopic techniques. *Australian Journal of Zoology* 40(2):143–152.

Smith, H.R. and G.A.J. Worthy. 2006. Stratification and intra- and inter-specific differences in fatty acid composition of common dolphin (*Delphinus* sp.) blubber: Implications for dietary analysis. *Comparative Biochemistry and Physiology Part B: Biochemistry and Molecular Biology* 143(4):486–499.

Stevens, E.D. 1973. The evolution of endothermy. *Journal of Theoretical Biology* 38:597–611.

Tattersall, G.J. and V. Cadena. 2010. Insights into animal temperature adaptations revealed through thermal imaging. *Imaging Science Journal* 58(5):261–268.

Trites, A.W. 1990. Thermal budgets and climate spaces: The impact of weather on the survival of Galapagos (*Arctocephalus galapagoensis* Heller) and northern fur seal pups (*Callorhinus ursinus* L.). *Functional Ecology* 4:753–768.

Walker, K.A., J.W. Davis, and D.A. Duffield. 2008. Activity budgets and prey consumption of sea otters (*Enhydra lutris kenyoni*) in Washington. *Aquatic Mammals* 34(4):393–401.

Williams, T.M. 1990. Heat-transfer in elephants—Thermal partitioning based on skin temperature profiles. *Journal of Zoology* 222:235–245.

Williams, T.M. and R.W. Davis. 1990. *Sea Otter Rehabilitation Program: 1989 Exxon Valdez Oil Spill.* Galveston, TX: International Wildlife Research.

Willis, K. and M. Horning. 2005. A novel approach to measuring heat flux in swimming animals. *Journal of Experimental Marine Biology and Ecology* 315(2):147–162.

Willis, K., M. Horning, D.A.S. Rosen, and A.W. Trites. 2005. Spatial variation of heat flux in Steller sea lions: Evidence for consistent avenues of heat exchange along the body trunk. *Journal of Experimental Marine Biology and Ecology* 315(2):163–175.

Yeates, L.C., T.M. Williams, and T.L. Fink. 2007. Diving and foraging energetics of the smallest marine mammal, the sea otter (*Enhydra lutris*). *Journal of Experimental Biology* 210(11):1960–1970.

section three

Reproduction

chapter ten

Post-partum

Daniel E. Crocker and Birgitte I. McDonald

Contents

10.1 Challenge of parturition for marine mammals

As air-breathing endotherms living in marine habitats, marine mammals have had to overcome the problem of giving birth in an environment that is potentially challenging to a newborn. As described in previous chapters, the marine environment presents a suite of physiological challenges that are magnified in neonates. For example, neonates have high surface-to-volume ratios and thin blubber layers, resulting in higher heat loss compared to adults (Chapter 9). As a consequence of these challenges, marine mammals give birth to a single, large, *precocial* young, with many species exhibiting rapid growth and development. Across marine mammal species, offspring birth mass increases as a power function of maternal mass, called the *scaling exponent*. This scaling exponent is 0.85 (using data from 58 species), indicating that smaller species give birth to relatively large offspring for their body size (Figure 10.1).

The transition to nutritional independence is a critical stage in marine mammal development and survival because young animals lack the skills and the physiology (e.g., oxygen stores, Chapter 2) to efficiently acquire food. In mammals, lactation is the most energetically

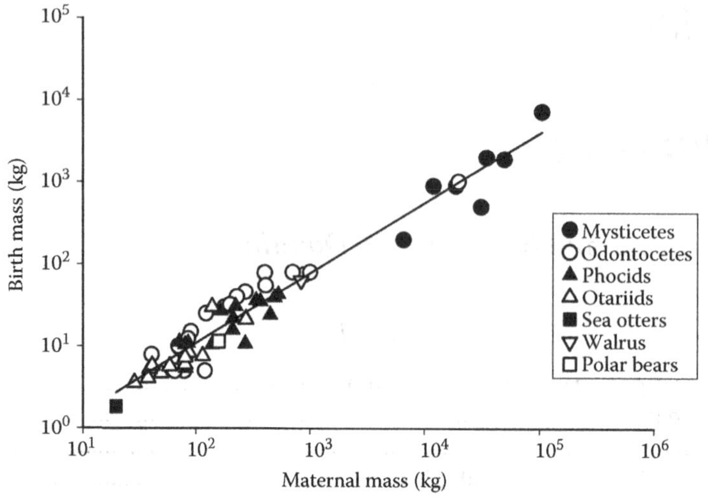

Figure 10.1 The relationship between body mass and offspring birth mass in selected marine mammal species where data or estimates are available. The slope of the line on this log–log plot, or the scaling exponent, is 0.85. Because the slope of this line is less than 1, it shows that smaller marine mammals produce larger offspring relative to their body size. (Modified with additional data from Costa, D.P., Energetics 2009, in *Encyclopedia of Marine Mammals*, Perrin, W.F., Wursig, B., and Thewissen, J.G.M. (eds.), Academic Press, San Diego, CA, 2009.)

demanding component of reproduction and can buffer the transition to independence by providing the offspring with energy stores, or extending the duration of the developmental period. Marine mammals exhibit a diverse range of lactation strategies that are influenced by size and environment. Periods of parental investment range widely in marine mammals from 4 days to several years (Figure 10.2). Cetaceans have evolved a suite of behavioral and anatomical

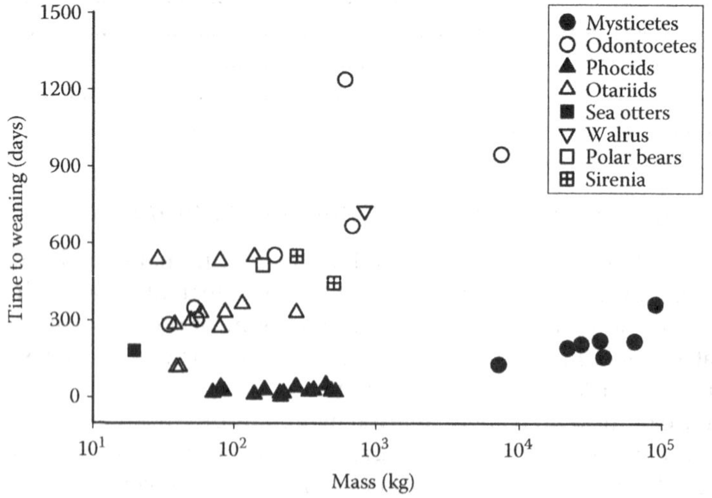

Figure 10.2 The relationship between body mass and time to weaning in selected marine mammal species where data or estimates are available. Mysticete whales and phocid seals have shorter lactation periods for their body size when compared to other marine mammals. (Modified with additional data from Costa, D.P., Energetics 2009, in *Encyclopedia of Marine Mammals*, Perrin, W.F., Wursig, B., and Thewissen, J.G.M. (eds.), Academic Press, San Diego, CA, 2009.)

adaptations that allow them to give birth, suckle, and nurture their young in water. In contrast, pinnipeds leave the marine environment to reproduce and give birth on land or ice.

Underneath this broad difference, there exists a diversity of parental strategies and physiological capabilities that shape the movements, breeding systems, foraging strategies, and life-history patterns of marine mammals. In some species (e.g., some mysticetes and many phocids), movement to an appropriate breeding environment necessitates a *capital breeding* strategy, where stored body reserves are accumulated between breeding episodes and then used for reproduction. Other species (e.g., odontocetes and otariids) use an *income breeding* strategy, where offspring are provisioned using resources acquired from foraging during the lactation period (Figure 10.3).

10.2 Knowledge by order

10.2.1 Cetaceans

The ability to give birth and nurse in the ocean was likely a key feature in allowing the success and wide distribution of cetacean species. Unfortunately for scientists, aquatic lactation has also made the post-partum physiology of cetaceans extremely difficult to study, and much of what is known about lactation physiology is based on carcasses from whaling or strandings and from studies of captive animals. For many species, birth has never been observed and knowledge of reproductive and parental behavior is fragmentary or based on comparisons with other species. Despite these limitations, field studies provide some important insights into post-partum behavior and physiology. Because cetaceans exhibit low adult mortality and *fecundity* (birth rates), but high rates of infant mortality, the period of parental care is critical to offspring survival and population health. Many of the maternal traits that support calf survival are behavioral and social, including communication and nursing behavior, but important differences in lactation physiology are evident that influence the life-history strategies of cetaceans.

10.2.1.1 Lactation strategies

The duration of lactation varies widely both between and within species of cetaceans, with the greatest difference between suborders. However, for many species the duration of lactation is estimated from just a few samples, and the methods used to determine lactation duration (i.e., milk in the stomach from harvested individuals or behavioral observations of known individuals) are prone to error. Cetacean offspring can travel with their mothers, suckling frequently with only brief periods of maternal absence. Offspring suckle from two mammary teats located in slits on either side of the genital opening. The calf wraps the tip of its tongue around the nipple from below and pushes against the roof of the mouth. The base of the tongue creates suction by moving up and down. The mother uses muscular contraction of both the myoepithelial cells of the mammary gland and surrounding cutaneous muscles to cause milk ejection.

Although both baleen and toothed whales nurse in the water, they exhibit different lactation strategies. Lactation duration ranges from around 5 months in mysticetes up to 3–6 years in some odontocetes. The mysticete whales exhibit a capital breeding strategy resulting in a short lactation period relative to their body size (Figure 10.1). Most species lactate for 5–8 months, typically weaning their calves during the summer months when food availability is high. Mysticetes have compartmentalized breeding to a distinct part of the year, requiring high intensity lactation and offspring growth. In the best-studied mysticetes, individuals forage in rich polar waters during summer months, and then move to

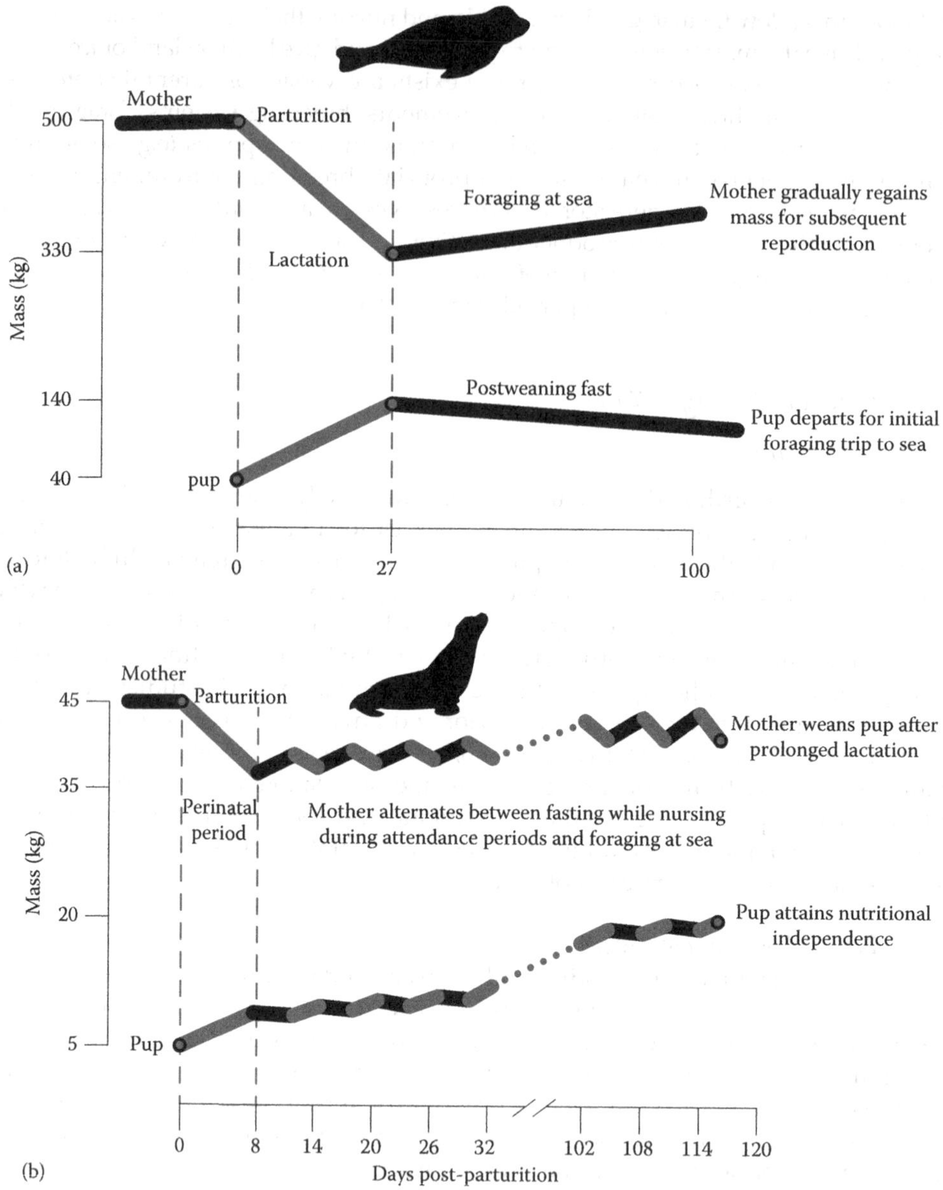

Figure 10.3 Schematic showing the difference between (a) capital breeding strategy (fasting during lactation) used by some phocids and mysticetes and (b) income breeding strategy (foraging/fasting cycle) used by otariids and some phocids. (From Champagne, C.D. et al., Fasting physiology of the pinnipeds: The challenges of fasting while maintaining high energy expenditure and nutrient delivery for lactation, in *Comparative Physiology of Fasting, Starvation, and Food Limitation*, McCue, M.D. (ed.), Springer, Berlin, Germany, 2012. With permission.)

warmer waters for parturition, nursing, and mating. Feeding may be essentially absent at the birthing locations. It has been hypothesized that these breeding areas may provide environments with warmer and calmer water for the calves, which may lower thermoregulatory costs, thereby increasing growth efficiency. However, more recent studies indicate that baleen whale mothers and their calves are thermal-neutral in cold water, and the migration may be to avoid predation on young calves. An exception to this general pattern is the bowhead whale (*Balaena mysticetus*) that calves at high latitudes in the summer and has an extended lactation period (12 months) compared to other mysticetes (Oftedal 1997). The life-history pattern of most mysticetes suggests that prey ingestion only overlaps slightly with nursing in offspring, indicating a relatively abrupt weaning compared to odontocetes.

Odontocetes exhibit an income breeding strategy, with mothers foraging during lactation. They nurse their calves for 6 months to several years, with lactation duration tending to increase with body size. Within species, there are records of extraordinarily long lactation periods for individuals, exceeding a decade, though these records may reflect artifacts of the observation methods employed. Long periods of parental investment may facilitate training of young in migration, foraging, and social behavior. Weaning in odontocetes is more gradual than mysticetes, and offspring may remain with the mother in social groups after weaning. In many species, females become pregnant while still lactating and may provision their young for a significant portion of gestation. This may allow females to "bet-hedge" against loss of the developing fetus, with the option of continuing investment in the current offspring.

The duration of lactation is influenced by several factors in cetaceans (Whitehead and Mann 2000). Cetacean young do not fast after weaning. Odontocetes show a gradual transition from nursing to foraging, while mysticetes have only a brief period where ingestion of solid food overlaps with nursing. The period of maternal investment may reflect (1) foraging strategies and the time required for the offspring to develop sufficiently to find and catch food; (2) the fitness consequences of continued parental investment; (3) the impact of continued lactation on the next breeding attempt; and (4) occurrence of *weaning conflict* between the mother and calf. Weaning conflict arises from the fact that the mother and offspring may be selected to desire different levels of investment in current offspring (Trivers 1974).

10.2.1.2 Investment

Though data from individual species are sparse, the milk of cetaceans has the highest fat and energy content of any mammalian taxa except for the pinnipeds (Oftedal and Iverson 1995) (Table 10.1). Fat content varies across lactation but is inversely related to duration of parental investment, being ~30%–50% at mid-lactation in mysticetes and ~10%–30% at mid-lactation in odontocetes (Oftedal 1997). While data are sparse, studies suggest that milk fat content peaks near mid-lactation. The high energy cost of lactation is reflected in thinner blubber layers in lactating females when compared to non-lactating individuals. The high milk energy outputs of mysticete whales represent the most extreme examples of parental investment in nature. A pregnant blue whale (*Balaenoptera musculus*) may weigh 119,000 kg of which as much as 45,000 kg is blubber. During lactation the mother converts most of this blubber to milk, allowing the calf to gain 17,000 kg by weaning (Lockyer 1981). Most odontocetes have smaller body reserves than mysticetes and their food resources are more varied in time and space. This may limit their ability to maintain the high milk energy outputs of large baleen whales. The duration of lactation in cetaceans can be thought to ultimately reflect the spatial and temporal characteristics of the food resources used by the population and whether sufficient body reserves are accrued to permit rapid energy transfer to the offspring in milk.

10.2.2 Pinnipeds

In contrast to the cetaceans, pinnipeds give birth on land, despite spending most of their lives in the water. Most neonates require some period of time on land to develop before being able to cope with life in the sea. This requires use of an appropriate breeding site,

Table 10.1 Lactation duration and milk composition for selected cetaceans

	Lactation duration	Milk composition		
Species	Months	% fat	% protein	% sugar
Blue whale	6–7	39	11	1.3
Bryde's whale	6–7	30	15	
Fin whale	6–7	33	11	2.4
Humpback whale	10–11	44	9	0.7
Gray whale	7–8	53	6	
Minke whale	5–6	30	14	1.4
Bottlenose dolphin	19	24	11	1.1
Common dolphin	6	30	10	
Harbor porpoise	8	46	11	
Sperm whale	25	26	9	0.1
Spinner dolphin	15–19	26	7	

Source: Data from Oftedal, O.T., *J. Mammary Gland Biol. Neoplasia*, 2(3), 205, 1997.
Note: Milk composition is from about mid-lactation.

away from potential predators, and sometimes distant from their patchy food resources. The strategy for managing the separation of marine feeding and terrestrial birth leads to the major differences between individual phocid and otariid species reproductive biology.

10.2.2.1 Lactation strategies

The general pattern in phocids is that the majority of parental investment is derived from stored body reserves, a capital breeding strategy. This strategy is associated with a short lactation period for their body size (Figure 10.1). In some species (e.g., harbor [*Phoca vitulina*], gray [*Halichoerus grypus*], bearded [*Erignathus barbatus*], harp [*Pagophilus groenlandicus*], ringed [*Pusa hispida*], and Weddell seals [*Leptonychotes weddellii*]), individuals may feed during lactation; but this tendency varies individually with body reserves and the proximity of food resources (Lydersen et al. 1995; Lydersen and Kovacs 1996). Some of the smallest phocids, like harbor seals and ringed seals, feed frequently throughout lactation (Lydersen 1995; Bowen et al. 2001). In phocid species that do forage during lactation, the foraging trip is usually short, lasting just a few hours. This capital breeding strategy is analogous to the short lactation period relative to body size described previously for fasting mysticete whales. The phocid lactation period varies between 4 days in the hooded seal to up to 6–7 weeks in the Weddell seal. Weaning is often abrupt and occurs when the mother abandons the pup to return to sea. In phocid species where the mother feeds during lactation, weaning may be less abrupt and pups may spend significant time in the water prior to weaning. The most extreme example of the abbreviated phocid lactation system is the hooded seal (*Cystophora cristata*), where females lactate for only 3–4 days. In this short time, the pup can double its birth mass receiving more than 10 L of milk per day containing an average fat content of 59% (Lydersen et al. 1997). The abbreviated duration and complete separation of lactation from foraging allows many phocid seals to use highly dispersed food resources and undergo long foraging migrations to distant feeding areas.

Within capital breeding phocids, there is wide variation in the postweaning behavior and physiology of the pups. In some species, like northern elephant seals (*Mirounga angustirostris*), the pup remains at the rookery and undergoes a period of extended fasting after weaning. During this period, the pup begins to develop the physiology needed to dive to appropriate depths for foraging, including development of oxygen stores and control of the cardiovascular system (Burns et al. 2004; Tift et al. 2013). The size of the weaned pup is often an important determinant of the duration of the postweaning fast and the onset of independent foraging (Noren et al. 2003). In many species, pups undergo a long migration to foraging areas after departing the breeding site. In these cases, pup body reserves may still be critical to allowing thermoregulation in cold water and to allow time for development of foraging skills.

Similar to phocids, otariid mothers also fast from food or water while on-shore. However, they only remain with their pups for approximately a week, called the *perinatal period*, before returning to sea to forage. The duration of the perinatal period varies with species and environmental conditions, but is typically 5–9 days. The female returns to her pup, which has been fasting during her absence, and suckles it for 1–3 days. This pattern of intermittent suckling and foraging continues for 4 months to up to several years. In some species, like subantarctic fur seals (*Arctocephalus tropicalis*), these fasting periods for the pup can be greater than 30 days late in lactation (Verrier et al. 2012). The duration of the period of parental care varies with latitude, being shortest in the polar species and longest in the equatorial Galápagos fur seal (*Arctocephalus galapagoensis*) and sea lion (*Zalophus wollebacki*). The need to return frequently to provision their offspring limits otariids to foraging distances within tens to hundreds of kilometers of their breeding rookeries. This constraint places a priority on maximizing the efficiency of foraging in order to minimize the time that the pup is left fasting on-shore. For this reason, otariids are often associated with highly productive coastal regions of ocean upwelling. To maximize foraging success during this short period, otariids exhibit high levels of foraging effort and energy expenditure for locomotion when compared to phocids. The importance of successful foraging is evident in the composition of otariid milk, which increases in fat content with the duration of the foraging trip (Costa 1991).

Walruses (*Odobenus rosmarus*) have many reproductive features in common with phocids and otariids; however, they exhibit one critical difference. Pups are born on ice or on land, but unlike the other pinnipeds, walruses nurse while at sea. The ability to feed at sea allows pups to accompany mothers on short foraging trips. The perinatal fasting period is only a few days in walrus and the milk they produce is relatively low in fat (~26%) compared to other pinnipeds. Use of an aquatic lactation strategy has reduced the importance of fasting metabolism and lactation efficiency in walrus. The ability of young pups to move around in the marine environment with their mothers provides them the opportunity to learn about foraging strategies and locations in a way that is not possible in the other pinnipeds. As a result, weaning takes place much more slowly in walrus, with the period of maternal investment lasting as long as two years.

10.2.2.2 Investment and lactation metabolism

The importance of fasting ability on the breeding strategies used by pinnipeds is reflected in the strong impact of body reserves at parturition on reproductive effort. A variety of studies have shown strong impacts of maternal mass and fat content on the level of maternal investment in pinnipeds (e.g., Crocker et al. 2001). Interspecific growth rates are correlated with maternal body size (Kovacs and Lavigne 1986). Additionally, animals with larger fat reserves are able to invest more in their offspring. This creates an important link between foraging success and maternal investment and leads to strong impacts of ocean climate on the magnitude of investment or rates of natality in many species (Le Boeuf and Crocker 2005). Females that expend too much energy in 1 year may experience reduced fecundity or lower breeding success in the following year if they are not able to recover those reserves from foraging.

The different lactation strategies exhibited by phocids and otariids result in different energetic investment in their offspring. The larger body size of phocids allows them to store more energy as fat, enabling longer fasting durations and more rapid rates of milk energy delivery. Most phocid species that fast during lactation lose 30%–40% of body mass during breeding. Because of the abbreviated nature of phocid lactation, pup maintenance costs are low and the bulk of the energy given in milk can be used for growth (>80% in elephant seals and gray seals). Additionally, the short lactation period results in a lower maternal *metabolic overhead*. Maternal metabolic overhead is the proportion of a female's energy expenditure that goes toward meeting her own metabolic needs. These costs can include maintenance metabolism but also the energy costs of milk synthesis and maternal behavior. By compressing the period of parental investment, phocid seals increase the efficiency of lactation, increasing the proportion of energy expenditure given to their pups. In addition, the rapid energy delivery means that the pups can store more of the energy obtained as body reserves. In contrast, as a result of the intermittent feeding strategy exhibited by otariids, only 20%–50% of ingested milk energy is available for growth (McDonald et al. 2012). Additionally, the longer lactation period requires otariid mothers to expend more energy to meet her and her pup's maintenance metabolism requirements prior to nutritional independence. Thus, despite smaller body size at weaning, otariid pups may receive greater absolute maternal investment when compared to the mother's body size. For this reason, offspring behavior can also have important impacts on how maternal energy is allocated in otariids (McDonald et al. 2012).

Pinnipeds have the most energy dense milk of all mammals, allowing for rapid transfer of energy to the pups (Table 10.2). Their milk is high in fat and protein content (as high as 61% fat in hooded seals and 14% protein in northern fur seals—*Callorhinus ursinus*) but very low in carbohydrate content (<1%). By comparison, typical whole milk from a dairy cow is ~4% fat, ~3% protein, and ~5% sugars. Species with shorter lactation durations produce milk with higher fat and energy contents in order to provision the pup more quickly. In many species, milk composition changes dramatically over lactation, increasing in fat content with time onshore. There is a strong negative association between milk fat content and the time spent onshore with the pup in both phocids and otariids (Costa 1991). Additionally, otariid milk has increased lipid content following longer foraging trips.

In order to produce such energy dense milk while fasting, pinnipeds must be able to rapidly mobilize stored body fat and protein from muscle and vital body organs and deliver it to the mammary gland, while preventing the mobilized protein from being burned as fuel for metabolism or used to make sugars (see Chapter 8). Pinniped milk also contains significant amounts of water. Although the high energy content reduces the water needed for milk production, the bulk of this water must be provided through *metabolic water production* from maternal metabolism. This water is produced largely from the oxidation of fats for energy. The metabolic adaptations for extended fasting are covered in detail in Chapter 8. Here, we consider the role that nutrient mobilization and fasting plays in determining the ability of pinnipeds to produce such energy dense milk.

The onshore accessibility of pinnipeds for study has allowed a greater level of scientific investigation of lactation physiology than is possible in free-ranging cetaceans. In addition to numerous studies on milk composition and production, recent work has examined the endocrine and cellular processes associated with nutrient mobilization for lactation. The primary determinant of nutrient delivery to the mammary gland is mobilization of fatty acids from stored triglycerides in adipose tissue, or *lipolysis*. Fasting and lactating

Table 10.2 Lactation variables for selected pinnipeds

	Lactation duration	Maternal foraging	Milk composition		Pup growth
Species	Days	% Time at sea	% Fat	% protein	kg/day
Bearded seal	24	84	47	10	3.3
Crabeater seal	17		51	11	4.2
Hooded seal	4	0	59	5	7.0
Harbor seal	25	55	50	9	0.6
Harp seal	12	71	50	8	2.2
Gray seal	16	0	54	7	2.5
N. elephant seal	26	0	55	11	4.0
Ringed seal	39	69	38	10	0.4
S. elephant seal	23	0	40	10	3.4
Weddell seal	50	40	48	8	2.0
Australian sea lion	532	60	31	10	0.11
Ca. sea lion	332	73	36	9	0.12
S.A. sea lion	548	59	32	10	0.21
Stellar sea lion	332	76	22	9	0.35
Antarctic fur seal	117	71	41	10	0.08
Galápagos fur seal	540	50	29	12	0.07
Subantarctic fur seal	300	80	43	12	0.04
N. fur seal	118	73	42	14	0.07
Walrus	730	na	26	8	0.45

Source: Data from Schulz and Bowen (2004).

pinnipeds have some of the highest circulating levels of fatty acids found in nature. By the end of lactation, northern elephant seals have plasma fatty acid levels that average over 3 mM, a concentration that would be considered harmful (dyslipidemia) in humans (Crocker et al. 2014a). These high levels of fatty acids not only support maternal maintenance energy needs, but also facilitate rapid uptake and use by the mammary gland. The mechanisms that allow such rapid mobilization of stored fats have begun to be elucidated in seals and appear to vary widely, even between species with similar life-history patterns. This suggests that lipid mobilization during fasting is under strong evolutionary selection pressure in seals.

The primary endocrine features associated with lactation in seals are a profound reduction in the hormone insulin and elevation in the hormone cortisol. While insulin's primary role is the regulation of carbohydrate metabolism, it also exerts strong anti-lipolytic effects. This means that insulin promotes the production of new fats and inhibits the mobilization of stored fats. Low insulin levels appear critical to fat mobilization in seals, resulting in a decline in insulin release over lactation when compared to other life-history stages (Fowler et al. 2008). Correlative analysis suggests that low insulin levels are the primary determinant of fatty acid mobilization during lactation in some species. The hormone cortisol helps mobilize fat and protein in most species of animals and usually also leads to increased use of mobilized proteins for energy metabolism and for making sugars, features that might lead to vital organ damage during extended fasting. Several species of phocids and otariids elevate plasma cortisol concentrations over their lactation periods. This elevation is strongly associated with both circulating fatty acid levels and milk fat content in some species. However, in contrast to most terrestrial mammals, the seals are able to avoid the protein wasting effects of cortisol. Protein sparing is efficient in lactating seals, with protein oxidation providing a maximum of 4%–8% of energy expenditure for various species as measured from the production of urea or changes in body composition (Crocker et al. 1998). Some investigations have suggested that elevations in cortisol may serve as a re-feeding signal in otariids and help initiate the return to sea to forage (Guinet et al. 2004). Interestingly growth hormone, the primary regulator of lipolysis, milk output, and fat content in dairy cattle, appears to play a more minor role in seals.

Several enzymes are important in the process of lipolysis (release of fatty acids from triglycerides). Direct measurements of lipolysis rates during lactation in elephant seals have shown that they are uncoupled from the levels of circulating fatty acids during lactation (Houser et al. 2007). The enzymes that allow adipose tissue to reuptake fatty acids for storage, fatty acid translocase (CD36) and fatty acid transport protein 1 (FATP1), decrease with fasting in seals (Viscarra and Ortiz 2013). This suggests that reducing the reuptake of fatty acids by adipose tissue is an important feature that keeps fatty acid levels high to supply the mammary gland as fasting seals deplete their fat reserves while lactating. In most mammals, the primary enzyme responsible for lipolysis of fats is hormone-sensitive lipase (HSL). Surprisingly, HSL levels are very low in the blubber of lactating elephant seals. Instead, the enzyme *adipocyte triglyceride lipase* (ATGL) that removes one fatty acid from a triglyceride molecule has an increased importance when compared to other mammals (Fowler et al. 2015). In some species, like gray and harbor seals, the levels of an enzyme bound to mammary gland tissues called *lipoprotein lipase* (LPL) is important to allowing the mammary gland to uptake fatty acids from triglycerides in circulation. In these species, LPL levels increase in parallel with milk fat content (Iverson et al. 1995). In other species, like elephant seals, this enzyme plays a minor role (McDonald and Crocker 2006). These differences suggest that despite similar reproductive strategies, the various seal lineages may have evolved important metabolic differences in lactation physiology under strong evolutionary pressure for efficient lactation while fasting.

Since blubber is the source of fatty acids for milk synthesis, its composition and the way individual fatty acids are mobilized and used for metabolism can potentially influence the composition of milk. It has been reported in many different species of marine mammals that blubber layers are stratified from inner to outer layers (Strandberg et al. 2008). External layers have a higher proportion of medium chain (≤18 C) monounsaturated fatty acids (MUFA), possibly as a homeoviscous adaptation for the purpose of maintaining membrane fluidity at the low temperatures encountered at depth. Interior layers in phocid blubber are highly enriched in saturated fatty acids (SFA) and long-chain (≥20 C) MUFA. This inner layer is heavily metabolized during fasting and lactation (Fowler et al. 2014). In phocids, the mobilization of specific fatty acids from blubber, and their incorporation into milk, conforms to biochemical predictions based on the number of carbons and saturated bonds. Long-chain (>20 C) MUFA are the least mobilized and polyunsaturated fatty acids (PUFA) and SFA are more highly mobilized from the blubber. In the mammary gland of terrestrial mammals, fatty acid synthesis from glucose and ketones results in short and medium chain fatty acids containing fewer than 12 carbons. However, these short- and medium-length fatty acids are usually not detected in seal milk, suggesting that there is little de novo lipid synthesis in the mammary gland. In other words, plasma fatty acid delivery and uptake by the mammary gland is responsible for milk fat content. PUFA availability to the developing pup's muscle tissue may contribute to the development of oxidative capabilities for diving and provides non-shivering thermoregulatory benefits (Trumble et al. 2010). The majority of long-chain MUFA mobilized is directed to milk synthesis and the mother may preferentially use PUFA and SFA for her own metabolism. The proportion of long-chain MUFA in milk increases late in lactation in several species. The MUFA delivered in milk may help pups establish a thermoregulatory blubber layer with optimal characteristics for energy density and thermoregulation.

The accessibility of pinnipeds when lactating onshore has allowed the use of cutting-edge techniques to study the metabolic adaptations associated with lactating while fasting. In some cases, tracer techniques have been used to study metabolism in lactating pinnipeds that have not been used in any other wildlife species. These studies demonstrate a remarkable ability to maintain metabolic homeostasis despite extraordinarily high rates of energy mobilization and loss. For example, the combined energy loss for metabolism and milk production in lactating elephant seals is nearly six times the predicted standard metabolic rate for a mammal of similar size (Crocker et al. 2001). Many of the metabolic features that enable these feats of high-energy expenditure fasting are also found in non-lactating conspecifics, but several studies have revealed important alterations of maternal metabolism and physiology associated with lactation. These alterations include changes in the release of hormones that regulate metabolism and evidence for changes in the tissue responses to regulatory hormones. These studies also provide evidence that metabolic features are strongly impacted by the amount of body fat, or *adiposity*, of lactating females. For example, lactating females release less insulin and exhibit reduced glucose clearance when compared to fasting pups. These changes can vary directly with the depletion of adipose tissue reserves while lactating. So in contrast to humans and most mammals, where being fat is associated with reduced insulin sensitivity, seals appear to develop more diabetic-like features as they deplete their body reserves while lactating. Similarly, responses to metabolic hormones like glucagon, which typically increases production of sugars from stored body protein and release of fatty acids from blubber, are altered in direct relation to depletion of body fat (Crocker et al. 2014). Responses that would hinder the ability to continue fasting and producing milk, like releasing insulin and making sugars from muscle or vital organs get down-regulated, while responses that facilitate milk production and fasting, like enhanced lipolysis, get upregulated. This represents a remarkable example of metabolic adaptation for rapid reorganization of systemic and tissue metabolism to provide appropriate nutrients to the mammary gland, while protecting mothers from harmful effects of lactating while fasting.

10.2.2.3 *Mammary gland physiology*

One aspect of pinniped lactation physiology that has received recent attention is the unique physiological requirements of the mammary gland. The mammary gland of all mammals undergoes a complex set of changes during lactation that includes proliferation and differentiation of cells, secretion and ultimately the death and regression of mammary cells after weaning in a process known as *involution*. The unusual composition of pinniped milk and the ability of the otariid mammary gland to sustain function despite long interruptions in suckling provide an important comparative model with which to better understand the regulation of involution in other species.

The milk sugar lactose has been detected in only trace concentrations in phocid milk and is absent from the milk of otariids and odobenids. From the perspective of lactating while fasting, the lack of carbohydrates in milk helps pinnipeds avoid the use of proteins from vital organs for carbohydrate synthesis. However, the low levels of milk lactose may also reflect a mammary gland that is highly adapted for pinniped lactation systems. Lactose is an important determinant of milk water content as it is unable to diffuse through the mammary cell membrane and draws water into the mammary alveoli by osmosis. Low lactose levels are associated with the ability to concentrate milk solids, facilitating the high lipid and protein content of the milk. Thus, low lactose levels may help promote the high fat and protein contents in pinniped milk.

The low lactose levels may also be a result of a mutation in the alpha-lactalbumin gene (LALBA) that normally encodes a protein that plays an important role in lactose synthesis. Mutations in the LALBA gene may help preserve mammary gland functionality during intermittent suckling (Sharp et al. 2008). During their long inter-suckling foraging bouts, otariid mammary glands do not undergo involution and remain active and ready to suckle the pup when the female returns to shore (Figure 10.4). In other mammals, the termination of suckling is associated with accumulation of milk in the gland. Prolonged exposure to factors in the milk regulates the mammary epithelium, causing downregulation of expression of milk protein genes. This downregulation is followed by involution during which mammary cells die via apoptosis. The α-lactalbumin protein encoded by the LALBA gene

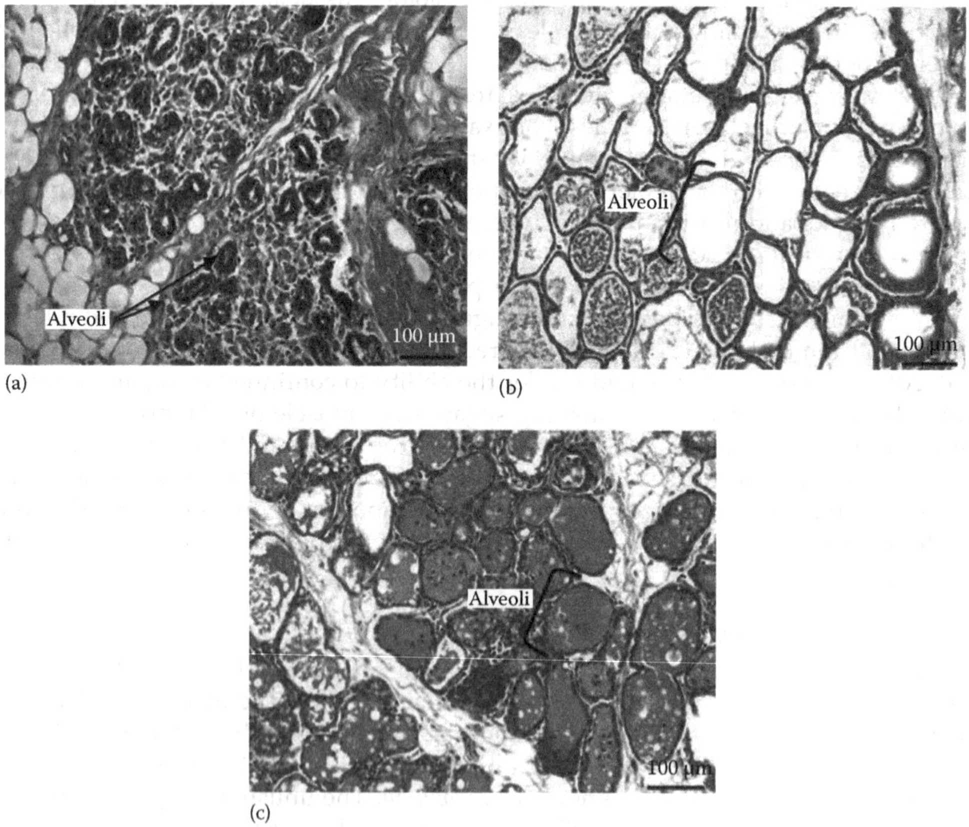

Figure 10.4 **(See color insert.)** Mammary gland morphology of an otariid. Histological sections of a mammary gland from (a) pregnant, (b) lactating onshore, and (c) lactating while foraging at-sea fur seals. Fat globules within alveoli appear as white bodies. Pink staining within alveoli represents milk components. Residence of milk components in alveoli would lead to involution in most species but otariids have evolved the ability to maintain mammary function in absence of suckling. (From Sharp, J.A. et al., *J. Mammary Gland Biol. Neoplasia*, 12(1), 47, 2007. With permission.)

has been identified as an important factor causing apoptosis and involution in mice and human mammary epithelial cells. Mutations in this gene identified in several otariids lead to lack of production of functional α-lactalbumin and may protect the otariid mammary gland from involution while foraging. In contrast, phocids lack this mutation and have shown involution of the mammary gland within 24 hours of weaning (Reich and Arnould 2007). The mutation of the LALBA gene in otariids and walruses may have been an important factor in allowing the divergence of pinniped life-history strategies between capital and income breeding strategies.

10.2.3 Sea otters

Although they are less specialized for the marine environment when compared to other marine mammals, sea otters (*Enhydra lutris*) spend most of their time at sea, including the period of maternal investment. Sea otters give birth, lactate, and rear their young in shallow coastal waters. Females give birth to a single pup and lactate for an average of 6 months. Along the income/capital breeding continuum, sea otters likely represent the extreme level of the income strategy found among marine mammals. The high surface-to-volume ratios of sea otters cause elevated heat loss and contribute to them having the highest metabolic rates for their body size among marine mammals (Costa and Williams 1999). Because of this, they must consume 20%–25% of their body mass in food each day (Costa and Kooyman 1982) and spend as much as 50% of their time foraging, depending on reproductive status and prey availability (Tinker et al. 2008). Sea otters lack the typical blubber layer of marine mammals and are dependent on increased food intake to supply the nutrients for lactation. The increased energy cost of lactation imposes a significant energetic burden on sea otter mothers with energy demands nearly 100% higher than pre-pregnancy levels by weaning (Thometz et al. 2014). Despite a lack of fat reserves, females often lose mass during lactation (Monson et al. 2000).

Female sea otters breed annually, irrespective of body condition or environmental factors. This "bet-hedging" strategy allows an otter to abandon pups early in development if she is not able to meet the energy cost of lactation through increased foraging (Monson et al. 2000). Similarly, females make decisions about when to wean their pup that are influenced by resource availability. Female sea otters must balance the benefits of extending lactation to the survival to their current offspring against potential loss of future reproduction from the health impacts of nutritional deficiencies. In areas of high otter population density, reductions in prey resource availability may make it difficult for mothers to complete lactation. These energetic constraints have been implicated in poorer body condition and higher mortality rates among sea otters nearing the end of lactation on the central California Coast, termed *end-lactation syndrome* (Thometz et al. 2014).

10.2.4 Polar bears

Polar bears (*Ursus maritimus*) are considered marine mammals because they move over a wide range of sea ice, enter the water to move between ice floes and land and rely on the ocean for food. This less marine-adapted lifestyle is reflected in their reproductive and lactation strategy. The post-partum physiology of polar bears has many similarities to that of brown bears. Polar bear offspring are unique among marine mammals in that they are *altricial*, being initially dependent on their mother for thermoregulation.

Pregnant females enter dens on land or, more rarely, on stable ice, and fast through the period of gestation and for the first several months of lactation. After a short 4-month gestation period, polar bears give birth to a small (~0.5 kg) cub while in their dens. The combined fasting period can be up to 8 months and females lose over 40% of the body mass during denning. Like other fasting adapted marine mammals, the bulk of energy reserves used are stored fat (Atkinson and Ramsay 1995). Fatter females produce larger cubs at the time of emergence. Like many pinnipeds, body fat content has direct impacts on the ability to spare protein from vital body organs while fasting (Atkinson et al. 1996). Despite these similarities, there are several important differences between polar bear blubber and that of other marine mammals. Polar bear adipose tissue does not exhibit the vertical stratification of fatty acid composition found in some cetaceans and pinnipeds, and does not undergo changes in fatty acid composition across lactation (Thiemann et al. 2006). In contrast to several pinnipeds, polar bear blubber undergoes dramatic changes in total lipid content as stored fats are used for milk synthesis. For example, a polar bear female's blubber declines from 78% to 62% lipid across the denning period. In contrast, an elephant seal undergoes significant loss of blubber during lactation but the remaining adipose tissue has a relatively constant lipid content of ~90%.

During denning, polar bear females have low metabolic rates that are less that resting metabolism, despite the costs of milk synthesis. This shows that denning is a good strategy for maximizing lactation efficiency. After emergence, mother and cub may stay in the denning area for a while before they make the trek to the sea ice to hunt seals. Polar bears have an extremely long lactation period ranging from 1.5 to 2.5 years, depending on environmental characteristics and the ability of the cub to hunt independently. Peak lactation is thought to occur during the denning period or soon after bears emerge from the den. Polar bear milk composition changes throughout denning and feeding, declining in milk fat content from as high as 36% fat and increasing in protein content after females begin to feed (Derocher et al. 1993) (Table 10.1). Unlike other marine mammals, polar bear milk has substantial carbohydrate content (as high as 5%), that declines across the lactation period (Arnould and Ramsay 1994). As the cubs grow, milk energy production lessens and females with older cubs may cease lactation during seasonal periods of fasting. Unlike many pinnipeds, the blubber of female polar bears has a fairly consistent fatty acid composition across lactation.

10.3 Toolbox

The study of lactation and post-partum reproduction in marine mammals has focused largely on pinnipeds because they are accessible on land while nursing their young. These studies have investigated how females invest energy in their offspring, how the offspring allocate the acquired energy toward growth and development, and the metabolic features that enable production of high energy density milk while fasting. These studies use a combination of techniques that allow estimation of energy expenditure and energy intake, ranging from determining mass and body condition, to using data loggers to measure movement and behavior. As technology advances, we will continue to increase our knowledge about post-partum physiology, particularly in the groups such as cetaceans, where data are lacking.

10.3.1 Mass and body condition

In the 1970s and 1980s, it was recognized that phocids provide an ideal study system in which to investigate *reproductive effort* and the potential cost of reproduction because the physiology of lactation in phocids is significantly impacted by constraints resulting from the temporal separation of foraging and parental investment. This makes it possible to look at reproductive effort without the confounding variable of food intake. The earliest studies investigating reproductive effort used changes in mass across lactation as an estimate of energy investment in pups. Although this seems like a simple approach, when working with large wild animals it is important to develop techniques to safely chemically or physically immobilize the animal allowing for a basic measurement, like mass. Even when safely immobilized, the large body size of some phocids provided distinct challenges. For example, in order to obtain mass measurements from large female phocids (250–600 kg), researchers must lift the animal using a winch attached to a scale hanging from a 3 m tall tripod. For large male pinnipeds, researchers have even gotten animals to move across truck scales.

Although these studies were important to the emerging field of life-history studies, it was recognized that mass just gives an approximation of energy investment, as not all mass is created equal (i.e., fat has twice the energy density of protein). Researchers started to use techniques that allowed them to calculate energy expenditure (mothers) or gain (offspring) from changes in body composition and mass by converting the mass of tissue lost (or gained) to an energy equivalent based on the energy content of lipid and protein. This allowed for more precise estimates of total post-partum reproductive effort (milk output and *metabolic overhead*) and to calculate transfer efficiency to offspring (energy gained by pup/energy lost by mother).

Obtaining precise body composition measurements in live animals is difficult. The most widely used techniques in field studies measure fat content and then partition the animal's mass into two compartments—fat and fat-free mass (often referred to as lean body mass). A number of approaches are used to estimate the body composition of marine mammals, with the size and accessibility of the animal determining what method is used. One technique, often considered the gold standard for determining body composition of live animals, is the *isotope dilution method* (Bowen and Iverson 1998). Total body water is determined by injecting a known quantity of labeled water (usually ^2H_2O or ^3H_2O) into the animal, allowing the isotope to equilibrate with the animal's body water, and taking a blood or urine sample to determine the isotope dilution space. The estimate of total body water is combined with estimates of tissue hydration state to derive total lipid mass. This is similar to how body composition bathroom scales calculate your percent fat from estimates of total body water based on electrical conductivity.

Morphometric measurements including length, girth, and blubber depth can also be used to estimate body composition. One such approach is the *truncated cones method*, where body composition is estimated by combining morphometrics with ultrasonic measures of blubber thickness at various sites on the body. These measurements are used to model the seal as a series of truncated cones, allowing calculation of blubber and non-blubber compartments of each cone, which can then be summed to estimate total body composition (Gales and Burton 1987). This technique has been shown to be an accurate estimate of lipid content in a wide variety of phocid seals. Since this method is based on estimating the volume of the subdermal blubber layer, it works best in species, like some phocids, whose fat reserves are largely subcutaneous. However, it may be less accurate in species that store significant lipid inside the body cavity or within muscle.

The size of some marine mammals, especially the larger odontocetes and mysticete whales, does not allow researchers to use the methods described above, so they rely on photogrammetric techniques for estimating size and condition. *Photogrammetry* is the technique of making measurements based on photographs. Two-dimensional photogrammetry has been used to estimate size and mass in several species of pinnipeds and cetaceans, but the accuracy of the mass measurements were often greatly impacted by camera angle and animal position. Three-dimensional photogrammetry, which uses several photographs to create a 3D model of the animal, has improved the accuracy and precision of photogrammetric mass estimates (Waite et al. 2007). Estimates of whale size and mass have been obtained using aerial stereophotogrammetry. For large animals like whales, these types of tools are currently the only way to obtain mass and condition information on wild animals.

10.3.2 *Field metabolic rate*

Although changes in mass and condition can be used to investigate energy transfer in fasting animals, these techniques are not appropriate for animals that feed during lactation like otariids and odontocetes. In non-fasting species, it is possible to measure field metabolic rate over short periods of time (~7–10 days), and this, combined with milk energy output (method discussed below), can be used to investigate lactation and post-partum reproductive effort.

Field metabolic rate can be measured using the *doubly labeled water* (DLW) method (Costa 1987; Speakman 1997). The basis of the DLW method is to follow the decline in enrichment of the isotopes of oxygen (^{18}O) and hydrogen (^{2}H or ^{3}H) in the body water. CO_2 production is measured by injecting known amounts of $^{2}H_2O$ (or $^{3}H_2O$) and $H_2{}^{18}O$ into the animal. An initial blood sample is taken after the isotopes equilibrate with the animal's body water (as described above) and is followed by a final blood sample at the end of the study period, 7–10 days later depending on animal size and metabolic rate. The decline in the hydrogen isotope is a measure of total water influx (TWI), which is composed of metabolic water production (MWP) and water consumed in the food (i.e., milk or fish). The ^{18}O isotope declines as a function of both water flux and CO_2 production. The difference between the rates of decline of these two isotopes is proportional to the animal's CO_2 production. Energy expenditure can then be calculated from CO_2 production using an appropriate conversion factor depending on the diet of the female or pup (Costa 1987).

10.3.3 *Milk composition and intake*

Knowledge on both the proximate composition of milk and rate of milk production is key to understanding the intra- and interspecific variation in the patterns of energy transfer observed in nature and described in this chapter. However, obtaining milk samples from large and/or difficult to access species can be challenging, and accurate estimates of milk production may be impractical. Despite this challenge, milk composition has been determined for many marine mammals. From these studies, we know that milk composition can change substantially over the course of lactation and attendance bouts, so it is important to consider the timing of sampling when studying milk composition and intake. In pinnipeds, an intramuscular injection of oxytocin is usually administered to help initiate milk let down before sampling. The lipid, protein, water, and ash components of the milk are measured independently and in duplicate following standard protocols (reviewed in Oftedal and Iverson 1995). Lipids are typically extracted using organic solvents and protein content is usually measured based on the nitrogen content of milk. The water content of milk is easily measured using oven drying. Carbohydrates are often not analyzed because of their established minor contribution to marine mammal milk. If measured accurately, the sum of the individual milk components should total ~100% of the initial sample mass.

The total energy content can be estimated using standard values for the energy density of lipid and protein. Alternatively, if the primary question is the energy content of milk, gross energy content can be determined with high precision using bomb calorimetry of the dry material and water content data.

Milk intake can be measured using labeled water techniques if one assumes that the only influx of water is from MWP and milk intake (Oftedal and Iverson 1987). Because of this assumption, this method is only appropriate when offspring are solely dependent on milk. Milk consumption rates are calculated using the water influx rates (calculated from the decline in labeled H as described above), milk water content, and MWP, ideally derived independently from the metabolic rate of each pup. With these values, milk intake is calculated using the equation

$$\text{Milk intake rate} = \frac{(\text{TWI} - \text{TWP})}{\text{Milk water content}}$$

Milk intake is then converted to energy consumed using the energy content of milk and can be combined with data on maternal energy expenditure and pup energy storage to create a lactation energy budget.

10.3.4 Data loggers

As described in previous chapters, many aspects of biomechanics, diving physiology, and foraging ecology of marine mammals have been uncovered with the use of data loggers. Given the enormous developments in these tags in recent years, we are now able to measure fine scale movements, along with physiological parameters, of behaviors in species that are difficult to observe. Most of what we know about lactation and reproductive effort in marine mammals is based on pinniped studies, but as technology advances we may be able to address more of these questions in cetaceans with the creative use of data loggers.

For example, stomach temperature data loggers have primarily been used in adult animals to investigate at sea feeding behavior but have also been used to observe suckling behavior in harbor seals (Hedd et al. 1995). This approach found that harbor seals commonly suckled in water (which would often be missed with behavioral observations), and that pups started ingesting prey while they were still nursing, suggesting that weaning may be a gradual process in this species (Schreer et al. 2010). This method can be used to study suckling behavior and the transition to nutritional independent in marine mammals.

Time depth recorders (TDRs) revolutionized our ability to study marine mammals at sea, and now many modern tags include accelerometers and magnetometers that measure fine scale movement that can be used to estimate energy expenditure through measures such as overall dynamic body acceleration and stroke/fluke rate. The use of these tags on lactating females provides information on time activity budgets (both at sea and on land) and improves our understanding of energy allocation during lactation. Additionally, accelerometers have also been used to pick up behavioral signals such as feeding events. With visual calibration of animals instrumented with accelerometers, it may be possible to identify other behaviors, such as suckling, resulting in a better understanding of reproductive effort in animals that are difficult to observe. Similarly, acoustic/behavior tags, like the D-Tag, have typically been used to study echolocation of odontocetes and behavioral responses of cetaceans to noise, but they have great potential to increase our knowledge on post-partum reproductive effort and behavior in cetaceans. For example, it may be possible to hear when a calf is suckling if a tag is placed near the head of the calf. If mothers/calf pairs are instrumented at different stages of lactation, it will be possible to estimate how much time the calf spends suckling and how this changes as lactation progresses.

10.3.5 Metabolism

The full scope of biomedical tools for the study of metabolism are available for the study of lactation in marine mammals, although their applicability is currently limited to pinnipeds that are able to be chemical immobilized without substantial handling artifacts (Champagne et al. 2012). The tools include the use of antibody-based techniques to measure protein expression in tissue and circulating levels of chemical messengers including hormones and adipokines. Western blots, enzyme immunoassays (EIA), enzyme-linked immunosorbent assays (ELISA), and radioimmunoassays are used to quantify the concentrations of proteins and chemical messengers. The use of omics tools including genomics, transcriptomics, and proteomics can be used to understand the changes in gene expression in specific tissues of fasting and lactating pinnipeds, usually blubber and muscle. The systemic impacts of these changes in gene expression can be studied using metabolomics, which attempts to measure as many of the metabolites in circulation or tissue as possible (Champagne et al. 2013). Finally, metabolic tracers can be used to quantify metabolite flux during lactation. A variety of techniques have been developed that use turnover of radioisotope or stable isotope-labeled metabolites to investigate whole-body metabolism. Several of these tracer methodologies that were developed for biomedical studies have been used for the first time in wildlife systems in studies on lactating pinnipeds (Champagne et al. 2006).

10.3.6 Detection of weaning

One challenge for some marine mammal species that slowly transition offspring from milk to nutritional independence is the ability to detect weaning. One method that can be used to study this transition is to compare the stable isotope values of C and N in milk and plasma between mothers and offspring. When milk is their only food source, stable isotopes show that offspring are feeding at a higher trophic level than their mothers. As offspring transition to other food sources these differences disappear. In this way, stable isotopes can give insight into the timing of weaning transition to other food sources in marine mammals (Polischuk et al. 2001).

10.4 Unsolved mysteries

The ability of the mother to rapidly produce such large quantities of high fat milk must be matched by the digestive and assimilation capabilities of the pup to process and use these nutrients. Maximum energy assimilation rates by the digestive and metabolic systems of animals have been theorized to constrain energy expenditure and growth (Weiner 1992). A limit of 7× basal metabolic rate (BMR) has been hypothesized as an upper limit to assimilation of dietary energy from free-ranging energy budgets (Hammond and Diamond 1997) and a mammalian record of 7.7× BMR was reported for mice under conditions of extreme cold exposure (Johnson and Speakman 2001).

An average elephant seal pup assimilates energy at a rate of 8.3× BMR during suckling (Crocker et al. 2001) and a hooded seal pup is likely the highest rate of nutritional assimilation relative to body weight found in nature. These feats require an appropriate lipid digestion and intestinal uptake capacity, mechanisms to transport the digested lipids to adipose tissue for storage, and the ability to withstand extreme blood lipemia, or high concentrations of emulsified fat. The digestive adaptations of pinnipeds have yet to be studied.

Glossary

Adipocyte triglyceride lipase (ATGL): An enzyme that removes the first fatty acid from a triaglycerol molecule. This enzyme has been shown to be the predominant lipolytic enzyme in some marine mammal's blubber.

Adiposity: The degree of fatness of an animal. High adiposity is a feature of most marine mammals.

Altricial: A young animal born in an undeveloped state with limited mobility and requiring a high level of care and feeding by the parents compared to precocial young.

Capital breeding: A breeding strategy in which all or most of the energy used for breeding comes from stored body reserves. Capital and income breeding are two extremes on a continuous scale.

Doubly labeled water: Isotopically labeled water, in which both the oxygen and hydrogen molecules have been labeled with isotopes. The doubly labeled water method uses turnover of the tracer to measure CO_2 production and is an important technique for measuring the field metabolic rates of free-ranging animals.

Fecundity: The ability to produce offspring. Fecundity is usually defined as the rate of births for an individual or species.

Income breeding: A breeding strategy in which animals foraging during the period of reproduction. Capital and income breeding are two extremes on a continuous scale.

Involution: The shrinkage of an organ to its former size when inactive. Involution is usually used in reference to reproductive organs or mammary glands after reproduction.

Isotope dilution method: A technique using the dilution of isotopically labeled compounds to measure the quantity of the compound in an animal. This method most frequently refers to the use of isotopic labeled water to measure total body water, body composition, or milk intake.

Lipolysis: The breakdown of fats and other lipids by hydrolysis to release fatty acids. Lipolysis is the key process in mobilizing energy for fasting marine mammals.

Metabolic overhead: The proportion of a lactating female's energy expenditure that goes to her own maintenance metabolism instead of milk energy delivery to the pup or calf.

Metabolic water production: The water that is produced as a result of cellular respiration. For many marine mammals this is the only source of water for milk production and survival during fasting.

Perinatal period: The period immediately after birth. In otariids, this period refers to the initial period of lactation before the mother returns to sea to forage.

Photogrammetry: The use of photographs to make measurements of animals usually estimates of mass or surface area. This technique allows measurements to be made of animals without handling them.

Precocial: A young animal born in a more developed state with increased mobility and reduced need for parental care compared to altricial young.

Reproductive effort: The proportion of resources or energy that an organism expends on reproduction. Pinnipeds have become important experimental systems for measuring reproductive effort in wildlife biology.

Scaling exponent: The exponent of a power function that describes the relationship between two log transformed variables. Many physiological and behavioral variables in biology exhibit power functions with body size across species.

Truncated cones method: A method that combines morphometric measurements and ultrasound measures of blubber depth to model the blubber proportions of an animal. This method has been validated for use in many species of phocid seals.

Weaning conflict: The period where the offspring and parent are in conflict about the level of continued parental investment. Offspring may be selected to seek more resources from parents than that which maximizes their ability to invest in future reproduction.

References

Arnould, J.P.Y. and M.A. Ramsay. 1994. Milk production and milk consumption in polar bears during the ice-free period in western Hudson Bay. *Canadian Journal of Zoology* 72:1365–1370.

Atkinson, S.N., R.A. Nelson, and M.A. Ramsay. 1996. Changes in the body composition of fasting polar bears (*Ursus maritimus*): The effect of relative fatness on protein conservation. *Physiological Zoology* 69 (2):304–316.

Atkinson, S.N. and M.A. Ramsay. 1995. The effects of prolonged fasting of the body composition and reproductive success of female polar bears (*Ursus maritimus*). *Functional Ecology* 9:559–567.

Bowen, W.D., S.L. Ellis, S.J. Iverson, and D.J. Boness. 2001. Maternal effects on offspring growth rate and weaning mass in harbour seals. *Canadian Journal of Zoology* 79 (6):1088–1101.

Bowen, W.D. and S.J. Iverson. 1998. Estimation of total body water in pinnipeds using hydrogen-isotope dilution. *Physiological and Biochemical Zoology* 71 (3):329–332.

Burns, J.M., C.A. Clark, and J.P. Richmond. 2004. The impact of lactation strategy on physiological development of juvenile marine mammals: Implications for the transition to independent foraging. *International Congress Series* 1275:341–350.

Champagne, C.D., S.M. Boaz, M.A. Fowler, D.S. Houser, D.P. Costa, and D.E. Crocker. 2013. A profile of carbohydrate metabolites in the fasting northern elephant seal. *Comparative Biochemistry and Physiology Part D: Genomics and Proteomics* 8 (2):141–151.

Champagne, C.D., D.E. Crocker, M.A. Fowler, and D.S. Houser. 2012. Fasting physiology of the pinnipeds: The challenges of fasting while maintaining high energy expenditure and nutrient delivery for lactation. In *Comparative Physiology of Fasting, Starvation, and Food Limitation*, ed. M.D. McCue. Springer, Berlin, Germany.

Champagne, C.D., D.S. Houser, and D.E. Crocker. 2006. Glucose metabolism during lactation in a fasting animal, the northern elephant seal. *American Journal of Physiology: Regulatory, Integrative and Comparative Physiology* 291 (4):R1129–R1137.

Costa, D.P. 1987. Isotopic methods for quantifying material and energy intake of free-ranging marine mammals. In *Approaches to Marine Mammal Energetics*, eds. A.C. Huntley, D.P. Costa, G.A.J. Worthy, and M.A. Castellini. Society for Marine Mammalogy, Lawrence, KS.

Costa, D.P. 1991. Reproductive and foraging energetics of pinnipeds: Implications for life history patterns. In *The Behavior of Pinnipeds*, ed. D. Renouf. Chapman & Hall, London, UK.

Costa, D.P. 2009. Energetics 2009. In *Encyclopedia of Marine Mammals*, eds. W.F. Perrin, B. Wursig, and J.G.M Thewissen. Academic Press, San Diego, CA.

Costa, D.P. and G.L. Kooyman. 1982. Oxygen consumption, thermoregulation, and the effect of fur oiling and washing on the sea otter, *Enhydra lutris*. *Canadian Journal of Zoology* 60 (11):2761–2767.

Costa, D.P. and T.M. Williams. 1999. Marine mammal energetics. In *Biology of Marine Mammals*, eds. J.E. Reynolds and S.A. Rommel. Smithsonian Institution Press, Washington, DC, pp. 176–217.
Crocker, D.E., C.D. Champagne, M.A. Fowler, and D.S. Houser. 2014a. Adiposity and fat metabolism in lactating and fasting northern elephant seals. *Advances in Nutrition* 5 (1):57–64.
Crocker, D.E., M.A. Fowler, C.D. Champagne, A.L. Vanderlugt, and D.S. Houser. 2014b. Metabolic response to a glucagon challenge varies with adiposity and life-history stage in fasting northern elephant seals. *General and Comparative Endocrinology* 195:99–106.
Crocker, D.E., P.M. Webb, D.P. Costa, and B.J. Le Boeuf. 1998. Protein catabolism and renal function in lactating northern elephant seals. *Physiological Zoology* 71 (5):485–491.
Crocker, D.E., J.D. Williams, D.P. Costa, and B.J. Le Boeuf. 2001. Maternal traits and reproductive effort in northern elephant seals. *Ecology* 82 (12):3541–3555.
Derocher, A.E., D. Andriashek, and J.P.Y. Arnould. 1993. Aspects of milk composition and lactation in polar bears. *Canadian Journal of Zoology* 71:561–567.
Fowler, M.A., C.D. Champagne, D.S. Houser, and D.E. Crocker. 2008. Hormonal regulation of glucose clearance in lactating northern elephant seals (*Mirounga angustirostris*). *Journal of Experimental Biology* 211 (18):2943–2949.
Fowler, M.A., D.P. Costa, D.E. Crocker, W. Shen, and F.B. Kraemer. 2015. Adipose triglyceride lipase, not hormone sensitive lipase, is the primary lipolytic enzyme in fasting elephant seals (*Mirounga angustirostris*). *Physiological and Biochemical Zoology* 88:284–294.
Fowler, M.A., C. Debier, E. Mignolet, C. Linard, D.E. Crocker, and D.P. Costa. 2014. Fatty acid mobilization and comparison to milk fatty acid content in northern elephant seals. *Journal of Comparative Physiology B* 184 (1):125–135.
Gales, N.J. and H.R. Burton. 1987. Ultrasonic measurement of blubber thickness of the southern elephant seal, *Mirounga leonina* (Linn.). *Australian Journal of Zoology* 35:207–217.
Guinet, C., N. Servera, S. Mangin, J.-Y. Georges, and A. Lacroix. 2004. Change in plasma cortisol and metabolites during the attendance period ashore in fasting lactating subantarctic fur seals. *Comparative Biochemistry and Physiology Part A: Molecular & Integrative Physiology* 137 (3):523–531.
Hammond, K.A. and J. Diamond. 1997. Maximal sustained energy budgets in humans and animals. *Nature* 386 (6624):457–462.
Hedd, A., R. Gales, and D. Renouf. 1995. Use of temperature telemetry to monitor ingestion by a harbour seal mother and her pup throughout lactation. *Polar Biology* 15 (3):155–160.
Houser, D.S., C.D. Champagne, and D.E. Crocker. 2007. Lipolysis and glycerol gluconeogenesis in simultaneously fasting and lactating northern elephant seals. *American Journal of Physiology: Regulatory, Integrative and Comparative Physiology* 293 (6):R2376–R2381.
Iverson, S.J., M. Hamosh, and W.D. Bowen. 1995. Lipoprotein lipase activity and its relationship to high milk fat transfer during lactation in grey seals. *Journal of Comparative Physiology B* 165:384–395.
Johnson, M.S. and J.R. Speakman. 2001. Limits to sustained energy intake V. Effect of cold-exposure during lactation in *Mus musculus*. *Journal of Experimental Biology* 204 (11):1967–1977.
Kovacs, K.M. and D.M. Lavigne. 1986. Maternal investment and neonatal growth in phocid seals. *Journal of Animal Ecology* 55:1035–1051.
Le Boeuf, B.J. and D.E. Crocker. 2005. Ocean climate and seal condition. *BMC Biology* 3 (1):9.
Lockyer, C. 1981. Growth and energy budgets of large baleen whales from the Southern Hemisphere. *Mammals in the Seas* 3:379–487.
Lydersen, C. 1995. Energetics of pregnancy, lactation and neonatal development in ringed seals (*Phoca hispida*). In *Whales, Seals, Fish and Man*, eds. A.S. Blix, L. Walloe, and O. Ulltang. Elsevier Science B.V., Amsterdam, the Netherlands.
Lydersen, C., M.O. Hammill, and K.M. Kovacs. 1995. Milk intake, growth and energy consumption in pups of ice-breeding grey seals (*Halichoerus grypus*) from the Gulf of St. Lawrence, Canada. *Journal of Comparative Physiology B* 164:585–592.
Lydersen, C. and K.M. Kovacs. 1996. Energetics of lactation in harp seals (*Phoca groenlandica*) from the Gulf of St. Lawrence, Canada. *Journal of Comparative Physiology B* 166:295–304.
Lydersen, C., K.M. Kovacs, and M.O. Hammill. 1997. Energetics during nursing and early postweaning fasting in hooded seal (*Cystophora cristata*) pups from the Gulf of St. Lawrence, Canada. *Journal of Comparative Physiology B* 167:81–88.

McDonald, B.I. and D.E. Crocker. 2006. Physiology and behavior influence lactation efficiency in northern elephant seals (*Mirounga angustirostris*). *Physiological and Biochemical Zoology* 79 (3):484–496.

McDonald, B.I., M.E. Goebel, D.E. Crocker, and D.P. Costa. 2012. Biological and environmental drivers of energy allocation in a dependent mammal, the Antarctic fur seal pup. *Physiological and Biochemical Zoology* 85 (2):134–147.

Monson, D.H., J.A. Estes, J.L. Bodkin, and D.B. Siniff. 2000. Life history plasticity and population regulation in sea otters. *Oikos* 90 (3):457–468.

Noren, D.P., D.E. Crocker, T.M. Williams, and D.P. Costa. 2003. Energy reserve utilization in northern elephant seal (*Mirounga angustirostris*) pups during the postweaning fast: size does matter. *Journal of Comparative Physiology B* 173:443–454.

Oftedal, O.T. 1997. Lactation in whales and dolphins: Evidence of divergence between baleen- and toothed-species. *Journal of Mammary Gland Biology and Neoplasia* 2 (3):205–230.

Oftedal, O.T. and S.J. Iverson. 1987. Hydrogen isotope methodology for measurement of milk intake and energetics of growth in suckling young. In *Approaches to Marine Mammal Energetics*, eds. A.C. Huntley, D.P. Costa, G.A.J. Worthy, and M.A. Castellini. Society for Marine Mammalogy, Lawrence, KS.

Oftedal, O.T. and S.J. Iverson. 1995. Comparative analysis of nonhuman milks. A. Phylogenetic variation in the gross composition of milks. In *Handbook of Milk Composition*, ed. R.G. Jensen. Academic Press, San Diego, CA, pp. 749–789.

Polischuk, S.C., K.A. Hobson, and M.A. Ramsay. 2001. Use of stable-carbon and-nitrogen isotopes to assess weaning and fasting in female polar bears and their cubs. *Canadian Journal of Zoology* 79 (3):499–511.

Reich, C.M. and J.P.Y. Arnould. 2007. Evolution of Pinnipedia lactation strategies: A potential role for α-lactalbumin? *Biology Letters* 3 (5):546–549.

Schreer, J.F., J.L. Lapierre, and M.O. Hammill. 2010. Stomach temperature telemetry reveals that harbor seal (*Phoca vitulina*) pups primarily nurse in the water. *Aquatic Mammals* 36 (3):270.

Schulz, T.M. and W.D. Bowen. 2004. Pinniped lactation strategies: Evaluation of data on maternal and offspring life history traits. *Marine Mammal Science* 20(1):86–114.

Sharp, J.A., C. Lefèvre, and K.R. Nicholas. 2008. Lack of functional alpha-lactalbumin prevents involution in Cape fur seals and identifies the protein as an apoptotic milk factor in mammary gland involution. *BMC Biology* 6 (1):48.

Sharp, J.A., C. Lefevre, A.J. Brennan, and K.R. Nicholas. 2007. The fur seal—A model lactation phenotype to explore molecular factors involved in the initiation of apoptosis at involution. *Journal of Mammary Gland Biology and Neoplasia* 12 (1):47–58.

Speakman, J. 1997. *The Doubly Labeled Water Method*. Chapman & Hall, London, UK.

Strandberg, U., A. Käkelä, C. Lydersen et al. 2008. Stratification, composition, and function of marine mammal blubber: The ecology of fatty acids in marine mammals. *Physiological and Biochemical Zoology* 81 (4):473–485.

Thiemann, G.W., S.J. Iverson, and I. Stirling. 2006. Seasonal, sexual and anatomical variability in the adipose tissue of polar bears (*Ursus maritimus*). *Journal of Zoology* 269 (1):65–76.

Thometz, N.M., M.T. Tinker, M.M. Staedler, K.A. Mayer, and T.M. Williams. 2014. Energetic demands of immature sea otters from birth to weaning: Implications for maternal costs, reproductive behavior and population-level trends. *Journal of Experimental Biology* 217 (12):2053–2061.

Tift, M.S. E.L. Ranalli, D.S. Houser, R.M. Ortiz, and D.E. Crocker. 2013. Development enhances hypometabolism in northern elephant seal pups (*Mirounga angustirostris*). *Functional Ecology* 27:1155–1165.

Tinker, T.M., G. Bentall, and J.A. Estes. 2008. Food limitation leads to behavioral diversification and dietary specialization in sea otters. *Proceedings of the National Academy of Sciences* 105 (2):560–565.

Trivers, R.L. 1974. Parent-offspring conflict. *American Zoologist* 14 (1):249–264.

Trumble, S.J., S.R. Noren, L.A. Cornick, T.J. Hawke, and S.B. Kanatous. 2010. Age-related differences in skeletal muscle lipid profiles of Weddell seals: Clues to developmental changes. *Journal of Experimental Biology* 213 (10):1676–1684.

Verrier, D., S. Atkinson, C. Guinet, R. Groscolas, and J.P.Y. Arnould. 2012. Hormonal responses to extreme fasting in subantarctic fur seal (*Arctocephalus tropicalis*) pups. *American Journal of Physiology: Regulatory, Integrative and Comparative Physiology* 302 (8):R929–R940.
Viscarra, J.A. and R.M. Ortiz. 2013. Cellular mechanisms regulating fuel metabolism in mammals: Role of adipose tissue and lipids during prolonged food deprivation. *Metabolism* 62 (7):889–897.
Waite, J.N., W.J. Schrader, J.E. Mellish, and M. Horning. 2007. Three-dimensional photogrammetry as a tool for estimating morphometrics and body mass of Steller sea lions (*Eumetopias jubatus*). *Canadian Journal of Fisheries and Aquatic Sciences* 64 (2):296–303.
Weiner, J. 1992. Physiological limits to sustainable energy budgets in birds and mammals: Ecological implications. *Trends in Ecology & Evolution* 7 (11):384–388.
Whitehead, H. and J. Mann. 2000. Female reproductive strategies of cetaceans. *Cetacean Societies: Field Studies of Dolphins and Whales*, eds. J. Mann, R.C. Connor, P.L. Tyack, and H. Whitehead. University of Chicago Press, Chicago, IL, pp. 219–246.

section four

Sensory systems

chapter eleven

Acoustics

Dorian Houser and Jason Mulsow

Contents

11.1 The big picture challenge and summary

Life as an aquatic mammal means exploiting an environment that is less conducive to the transmission of light than it is to sound. Light energy is attenuated rapidly by sea water such that in many oceans of the world little usable light exists beyond ~200 m depth. Even above this point, termed the euphotic zone, light attenuation limits the distance over which vision is useful, which is seldom more than tens of meters. On the other hand, the ocean is a favorable environment for sound transmission. Sound travels faster in water (~1500 m/s) than in air (~340 m/s) and is attenuated orders of magnitude less per unit of distance traveled. So, how does an animal whose ancestors were land-walking, primarily visual creatures adapt to an environment that is unfavorable to vision (see also Chapter 12)? Marine mammals must forage for food (often mobile and elusive), find mates, navigate the oceans, and avoid predators over scales where vision may be of limited use in the sea. To achieve these goals, marine mammals need to be able to detect and identify sound, localize sound sources, and project sound in an efficient manner, all within the constraints of the generalized mammalian bauplän.

11.2 Knowledge by order

11.2.1 Order Cetacea

11.2.1.1 Hearing

The auditory system of the terrestrial mammal is adapted for transducing changes in air pressure (sound waves) to neurological signals that are perceived as pitch and loudness, the perceptual correlates of the physical characteristics of sound, frequency and amplitude, respectively (Figure 11.1). The tympanic membrane, or eardrum, is the tissue that first contributes to this process. As sound vibrates the tympanic membrane, the vibrations move a chain of small bones called the ossicles, which consist of the malleus, incus, and stapes. The stapes connects to the oval window of the inner ear and is responsible for generating pressure waves within the fluid-filled cochlea. It is an elegantly designed system that functions well due to the presence of air on both sides of the tympanic membrane (i.e., the middle ear contains air fed to it from the Eustachian tubes). This changes when an animal is submerged in water. Now surrounded by water, the acoustic impedance differences between the air of the middle ear/external ear canal and the water are such that relatively little acoustic energy will be transduced into mechanical waves via the ossicular chain.

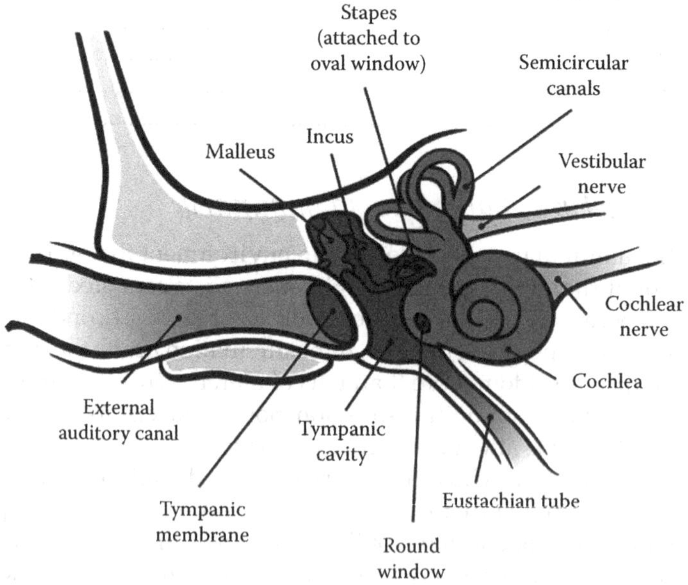

Figure 11.1 **(See color insert.)** A schematic of the outer, middle, and inner ear of terrestrial mammals (in this case based on the human anatomy). (Adapted from Chittka, L. and Brockmann, A., *PLoS Biol.*, 3(4), e137, 2005.) For terrestrial mammals in air, the outer and middle ear are both air-filled, and the middle ear ossicles (incus, malleus, and stapes) amplify sound vibrations from the tympanic membrane (eardrum) into the sensory organ of the fluid-filled cochlea via the oval window. Sound is then transmitted to the brain through the auditory nerve. When this terrestrial ear is submerged, the high acoustic impedance of external water in the outer ear relative to the low impedance of the air-filled middle ear results in decreased efficiency of sound transmission to the cochlea. Marine mammal auditory systems have adapted to various degrees to overcome this problem.

The cetaceans, which consist of the toothed and baleen whales, are undoubtedly the most derived of the marine mammals when it comes to the auditory system. Having an evolutionary history traced through ungulate ancestors, the Cetacea demonstrate adaptations to the auditory system and vestibular systems that are unique relative to other mammalian species. However, within the Cetacea there are dramatic differences between the suborders Odontoceti (toothed whales) and Mysticeti (baleen whales), with the former being the more derived of the two suborders.

The overall morphology of the cetacean head has resulted from selective pressures related to underwater movement efficiency (e.g., reduction in drag) and surface respiration. Cetaceans typically demonstrate a streamlined, fusiform shape (see Chapter 1) with a telescoping skull (Figure 11.2). The process of telescoping has led to a dorsal positioning of the nares and elongation of the rostrum, particularly the maxillary and mandibular bones. The ears in cetaceans were also modified such that the bony structures containing the middle and inner ear, the tympanic and periotic bulla (collectively termed the tympanoperiotic complex or the auditory bulla), migrated laterally from the skull (Oelschlager 1990). In the odontocetes, the bony separation from the skull is complete (except for sperm whales and some beaked whales), but some bony connections between the ear and the skull remain in mysticetes. The external pinnae, which are commonly found in terrestrial mammals and which are important to sound localization and amplification in air, are essentially absent in cetaceans although vestigial tissue structures surrounding the external meatus are present. The external meatus is largely diminished in cetaceans and appears as a pin-sized hole in the side of the cetacean head. The external auditory canal in odontocetes does not connect to the

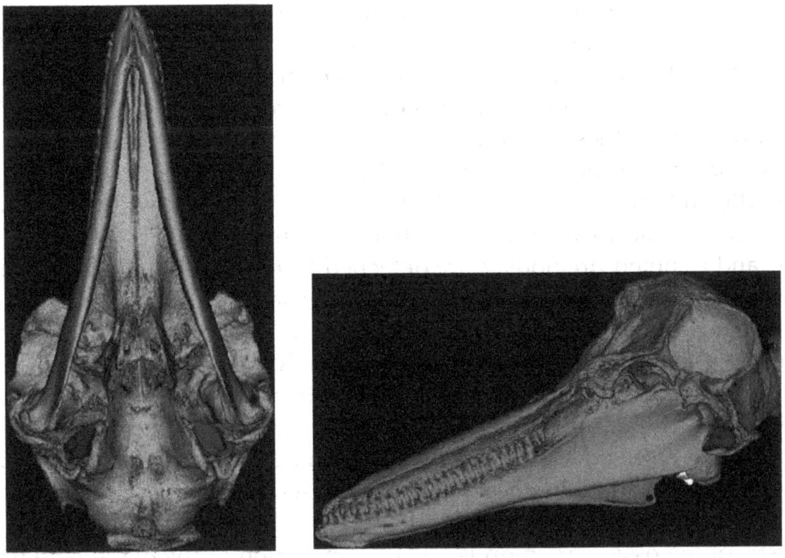

(a) (b)

Figure 11.2 **(See color insert.)** (a) Ventral perspective of the dolphin skull with the left (green) and right (red) auditory bullae displayed *in situ*. The projections are based on a CT scan of a bottlenose dolphin. (b) Lateral perspective of the same dolphin showing the relationship of the left bulla to the posterior terminus of the lower jaw. The bullae are detached from the skull and surrounded by tissue and airspaces (sinuses).

tympanic membrane and the canal is filled with cerumen and cellular debris. These factors contribute to arguments that the outer ear of the dolphin contributes relatively little to sound reception (Kellogg 1938; McCormick et al. 1970; Ridgway 2000). In contrast, the external ear canal of the mysticete whale clearly connects to an everted tympanic membrane.

Abundant anatomical, physiological, and behavioral evidence suggests that the acoustic pathway to the odontocete ear primarily is achieved via fatty channels that connect directly to the auditory bulla. The fats, commonly referred to as *acoustic fats*, are contained within the lower jaw of the dolphin and consist of wax esters that have acoustic impedances similar to that of water. A thin, bony region of the lower mandible, termed the *pan*, is in connection with mandibular fat bodies both internally and externally. The fat bodies extend posteriorly from the jaw to connect with the tympanic bone and middle ear, while another fat body also connects to the tympanic bone. Studies have demonstrated that sound passes into the jaw fats, and it is believed that the sound energy is directly coupled to the tympanic bone and middle ear via this path (Brill and Harder 1991). Odontocetes are particularly sensitive to ultrasonic frequencies (i.e., >25 kHz or so) along the jaw, and specifically in the region of the pan, suggesting the importance of this pathway in echolocation (see below). Indeed, hearing via this pathway is highly directional, that is, there is a narrow receiving beam for sound along the forward looking longitudinal axis of the animal (Au and Moore 1984). A narrow receiving beam is an important component of any sonar system and enables target localization while minimizing unwanted interferences and reverberation. At lower frequencies, within the communication range of odontocetes (<20 kHz), hearing is less directional and the greatest sensitivity to sounds within this frequency range appears to be more lateral with acoustically sensitive regions existing near the external auditory meatus (Popov et al. 2006). Nevertheless, the lack of a connection between the external ear canal and the tympanum in odontocetes suggests that the meatus and ear canal are vestigial and contribute little to sound reception. However, the presence of the posterolateral fat body adjoined to the tympanic might indicate another peripheral pathway, possibly tuned to lower frequency sound reception (Ketten 1994, 1997; Popov et al. 2008). Collectively, these findings are suggestive of different *modes of hearing* between echolocation and communication signals within the odontocetes.

The middle and inner ears of the odontocetes are contained within the tympanoperiotic complex or the auditory bulla. These bones are the densest bones in the body of an odontocete, and as noted previously, are detached from the skull (Figure 11.2). The bullae are suspended by ligamentous connections and partially surrounded both medially and dorsally by air, an anatomical arrangement that serves to acoustically isolate the ears. The isolation and presence of air between the bullae likely contributes to changes in received sound as it passes from one ear to the other and probably contributes to sound localization capabilities, that is, the air may impede sound conduction and contribute to spectral differences in the sound received at each ear (Houser et al. 2004; Mulsow et al. 2014).

The middle ear of the odontocete is characterized by ossicles that are more massive than in most terrestrial mammals and stiffened by ligaments and membranous sheaths. Characteristic of the ossicular chain is a fixed connection of the malleus to the tympanic bulla and the presence of a robust stapedial ligament attached to the head of the stapes. Although the tympanic membrane has a membranous attachment to the malleus, the lack of a connection between the external auditory canal and the tympanic membrane has led to much debate over the exact function of the ossicular chain in odontocete sound reception. Prior work in which the tympanic conus, malleus, and external auditory canal were disrupted had little effect on cochlear potentials (McCormick et al. 1970). This finding was used to argue that the motion of the bulla relative to the ossicular chain was the mechanism by which compressional waves were made within the cochlear fluids, not the action of the stapes on the oval window as would be caused by activation of the ossciular chain via motion of the tympanic membrane (McCormick et al. 1980). However, it has also been argued that the tympanum likely maintains the role of activating the ossicular chain, regardless of the tissue path to the tympanum as is observed in other mammals (Fleischer 1978). Little has changed in this debate, and it remains uncertain as to how sound is translated into compressional waves within the cochlear fluid.

The hearing range of odontocetes is broad and ultrasonic, and ranges from several hundred Hz to in excess of 160 kHz in some smaller odontocetes (Figure 11.3). There is a crude inverse relationship between the frequency range of hearing and size within the odontocetes, as larger odontocetes (e.g., killer whales, beaked whales) can have upper-frequency limits of hearing (80 to 100 kHz) nearly an octave less than some of the smaller odontocetes (160 kHz). The broad, ultrasonic frequency range of hearing in odontocetes is critical to their system of echolocation, that is, the ear of the odontocete is adapted to serve as an optimal receiver with respect to their sound transmission system (see below). Although the inner ear of the odontocete retains a general mammalian structure, it has highly derived features that enable ultrasonic hearing. Many structures within the cochlea are hypertrophied: high ganglion cell counts, a dense stria vascularis, and disproportionately large cochlear aqueducts (e.g., see Ketten 1992 for review). The basilar membrane, as

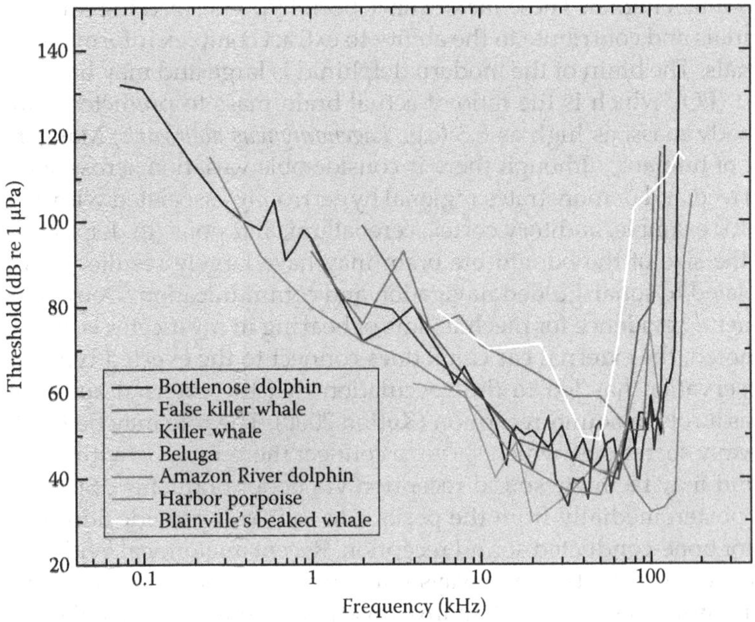

Figure 11.3 **(See color insert.)** Underwater audiograms for selected odontocete cetacean species. All thresholds were obtained using behavioral methods except those for the beaked whale species, which were obtained using auditory evoked potential methods. Note the extended range of high-frequency hearing in the harbor porpoise relative to the other mid-frequency cetaceans. Studies for which thresholds are shown: bottlenose dolphin (Johnson 1967), false killer whale (Thomas et al. 1988), killer whale (Szymanski et al. 1999), beluga (White et al. 1978), Amazon River dolphin (Jacobs and Hall 1972), harbor porpoise (Kastelein et al. 2002), and Blainville's beaked whale (Pacini et al. 2011).

in all mammals, serves as the frequency map of the inner ear and vibrates in response to changes in the pressure of the cochlear fluids caused by action of the stapes on the oval window. Differences in the thickness and width of the basilar membrane dictate, at least in large part, where along the basilar membrane length the maximum response to a particular frequency of stimulation occurs; the basilar membrane is narrower and thicker at the base, where resonant responses to higher frequencies occur, and wider and thinner at the apex, where resonant responses to lower frequencies occur. The ratio of the basilar membrane thickness to width (T/W) is one of several determinants of the frequency response of the membrane (Ketten 2000), the vibration of which activates hair cells along the organ of Corti and results in neural signals being transmitted via the auditory nerve to the brain. The T/W ratios along the basilar membranes of odontocetes are consistent with the maximum high and low frequencies reported in these species (Ketten and Wartzok 1990). The membrane itself is supported and stiffened by outer and inner bony laminae that buttress the membrane and increase its stiffness. The degree that the outer lamina buttresses the basilar membrane varies by species with more of the basilar membrane being supported in odontocetes producing echolocation signals with higher peak frequencies (e.g., *Phocoena phocoena*, >100 kHz) than those producing echolocation signals with lower peak frequencies (e.g., *Tursiops truncatus*, <100 kHz; Ketten 2000).

Auditory ganglion cells of the cochlear nerve are dense in odontocetes with calculated ratios of ganglion to hair cells of ~6.5–7.3:1 (Ketten 2000). In addition, auditory fiber diameters tend to be greater in odontocetes than in terrestrial mammals and the number of auditory cell counts is much higher. These factors have been hypothesized to reduce the latencies of action potentials and contribute to the ability to extract complex information from received acoustic signals. The brain of the modern delphinid is large and may have an encephalization quotient (EQ), which is the ratio of actual brain mass to predicted brain mass based upon total body mass, as high as 6.3 (e.g., *Lagenorhyncus albirostris*; Manger 2006). This is close to that of humans, although there is considerable variation across species. The brain is highly derived and demonstrates regional hypertrophy associated with acoustic processing centers; for example, auditory cortex, cerebellum, and pons (in delphinids). Indeed, the increase in the size of the odontocete brain may have largely resulted from the adaptive pressures related to sonar-guided navigation and communication (Oelschlager 2008).

Experimental evidence for mechanisms of hearing in mysticetes is absent, although as previously noted, the external ear canal does connect to the everted tympanic membrane and this observation has led to the speculation that the external auditory canal of the mysticete has a role in sound reception (Ketten 2000). The role may be limited to the production of waxy secretions, which serve to connect the tympanic with surrounding bony structures and may facilitate sound reception via bone conduction. Spongy bone flanges that extend posteromedially from the periotic to the skull provide additional anatomical arguments for bone-conducted sound reception. Recent anatomical evidence also suggests that the fatty sound reception pathways may not be unique to odontocetes but may also be present in mysticetes. Postmortem investigations of the minke whale (*Balaenoptera acutorostrata*) demonstrate that a fatty body exists lateral to the tympanoperiotic complex and connects with it, as is observed in odontocetes (Yamato et al. 2012). It connects directly to the ossicles and extends laterally to the blubber suggesting the possible presence of a lateral sound transmission pathway.

The ossicles of the middle ear in the mysticete whale are massive and loosely coupled, which is suggestive of specialization for low-frequency sound reception. In the cochlea, the apical basilar membrane thicknesses and widths are implicative of high-frequency hearing limits possibly as high as in humans (or higher) (Ketten 2000).

However, basal thickness are flaccid and the widths are broad, and the laminar supports are either poorly developed or absent, all of which further supports the idea of infrasonic specialization.

As with odontocetes, auditory fiber diameters, ganglion cell counts, and innervation densities are generally greater than observed in terrestrial mammals and presumably are related to the same adaptation of decreasing latencies and enabling the extraction of information from complex acoustic arrivals. The mysticete brain does not demonstrate the same degree of hypertrophy of brain auditory structures observed in the odontocetes, although the auditory cortex appears to have a greater cellular volume and other comparable specializations exist (Eriksen and Pakkenberg 2007). With an EQ < 1.0 for all mysticetes, brain size appears to be uncoupled from body size and deviates less from allometric predictions than it does in odontocete counterparts (Worthy and Hickie 1986; Boddy et al. 2012), which is likely driven by the evolution of underwater echolocation in odontocete species.

11.2.1.2 Sound production

The location of sound production by odontocetes, whether it occurred in the nasal passages or the larynx, was once hotly debated. However, nearly all data to date have provided evidence that sound production occurs in the nasal system of odontocetes. It is now commonly accepted that sound production in odontocetes is primarily achieved through the use of specialized structures located within the nasal passages and below the blowhole plug. These structures, termed the phonic lips, are believed to be used in producing the most typical of dolphin acoustic signals, mainly whistles, burst-pulses, and echolocation clicks (Cranford 2000). The odontocete nasal passages have been demonstrated to pressurize just prior to signal production (Ridgway et al. 1980; Amundin and Andersen 1983) and it is the passage of pressurized air across the phonic lips that is responsible for signal production. The phonic lips are associated with a series of air sacs (pre-maxillary, vestibular, and accessory) that support the pneumatic operation of signal production, although the exact role they play is not well understood.

Whistles are tonal signals, commonly frequency modulated, that are typically produced at fundamental frequencies below 20 kHz (but which may have harmonic components at higher frequencies). Whistles are believed to be used primarily for communication and social interaction, but not all species of odontocetes produce whistles (e.g., the sperm whale [*Physeter macrocephalus*]). Whistles demonstrate a more omnidirectional propagation from the animal relative to signals like echolocation clicks (although the higher the frequency of the signal, the more directionality it will display). Based upon the contours displayed on a spectrogram, whistles can roughly be grouped as constant frequency, upsweep, downsweep, concave, convex, or sinusoidal (multiple) (see Au and Hastings 2008, for a review). The role of the different signal types in communication is largely unknown, although the identification of individually distinctive whistles produced by dolphins and the context in which they are produced have been used as evidence for a class of *signature whistles* that dolphins utilize for identification and individual localization (King et al. 2013). The signature whistle hypothesis, however, does have its detractors (e.g., McCowan and Reiss 2001).

Burst pulses are short duration, broadband signals produced in rapid succession and which resemble squeaks, creaks, or groans. They are produced by all species of odontocete studied to date, although there is some deviation and specialization of the theme (e.g., sperm whale production of codas). Burst pulses are much less studied than whistles and will not be discussed extensively here. However, it is worth noting that burst pulses may be the primary means of acoustic communication in some species (e.g., harbor porpoises).

Echolocation clicks are short duration (tens to hundreds of μs), transient signals, although the exact type of signal varies by species. For example, the bottlenose dolphin, which is a delphinid, produces clicks that are broadband and may have half-power frequency bandwidths of greater than 85 kHz (Houser et al. 1999). From a sonar design perspective, such a signal provides considerable information about the targets that are ensonified by returning echoic information to the dolphin across a large band of frequencies. Conversely, the harbor porpoise, which is a phocoenid, produces narrower band signals with peak frequencies typically between 120 and 140 kHz (Au et al. 2006; Koblitz et al. 2012). It has been hypothesized that the smaller harbor porpoise produces high frequency, narrowband signals in order to avoid detection by one of its primary predators, the killer whale (*Orcinus orca*) (Koblitz et al. 2012); that is, the signal is produced above the killer whale's upper-frequency limit of hearing.

Echolocation clicks are projected through the melon, a specialized body of lipids that forms the forehead of the odontocete. The lipids are similar in composition to those that fill the lower jaw and which connect to the auditory bullae. These so-called *acoustic fats* form a structure that has a low sound-velocity core with a gradation of higher velocity fats surrounding it (Norris and Harvey 1974). The melon serves to collimate the projected echolocation signal and the echolocation beam has been found to be narrow and forward projected in all species studied (e.g., see Table 1 in Koblitz et al. 2012). Echolocation clicks are produced in series, or in trains, and for some species they can vary dynamically in frequency content. The echolocation beamwidth can also be varied to some extent and steered, much like one's vision can be changed by movement of the eyes but without movement of the head (Moore et al. 2008; Madsen et al. 2010). The ability to manipulate the frequency content of click and the beam structure is likely achieved through manipulation of the melon by underlying muscle in combination with changes in the shape and air volume of associated air sacs, which would act as reflectors; however, the exact process remains uncharacterized (Cranford 1992; Moore et al. 2008; Madsen et al. 2010). The melon and the acoustic fats of the lower jaw are also vascularized, yet they are metabolically inert structures requiring little in the way of oxygen or glucose delivery (Houser et al. 2004). Since the lipid density is dependent on temperature, and the density of the lipid will affect the bulk modulus, shear modulus, and sound speed through the acoustic fats, it has been proposed that the vasculature might serve to regulate the temperature of the acoustic fats and thus stabilize the propagation of sound under conditions of varying temperature (e.g., changes in water depth or seasonal temperature changes) (Houser et al. 2004).

The mechanisms by which mysticete whales produce sounds are unknown, and unlike odontocetes where considerable experimental evidence supports the phonic lips as the primary sound production mechanism, it is quite possible that mysticete sound production is laryngeal in origin. Mysticete whales produce a variety of call types that can roughly be categorized as songs, simple calls, complex calls, and knocks, grunts, pulses or clicks (Clark 1990). The duration of sounds produced by mysticetes may range from <100 ms (e.g., knocks and grunts) to several seconds (simple calls) and may contain dominant frequency content as low as 10 Hz and as high as 10 kHz. Many signals may be strung together in units and phrases to produce songs, as is probably most well known for the humpback whale (*Megaptera novaeangliae*). Mysticete vocalizations can be produced at high amplitudes and the source levels of some vocalizations have been observed to exceed (root-mean-squared) sound pressure levels of 180 dB (re 1 μPa); (decibels relative to a reference pressure of 1 micro-Pascal). There is considerable species variability in sound production by mysticetes and a summary of the frequency, duration, and source levels can be found in Au and Hastings (2008).

11.2.2 Order Carnivora

The marine carnivores comprise five familial groups: Phocidae (true or earless seals), Otariidae (sea lions and fur seals), Odobenidae (walruses), Mustelidae (sea otters), and Ursidae (polar bears). With regards to acoustic ecology, the defining characteristic of the marine carnivores is their amphibious nature; while all species forage on prey in the aquatic environment, each carries out some essential life-history functions in air. The degrees to which their auditory systems have modified the ear of their terrestrial ancestors for underwater function reflect the importance of underwater function relative to aerial function in the life of each species, as some trade-offs exist with specific adaptation for a particular physical medium.

11.2.2.1 The Pinnipeds: Families Phocidae, Otariidae, and Odobenidae

Of the five Carnivore families which contain marine species, the three pinniped families are the only ones in which all species are amphibious (although the degrees of marine versus terrestrial activity and adaptation varies greatly across pinniped species). The pinnipeds are thought to represent a monophyletic group, having evolved from a terrestrial dog or bear-like ancestor more than 25 million years ago in the Oligocene epoch. This long evolutionary history in the marine environment relative to the other marine carnivores has resulted in the greatest degree of adaptation of the auditory system (Thewissen and Nummela 2007). Auditory anatomy and processing capabilities are thought to be generally similar among otariid species. In contrast, the phocids display greater interspecies differences in these qualities, a result of their greater diversity in morphologies and ecological roles. Odobenids are thought to be more closely related to otariids but possess an auditory system that appears in many ways to be intermediate between those of otariids and phocids (Nummela 2007).

The external pinnae of the pinnipeds are notable for their degrees of reduction when compared to those of terrestrial carnivores. In the otariids, the pinnae have been reduced to small folds, while the phocids lack external pinnae altogether (Ramprashad et al. 1972; Repenning 1972). The reduction in pinna size is likely a result of selection for an increasingly hydrodynamic, fusiform shape when swimming underwater (see Chapter 1). This adaptation for underwater locomotion likely has ramifications, however, for pinnipeds while listening in air. In air, the pinnae direct and filter sound from the external environment into the ear canal toward the tympanic membrane. Their directional filtering properties also help individuals resolve ambiguities of the location of sound sources when large interaural differences are not present, for example, when sounds arrive from directly in front of or behind a subject, or when a sound source is moved only in the vertical plane (Yost 2007). Underwater, the impedances of the tissues of the pinnae are similar to water, and the absence of these structures is likely of relatively little importance to hearing.

The various morphologies of pinniped ears have marked impacts on species-specific hearing sensitivities (Figure 11.4). The ear canals of otariids are generally similar to those of terrestrial mammals (Ramprashad et al. 1972; Repenning 1972). In air, they appear to have hearing that is generally as sensitive as terrestrial carnivores. Hearing sensitivity is also quite good underwater, although otariids are not as sensitive as and lack the broad frequency range of odontocetes and phocids. These findings support the view that otariids are adapted primarily for aerial hearing (Reichmuth et al. 2013). Phocid species display a variety of outer ear morphologies that diverge from the basic mammalian construction. In the extreme case of the deep-diving elephant seals, the external auditory meatus is reduced to nearly the size of a pinhole, and there is probably not a continuous

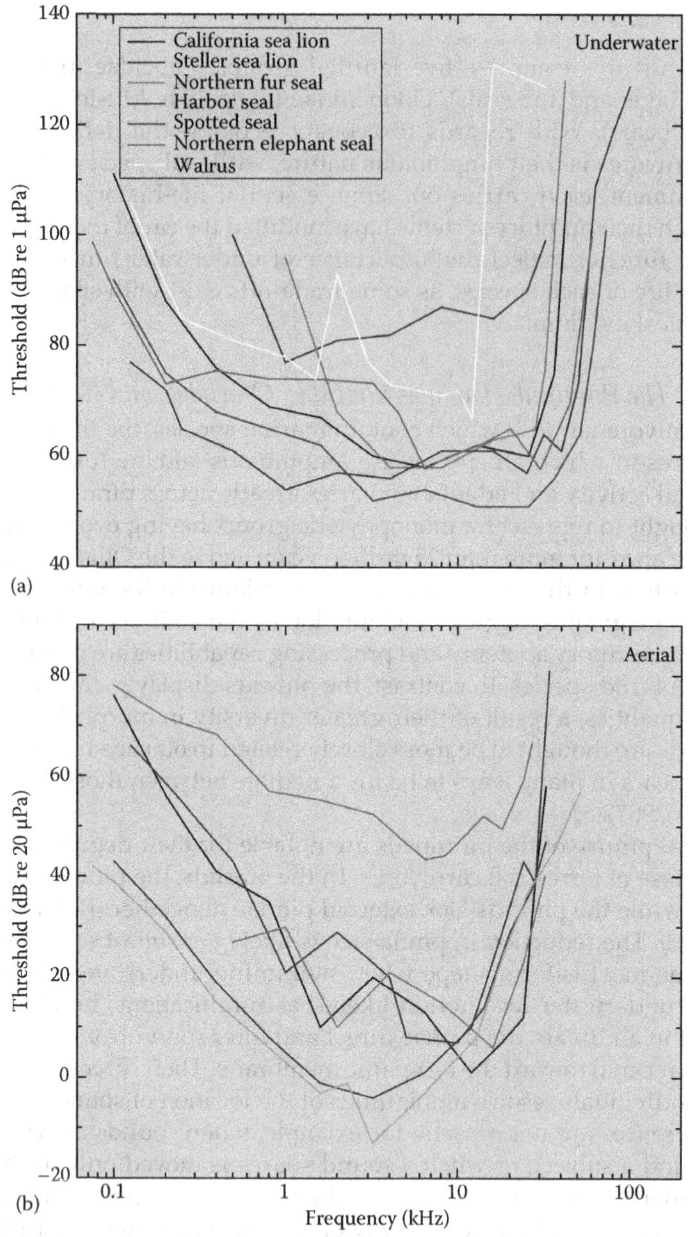

Figure 11.4 **(See color insert.)** Underwater (a) and aerial (b) hearing thresholds for selected pinniped species. All audiograms were obtained using behavioral methods. The two plots are shown with identical x-axes to emphasize the differences in frequency range of hearing between air and water. Note the difference in reference pressures for underwater (1 µPa) and aerial (20 µPa) thresholds. Studies for which underwater thresholds are shown: California sea lion (Reichmuth et al. 2013), Steller sea lion (Kastelein et al. 2005), northern fur seal (Moore and Schusterman 1987), harbor seal (Reichmuth et al. 2013), spotted seal (Sills et al. 2014), northern elephant seal (Kastak and Schusterman 1999), and walrus (Kastelein et al. 2002). Aerial studies: California sea lion (Reichmuth and Southall, 2012), Steller sea lion (Mulsow and Reichmuth 2010), northern fur seal (Moore and Schusterman 1987), harbor seal (Reichmuth et al. 2013), spotted seal (Sills et al. 2014), and northern elephant seal (Reichmuth et al. 2013).

airspace to the tympanic membrane due to a narrow ear canal that is filled with cellular debris. This modification for deep diving results in relatively poor aerial hearing sensitivity relative to terrestrial carnivores and other pinniped species, although underwater hearing appears to be sensitive across a wide frequency range (Kastak and Schusterman 1999). Many phocid species (e.g., harbor seals) possess musculature that can constrict the outer ear canal, presumably an adaptation for restricting the entrance of water into the ear canal when diving (Ramprashad et al. 1972; Repenning 1972). In contrast to the elephant seals, the ear canal can remain open in air, imparting hearing sensitivity in air that is comparable to that of terrestrial mammals (Reichmuth et al. 2013). These levels of adaptation reflect the relative degrees to which various pinniped species spend their time in aerial versus underwater environments: elephant seals are by far the *most aquatic*, otariids the *most terrestrial*, with harbor seals lying somewhere in-between (Kastak and Schusterman 1998, 1999).

The three middle ear ossicles of pinnipeds appear to function in a similar way to that of terrestrial mammals in air, with pressure differences between the ear canal and the middle ear transferred via the tympanic membrane and ossicular chain to a stapes that is moveable in the cochlear oval window (Ramprashad et al. 1972; Repenning 1972; Nummela 2007). While the mass and density of otariid middle ear ossicles are essentially similar to those of terrestrial carnivores, those of phocids (and to a lesser degree odobenids) have a relatively increased mass and density. Additionally present in the pinniped middle ear and outer ear canal is cavernous tissue that may engorge with blood upon diving in order to minimize pressure differences between the middle ear and the external environment (Odend'hal and Poulter 1966; Møhl 1967b, 1968; Repenning 1972).

The pinniped cochlea is essentially the same as that of other mammalian carnivores and notably different from that of the cetaceans that possess structural modifications for very high-frequency underwater hearing. Under water, the basal end of the cochlea determines high-frequency hearing limit in both otariids and phocids, although these limits are very different for the two families, with high-frequency hearing limits of approximately 40 and 100 kHz, respectively (Reichmuth et al. 2013). In air, the increased mass of the middle ear ossicles of phocids results in an inertial constraint on the transfer of high-frequency sound from the tympanic membrane to the inner ear, reducing the high-frequency limit to between 20 and 30 kHz (Hemilä et al. 2006). The only hearing data that exist for odobenids are underwater thresholds for one individual, which show a high-frequency limit of approximately 15 kHz (Kastelein et al. 2002). It is unknown whether this relatively low cutoff is representative of walruses in general, or just the subject of that study.

The sensory physiology of pinnipeds does not have the same extensive history of comparable research as odontocete cetaceans. This has resulted historically from the lower scientific interest in the capabilities of pinniped auditory nervous systems, as they lack the echolocation abilities and the high degree of specialization seen in odontocete cetaceans. Although some investigators suggested that pinnipeds might echolocate (Poulter 1963, 1966), electrophysiological and behavioral observations have demonstrated that pinnipeds do not possess a sophisticated biosonar system (Schusterman et al. 2000). For example, the auditory processing capabilities of odontocetes that are required for echolocation, such as the ability of the auditory nervous system to follow the envelopes of rapidly presented acoustic stimuli, are much less developed in pinnipeds and more akin to the capabilities of terrestrial carnivores (Bullock et al. 1971; Mulsow and Reichmuth 2007). Likewise, anatomical studies of the cerebral cortex have shown that auditory projection areas are essentially similar to other carnivores (Alderson et al. 1960; Popov et al. 1986).

The submersion of the pinniped head in water has profound implications for hearing. Upon submersion, the impedance mismatch between the external environment and biological tissues is reduced, resulting in an increased potential for *bone conduction* of sound to the cochlea along pathways other than that of the outer/middle ear (although work by Møhl and Ronald 1975 demonstrated best reception of sound near the opening of the ear canal in harbor seals). Underwater bone-conduction pathways and the 4–4.5 times increase in sound speed relative to air reduce the ability to detect and localize sound; however, anatomical evidence, including the partial separation of the auditory bullae from the skull, suggests anatomical modifications may aid pinnipeds in sound localization. Indeed, underwater sound localization experiments with pinnipeds have shown that, while inferior to the capabilities of odontocete cetaceans, pinnipeds are still able to acutely localize underwater sound (Gentry 1967; Moore and Au 1975; Babushina and Poliakov 2004; Bodson et al. 2006; Bodson et al. 2007). Many critical aspects of underwater hearing pathways in pinnipeds, however, remain poorly understood.

Most psychological measurements of auditory capabilities that are independent of the physical media in which they occur (i.e., at the level of cochlear and neural processes) have been limited to measurements with simple acoustic stimuli. Studies have shown that some features of auditory processing, such as the abilities to detect changes in tonal frequency and amplitude, are inferior to those of odontocetes and generally similar to those of other mammals (Møhl 1967; Moore and Schusterman 1976; Terhune and Ronald 1976; Schusterman and Moore 1978). In contrast, recent measurements have demonstrated that pinnipeds appear to be able to more efficiently extract tones from noise than other mammalian species, potentially an adaptation of the auditory system for function in noisy marine environments (Southall et al. 2000, 2003; Sills et al. 2014). Investigations utilizing more complex *real-world stimuli* have generally been limited to the realm of playback studies in field settings, but there is increasing interest in studying the perception of complex stimuli in controlled laboratory settings (e.g., Cunningham et al. 2014).

11.2.2.2 Family Mustelidae

Having diverged from non-marine ancestors approximately 3–5 million years ago, the sea otter has a much shorter evolutionary history in the marine environment than pinnipeds (Thewissen and Nummela 2007). Sea otters do, however, display some of the adaptations of the outer and middle ear seen in pinnipeds, including reductions in pinna size, increases in the size of middle-ear ossicles, and a thickening of the tympanic membrane (Solntseva 2007). Also like the pinnipeds, these adaptations may have been primarily driven by needs related to swimming and diving as opposed to hearing, although there are almost certainly effects of these morphological changes on auditory capabilities relative to non-marine species.

Behavioral measurements of hearing sensitivity with the sea otter have shown that aerial hearing is similar to that of otariid pinnipeds, while underwater hearing is apparently inferior to that of otariids, phocids, and most terrestrial mammals (Ghoul and Reichmuth 2014). Additionally, sea otters are apparently similar to terrestrial mammals in terms of their ability to detect tones in simultaneous noise (Ghoul and Reichmuth 2014). These findings reinforce the notion that due to their relatively recent transition to the marine environment, the hearing capabilities of sea otters are still primarily adapted for aerial function, although some adaptations of the ear for diving may have been at the cost of aerial hearing sensitivity.

11.2.2.3 Family Ursidae

Similar to sea otters, polar bears have a relatively short evolutionary history in the marine environment, having diverged from terrestrial bears <2 million years ago (Cahill et al. 2013). Little research has been conducted on the auditory system of the polar bear (as an aside, the polar bear is the only bear for which hearing data are available). The only known modifications to the auditory apparatus relative to that of terrestrial carnivores are prominent fur inside of the pinna and a reduction in pinna size relative to terrestrial bears, suggesting thermoregulatory adaptations for reducing heat loss in the extreme cold of the Arctic (Stirling 1988).

Behavioral and electrophysiological measurements of aerial hearing sensitivity have demonstrated that polar bears are sensitive to aerial sound over a frequency range similar to that of pinnipeds and mustelids: approximately 125 Hz to 32 kHz (Nachtigall et al. 2007; Owen and Bowles 2011). There are currently no data on underwater hearing for polar bears, but their capabilities are probably poor given the lack of apparent adaptations of the auditory system and the suspected low importance of underwater sound in communication, foraging, and navigation. Future measurements of underwater hearing in the polar bear would, however, be of interest in order to make amphibious comparisons in the *most terrestrial* marine mammal.

11.2.3 Order Sirenia

The sirenians comprise manatee species and the dugong. This order of marine mammals is relatively divergent from the marine carnivores and cetaceans in terms of evolution, sharing their closest phylogenetic relatives with elephants and hyraxes. The sirenians likely became distinct from terrestrial ancestors in the Eocene Epoch (along with cetaceans), and their auditory systems display the high level of modification of the terrestrial mammalian ear that is consistent with a fully aquatic existence (Thewissen and Nummela 2007).

Like cetaceans, sirenians lack pinnae and have a small pin-sized opening that leads to an ear canal that is probably occluded with cellular debris (Bullock et al. 1980; Ketten et al. 1992; Chapla et al. 2007). The middle-ear ossicles are large, also like cetaceans and phocids, and have a high bone density (Robineau 1969; Fleisher 1978; Chapla et al. 2007). Unlike the phocids, however, the middle ear likely remains air-filled without the engorgement of cavernous tissue. The malleus is strongly fused to the tympanic bone and the incus fused to the periotic bone, resulting in a middle ear chain that is not activated through a typical mammalian pathway via the tympanic membrane. A striking feature of the sirenian ear is that the malleus presses on the tympanic membrane, forcing it outwards. This feature is believed to be unique among mammals, although its function in aquatic hearing is unknown (Nummela 2007). The cochlea is essentially similar to that of other non-echolocating mammals (Ketten et al. 1992). Despite some knowledge of the morphological configuration of the sirenian ear, the manner in which sound is transferred to the inner ear is not currently understood. Some studies have suggested a pathway could exist through fatty tissue located in the zygomatic process (Bullock et al. 1980; Ketten et al. 1992; Chapla et al. 2007).

The auditory electrophysiological voltages that are recorded at the skin surface of manatees are small compared to those of odontocetes, a result of the lack of hypertrophy of neural structures (Klishin et al. 1990). Electrophysiological measures of auditory system function have shown that sirenians are roughly comparable to pinnipeds in terms of their ability to track the envelope of rapidly presented acoustic stimuli, a result that is not particularly surprising given the lack of echolocation abilities in both groups (Popov and Supin 1990; Mann et al. 2005). Underwater hearing threshold measurements with manatees have demonstrated low-frequency hearing limits near a few hundred hertz, with high-frequency hearing up to approximately 50–100 kHz (Klishin et al. 1990; Popov and Supin 1990; Gerstein et al. 1999; Mann et al. 2005; Gaspard et al. 2012). Like the pinniped species that have been examined, manatees appear to be better at detecting tonal acoustic stimuli in noise than terrestrial mammals (Gaspard et al. 2012). Together, these findings demonstrate that although sirenians have an auditory system that is adapted for a fully aquatic lifestyle, the lack of specializations seen in odontocetes is most likely a result of the absence of echolocation in this phylogenetic group.

11.3 Responses to environmental noise

Most marine mammals rely on sound to exploit their environment and engage in interactions that support life-history functions. Since the beginning of the industrial age, there has been a continuing growth in ocean noise. Increases in ocean noise have the potential to impact the ability of marine mammals to detect and identify important acoustic information, a process known as masking wherein the presence of one sound impedes the detection of another. The rate of increase in ocean noise dwarfs the evolutionary time scale over which marine mammals have evolved to deal with natural ocean noise conditions, suggesting that their ability to compensate for increased ocean noise is limited by the inherent plasticity of their sound production and reception systems. Human contributions to ocean noise result from shipping, commercial and recreational boating, military activities, green energy construction and operations (e.g., wind turbines), and seismic exploration. Of these, commercial shipping is likely the most persistent noise source in the ocean as commercial shipping continues 24 hours a day and it is distributed throughout the world's oceans. Because of the low-frequency nature of shipping noise (generally in the hundreds of hertz or less), mysticete whales are the most likely to be impacted by commercial shipping, particularly in regions with very active shipping lanes (Clark et al. 2009; Hatch et al. 2012). Nevertheless, there are many different opportunities for high-level acoustic activity resulting from human sources to impact sound reception in marine mammals and consideration should not be limited to mysticetes. Neither should the question be constrained to anthropogenic sound. Biological sources significantly contribute to ocean noise in some systems (e.g., snapping shrimp) and even marine mammals themselves may be significant contributors to ocean noise (e.g., song competition among whales). This then raises the question: How does a marine mammal respond when ocean noise interferes with its ability to capitalize on acoustic cues?

When in the presence of excessive noise, communication signals may be masked by the noise excess. In such a situation, there is a poor signal-to-noise ratio and the potential for a receiver (e.g., a listening whale) to detect and interpret the signal is reduced. When marine mammals face conditions in which noise limits their ability to communicate, they may change the characteristics of the signals they produce. In some instances, both the amplitude of the signal and the rate at which it is produced may be increased, a phenomenon known as the Lombard effect. This phenomenon, which is named for the man who discovered it, is found to occur in humans and other mammals, as well as in birds (Brumm and Zollinger 2013). Elevations in signal amplitude are produced in northern right whales and killer whales in response to increased ambient noise (Holt et al. 2008; Parks et al. 2011). Increases in the rate of signal production and amplitude, as well as shifts in the frequency of signals outside the spectrum of interfering noise, all serve to increase the signal-to-noise ratio of the caller's signal. In this regard, what limited evidence exists suggests that marine mammals deal with environmental noise in the same manner that terrestrial mammals do.

11.4 Tools and methods

The sound projector, microphone (for amphibious species), and hydrophone (underwater microphone) are mainstays in the arsenal of the marine mammal bioacoustician. Projectors and hydrophones are transducers, that is, they either change an electrical signal into motion (e.g., projector) or they change variations in pressure (i.e., the sound pressure) into an electrical signal (e.g., hydrophone). Many underwater transducers can do both, to some degree. There are many types of underwater projectors and receivers, and each of these has different transmission capabilities or sensitivities at different frequencies based upon their material composition and design. For this reason, proper calibration of a transducer is critical to any study that desires to quantify sound or project a desired level of sound. In line with the hydrophone is generally an amplifier for increasing the signal level and a filter. Filtering is often required in order to eliminate unwanted signals or to prevent aliasing, which is a problem that occurs when high-frequency signals are not digitally sampled at a high enough rate to adequately characterize the signal.

In passive acoustic monitoring, which is common to the study of free-ranging marine mammals, hydrophone arrays are used to record the sound pressure level of received marine mammal sounds. Historically, as few as one hydrophone has been used to characterize sounds produced by marine mammals. However, arrays of hydrophones, appropriately placed in space, can be used for more advanced procedures such as localizing a phonating animal (e.g., hyperbolic fixing). These techniques are useful for tracking vocal animals at sea, monitoring their acoustic behavior, and possibly assisting with estimates of population size (e.g., Marques et al. 2013; Stanistreet et al. 2013). In combination, this type of information provides valuable insight into the animal's ecology. With the increased awareness of the potential for anthropogenic sound to impact marine mammals, the techniques of localizing, tracking, and monitoring acoustic behavior have become important methods for relating particular types of sound exposure to changes in marine mammal behavior. For example, work with bottom-mounted hydrophone arrays has demonstrated that beaked whales cease the production of foraging-related echolocation signals when exposed to certain levels of sonar (Moretti et al. 2014).

Underwater sound transducers are also a mainstay of work with animals under human care. Indeed, almost all of the research on underwater hearing has involved the use of these tools in psychophysical research protocols. Psychophysics is the field of study related to how a physical quantity (e.g., sound frequency) translates to the subjective perception of the physical quantity. Psychophysical procedures involve training an animal to perform a behavior in response to a given stimulus of which one or more of its physical properties are varied. The process is behavioral in that the animal is asked to make a decision, with the question asked and the type of decision desired dictated by the experimental design. For example, until ~2005 when electrophysiological techniques became more common for studying hearing in odontocetes (mentioned later), information about the frequency range of hearing and hearing sensitivity were nearly always determined through psychophysical means. A simple form of a psychophysical hearing test relates the amplitude of a sound, typically measured as the sound pressure level, to an animal's ability to detect the sound. Procedures involve presenting a sound to an animal that is trained to station itself in a precise location within a calibrated sound field, or in the case of some pinnipeds, to wear headphones. If the animal hears the tone, it provides a response, such as producing a sound (e.g., whistle) or performing a behavior (e.g., touching a paddle). If no sound is heard, the animal either provides a different response (e.g., touches a different paddle) or remains quiet. In *staircase* procedures, designed to determine hearing thresholds, correct responses to sound detection result in the sound level, the physical quantity under study, being reduced before it is presented again. If sounds are presented, but are not detected, then sound levels are typically increased. Through this titration process, the minimum level of sound that is detectable, which is the threshold of detection, can be determined. If this process is performed across a broad range of frequencies, at least enough to cover the range of hearing, then a curve defining hearing sensitivity as a function of frequency can be derived. This is the audiogram, which provides the most fundamental piece of sensory information necessary to begin understanding marine mammal auditory perception and ecology, as well as the potential for anthropogenic sound to affect marine mammals.

There are many standardized psychoacoustic procedures that can be performed and a detailed discussion of the psychophysical procedures used in marine mammal research is beyond the scope of this chapter. Nevertheless, it is important to be aware that the use of these procedures is critical to exploring both sensory capability and inferring physiological function. Examples of auditory system information obtained with psychophysical methods include: quantification of hearing thresholds; interaural time and intensity differences (Moore et al. 1995); angular discrimination (Branstetter et al. 2003); minimum detectable frequency and amplitude differences (Herman and Arbeit 1972); auditory filter shape, critical ratios, and critical bands (Finneran et al. 2002; Lemonds et al. 2012); temporal auditory summation (Terhune 1988; Kastelein et al. 2010; Holt et al. 2012); receiving beam patterns (i.e., directionality of hearing; Au and Moore 1984); auditory masking (Branstetter et al. 2013); and others (e.g., see references from the Order Carnivora section).

Electrophysiological studies of marine mammal hearing dates back to the 1960s, when it was performed by a relatively small group of researchers (see Supin et al. 2001). However, with modern advances in computational capabilities and the miniaturization of technological components utilized for electrophysiological research, the ability to perform electrophysiological research with marine mammals has become more common both in the laboratory and in the field. Unlike psychophysical approaches to studying sensory capabilities, electrophysiological approaches to studying hearing monitor the neural responses of the auditory system to acoustic stimuli, either as the voltages generated by single units or as a conglomerate (summed) response. The presence and magnitude of neural responses can correlate with some psychophysical functions; indeed, one of the most common electrophysiological procedures performed in studying hearing sensitivity in marine mammals is the recording of the auditory evoked potential (AEP). Auditory evoked potentials are voltages generated by the brain in response to the detection of an acoustic stimulus, and the threshold of acoustic stimulation required to produce a measureable response has been shown to correlate with hearing sensitivity. The use of AEP methods to study hearing has greatly advanced the number of marine mammal subjects for which hearing tests have been performed (Figure 11.5) (Houser and Finneran 2006), and has permitted a number of species not maintained under human care to be tested when stranded or when rehabilitating (Nachtigall et al. 2005; Cook et al. 2006; Finneran et al. 2009).

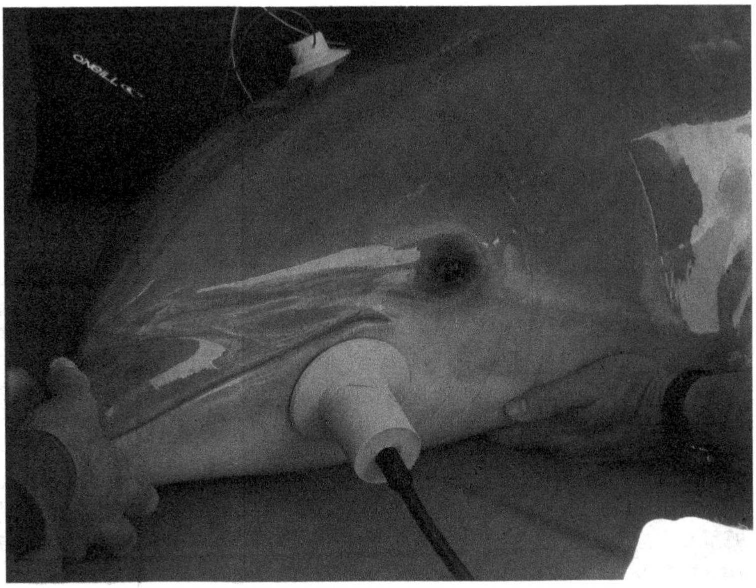

Figure 11.5 **(See color insert.)** A bottlenose dolphin undergoing a hearing test using auditory evoked potentials. The sound projector is attached to the lower jaw over the region of the *pan*, a thin portion of the lower mandible through which sound passes to the ear (also see Figure 11.2). An electrode placed on the dorsal surface of the animal over the brainstem, records the evoked responses produced by the brain in response to the projected sounds.

11.5 Lingering mysteries

One of the greatest mysteries related to marine mammal sound production and hearing is the frequency range of hearing and hearing sensitivity of any mysticete whale. To date, no mysticete whale has ever had its hearing effectively tested, although attempts have been made. For example, evoked potential methods were attempted on a gray whale calf (*Eschrichtius robustus*) in an effort to obtain some information about hearing in this species (Ridgway and Carder 2001). The attempt met with limited success and no frequency-specific information on hearing sensitivity was obtained. Other methods have also been used to try and assess what the large mysticetes can hear; these include anatomical modeling (Figure 11.6), looking for behavioral reactions to sound exposure, and determining the frequency range over which vocalizations are produced. Unfortunately, none of these methods can fully answer the question about the perceptual aspect of hearing. Although behavioral reactions may indicate that a whale heard a sound, whales may not respond to signals they hear due to contextual factors, such as their motivational state (e.g., greater desire to continue feeding during a feeding bout) or a lack of concern over the sound source. In other words, because an animal does not respond to a sound exposure cannot be interpreted as it not hearing a sound. It is also not possible to extrapolate the frequency range of hearing for an animal from the frequencies at which it vocalizes. Most mammals, including humans, hear across a much greater range of frequencies than that within which they vocalize.

Mysticetes provide an immense challenge to empirically determining hearing capabilities. Mysticetes are too large to maintain under human care for the time necessary to perform behavioral audiometry. For this reason, much of what we believe about mysticete

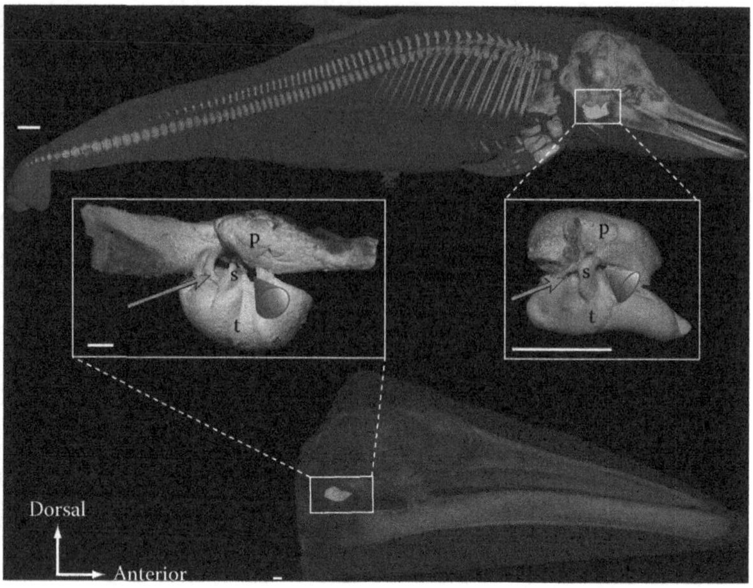

Figure 11.6 **(See color insert.)** Images of a pantropical spotted dolphin (top) and a minke whale (bottom) based on computerized tomography (CT) scans. (From Yamato, M. and Pyenson, N.D., *PLoS One*, 10(3), e0118582, 2015.) The tympanoperiotic complexes (identified in yellow, and as pictures from adult specimens) are also shown. The blue arrows indicate the tympanic apertures. The pink cones represent the *acoustic funnels*, through which sound is conducted through acoustic fats to the tympanoperiotic complex. Note the anterior orientation of the acoustic funnel in the odontocete relative to the mysticete, which is indicative of the forward-facing, highly directional echolocation system in odontocetes. These structures suggest a more lateral sound reception system in mysticetes, and have value in making predictions in the absence of physiological or psychological data.

hearing ranges has been derived from anatomical models (Houser et al. 2001; Parks et al. 2007; Tubelli et al. 2012). Unfortunately, without validation of the models, there remains much uncertainty as to their predictive capability. The use of AEP methods, previously mentioned, may provide one of the better opportunities for measuring hearing in mysticetes. However, this, too, is fraught with challenges. Evoked potential methods work well with odontocetes because they have a large auditory nerve, smooth skin, thin skull, thin blubber layers (for many species), a large brain:body mass ratio, and highly refined auditory processing. On the other hand, mysticetes have large and thick skulls, large blubber layers that attenuate electrical signals recorded at the body surface (i.e., blubber is a poor conductor of electricity), a largely unfavorable brain:body mass ratio, and a less derived auditory system. These factors either contribute to small amplitude AEPs or make the recording of the AEPs difficult, at least by non-invasive methods. Nevertheless, the first empirical measurements of hearing in a mysticete whale may likely have to be made through electrophysiological methods but may have to be performed on the smallest species (e.g., minke whale) or calves of relatively small species (e.g., gray whale calves) for early successes.

A related controversy exists with our knowledge of hearing in odontocetes. Although research primarily conducted on bottlenose dolphins has provided considerable information about the peripheral auditory pathway and how sound enters the dolphin head, it remains unclear as to how acoustic energy is conducted to the cochlea. Is the stapes vibrated by the malleus and incus such that it manipulates the oval window, as occurs in terrestrial mammals? Are the tympanic membrane and external acoustic meatus and canal vestigial, having lost their function in the transition to a fully aquatic lifestyle? Is hearing achieved by passage of the sound directly to the cochlear capsule, which produces differential motion between the cochlear capsule and the footplate of the stapes such that displacement of the cochlear fluids occurs?

Even with relatively extensive study, there remains much to be discerned about hearing in odontocetes, and marine mammals as a whole. Certainly, the return to the marine environment, either fully or amphibiously, has resulted in modifications to both sound production and sound reception systems. The degree of anatomical modification supporting these changes is broad, ranging from reductions of terrestrial traits (e.g., reduced but present pinna of some pinnipeds) to major rearrangements of the terrestrial bauplän (e.g., separation of the auditory bullae from the skull in cetaceans). How these changes relate to hearing capabilities and physiological function will remain research questions for years to come, and they will have an increasing importance related to conservation issues as it relates to human-caused sound in the world's oceans.

References

Alderson, A.M., E. Diamantopoulos, and C.B.B. Downman. 1960. Auditory cortex of the seal (*Phoca vitulina*). *Journal of Anatomy* 94:506–511.

Amundin, M. and S.H. Andersen. 1983. Bony nares air pressure and nasal plug muscle activity during click production in the harbour porpoise, *Phocoena phocoena*, and the bottlenosed dolphin, *Tursiops truncatus*. *Journal of Experimental Biology* 105 (1):275–282.

Au, W.W.L. and M.C. Hastings. 2008. *Principles of Marine Bioacoustics*. New York: Springer.

Au, W.W.L., R.A. Kastelein, K.J. Benoit-Bird, T.W. Cranford, and M.F. McKenna. 2006. Acoustic radiation from the head of echolocating harbor porpoises (*Phocoena phocoena*). *Journal of Experimental Biology* 209 (Part 14):2726–2733.

Au, W.W.L. and P.W.B. Moore. 1984. Receiving beam patterns and directivity indices of the Atlantic bottlenosed dolphin (*Tursiops truncatus*). *Journal of the Acoustical Society of America* 75 (1):255–262.

Babushina, E.S. and M.A. Poliakov. 2004. The underwater and airborne horizontal localization of sound by the northern fur seal. *Biofizika* 49 (4):723–726.

Boddy, A.M., M.R. McGowen, C.C. Sherwood, L.I. Grossman, M. Goodman, and D.E. Wildman. 2012. Comparative analysis of encephalization in mammals reveals relaxed constraints on anthropoid primate and cetacean brain scaling. *Journal of Evolutionary Biology* 25:981–994.

Bodson, A., L. Miersch, and G. Dehnhardt. 2007. Underwater localization of pure tones by harbor seals (*Phoca vitulina*). *Journal of the Acoustical Society of America* 122 (4):2263–2269.

Bodson, A., L. Miersch, B. Mauck, and G. Dehnhardt. 2006. Underwater auditory localization by a swimming harbor seal (*Phoca vitulina*). *Journal of the Acoustical Society of America* 120 (3):1550–1557.

Branstetter, B.K., S.J. Mevissen, L.M. Herman, A.A. Pack, and S.P. Roberts. 2003. Horizontal angular discrimination by an echolocating bottlenose dolphin *Tursiops truncatus*. *Bioacoustics* 14:15–34.

Branstetter, B.K., J.S. Trickey, K. Bahktiari, A. Black, H. Aihara, and J.J. Finneran. 2013. Auditory masking patterns in bottlenose dolphins (*Tursiops truncatus*) with natural, anthropogenic, and controlled noise. *Journal of the Acoustical Society of America* 133 (3):1811–1818.

Brill, R.L. and P.J. Harder. 1991. The effects of attenuating returning echolocation signals at the lower jaw of a dolphin (*Tursiops truncatus*). *Journal of the Acoustical Society of America* 89 (6):2851–2857.

Brumm, H. and S.A. Zollinger. 2013. Avian vocal production in noise. In *Animal Communication and Noise*, ed. H. Brumm. Berlin, Germany: Springer-Verlag.

Bullock, T.H., D.P. Domning, and R.C. Best. 1980. Evoked brain potentials demonstrate hearing in a manatee (*Trichechus inunguis*). *Journal of Mammalogy* 61 (1):130–133.

Bullock, T.H., S.H. Ridgway, and S. Nobuo. 1971. Acoustically evoked potentials in midbrain auditory structures in sea lions (Pinnipedia). *Zeitshchrift für Vergleichende Physiologie* 74:372–387.

Cahill, J.A., R.E. Green, T.L. Fulton et al. 2013. Genomic evidence for island population conversion resolves conflicting theories of polar bear evolution. *PLoS Genetics* 9:e1003345.

Chapla, M.E., D.P. Nowacek, S.A. Rommel, and V.M. Sadler. 2007. CT scans and 3D reconstructions of Florida manatee (*Trichechus manatus latirostris*) heads and ear bones. *Hearing Research* 228:123–135.

Chittka, L. and A. Brockmann. 2005. Perception space—The final frontier. *PLoS Biology* 3 (4):e137.

Clark, C.W. 1990. Acoustic behavior of mysticete whales. In *Sensory Abilities of Cetaceans*, eds. J. Thomas, and R. Kastelein. New York: Plenum Press.

Clark, C.W., W.T. Ellison, B.L. Southall et al. 2009. Acoustic masking in marine ecosystems: Intuitions, analysis, and implication. *Marine Ecology Progress Series* 395:201–222.

Cook, M.L.H., R.A. Varela, J.D. Goldstein et al. 2006. Beaked whale auditory evoked potential hearing measurements. *Journal of Comparative Physiology A* 192 (5):489–495.

Cranford, T.W. 1992. Functional morphology of the odontocete forehead: Implications for sound generation. Dissertation, Department of Biology, University of California Santa Cruz, Santa Cruz, CA.

Cranford, T.W. 2000. In search of impulse sound sources in odontocetes. In *Hearing by Whales and Dolphins*, eds. W.W.L. Au, A.N. Popper, and R.R. Fay. New York: Springer-Verlag.

Cunningham, K.A., B.L. Southall, and C. Reichmuth. 2014. Auditory sensitivity of seals and sea lions in complex listening scenarios. *Journal of the Acoustical Society of America* 136 3410–3421.

Eriksen, N. and B. Pakkenberg. 2007. Total neocortical cell number in the mysticete brain. *Anatomical Record: Advances in Integrative Anatomy and Evolutionary Biology* 290 (1):83–95.

Finneran, J.J., D.S. Houser, B. Mase-Guthrie, R.Y. Ewing, and R.G. Lingenfelser. 2009. Auditory evoked potentials in a stranded Gervais' beaked whale (*Mesoplodon europaeus*). *Journal of the Acoustical Society of America* 126 (1):484–490.

Finneran, J.J., C.E. Schlundt, D.A. Carder, and S.H. Ridgway. 2002. Auditory filter shapes for the bottlenose dolphin (*Tursiops truncatus*) and the white whale (*Delphinapterus leucas*) derived with notched noise. *Journal of the Acoustical Society of America* 112 (1):322–328.

Fleischer, G. 1978. Evolutionary principles of the mammalian middle ear. *Advances in Anatomy, Embryology and Cell Biology* 55:1–70.

Gaspard J.C. III, G.B. Bauer, R.L. Reep et al. 2012. Audiogram and auditory critical ratios of two Florida manatees (*Trichechus manatus latirostris*). *Journal of Experimental Biology* 215:1442–1447.

Gentry L. 1967. Underwater auditory localization in the California sea lion (*Zalophus californianus*). *Journal of Auditory Research* 7:187–193.

Gerstein, E.R., L. Gerstein, S.E. Forsythe, and J.E. Blue. 1999. The underwater audiogram of the West Indian manatee (*Trichechus manatus*). *Journal of the Acoustical Society of America* 105 (6):3575–3583.

Ghoul, A. and C. Reichmuth. 2014. Hearing in the sea otter (*Enhydra lutris*): Auditory profiles for an amphibious marine carnivore. *Journal of Comparative Physiology A* 200 (11):967–981.

Hatch, L.T., C.W. Clark, S.M. Van Parijs, A.S. Frankel, and D.W. Ponirakis. 2012. Quantifying loss of acoustic communication space for right whales in and around a U.S. National Marine Sanctuary. *Conservation Biology* 26 (6):983–994.

Hemilä, S., S. Nummela, A. Berta, and T. Reuter. 2006. High-frequency hearing in phocid and otariid pinnipeds: An interpretation based on inertial and cochlear constraints (L). *Journal of the Acoustical Society of America* 120 (6):3463–3466.

Herman, M. and W.R. Arbeit. 1972. Frequency difference limens in the bottlenosed dolphin: 1–70 KC/S. *Journal of Auditory Research* 2:109–120.

Holt, M.M., A. Ghoul, and C. Reichmuth. 2012. Temporal summation of airborne tones in a California sea lion (*Zalophus californianus*). *Journal of the Acoustical Society of America* 132 (5):3569–3575.

Holt, M.M., D.P. Noren, V. Veirs, C.K. Emmons, and S. Veirs. 2008. Speaking up: Killer whales (*Orcinus orca*) increase their call amplitude in response to vessel noise. *Journal of the Acoustical Society of America* 125 (1):EL27–EL32.

Houser, D.S. and J.J. Finneran. 2006. Variation in the hearing sensitivity of a dolphin population obtained through the use of evoked potential audiometry. *Journal of the Acoustical Society of America* 120 (6):4090–4099.

Houser, D.S., J.J. Finneran, D.A. Carder et al. 2004. Structural and functional imaging of bottlenose dolphin (*Tursiops truncatus*) cranial anatomy. *Journal of Experimental Biology* 207:3657–3665.

Houser, D.S., D.A. Helweg, and P.W.B. Moore. 1999. Classification of dolphin echolocation clicks by energy and frequency distributions. *Journal of the Acoustical Society of America* 106 (3):1579–1585.
Houser, D.S., D.A. Helweg, and P.W.B. Moore. 2001. A bandpass filter-bank model of auditory sensitivity in the humpback whale. *Aquatic Mammals* 27 (2):82–91.
Jacobs, D.W. and J.D. Hall. 1972. Auditory thresholds of a fresh water dolphin, *Inia geoffrensis* Blainville. *Journal of the Acoustical Society of America* 51 (2B):530–533.
Johnson, C.S. 1967. Sound detection thresholds in marine mammals. In *Marine Bioacoustics*, ed. W.N. Tavolga. Oxford, UK: Pergamon Press.
Kastak, D. and R.J. Schusterman. 1998. Low-frequency amphibious hearing in pinnipeds: Methods, measurements, noise, and ecology. *Journal of the Acoustical Society of America* 103 (4):2216–2228.
Kastak, D. and R.J. Schusterman. 1999. In-air and underwater hearing sensitivity of a northern elephant seal (*Mirounga angustirostris*). *Canadian Journal of Zoology* 77 (11):1751–1758.
Kastelein, R.A., P. Bunskoek, M. Hagedoorn, W.W.L. Au, and D. de Haan. 2002a. Audiogram of a harbor porpoise (*Phocoena phocoena*) measured with narrow-band frequency-modulated signals. *Journal of the Acoustical Society of America* 112 (1):334–344.
Kastelein, R.A., L. Hoek, C.A. de Jong, and P.J. Wensveen. 2010. The effect of signal duration on the underwater detection thresholds of a harbor porpoise (*Phocoena phocoena*) for single frequency-modulated tonal signals between 0.25 and 160 kHz. *Journal of the Acoustical Society of America* 128 (5):3211.
Kastelein, R.A., P. Mosterd, B. van Santen, M. Hagedoorn, and D. de Haan. 2002b. Underwater audiogram of a Pacific walrus (*Odobenus rosmarus divergens*) measured with narrow-band frequency-modulated signals. *Journal of the Acoustical Society of America* 112 (5):2173–2182.
Kastelein, R.A., R. van Schie, W.C. Verboom, and D. de Haan. 2005. Underwater hearing sensitivity of a male and a female steller sea lion (*Eumetopias jubatus*). *Journal of the Acoustical Society of America* 118 (3):1820–1829.
Kellogg, R. 1938. Adaptation of structure to function in whales. *Cooperation in Research* 501:649–682.
Ketten, D.R. 1992. The marine mammal ear: Specializations for aquatic audition and echolocation. In *The Evolutionary Biology of Hearing*, eds. D. Webster, R. Fay, and A. Popper. New York: Springer-Verlag.
Ketten, D.R. (1994). Functional analyses of whale ears: Adaptations for underwater hearing. *IEEE Proceedings in Underwater Acoustics* 1:264–270.
Ketten, D.R. (1997). Structure and function in whale ears. *Bioacoustics* 8:103–135.
Ketten, D.R. 2000. Cetacean ears. In *Hearing by Whales and Dolphins*, eds W. Au, A.N. Popper, and R.R. Fay. New York: Springer-Verlag.
Ketten, D.R., D.K. Odell, and D.P. Domning. 1992. Structure, function, and adaptation of the Manatee ear. In *Marine Mammal Sensory Systems*, eds. J.A. Thomas, R.A. Kastelein, and A.Y. Supin. New York: Plenum Press.
Ketten, D.R. and D. Wartzok. 1990. Three-dimensional reconstructions of the dolphin ear. In *Sensory Abilities of Cetaceans: Laboratory and Field Evidence*, eds. J.A. Thomas and R.A. Kastelein. New York: Plenum Press.
King, S.L., L.S. Sayigh, R.S. Wells, W. Fellner, and V.M. Janik. 2013. Vocal copying of individually distinctive signature whistles in bottlenose dolphins. *Proceedings of the Royal Society B* 280:20130053.
Klishin, V.O., R. Pezo Diaz, V.V. Popov, and A.Y. Supin. 1990. Some characteristics of hearing of the Brazilian manatee, *Trichechus inunguis*. *Aquatic Mammals* 16 (3):139–144.
Koblitz, J.C., M. Wahlberg, P. Stilz, P.T. Madsen, K. Beedholm, and H.U. Schnitzler. 2012. Asymmetry and dynamics of a narrow sonar beam in an echolocating harbor porpoise. *Journal of the Acoustical Society of America* 131 (3):2315–2324.
Lemonds, D.W., W.W.L. Au, S.A. Valchos, and P.E. Nachtigall. 2012. High-frequency auditory filter shape for the Atlantic bottlenose dolphin. *Journal of the Acoustical Society of America* 132 (2):1222–1228.
Madsen, P.T., D. Wisniewska, and K. Beedholm. 2010. Single source sound production and dynamic beam formation in echolocating harbour porpoises (*Phocoena phocoena*). *Journal of Experimental Biology* 213:3105–3110.
Manger, P.R. 2006. An examination of cetacean brain structure with a novel hypothesis correlating thermogenesis to the evolution of a big brain. *Biological Reviews* 81:293–338.

Mann, D.A., D.E. Colbert, J.C. Gaspard et al. 2005. Temporal resolution of the Florida manatee (*Trichechus manatus latirostris*) auditory system. *Journal of Comparative Physiology A* 191 (10):903–908.

Marques, T.A., L. Thomas, S.W. Martin et al. 2013. Estimating animal population density using passive acoustics. *Biological Review* 88:287–309.

McCormick, J.G., E.G. Wever, J. Palin, and S.H. Ridgway. 1970. Sound conduction in the dolphin ear. *Journal of the Acoustical Society of America* 48 (6):1418–1428.

McCormick, J.G., E.G. Wever, S.H. Ridgway, and J. Palin. 1980. Sound reception in the porpoise as it relates to echolocation. In *Animal Sonar Systems*, eds. R.G. Busnel and J.F. Fish. New York: Plenum Press.

McCowan, B. and D. Reiss. 2001. The fallacy of 'signature whistles' in bottlenose dolphins: A comparative perspective of 'signature information' in animal vocalizations. *Animal Behavior* 62:1151–1162.

Møhl, B. 1967a. Frequency discrimination in the common seal and a discussion of the concept of upper hearing limit. In *Underwater Acoustics*, ed. V. Albers. New York: Plenum Press.

Møhl, B. 1967b. Seal ears. *Science* 157 (3784):99.

Møhl, B. 1968. Auditory sensitivity of the common seal in air and water. *Journal of Auditory Research* 8:27–38.

Møhl, B. and K. Ronald. 1975. The peripheral auditory system of the Harp seal, *Pagophilus groenlandicus* (Erxleben, 1777). In *Biology of the Seal*, eds. K. Ronald and A.W. Mansfield. Charlottenlund Slot, Denmark: Conseil International Pour L'exploration.

Moore, P.W., L.A. Dankiewicz, and D.S. Houser. 2008. Beamwidth control and angular target detection in an echolocating bottlenose dolphin (*Tursiops truncatus*). *Journal of the Acoustical Society of America* 124 (5):3324–3332.

Moore, P.W.B. and W.W.L. Au. 1975. Underwater localization of pulsed pure tones by the California sea lion (*Zalophus californianus*). *Journal of the Acoustical Society of America* 58 (3):721–727.

Moore, P.W.B., D.A. Pawloski, and L. Dankiewicz. 1995. Interaural time and intensity difference thresholds in the bottlenose dolphin (*Tursiops truncatus*). In *Sensory Systems of Aquatic Mammals*, eds. R.A. Kastelein, J.A. Thomas, and P.E. Nachtigall. Woerden, the Netherlands: De Spil Publishers.

Moore, P.W.B. and R.J. Schusterman. 1976. Discrimination of pure-tone intensities by the California sea lion. *Journal of the Acoustical Society of America* 60 (6):1405–1407.

Moore, P.W.B. and R.J. Schusterman. 1987. Audiometric assessment of northern fur seals, *Callorhinus ursinus*. *Marine Mammal Science* 3 (1):31–53.

Moretti, D., L. Thomas, T. Marques et al. 2014. A risk function for behavioral disruption of Blainville's beaked whales (*Mesoplodon densirostris*) from mid-frequency active sonar. *PLoS One* 9 (1):e85064.

Mulsow, J., J.J. Finneran, and D.S. Houser. 2014. Interaural differences in the bottlenose dolphin (*Tursiops truncatus*) auditory nerve response to jawphone stimuli. *Journal of the Acoustic Society of America* 136 (3):1402.

Mulsow, J. and C. Reichmuth. 2007. Electrophysiological assessment of temporal resolution in pinnipeds. *Aquatic Mammals* 33 (1):122–131.

Mulsow, J.L. and C. Reichmuth. 2010. Psychophysical and electrophysiological aerial audiograms of a Steller sea lion (*Eumetopias jubatus*). *Journal of the Acoustical Society of America* 127 (4):2692–2701.

Nachtigall E., M.M.L. Yuen, T.A. Mooney, and K.A. Taylor. 2005. Hearing measurements from a stranded infant Risso's dolphin, *Grampus griseus*. *Journal of Experimental Biology* 208:4181–4188.

Nachtigall, P.E., A.Y. Supin, M. Amundin et al. 2007. Polar bear *Ursus maritimus* hearing measured with auditory evoked potentials. *Journal of Experimental Biology* 210 (7):1116–1122.

Norris, K.S. and G.W. Harvey. 1974. Sound transmission in the porpoise head. *Journal of the Acoustical Society of America* 56 (2):659–664.

Nummela, S. 2007. Hearing in aquatic mammals. In *Sensory Evolution on the Threshold: Adaptations in Secondarily Aquatic Vertebrates*, eds. J.G.M. Thewissen and S. Nummela. Berkeley, CA: University of California Press.

Odend'hal, S. and T.C. Poulter. 1966. Pressure regulation in the middle ear cavity of sea lions: A possible mechanism. *Science* 153:768–769.

Oelschlager, H.A. 1990. Evolutionary morphology and acoustics in the dolphin skull. In *Sensory Abilities of Cetaceans: Laboratory and Field Evidence*, eds. J.A. Thomas and R.A. Kastelein. New York: Plenum Press.

Oelschlager, H.H.A. 2008. The dolphin brain—A challenge for synthetic neurobiology. *Brain Research Bulletin* 75:450–459.

Owen, M.A. and A.E. Bowles. 2011. In-air auditory psychophysics and the management of a threatened carnivore, the polar bear (*Ursus maritimus*). *International Journal of Comparative Psychology* 24:244–254.

Pacini, A.F., P.E. Nachtigall, C.T. Quintos et al. 2011. Audiogram of a stranded Blainville's beaked whale (*Mesoplodon densirostris*) measured during auditory evoked potentials. *Journal of Experimental Biology* 214:2409–2415.

Parks, S.E., M. Johnson, D. Nowacek, and P.L. Tyack. 2011. Individual right whales call louder in increased environmental noise. *Biology Letters* 7:33–35.

Parks, S.E., D.R. Ketten, J.T. O'Malley, and J. Arruda. 2007. Anatomical predictions of hearing in the North Atlantic right whale. *The Anatomical Record* 290:734–744.

Popov, V.V., T.F. Ladygina, and A.Y. Supin. 1986. Evoked potentials of the auditory cortex of the porpoise, *Phocoena phocoena*. *Journal of Comparative Physiology A* 158 (5):705–711.

Popov, V.V. and A.Y. Supin. 1990. Electrophysiological studies of hearing in some cetaceans and a manatee. *Sensory Abilities of Cetaceans*, eds. J.A. Thomas and R.A. Kastelein, New York: Plenum Press, 405–415.

Popov, V.V., A.Y. Supin, V.O. Klishin, and T.N. Bulgakova. 2006. Monaural and binaural hearing directivity in the bottlenose dolphin: Evoked-potential study. *Journal of the Acoustical Society of America* 119 (1):636–644.

Popov, V.V., A.Y. Supin, V.O. Klishin, and M.B. Tarakanov. 2008. Evidence for double acoustic windows in the dolphin, *Tursiops truncatus*. *Journal of the Acoustical Society of America* 123(1): 552–560.

Poulter, T.C. 1963. Sonar signals of the sea lion (*Zalphus californianus*). *Science* 139:753–754.

Poulter, T.C. 1966. The use of active sonar by the California sea lion, *Zalophus californianus*. *Journal of Auditory Research* 6:165–173.

Ramprashad, F., S. Corey, and K. Ronald. 1972. Anatomy of the seal ear *Pagophilus groenlandicus* (Erxlebel 1777). In *Functional Anatomy of Marine Mammals*, ed. R.J. Harrison. London, UK: Academic Press.

Reichmuth, C., M.M. Holt, J. Mulsow, J.M. Sills, and B.L. Southall. 2013. Comparative assessment of amphibious hearing in pinnipeds. *Journal of Comparative and Physiology A* 199 (6):491–507.

Reichmuth, C. and B.L. Southall. 2012. Underwater hearing in California sea lions (*Zalophus californianus*): Expansion and interpretation of existing data. *Marine Mammal Science* 28 (2):358–363.

Repenning, C.A. 1972. Underwater hearing in seals: Functional morphology. In *Functional Anatomy of Marine Mammals*, ed. R.J. Harrison. London, UK: Academic Press.

Ridgway, S.H. 2000. The auditory central nervous system of dolphins. In *Hearing by Whales and Dolphins*, eds. W.W.L. Au, R.R. Fay, and A.N. Popper. New York: Springer.

Ridgway, S.H. and D.A. Carder. 2001. Assessing hearing and sound production in cetaceans not available for behavioral audiograms: Experiences with sperm, pygmy sperm, and gray whales. *Aquatic Mammals* 27 (3):267–276.

Ridgway, S.H., D.A. Carder, R.F. Green, A.S. Gaunt, S.L.L. Gaunt, and W.E. Evans. 1980. Electromyographic and pressure events in the nasolaryngeal system of dolphins during sound production. In *Animal Sonar Systems*, eds. R.G. Busnel and J.F. Fish. New York: Plenum Press.

Robineau, D. 1969. Les osselets de l'ouie de la Rhytine. *Mammalia* 29:412–425.

Schusterman, R.J., D. Kastak, D.H. Levenson, C.J. Reichmuth, and B.L. Southall. 2000. Why pinnipeds don't echolocate. *Journal of the Acoustical Society of America* 107 (4):2256–2264.

Schusterman, R.J. and P.W. Moore. 1978. The upper limit of underwater auditory frequency discrimination in California sea lion. *Journal of the Acoustical Society of America* 63 (5):1591–1595.

Sills, J.M., B.L. Southall, and C. Reichmuth. 2014. Amphibious hearing in spotted seals (*Phoca largha*): Underwater audiograms, aerial audiograms and critical ratio measurements. *Journal of Experimental Biology* 217 (5):726–734.

Solntseva, G.N. 2007. *Morphology of the Auditory and Vestibular Organs in Mammals, with Emphasis on Marine Species*. Sofia, Bulgaria: Pensoft Publishers & Brill Academic Publishers.

Southall, B.L., R.J. Schusterman, and D. Kastak. 2000. Masking in three pinnipeds: Underwater, low-frequency critical ratios. *Journal of the Acoustical Society of America* 108 (3):1322–1326.

Southall, B.L., R.J. Schusterman, and D. Kastak. 2003. Auditory masking in three pinnipeds: Aerial critical ratios and direct critical bandwidth measurements. *Journal of the Acoustical Society of America* 114 (3):1660–1666.

Stanistreet, J.E., D. Risch, and S.M. Van Parijs. 2013. Passive acoustic tracking of singing humpback whales (*Megaptera novaeangliae*) on a northwest Atlantic feeding ground. *PLoS One* 8 (4):e61263.

Stirling, I. 1988. *Polar Bears*. Ann Arbor, MI: University of Michigan Press.

Supin, A.Y., V.V. Popov, and A.M. Mass. 2001. *The Sensory Physiology of Aquatic Mammals*. Boston, MA: Kluwer Academic Publishers.

Szymanski, M.D., D.E. Bain, K. Kiehl, S. Pennington, S. Wong, and K.R. Henry. 1999. Killer whale (*Orcinus orca*) hearing: Auditory brainstem response and behavioral audiograms. *Journal of the Acoustical Society of America* 106 (2):1134–1141.

Terhune, J.M. 1988. Detection thresholds of a harbour seal to repeated underwater high-frequency, short-duration sinusoidal pulses. *Canadian Journal of Zoology* 66:1578–1582.

Terhune, J.M. and K. Ronald. 1976. The upper frequency limit of ringed seal hearing. *Canadian Journal of Zoology* 54:1226–1229.

Thewissen, J.G.M. and S. Nummela. 2007. On becoming aquatic. In *Sensory Evolution on the Threshold: Adaptations in Secondarily Aquatic Vertebrates*, eds. J.G.M. Thewissen and S. Nummela. Berkeley, CA: University of California Press.

Thomas, J., N. Chun, W. Au, and K. Pugh. 1988. Underwater audiogram of a false killer whale (*Pseudorca crassidens*). *Journal of the Acoustical Society of America* 84:936–940.

Tubelli, A.A., A. Zosuls, D.R. Ketten, M. Yamato, and D.C. Mountain. 2012. A prediction of the minke whale (*Balaenoptera acutorostrata*) middle-ear transfer function. *Journal of the Acoustical Society of America* 132 (5):3263–3272.

White, M.J. Jr., J. Norris, D.K. Ljungblad, K. Baron, and G. di Sciara. 1978. *Auditory Thresholds of Two Beluga Whales (Delphinapterus leucas)*. San Diego, CA: Hubbs Sea World Research Institute.

Worthy, G.A. and J.P. Hickie. 1986. Relative brain size in marine mammals. *The American Naturalist* 128:445–459.

Yamato, M., D.R. Ketten, J. Arruda, S. Cramer, and K. Moore. 2012. The auditory anatomy of the minke whale (*Balaenoptera acutorostrata*): A potential fatty sound reception pathway in a baleen whale. *The Anatomical Record* 295 (6):991–998.

Yamato, M. and N.D. Pyenson. 2015. Early development and orientation of the acoustic funnel provides insight into the evolution of sound reception pathways in cetaceans. *PLoS One* 10 (3):e0118582.

Yost, W.A. 2007. *Fundamentals of Hearing: An Introduction*. San Diego, CA: Academic Press.

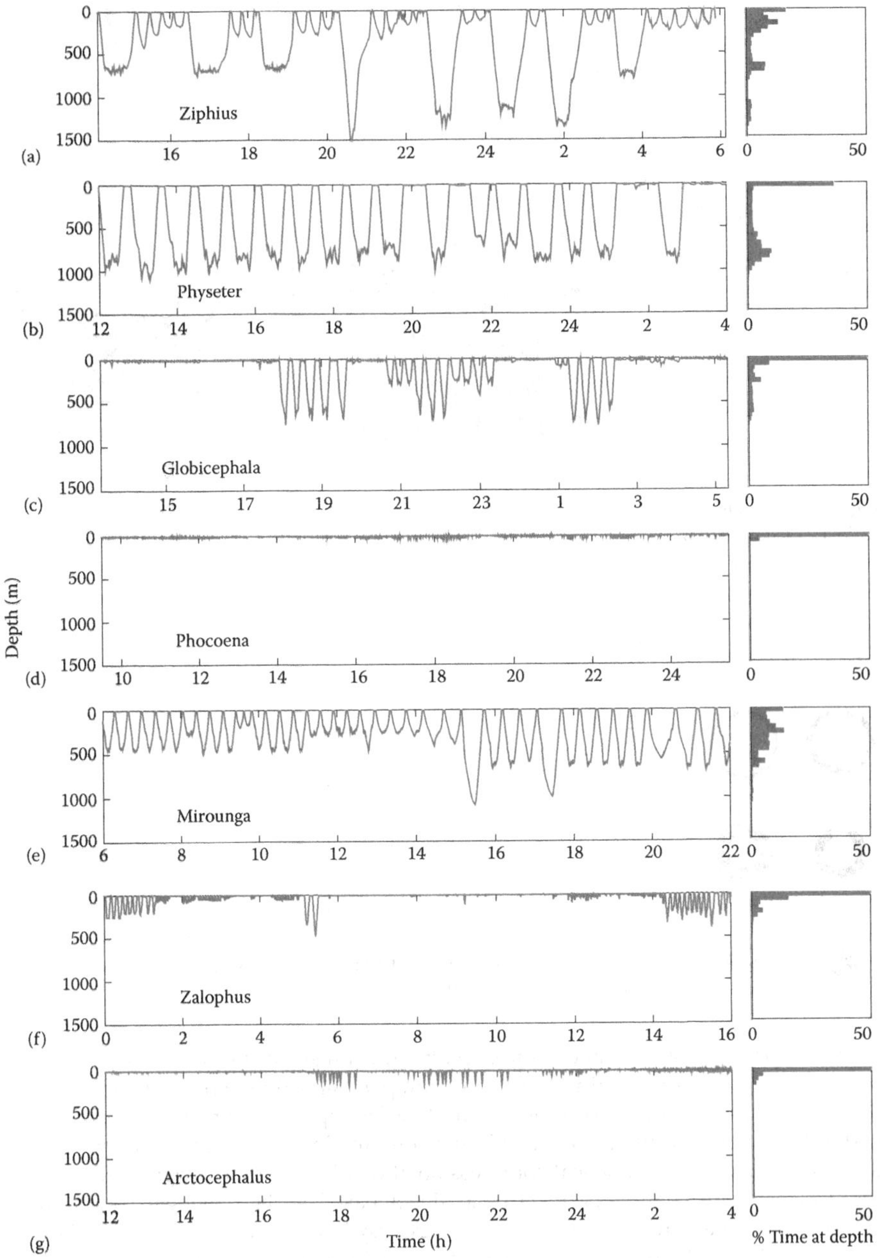

Figure 4.1 Dive traces (left) and frequency histograms (right) showing time at depth (in 50 m depth intervals) for various marine mammal species. Dive traces are plotted to identical scales: 1500 m depth over a 16 h time period, to illustrate the differences in the use of depth and patterns of diving between species. (a) Cuvier's beaked whale (Ziphiidae); Mediterranean, September 2003; (b) sperm whale (Physeteridae), Azores, August 2010; (c) short-finned pilot whale (Delphinidae), Canary Islands, October 2004; (d) harbor porpoise (Phocoenidae), Jutland Peninsula, Denmark, October 2012; (e) northern elephant seal (Phocidae), eastern Pacific, March 2014; (f) California sea lion (Otariidae), San Nicholas Island, California, November 2006; and (g) Antarctic fur seal (Otariidae), South Georgia, December 2001. (Data courtesy of (a) M. Johnson, (b) C. Oliveira, (c) N. Aguilar and M. Johnson, (d) D. Wisniewska (e) D. Costa, (f) D. Costa, (g) S. Hooker.)

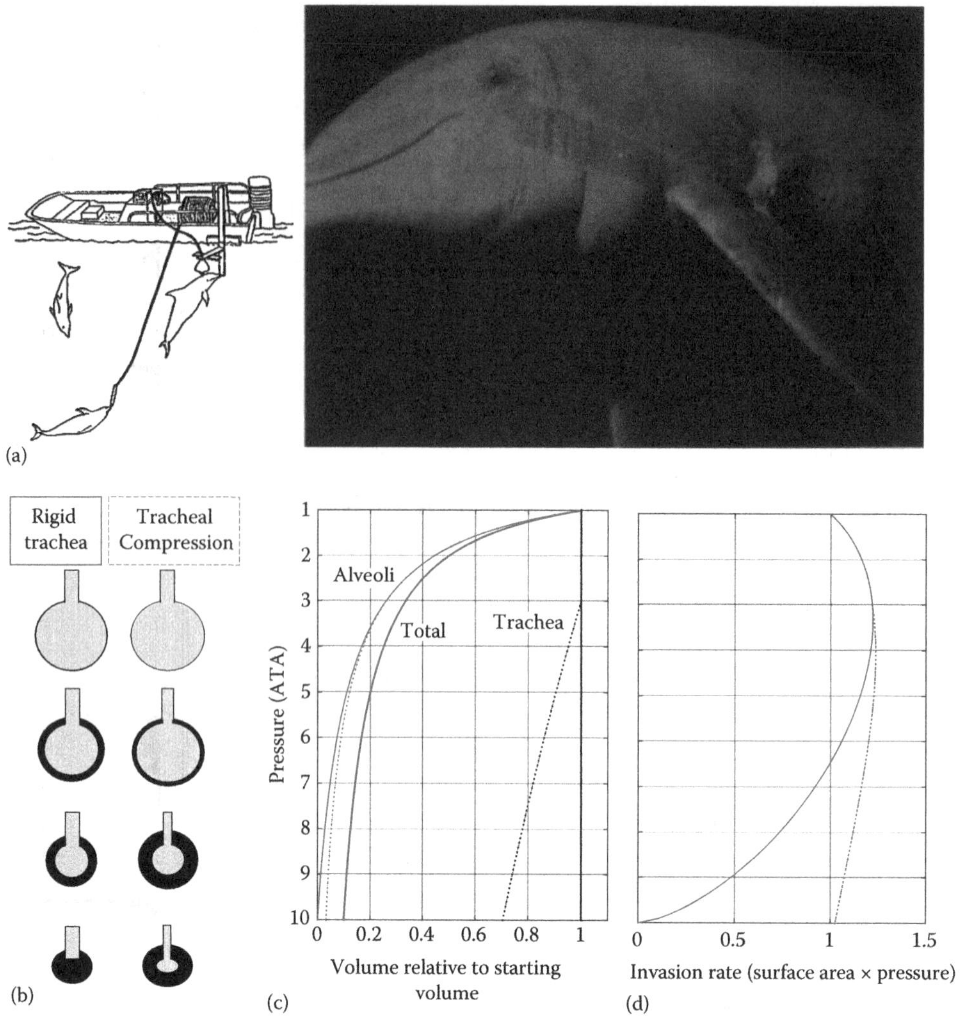

Figure 4.2 Compression of lung and thorax results in changes in invasion/diffusion rate. (a) Chest compression of a bottlenose dolphin (Tuffy), photographed at 300 m depth, with experimental setup shown on the left. Thoracic collapse is visible behind the left flipper. (Image copyright U.S. Navy; details published in Ridgway et al. 1969). (b) Graded lung compression showing illustrations of two models: (i) rigid trachea (solid) and (ii) tracheal compression at depths less than alveolar collapse (dotted). (c) Modeled relative volume changes of trachea and alveoli based on these models. Total volume (trachea + alveoli = 1 l) reduces according to Boyle's law (thick gray line). Volume changes are shown based on a rigid trachea (solid black line, 0.1 l), and assuming that the tracheal volume decreases exponentially after 3 ATA (dotted black line), with resulting changes to the alveolar volume from 0.9 l at the surface, decreasing more quickly with a rigid trachea (gray solid line) than with a compressing trachea (dotted gray line). (d) The resulting invasion rate (alveolar surface area × pressure) (sensu Scholander 1940). The invasion rate is plotted from 1 at the surface breath-hold, to 0 at the depth of alveolar collapse. Fick's law states that the diffusion rate is proportional to surface area × pressure/membrane thickness, thus the invasion rate divided by the membrane thickness would give the diffusion rate (i.e., this would further reduce the invasion rate). It can be seen that near the surface, the invasion rate increases with a pressure increase, but that at deeper depths this reduces to the depth of alveolar collapse. The effect of tracheal compression at depths shallower than alveolar collapse effects deeper alveolar collapse and increases the range of depths which have a high invasion (and diffusion) rate.

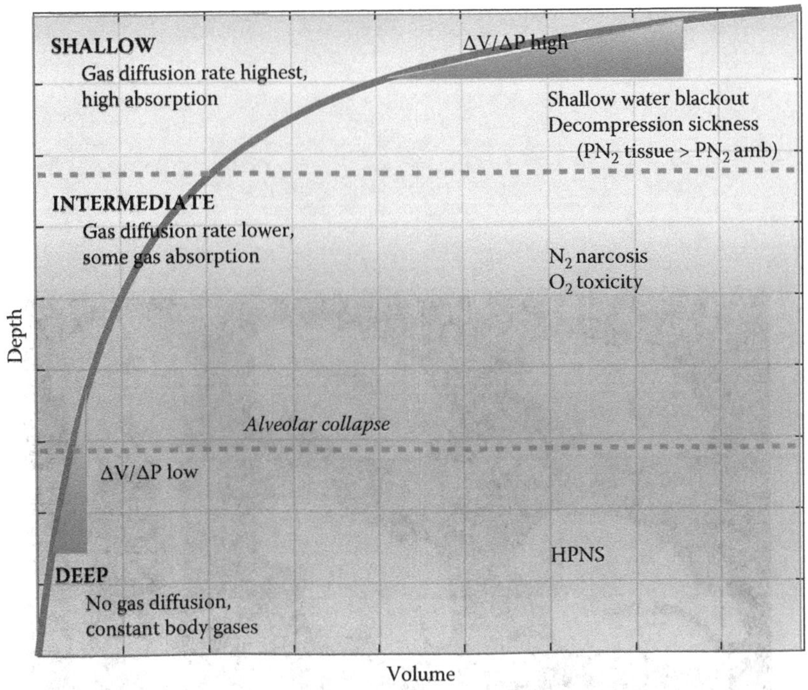

Figure 4.3 Risks of different diving-related problems are related to pressure and gas diffusion. The water column can be divided into a shallow, intermediate, and deep region. In the shallow region, the rate of change of volume (ΔV/ΔP) is high, and there is high gas diffusion (see Figure 4.2). In this region, gases are exchanged and animals may be at risk of gas bubble disease when blood and tissue PN_2 exceeds ambient pressure. In addition, shallow-water blackout may also occur in this region due to the rapid changes in volume. In the intermediate region, a reduction in the alveolar surface area and thickening of the alveolar membrane reduces gas exchange. The N_2 and O_2 taken up may cause nitrogen narcosis and increase the risk for O_2 toxicity. Once the alveoli collapse in the deep region, no further gas is exchanged and as pressure increases, animals may be more at risk of HPNS.

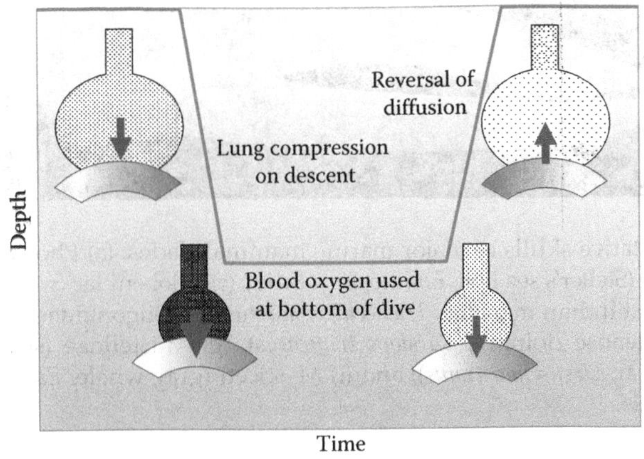

Figure 4.4 Mechanism underlying shallow-water blackout. Alveolar O_2 pressure increases with ambient pressure on descent, increasing diffusion into blood. Arterial O_2 is increased at the bottom of a shallow dive, and then decreases (due to metabolism) over the course of the dive. During ascent, alveolar O_2 pressure decreases as the lung volume increases with decreasing ambient pressure. This can cause a reversal of the diffusion gradient, pulling O_2 from the blood into the lungs, causing a transient decrease in arterial O_2 and leading to blackout. Darker shading of the alveoli indicates higher partial pressure. Darker vessel shading indicates higher blood gas tension. The arrows show direction of net diffusion and the greater arrow thickness indicates greater diffusion rate.

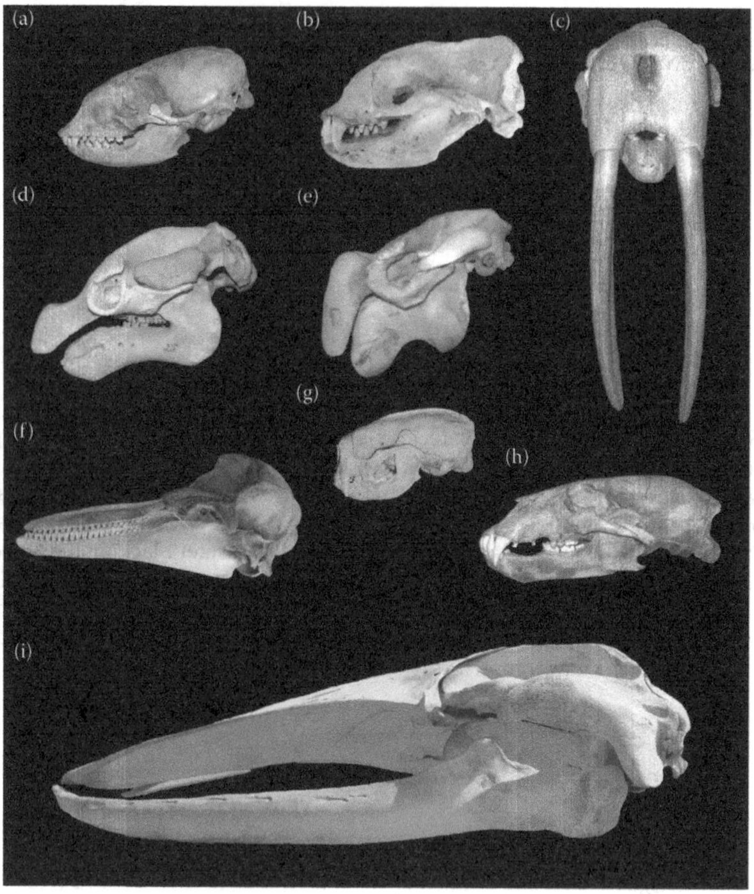

Figure 5.1 Representative skulls of major marine mammal clades. (a) Phocidae (harbor seal, *Phoca vitulina*); (b) Otariidae (Steller's sea lion, *Eumetopias jubatus*); (c) Odobenidae (walrus, *Odobenus rosmarus*); (d) Trichechidae (West Indian manatee, *Trichechus manatus*); (e) Dugongidae (dugong, *Dugong dugon*); (f) Odontoceti (bottlenose dolphin, *Tursiops truncatus*); (g) Mustelidae (sea otter, *Enhydra lutris*); (h) Ursidae (polar bear, *Ursus maritimus*); and (i) Mysticeti (gray whale, *Eschrichtius robustus*). Skull images are not to scale.

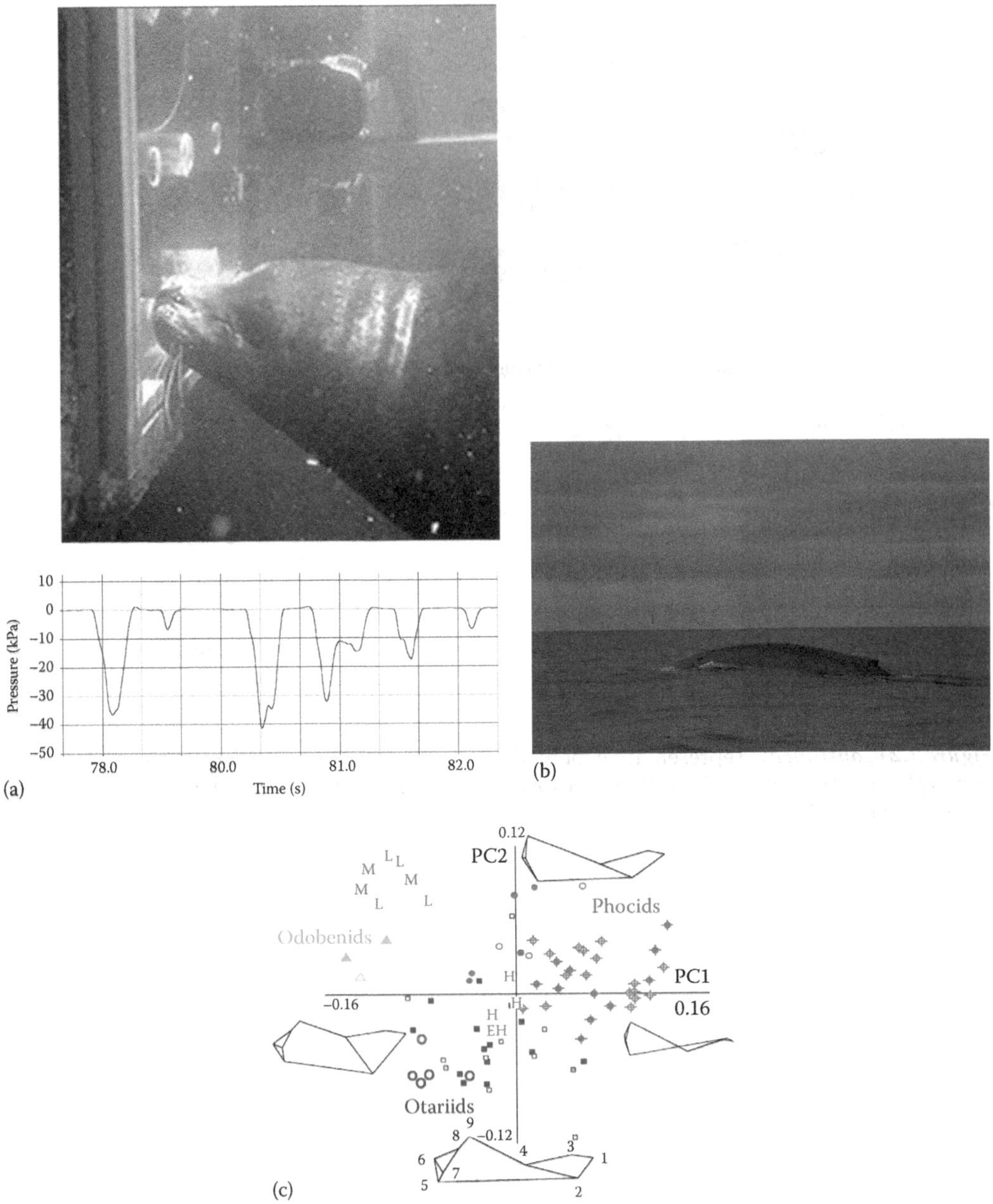

Figure 5.2 Representative methods of marine mammal feeding studies. (a) Functional performance, kinematics, and biomechanics. (From Marshall, C.D. et al., *PLoS ONE*, 9, e86710, 2014, doi: 10.1371/journal.pone.0086710.) (b) Animal-borne tags on free-ranging animals. (Photo by Jeremy Goldbogen. Unpublished video tag from: Jeremy Goldbogen, Dave Cade, Ari Friedlaender, and John Calambokidis. NMFS Permits: #14534-2 and 14809.) (c) Comparative and integrative morphology, biomechanics, and geometric morphometrics. (From Jones, K.E. et al., *Anat. Rec.*, 296, 1049, 2013.)

(*Continued*)

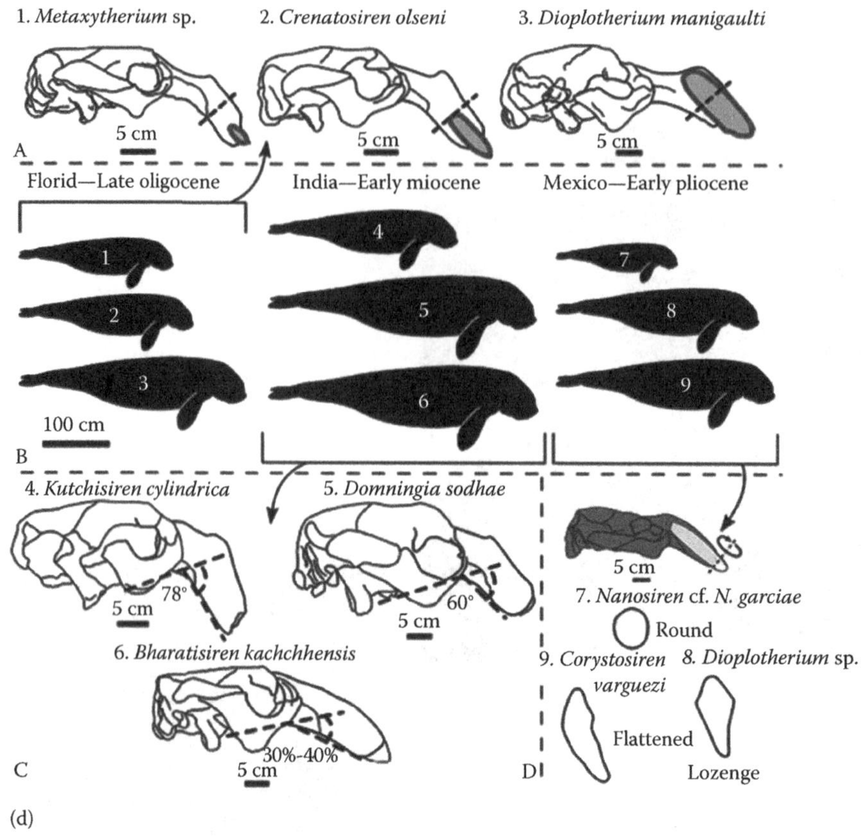

***Figure 5.2* (*Continued*)** Representative methods of marine mammal feeding studies. (d) Integration of morphology and paleoecology. (From Velez-Juarbe, J. et al., *PLoS ONE*, 7, e31294, 2012, doi: 10.1371/journal.pone.0031294.)

Figure 6.1 An Antarctic krill swarm. (Photo by Steve Nicol.)

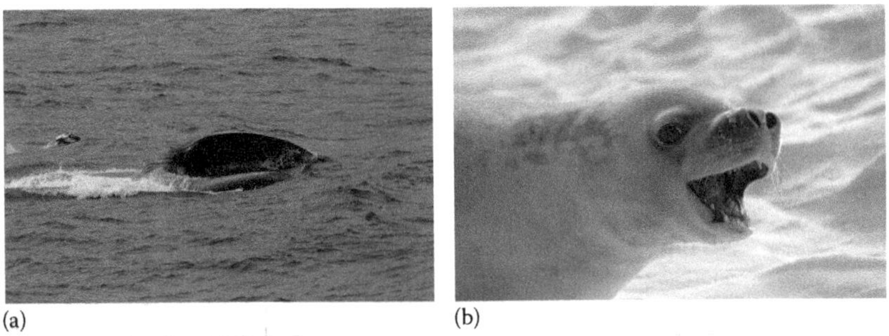

Figure 6.2 (a) Pygmy blue whale, feeding. Showing baleen, Portland, Victoria, Australia. (Photo by Paul Enser, Australian Antarctic Division, Kingston, South Australia, Australia.) (b) Crabeater seal head and shoulders illustrating the heavily cusped post-canine teeth used by this species to sieve Antarctic krill. (Photo by Mark A. Hindell.)

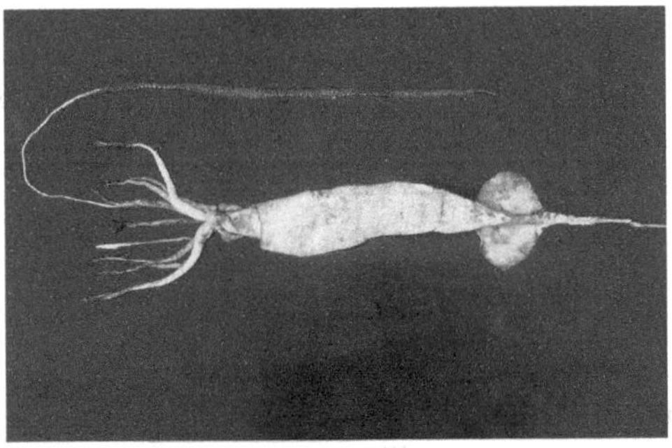

Figure 6.3 Direct observation of sub-surface feeding events is rare. Sub-surface feeding on bottom-dwelling or mid-water prey, such as squid as shown here, are greatly underestimated in this method (*Chiroteuthis* sp.). (Photo by Glenn Jacobson, Australian Antarctic Division © Commonwealth of Australia.)

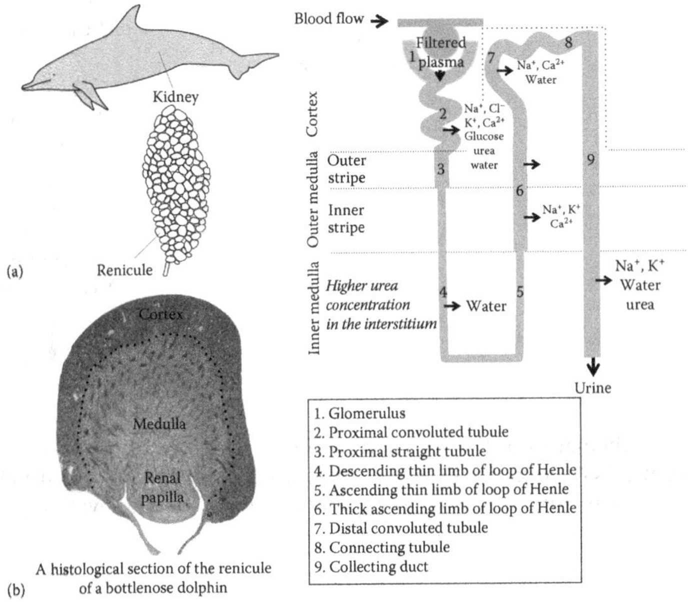

Figure 7.2 Kidney structure. The kidneys of marine mammals are lobulated and composed of hundreds to thousands of renicules (a). Each renicule possesses its independent cortex and medulla that empty into a common renal pelvis (b).

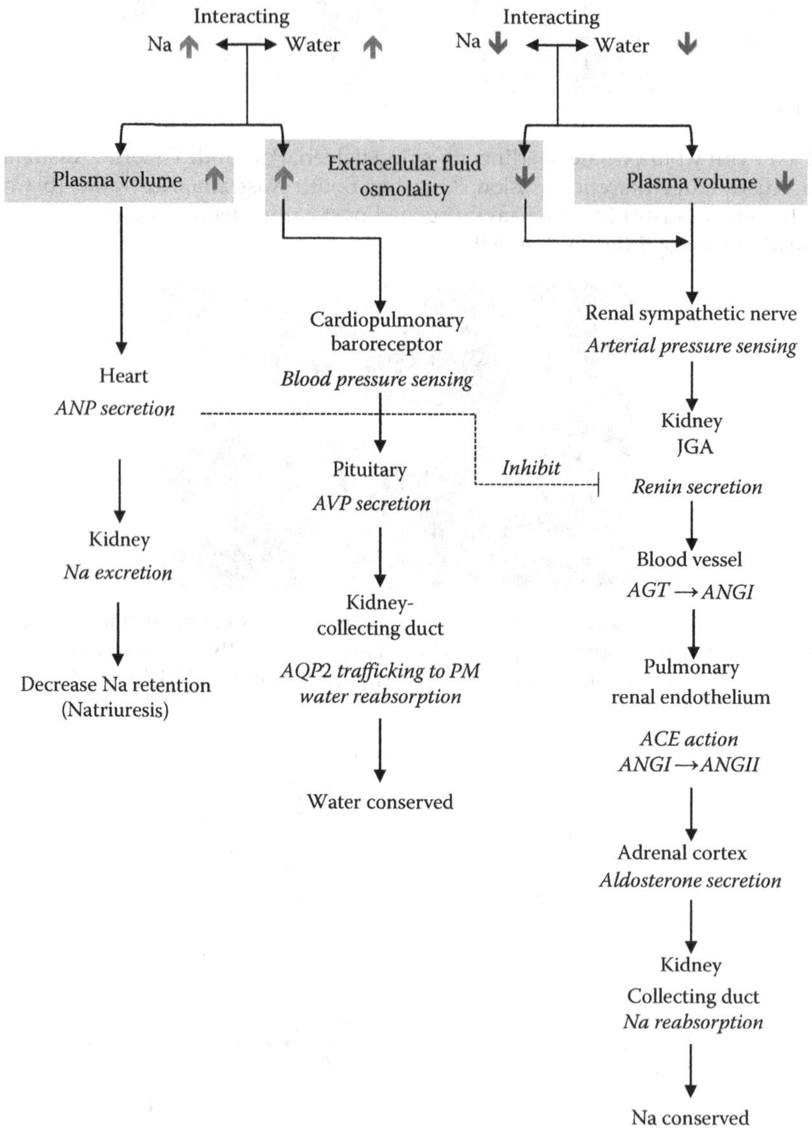

Figure 7.3 General schematics of the functions of osmoregulatory hormones. (ACE, angiotensin I-converting enzyme; AGT, angiotensinogen; ANG, angiotensin; ANP, atrial natriuretic peptide; AVP, arginine vasopressin; juxta-glomerular apparatus cells in the kidney; PM, plasma membrane.)

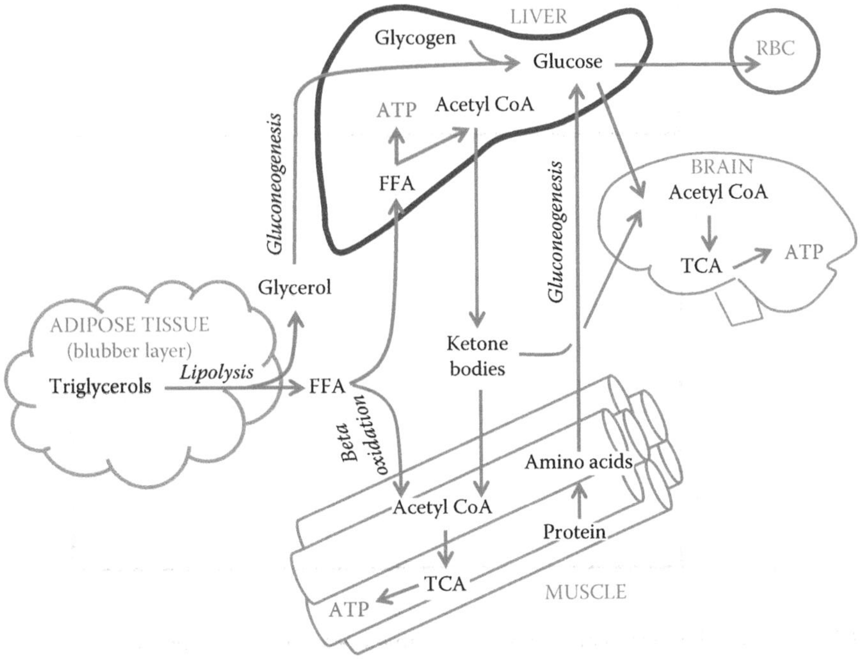

Figure 8.2 Summary of some of the major metabolic pathways related to energy production during a fast. Major pathways are illustrated for catabolism of lipids from adipose tissue and protein from muscles, as well as production of glucose in the liver from either stored glycogen or from gluconeogenesis from other precursors. Triacylglycerols catabolized via lipolysis from adipose tissues form the major source of energy during fasting. The resulting free fatty acids (FFA) are transformed via beta-oxidation into acetyl coenzyme A (acetyl CoA), which produces energy (ATP) at target tissues via the tricarboxylic acid (TCA) cycle. Some of the FFA that enter the liver are oxidized directly, but most are converted to ketone bodies (acetoacetate and β-hydroxybutyrate), which are released into the blood (the liver cannot use ketone bodies itself). The ketone bodies are oxidized in the mitochondria of target cells into acetyl-CoA, which enters the TCA cycle. Certain tissues require glucose to function. This can be originally met by liver glycogen stores, which are rapidly depleted. New glucose can be formed in the liver (gluconeogenesis) from the smaller glycerol segment of triacylglycerols and from amino acids freed via protein catabolism (proteolysis). The central nervous system can derive some of its energy from oxidation of ketone bodies, but red blood cells are obligate glucose consumers. In the later stages of fasting, protein becomes a more dominant metabolic substrate. (Adapted from Lieberman, M. and Marks, A.D., *Marks' Basic Medical Biochemistry: A Clinical Approach*, Lippincott Williams & Wilkins, Baltimore, MD, 2009.)

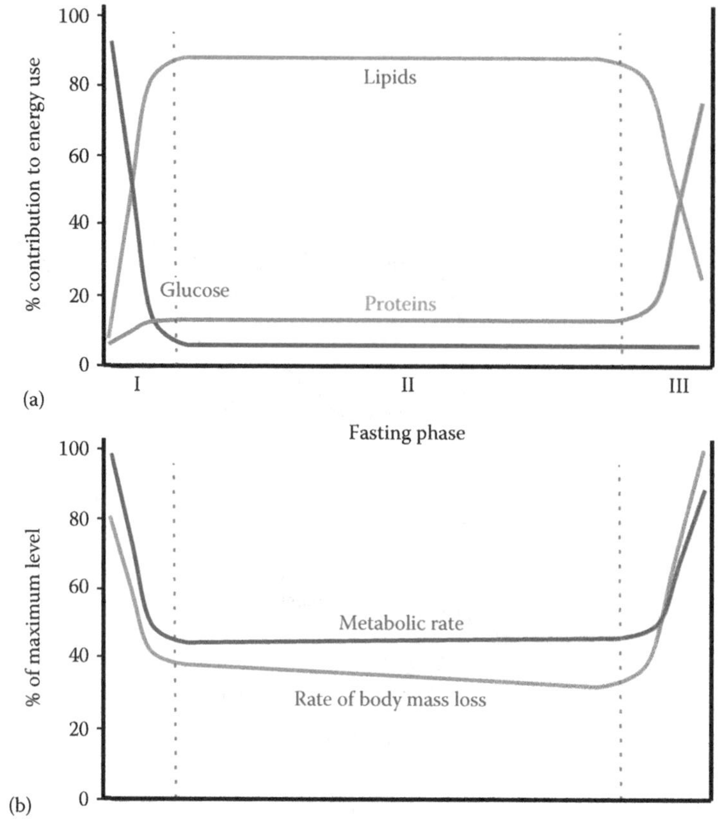

Figure 8.3 Predicted changes in physiological parameters according to the classical three phases of fasting as detailed by Cherel et al. (1988). Significant changes include a shift in metabolic fuel use (a) and rates of body mass loss and mass-specific metabolic rates (b).

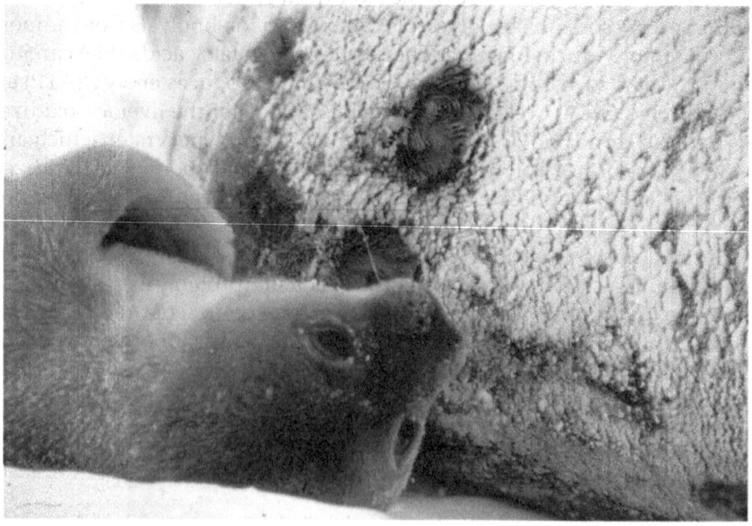

Figure 9.2 A young Weddell seal pup and nursing mother. Note the lanugo coat of the pup. The fur of the mother is kept cool and snow can sit on the surface without melting, except where the pup has been nursing. (Photo by M. Castellini, MMPA Permit #801.)

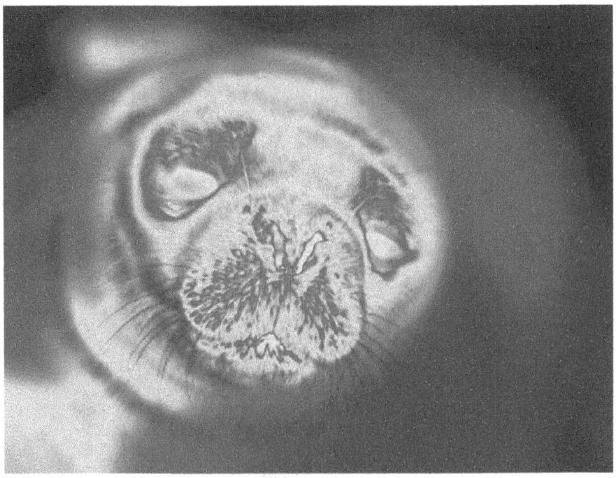

Figure 9.3 An infrared image reveals the hot (red) and cold (blue) surface temperatures on the face of a Weddell seal pup. Most of the pup surface remains cool, with the exception of the un-insulated eyes, lightly insulated head, and highly vascularized muzzle. (Photo by J. Mellish, NMFS 15748.)

Figure 9.4 A group of resting Weddell seals on the summer sea ice of McMurdo Sound, Antarctica. These animals can face winter air temperatures in this region as low as –50°C, before factoring in the effects of convection, or *wind chill*. (Photo by D. Uhlmann, NMFS 1034–1854.)

Figure 9.5 A Weddell seal mother watches her young pup explore the comparatively stable –2°C water of McMurdo Sound, Antarctica. (Credit to H. Kaiser, NMFS 15748.)

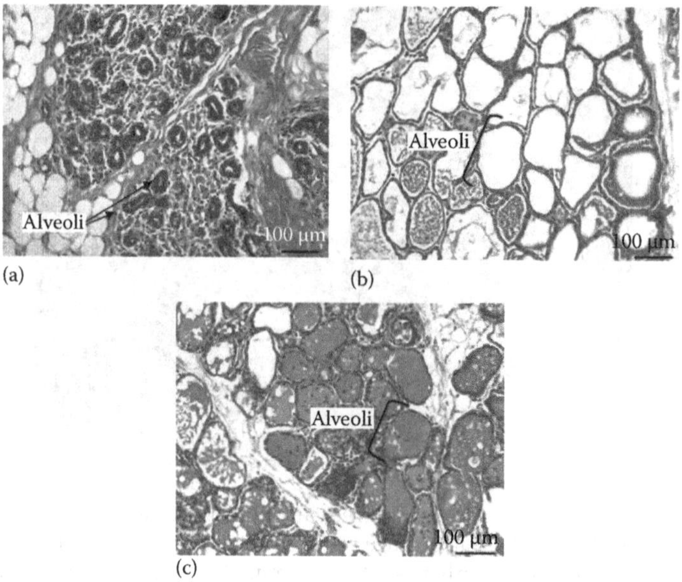

Figure 10.4 Mammary gland morphology of an otariid. Histological sections of a mammary gland from (a) pregnant, (b) lactating onshore, and (c) lactating while foraging at-sea fur seals. Fat globules within alveoli appear as white bodies. Pink staining within alveoli represents milk components. Residence of milk components in alveoli would lead to involution in most species but otariids have evolved the ability to maintain mammary function in absence of suckling. (From Sharp, J.A. et al., *J. Mammary Gland Biol. Neoplasia*, 12(1), 47, 2007. With permission.)

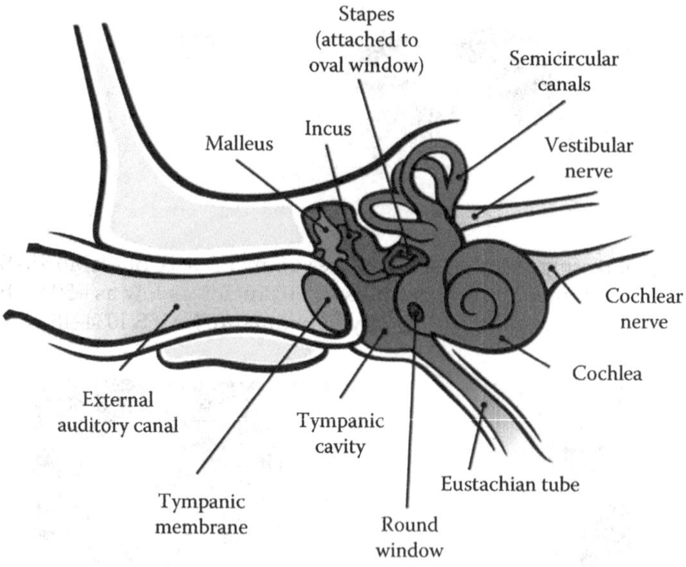

Figure 11.1 A schematic of the outer, middle, and inner ear of terrestrial mammals (in this case based on the human anatomy). (Adapted from Chittka, L. and Brockmann, A., *PLoS Biol.*, 3(4), e137, 2005.) For terrestrial mammals in air, the outer and middle ear are both air-filled, and the middle ear ossicles (incus, malleus, and stapes) amplify sound vibrations from the tympanic membrane (eardrum) into the sensory organ of the fluid-filled cochlea via the oval window. Sound is then transmitted to the brain through the auditory nerve. When this terrestrial ear is submerged, the high acoustic impedance of external water in the outer ear relative to the low impedance of the air-filled middle ear results in decreased efficiency of sound transmission to the cochlea. Marine mammal auditory systems have adapted to various degrees to overcome this problem.

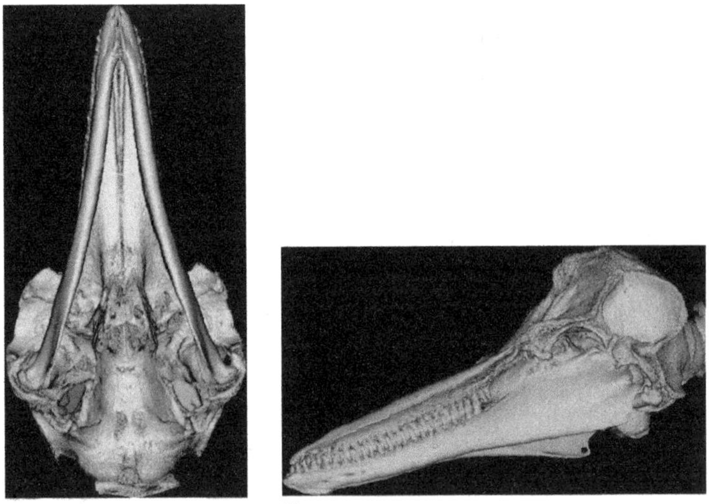

(a) (b)

Figure 11.2 (a) Ventral perspective of the dolphin skull with the left (green) and right (red) auditory bullae displayed *in situ*. The projections are based on a CT scan of a bottlenose dolphin. (b) Lateral perspective of the same dolphin showing the relationship of the left bulla to the posterior terminus of the lower jaw. The bullae are detached from the skull and surrounded by tissue and airspaces (sinuses).

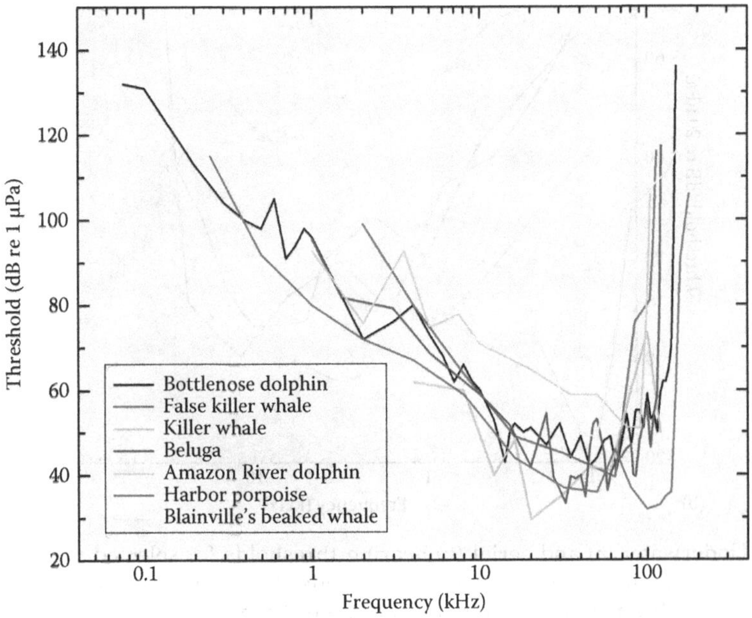

Figure 11.3 Underwater audiograms for selected odontocete cetacean species. All thresholds were obtained using behavioral methods except those for the beaked whale species, which were obtained using auditory evoked potential methods. Note the extended range of high-frequency hearing in the harbor porpoise relative to the other mid-frequency cetaceans. Studies for which thresholds are shown: bottlenose dolphin (Johnson 1967), false killer whale (Thomas et al. 1988), killer whale (Szymanski et al. 1999), beluga (White et al. 1978), Amazon River dolphin (Jacobs and Hall 1972), harbor porpoise (Kastelein et al. 2002), and Blainville's beaked whale (Pacini et al. 2011).

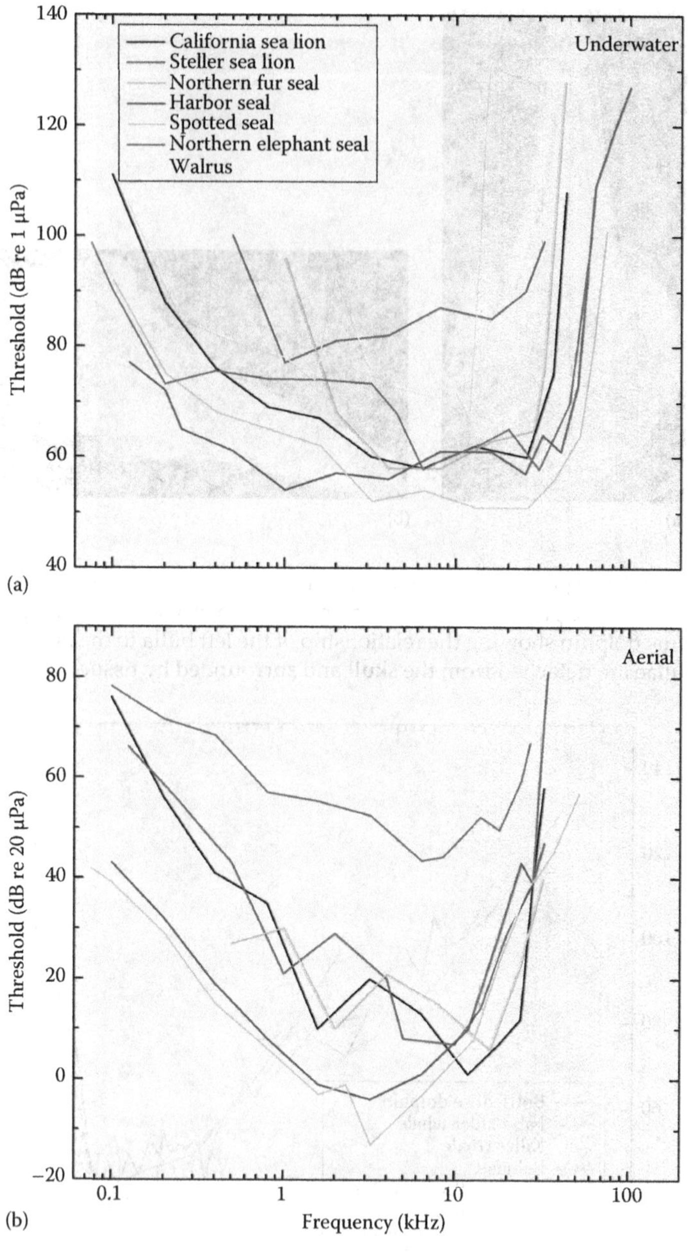

Figure 11.4 Underwater (a) and aerial (b) hearing thresholds for selected pinniped species. All audiograms were obtained using behavioral methods. The two plots are shown with identical x-axes to emphasize the differences in frequency range of hearing between air and water. Note the difference in reference pressures for underwater (1 μPa) and aerial (20 μPa) thresholds. Studies for which underwater thresholds are shown: California sea lion (Reichmuth et al. 2013), Steller sea lion (Kastelein et al. 2005), northern fur seal (Moore and Schusterman 1987), harbor seal (Reichmuth et al. 2013), spotted seal (Sills et al. 2014), northern elephant seal (Kastak and Schusterman 1999), and walrus (Kastelein et al. 2002). Aerial studies: California sea lion (Reichmuth and Southall, 2012), Steller sea lion (Mulsow and Reichmuth 2010), northern fur seal (Moore and Schusterman 1987), harbor seal (Reichmuth et al. 2013), spotted seal (Sills et al. 2014), and northern elephant seal (Reichmuth et al. 2013).

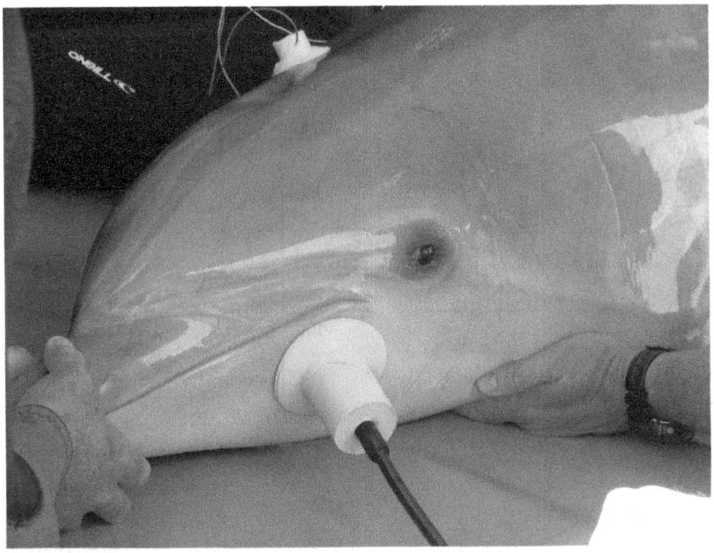

Figure 11.5 A bottlenose dolphin undergoing a hearing test using auditory evoked potentials. The sound projector is attached to the lower jaw over the region of the *pan*, a thin portion of the lower mandible through which sound passes to the ear (also see Figure 11.2). An electrode placed on the dorsal surface of the animal over the brainstem, records the evoked responses produced by the brain in response to the projected sounds.

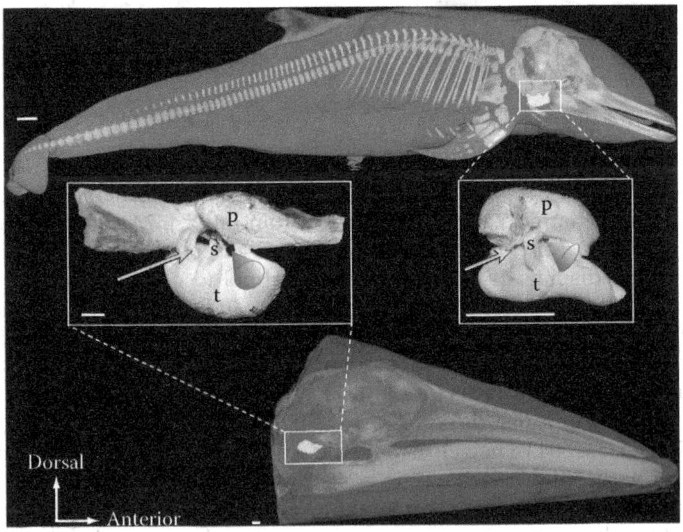

Figure 11.6 Images of a pantropical spotted dolphin (top) and a minke whale (bottom) based on computerized tomography (CT) scans. (From Yamato, M. and Pyenson, N.D., *PLoS One*, 10(3), e0118582, 2015.) The tympanoperiotic complexes (identified in yellow, and as pictures from adult specimens) are also shown. The blue arrows indicate the tympanic apertures. The pink cones represent the *acoustic funnels*, through which sound is conducted through acoustic fats to the tympanoperiotic complex. Note the anterior orientation of the acoustic funnel in the odontocete relative to the mysticete, which is indicative of the forward-facing, highly directional echolocation system in odontocetes. These structures suggest a more lateral sound reception system in mysticetes, and have value in making predictions in the absence of physiological or psychological data.

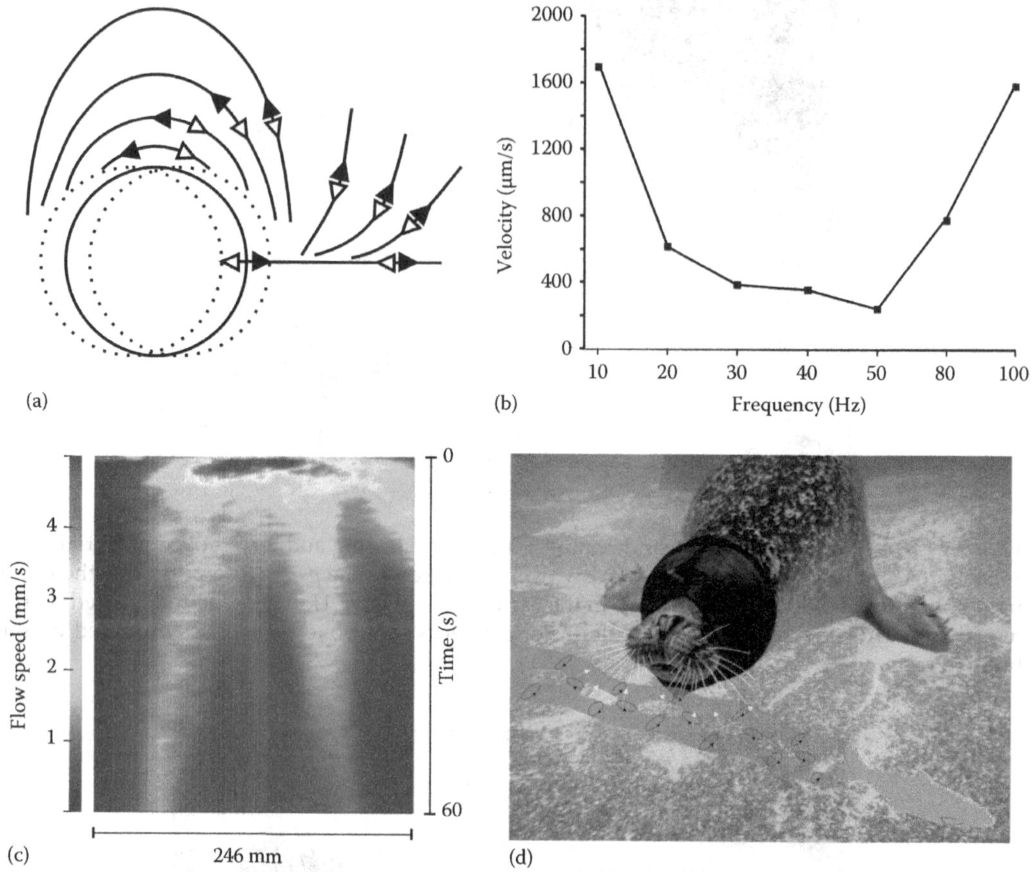

Figure 12.2 Hydrodynamic flow perception. (a) Water movements generated by a dipole as usually used to examine hydrodynamic receptors. (Modified from Tautz, J., *Medienbewegung in der Sinneswelt der Arthropoden—Fallstudien zu einer Sinnesökologie,* Gutsav Fischer Verlag, Stuttgart, Germany, 1989.) Movement vectors are displayed in the vicinity of a dipol (oscillating sphere), the amplitude of oscillation is indicated by the arrows and the dotted lines. (b) Performance of a harbor seal during hydrodynamic flow testing. (Modified from Dehnhardt, G. et al., *Nature*, 394, 235, 1998.) The seal's performance is depicted as velocity threshold (in μm/s) as a function of the frequency (in Hz) of the stimulus—see (a). The seal achieved its best threshold performance of 245 μm/s at 50 Hz. (c) Velocity profile of the wake behind a fish (*Lepomis gibbosus*). (Modified from Hanke, W. and Bleckmann, H., *J. Exp. Biol.*, 207, 1585, 2004.) The velocity profile was obtained from particle image velocimetry (PIV) (see Section 12.3.3). Flow speeds within a 246 mm region of interest are color-coded from blue to red corresponding to water velocities from 0 to 5 mm/s. The fish swam from the bottom to the top of the figure, thus the flow speeds at the top are measured in the wake directly behind the fish, whereas the flow velocities at the bottom of the figure depict the flow speeds that persist after 60 s. (d) A visually masked harbor seal is encountering a hydrodynamic trail of a fish. The flow direction of the water particles in the hydrodynamic trail is indicated by arrows. (Modified from Bleckmann, H., Reception of hydrodynamic stimuli in aquatic and semi-aquatic animals, in *Progress in Zoology,* Rathmayer, W. (ed.), Gustav Fischer Verlag, Stuttgart, Germany, 1994.) The seal can gain information on the 3D structure of the hydrodynamic trail by multiple point-to-point measurements (indicated by yellow arrows).

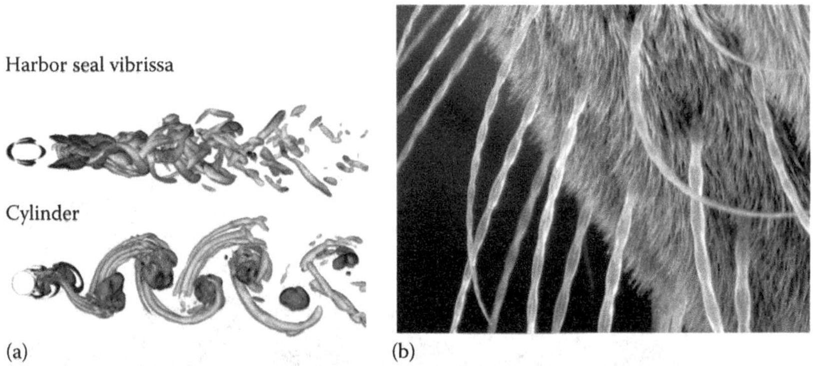

Figure 12.4 Morphology of harbor seal vibrissae and their behavior in flow. (a) Wake flow behind a vibrissae of a harbor seal and behind a circular cylinder as obtained from numerical simulations. (Modified after Hanke, W. et al., *J. Exp. Biol.*, 213, 2665, 2010.) Behind a vibrissa, a complex 3D vortex structure generates downstream from the vibrissa leaving a gap between the vibrissa and the region with fluctuating vortex flows. Furthermore, the complex vortex structure is not stable over time. In contrast, behind a circular cylinder, primary vortices regularly shed directly from the cylinder (Kármán street) and largely persist over time. (b) Close-up view of a snout of a harbor seal showing the characteristic morphology of the vibrissae of most phocids. Most phocids possess vibrissae that are flattened and possess an undulatory shape.

Figure 13.1 Bottlenose dolphin (*Tursiops truncatus*) affected by lacaziosis in Sarasota Bay, Florida. Note the raised, bumpy lesions that may affect large areas of the body. (Photo courtesy of Sarasota Dolphin Research Program, obtained under National Marine Fisheries Service Scientific Research Permit No. 15543.)

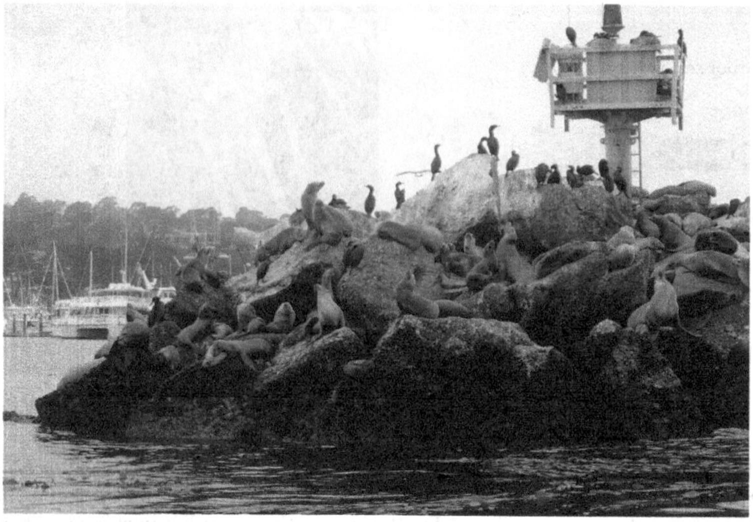

Figure 13.2 California sea lions (*Zalophus californianus*) haul out, or congregate on land, in large groups on a jetty in Monterey Bay, California. (Photo courtesy of The Marine Mammal Center, Sausalito, California.)

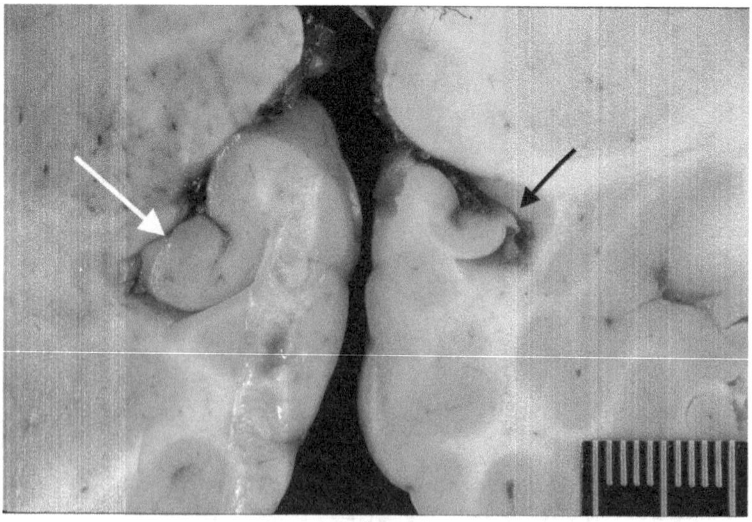

Figure 13.3 Domoic acid specifically targets the hippocampus in the brain. Left: A normal California sea lion brain section (the hippocampus is noted with the white arrow). Right: A California sea lion's brain affected by domoic acid exposure. Note the shrunken hippocampus (black arrow). (Photo courtesy of The Marine Mammal Center, Sausalito, California.)

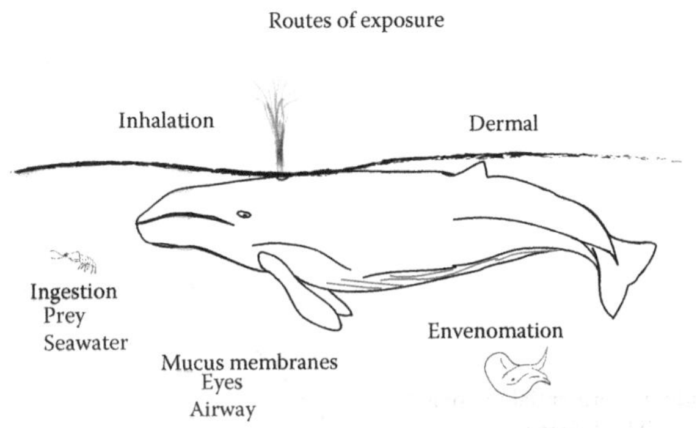

Figure 14.1 A conceptual figure displaying the potential routes of exposure of marine mammals to poisons. Some of these pathways are often overlooked, including the exposure of mucosal membranes to poisons such as brevitoxins or volatile organic compounds. Envenomation may be rare, although some marine mammals are known to feed on venomous animals (i.e., killer whales and stingrays) it has been hypothesized as a potential cause of mortality in some marine mammals. (From Duignan, P.J. et al., *Aquat. Mamm.*, 17, 143, 2000.)

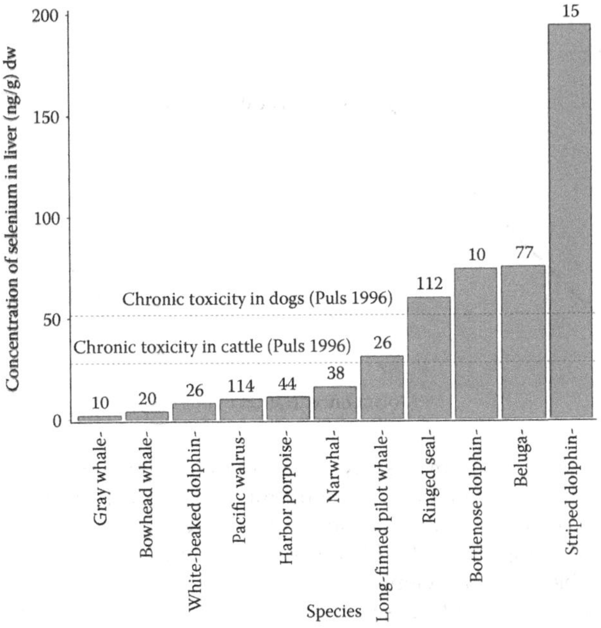

Figure 14.2 Marine mammals display a large variation in selenium (Se) concentrations in the liver. Mean concentrations and thresholds of toxicity are presented in dry weight (dw) assuming a mean water weight of 75% as reported in Das et al. (2003) and Puls (1994). The number of individuals for which the mean concentration is displayed is reported above each bar. Data are reported in the following references by increasing the concentration of Se (left to right in the figure); gray whale (*Eschrichtius robustus*) (Varanasi et al. 1994), bowhead whale (*Balaena mysticetus*) (Krone et al. 1999), white-beaked dolphin (*Lagenorhynchus albirostris*) (Muir et al. 1988), Pacific walrus (*Odobenus rosmarus divergens*) (Wagemann and Stewart 1994), harbor porpoise (*Phocoena phocoena*) (Paludan-Müller et al. 1993), harbor seal (*Phoca vitulina*), narwhal (*Monodon monoceros*) (Wagemann et al. 1983), long-finned pilot whale (*Globicephala melaena*) (Muir et al. 1988), ringed seal (*Pusa hispida*) (Smith and Armstrong 1978), bottlenose dolphin (*Tursiops truncatus*) (Kuehl and Haebler 1995), beluga (*Delphinapterus leucas*) (Wagemann et al. 1996), and striped dolphin (*Stenella coeruleoalba*) (Itano et al. 1984). (Adapted from Das, K. et al., Heavy metals in marine mammals, in *Toxicology of Marine Mammals*, Taylor & Francis, New York, 2003, pp. 135–167.)

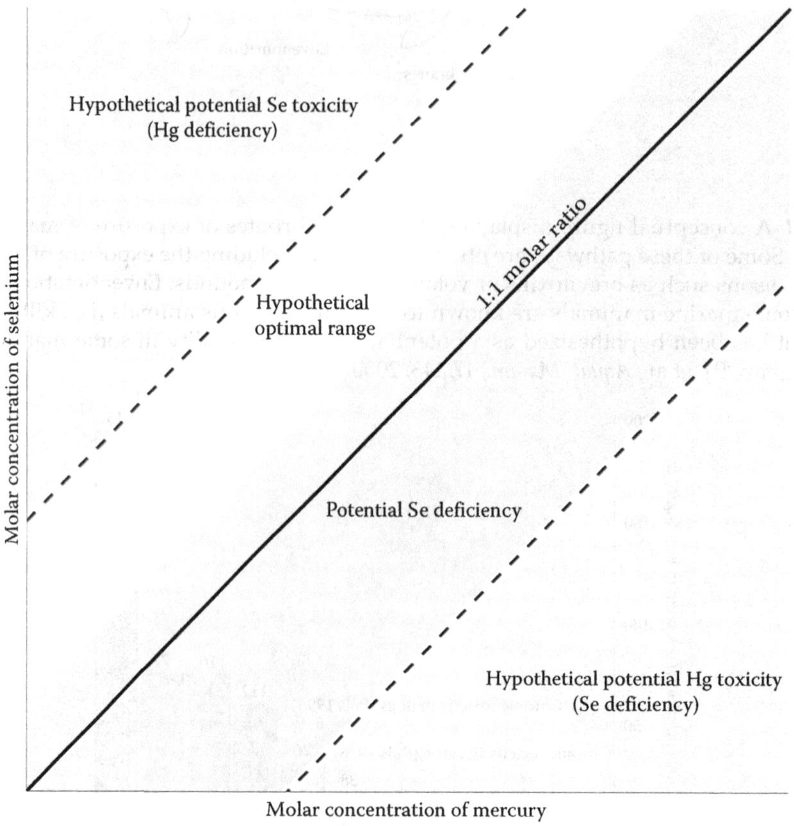

Figure 14.3 Mercury and selenium display a multi-faceted antagonistic relationship in marine mammals, including the formation of an insoluble mineral (HgSe, tiemmanite) occasionally resulting in a near 1:1 molar correlation (i.e., Koeman et al. 1973; Koeman et al. 1975; Brookens et al. 2007). As noted by Khan and Wang (2009) the 1:1 molar ratio can potentially be viewed as Se deficient if Hg–Se is not bioavailable, wherein selenium is unable to participate in Se-dependent detoxification processes (i.e., glutathione peroxidase, selenoproteins). We emphasize that, somewhat counterintuitively, with high concentrations of Se seen in marine mammals (see Figure 14.2) that potential selenosis could occur in marine mammals as a result of Hg deficiency. (Adapted from Khan, M. and Wang, F., *Environ. Toxicol. Chem.*, 28, 1567, 2009.)

chapter twelve

Visual and hydrodynamic flow perception

Frederike D. Hanke and Guido Dehnhardt

Contents

12.1 The big picture challenge

Marine mammals are generally very mobile species. Whereas some species undertake large-scale movements, most marine mammals move on small scales within, for example, a home range. These movements are required for successful foraging or breeding. Consequently, marine mammals regularly experience sometimes even tremendous changes in the environment, and these changes occur when moving horizontally but also vertically during dives. However, under all conditions, the animals have to rely on their senses as interfaces between environment and behavior. While they generally assist orientation, the senses help to collect important information about general features of the environment, predators, prey, or mating partners. Although an individual's senses most likely interact to form a multimodal representation of the environment (Dehnhardt 2002), it may happen under certain conditions that only a few or, in the most extreme case, only one sensory system is providing reliable information.

Among the senses, *vision* is of paramount importance from a human perspective. In contrast, in marine mammals, the role of the visual system for underwater orientation is controversially discussed. Many authors have speculated that vision cannot play a major role underwater due to low light levels that seem to dominate the visual environment of marine mammals. Low light levels are encountered at night but also when diving to deep waters even during the day when marine mammals experience perfect light conditions at the water surface to which they have to return for breathing. When diving, low light levels are caused by sunlight getting more and more absorbed and scattered with depth. Absorption and scattering is increased if particles are dissolved in the water which especially characterizes coastal waters or rivers. Under these dim light conditions, are marine mammals able to use vision? Which role could vision nevertheless play in their visual environment?

Marine mammals possess well-developed eyes with one exception which will be highlighted later in this chapter (for a review, see for example, Mass and Supin 2007; Kröger and Katzir 2008; Reuter and Peichl 2008; Hanke et al. 2009; Mass and Supin 2009). In most marine mammals, many adaptations have evolved to increase sensitivity comparable to adaptations found, for example, in crepuscular or nocturnal animals. Most marine mammals have large eyes and large ranges of *pupillary opening*, the *retina* is dominated by light-sensitive *rods*, visual pigments possess maximum sensitivity in blue light, giant ganglion cells pool visual information, and *tapeta* line the fundus (see Section 12.2.1). Moreover, diving straight down from the potentially very bright water surface to deep waters, the eyes of marine mammals also quickly and efficiently adapt to the light intensities that dominate the light conditions at the depth at which they are foraging (Levenson and Schusterman 1999). Some marine mammals such as elephant seals or beaked (Ziphiidae) and sperm (*Physeter macrocephalus*) whales even dive down to the deep sea and hunt deep-sea squid and fish at more than 1500 m depth. Although sunlight is never reaching these depths, the deep sea is not necessarily completely dark. *Bioluminescent* point-like flashes contribute light for vision in the deep sea (Warrant and Locker 2004). This visual environment has shaped the eyes of deep-diving marine mammals. In addition to a very fast *dark adaptation* (Levenson and Schusterman 1999) mediated by large pupillary openings (Levenson and Schusterman 1997), the rod pigment, the *rhodopsin*, of deep divers is highly blue-shifted and thus maximally sensitive to light of 480–486 nm wavelength (Lythgoe and Dartnall 1970; McFarland 1971; Fasick and Robinson 2000; Southall et al. 2002; Levenson et al. 2006). These blue-shifted pigments might be a response to bluish light that penetrates best into deep waters and to bluish bioluminescence. Moreover, the blue pigments have a reduced noise levels which helps to detect a signal and thus is required to gain a reasonably good image when photons are scarce and consequently noise is high (Reuter and Peichl 2008). Furthermore, the number of axons within the optic nerves of deep-diving animals is highest among the marine mammals with, for example, 750,000 axons in southern elephant seals (*Mirounga leonina*; Pütter 1903) and 1,000,000 in the pygmy sperm whale (*Kogia breviceps*; Dawson 1980). The putatively resulting high resolution might enable these animals to detect small bioluminescent sources even at distance (Kröger and Katzir 2008).

Marine mammals encounter dim light conditions when diving deep or at night, however, also when, even under perfect light conditions, dissolved particles additionally cause absorption and scattering (Jerlov 1976; Mobley 1994). With increasing turbidity, the underwater visual acuity drops drastically as shown for harbor seals (*Phoca vitulina*) (Weiffen et al. 2006). Thus, the detection of small prey items is limited. However, visual orientation underwater requires more than just object detection in a foraging context. Recently, researchers came up with the idea that the particles that seem to render vision in murky waters difficult, if not impossible, create a perfect *optic flow* environment when passed by an animal (Gläser et al. 2014). Optic flow is defined as the pattern of visual motion elicited on the retina of a moving observer (Gibson 1950). When moving forward, an optic flow pattern is generated in which all motion vectors emanate from the point of heading, the *focus of expansion* (FOE) (Figure 12.1a). The interpretation of optic flow patterns may allow an animal to avoid colliding with obstacles in its surround

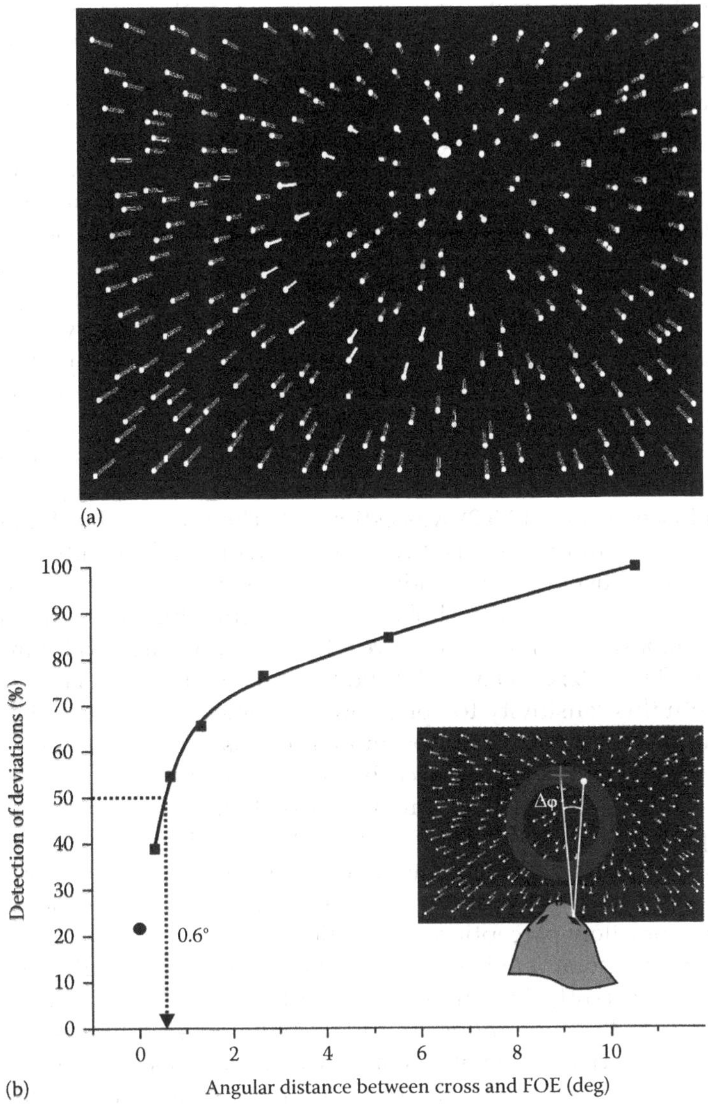

Figure 12.1 Optic flow perception. (a) Optic flow simulation mimicking a forward movement on a straight path through a cloud of particles. The motion vectors emanate from one point, the focus of expansion (FOE; large white dot), that indicates the direction of movement. (b) Performance of a harbor seal during optic flow testing. (Modified from Gläser, N. et al., *PloS One*, 9(7), e103555, 2014.) The seal's performance is depicted in detection of deviations (delta phi) between a cross superimposed on the simulation and the actual FOE (in %) as a function of the angular distance between the cross and the FOE (delta phi; in deg). The seal was still able to detect deviations of 0.6° between a cross and FOE (50% threshold). The inset shows the optic flow projection used for testing optic flow perception in a harbor seal. The gray ring was shading the region in which the FOE (white dot, not visible during experiments) could be programmed. A green cross was superimposed that could either match or deviate by an angular distance (delta phi) from the FOE.

(Regan and Gray 2000), estimate its heading, thus the direction it is traveling to (Warren and Hannon 1988), to locomote precisely toward a goal (Gibson 1950; Cutting et al. 1992; Warren et al. 2001) or to assess the distance that it has just traveled (Lappe et al. 2007). After the documentation of motion vision (Hanke et al. 2008) and high motion sensitivity in harbor seals (Weiffen et al. 2014; see Section 12.2.2), optic flow perception was recently investigated for the first time in a marine mammal (Gläser et al. 2014). In the optic flow experiment, a harbor seal was presented with an underwater optic flow projection which simulated a forward movement through a cloud of particles represented by 500 white dots moving out of the FOE toward the observer on straight paths (Figure 12.1). The FOE could be on a numerous positions which were, however, covered by a mask (Figure 12.1b, inset) in order to force the animal to rely on the global optic flow instead of on local motion signals at the FOE. Thus, the FOE and its vicinity were not visible to the seal. Additionally, a cross was superimposed on the flow pattern (Figure 12.1b, inset). It could either directly mark the FOE or could deviate from the FOE by a preset angle of deviation. In the first experimental condition, the seal correctly responded if it touched the cross on the projection; in the second condition, it was required to turn away from the projection in order to receive a reward. By varying the angle of deviation, the 50% threshold (see Section 12.3.2) was estimated. The seal quickly found access to the complex optic flow stimuli and was directly going with the flow indicating that the flow pattern created an illusion of self-motion. The seal could still detect a deviation of 0.6° between cross and FOE (50% threshold) (Figure 12.1b) which is comparable to the best performances published so far in literature which was documented for monkeys (Britten and van Wezel 1998, 2002; Gu et al. 2010) and humans (Warren and Hannon 1988; Warren et al. 1988). With this sensitivity to optic flow, harbor seals can most likely rely on optic flow information for goal-directed locomotion or assessing parameters important for *path integration* such as traveled distance. In conclusion, with the documented sensitivity to optic flow fields, harbor seals and maybe also other marine mammals can probably benefit from particle load in the water contrary to the notion that particles only impede underwater vision. Thus researchers will be forced to rethink visual orientation on the basis of these findings, and numerous research questions will probably entail this first experiment on optic flow perception (see Section 12.4).

However, under some environmental conditions, vision might indeed be almost impossible and constantly impaired in contrast to the situations outlined so far in which particle load changes temporally and spatially. River dolphins (Platanistidae and Lipotidae) inhabiting riverine habitats with extreme amounts of particles constantly dissolved in the water. As expected, these species have reduced eyes (Herald et al. 1969; Pilleri 1974). In the most extreme case, in the Indus River dolphin (*Platanista gangetica),* the eyes are very small and immobile with only a small eyelid opening, the lens is lacking, the *ciliary* body is atrophied, and the *tapetum* is almost absent. The optic nerves only contain low fiber numbers as also found in the Amazon River dolphin (*Inia geoffrensis*) that only possesses approximately 15,000 axons in the optic nerve (Morgane and Jacobs 1972; Mass and Supin 1989). This highly contrasts with, for example, the bottlenose dolphin (*Tursiops truncatus*) with its well-developed eyes from which optical information within 160,000–180,000 axons is transmitted to the brain (Morgane and Jacobs 1972; Dawson et al. 1982). On the other hand, Waller observed behaviors in *Platanista* (Waller 1983) comparable to visually guided behavior generally found in dolphins such as visual inspection and a well-developed retina (Waller 1982). In conclusion, it remains to be investigated if the eyes of the river dolphins only function to detect and gather light (Herald et al. 1969) and determine the direction of light (Herald et al. 1969; Pilleri 1974), the only visual tasks that can most likely be fulfilled in very turbid waters, or if they can use their eyes for more sophisticated visual tasks. Generally, all marine mammals might sometimes face the problem that vision does not provide any reliable data about the environment, the question arises which alternative sense marine mammals could rely on to replace or complement vision in dark and murky waters?

Please note that, beyond vision, the focus is laid on hydrodynamic flow perception in this chapter as marine mammal acoustics is elaborated profoundly and extensively in Chapter 11. Readers interested in chemoreception in marine mammals are relegated to existing reviews covering this field of research (see for example, Dehnhardt 2002).

Marine mammals are foraging in the *benthic* (at the sea bottom) as well as the *pelagic* (open water) zone. At the sea bottom, prey items might be hard to detect visually as many benthic fish are hiding in the mud or are cryptic. If vision is additionally impaired due to low light levels and turbidity, marine mammals have to adopt alternative strategies for the detection of prey. When hunting in the benthic zone, harbor seals were observed to cruise over or even dig into the sea bottom looking for prey (Bowen et al. 2002), and also dolphins (Rossbach and Herzing 1997; Smolker et al. 1997; de Gurjao et al. 2003; Mann et al. 2008), walruses (Odobenidae; cited after Kastelein and van Gaalen 1988; Kastelein et al. 1990), and sea otters (*Enhydra*) (Hines and Loughlin 1980) seem to have specific benthic feeding tactics. In these conditions, marine mammals might come in direct contact with the prey and could even actively touch the prey item. *Active touch* is mediated by claws or by vibrissae in the facial region (for a review, see for example, Dehnhardt 2002; Dehnhardt and Mauck 2008; Dehnhardt et al. 2014). Almost all marine mammals possess facial vibrissae (see Section 12.2.3). In a number of marine mammal species, these vibrissae are able to identify objects and specific object parameters (see Section 12.2.4) and could consequently be used during benthic foraging. However, how can marine mammals also detect and hunt prey in the pelagic zone with their vibrissae?

Every fish that is swimming through the water is generating water disturbances and this way leaves a so-called *hydrodynamic trail* behind itself (Figure 12.2c,d; Hanke et al. 2000; Hanke and Bleckmann 2004). On the basis of the finding that harbor seals are sensitive to vibrations when a rod generating the vibrations was directly in contact with the vibrissae (Renouf 1979; Mills and Renouf 1986), a hydrodynamic function of the vibrissae of harbor seals was established when a visually and acoustically masked seal was asked to sense dipole water movements generated by a sinusoidally oscillating sphere (Figure 12.2a; Dehnhardt et al. 1998). Hydrodynamic dipole stimuli are best suited for the physiological characterization of the sensory system in terms of *detection* and/or *difference thresholds* (see Section 12.3.2). The seal achieved the lowest detection threshold for dipole water movements with velocities of 245 μm/s at 50 Hz (Figure 12.2b; Dehnhardt et al. 1998), which compares well to the sensitivity of the *lateral line* of some decapods and marine teleosts (Bleckmann 1994). The detection thresholds of a California sea lion (*Zalophus californianus*) measured at 20 and 30 Hz was even lower than the highest sensitivity of the harbor seal (Dehnhardt and Mauck 2008), and the manatee's (*Trichechus manatus latirostris*) exceptional sensitivity reached 1 μm/s at 150 Hz (Gaspard et al. 2013). Thus, the vibrissae of the marine mammals tested so far indeed function as very sensitive hydrodynamic receptors.

However, to show that an animal can indeed use a certain type of sensory information as suggested by psychophysical studies, experiments designed in a sensory ecology approach are required. If a seal wants to use its vibrissae for prey tracking, it has to be able to not only detect hydrodynamic events but also to actively follow hydrodynamic trails (Figure 12.2c,d). The ability of harbor seals to track such hydrodynamic trails was

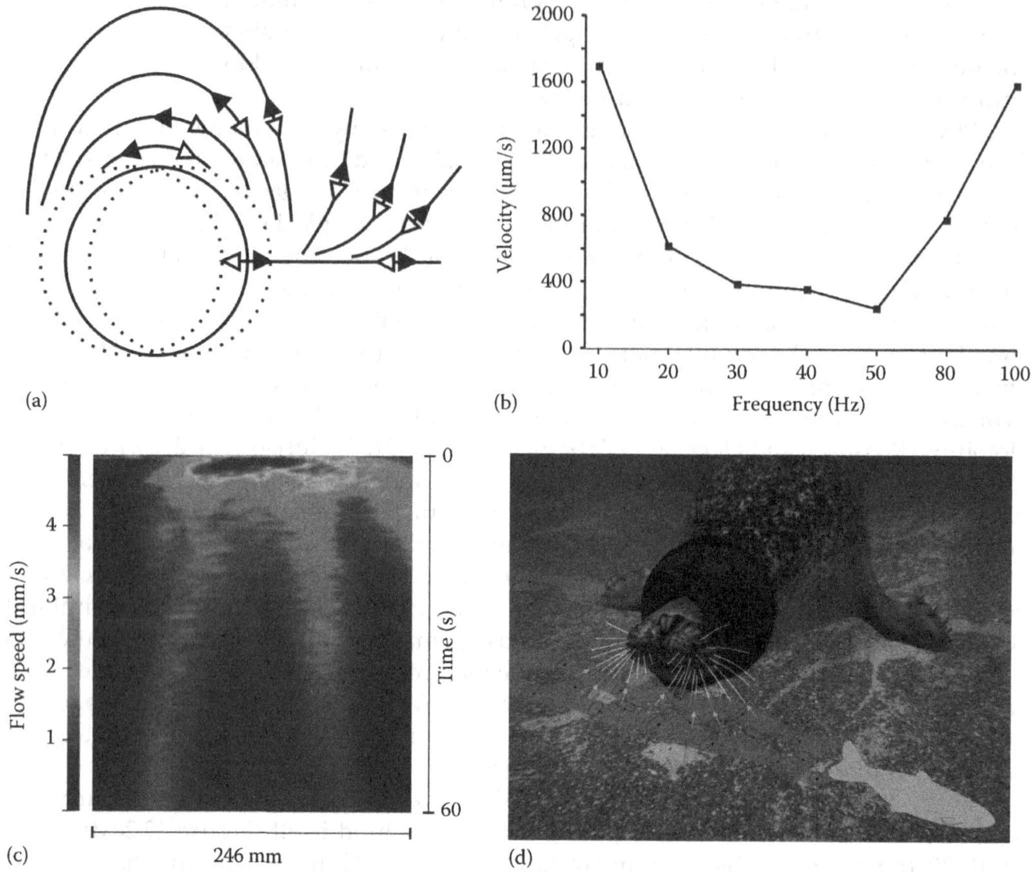

Figure 12.2 **(See color insert.)** Hydrodynamic flow perception. (a) Water movements generated by a dipole as usually used to examine hydrodynamic receptors. (Modified from Tautz, J., *Medienbewegung in der Sinneswelt der Arthropoden—Fallstudien zu einer Sinnesökologie,* Gutsav Fischer Verlag, Stuttgart, Germany, 1989.) Movement vectors are displayed in the vicinity of a dipol (oscillating sphere), the amplitude of oscillation is indicated by the arrows and the dotted lines. (b) Performance of a harbor seal during hydrodynamic flow testing. (Modified from Dehnhardt, G. et al., *Nature,* 394, 235, 1998.) The seal's performance is depicted as velocity threshold (in μm/s) as a function of the frequency (in Hz) of the stimulus—see (a). The seal achieved its best threshold performance of 245 μm/s at 50 Hz. (c) Velocity profile of the wake behind a fish (*Lepomis gibbosus*). (Modified from Hanke, W. and Bleckmann, H., *J. Exp. Biol.,* 207, 1585, 2004.) The velocity profile was obtained from particle image velocimetry (PIV) (see Section 12.3.3). Flow speeds within a 246 mm region of interest are color-coded from blue to red corresponding to water velocities from 0 to 5 mm/s. The fish swam from the bottom to the top of the figure, thus the flow speeds at the top are measured in the wake directly behind the fish, whereas the flow velocities at the bottom of the figure depict the flow speeds that persist after 60 s. (d) A visually masked harbor seal is encountering a hydrodynamic trail of a fish. The flow direction of the water particles in the hydrodynamic trail is indicated by arrows. (Modified from Bleckmann, H., Reception of hydrodynamic stimuli in aquatic and semi-aquatic animals, in *Progress in Zoology,* Rathmayer, W. (ed.), Gustav Fischer Verlag, Stuttgart, Germany, 1994.) The seal can gain information on the 3D structure of the hydrodynamic trail by multiple point-to-point measurements (indicated by yellow arrows).

investigated in a number of experiments. Hydrodynamic trails up to 40 m length were generated mainly by remote-controlled miniature submarines (Dehnhardt et al. 2001; Wieskotten et al. 2010a), but additionally by conspecifics (Schulte-Pelkum et al. 2007) or artificial fish (Kilian 2010) for comparison. All experiments revealed the seal's extraordinary trail-following abilities. They could not only follow the trails with high precision irrespective of trail generator, but even did so when they were allowed to follow the trail after the trail had aged by more than 20 s (Wieskotten et al. 2010a). Mimicking *burst-and-glide* swimming of some fish (Videler and Weihs 1982; Blake 1983; Hinch et al. 2002; Standen et al. 2004), the seals even performed well when the trail consisted of a phase in which the submarine was actively swimming and a phase in which it was only gliding (Wieskotten et al. 2010a). However, if the seal was unsuccessful in following this kind of trail, it lost the trail mainly in the transition zone between active propelled and gliding phase indicating that burst-and-glide swimming might be an effective antipredator strategy. In 2011, a comparative hydrodynamic trail-following study was conducted with a California sea lion (Gläser et al. 2011). The sea lion followed the trails successfully, however, its performance broke down rapidly when a delay was introduced between trail generation and start of trail-following. These differences might originate in the different morphology of the vibrissae of seals and sea lions (see Section 12.2.4).

Before actually starting a trail-following event, it is essential to know, for example, in which direction the fish that had generated the trail is swimming. Additionally, if a trail follower was able to analyze a hydrodynamic trail in detail, reading a generator's features such as size or form out of the trail, it could optimize foraging and could only go for the best prey item or it could also decide to swim into the opposite direction if the trail had been generated by a predator. These suggestions are based on the fact that the hydrodynamic trails generated by different fish species differ due to the fish's different size, shape, or swimming style (Hanke and Bleckmann 2004). Wieskotten et al. (2010b, 2011) approached these questions with harbor seals and generated hydrodynamic trails by a multitude of paddles. All these experiments were conducted in a box which enabled the generation of well-defined and measurable hydrodynamic events in calm waters. The seal indicated movement direction with high precision and with a contact time of vibrissae with the hydrodynamic trail of less than 0.5 s (Wieskotten et al. 2010b). Again, the seal was reliably answering even when the hydrodynamic event was up to 35 seconds old. Additionally, the seal discriminated objects differing in size and shape with high precision (Wieskotten et al. 2011). In conclusion, harbor seals can gain a multitude of information from reading a hydrodynamic trail.

With *particle image velocimetry* (PIV) (see Section 12.3.3), the seal's behavior during hydrodynamic perception as well as the generated trails were measured and analyzed which revealed various parameters that the seals could have potentially used to make their decisions. Among those, the vortices included in the trails seem to provide powerful information. This finding led to a series of hydrodynamic experiments in which it was and still is investigated which kind of information harbor seals can gain from single vortices (Dehnhardt et al. 2014). The results indicate that harbor seals are able to assess the travel direction of a single vortex ring irrespective of position of stimulation at the vibrissae, ipsi- or contralateral vibrissal pad. Furthermore, when presented with two vortex rings in succession they can judge these vortices on the basis of size (Krüger et al. 2014). In conclusion, the harbor seal's documented sensitivity to hydrodynamic events allows the seal to detect and follow prey even in dark and murky waters. Moreover, when encountering a hydrodynamic event, seals can interpret the events with high precision for example, comparable to human trackers reading and interpreting the traces left behind by animals or other humans.

With this knowledge in mind and returning to *benthic prey*, is there any hydrodynamic information that a marine mammal could use to detect even benthic prey? Every animal irrespective of lifestyle has to breathe. The *breathing current* expelled from the gills is causing water movements that are within the perception range of the vibrissal system of seals and sea lions (Bublitz 2010). Moreover, harbor seals seem to be able to localize artificially generated breathing currents over a large area (Niesterok et al. 2014). Thus the hydrodynamic detection of breathing currents of benthic fish might answer the question raised by Bowen et al. (2002), namely, which cues seals and marine mammals in general can rely on for a successful detection of benthic prey.

12.2 Knowledge by order

12.2.1 The visual system: Adaptation to low light levels

The marine mammal eye is of general vertebrate eye bauplan (Walls 1942), however, specific adaptations to the visual ecology and lifestyle of a particular species are apparent. A common trait of the eyes of marine mammals is that they have evolved many adaptations to increase sensitivity as a consequence of low light levels dominating their visual environment—important to mention, the eyes of manatees, sea otters, and polar bears (*Ursus maritimus*) share only some of these adaptations probably as manatees and polar bears inhabit light-rich environments and sea otters are phylogenetically young semi-aquatic animals with eyes still largely resembling eyes of terrestrial mammals.

First, an effective means of increasing sensitivity is to increase pupil diameter. And as the pupil diameter cannot exceed eye size, a further increase in sensitivity can only be achieved by increasing eye size (Land and Nilsson 2002). And indeed, most marine mammals possess large eyes; the largest extant eyes in vertebrates can actually be found in cetaceans (Walls 1942). Their eyes are also larger in comparison to their terrestrial relatives (Debey and Pyenson 2013). And moreover, among the pinnipeds, the pupil can dilate up to an area of 422 mm^2 as measured in northern elephant seals (*Mirounga angustirostris*) in dim light (Levenson and Schusterman 1997) which is more than five times larger than the fully dilated pupil of a young human (Rogers 2011). Second, the retina is dominated by *rods*. Generally, *rods* are more sensitive to light than *cones* as they contain a very light-sensitive pigment, the rhodopsin, they collect photons over longer time periods (temporal summation), and they pool photons over a larger area (spatial summation) due to a higher convergence. The ringed seal (*Phoca hispida*) retina only contains approximately 1.5% cones with a mean cone/rod ratio of 1:64 (Peichl and Moutairou 1998) in comparison to approximately 5% cones and a cone/rod ratio of 1:19 in humans (Jonas et al. 1992). Pronouncedly in whales, the visual information is spatially pooled in large receptive fields of giant ganglion cells that also have a very low density across the retina (Mass and Supin 2007; Reuter and Peichl 2008). And third, a thick tapetum (Pütter 1903; Walls 1942; Nagy and Ronald 1970; Jamieson and Fisher 1971; Young et al. 1988) can be found behind the inversely organized retina. This thick cell layer is reflecting photons that were not absorbed during the first passage of the retina and thus enables photon absorption during the second passage of the retina. According to Walls (1942), the tapeta of seals are the thickest tapeta found in the animal kingdom. In some marine mammals (see for example, Johnson 1901; Dawson et al. 1987), the tapetum backs the whole ocular fundus and is not restricted to the ventral fundus as in terrestrial carnivores. This way, the tapeta can increase the probability of photon absorbance irrespective of the direction from which the photon is entering the eye. The omnidirectionality of the tapeta might have evolved in marine mammals as a response to the 3D underwater world in which a marine mammal can translate but also rotate in any orientation and in which photons thus might arrive at the eye from above, below, or from the sides.

Marine mammal eyes not only possess adaptations to increase sensitivity, but their eyes can also quickly adapt to low light levels. Levenson and Schusterman (1999) could show that pinnipeds can dark-adapt within several minutes in comparison to human eyes that require more than 20 min to reach maximum sensitivity in darkness. Furthermore, the rate of dark adaptation seems to correlate with diving depth as the northern elephant seal dark adapts within only 6 min which corresponds with the time required to reach depths of approximately 1500 m to which they were found to regularly dive (Le Boeuf and Laws 1994). In contrast, it took the eyes of the shallow-diving harbor seals 18 min to dark-adapt and humans even 22 min (Levenson and Schusterman 1999). It is also noteworthy that a positive relationship between diving depth and orbit size has also been discovered recently (Debey and Pyenson 2013).

12.2.2 *Basis for optic flow perception*

Among the marine mammals, motion vision has so far only been investigated in harbor seals. The first optic flow experiment in harbor seals (Gläser et al. 2014) was based on two motion vision studies. Research in this field started in 2008, when Hanke et al. (2008) investigated *optokinetic eye movements* in harbor seals (Figure 12.3a). The optokinetic nystagmus describes a basic motion stabilizing reflex consisting of a *pursuit eye movement* during which the eye follows the whole-field motion stimulus. Optimally, the visual motion is canceled during the pursuit eye movement which is the prerequisite for a sharp image on the retina. In order to re-center the eye in the orbit, the eye performs a fast *saccade* against stimulus direction after the pursuit eye movement. Pursuit eye movement and saccade alternate as long as there is a visual motion stimulus. Hanke et al. (2008) showed that harbor seals possess optokinetic eye movements and that the eye stabilizes image motion equally well irrespective of stimulus movement direction. The latter phenomenon has never been described before and might be an adaptation to the low structured, 3D underwater world in which harbor seals operate and in which important visual information might arrive at the eye from any direction. In conclusion, this study showed that harbor seals are indeed able to perceive visual motion.

In a follow-up experiment, it was determined how sensitive harbor seals are in respect to global motion stimuli (Weiffen et al. 2014). For his purpose, the seal was presented with large *random dot* displays in which a specific number of dots are displayed in the display area (Figure 12.3b). At the extremes, these dots move either randomly (0% coherence) or all dots move into one direction (100% coherence). A threshold of motion sensitivity is achieved by varying the coherence of the dots. When presented with random dot displays, the seal quickly learned the task with these complex stimuli, and it was able to detect the display with *coherent motion* out of two displays with the lowest threshold amounting to 4.7% coherence (Weiffen et al. 2014), a sensitivity to coherent motion equivalent to the highest sensitivity assessed so far in humans and monkeys (see for example, Newsome and Paré 1988). However, the experimental conditions used to assess this threshold might have allowed the seal to use secondary cues that were eliminated in the second phase of the experiment. In this phase, the seal's threshold was determined at 23% coherence.

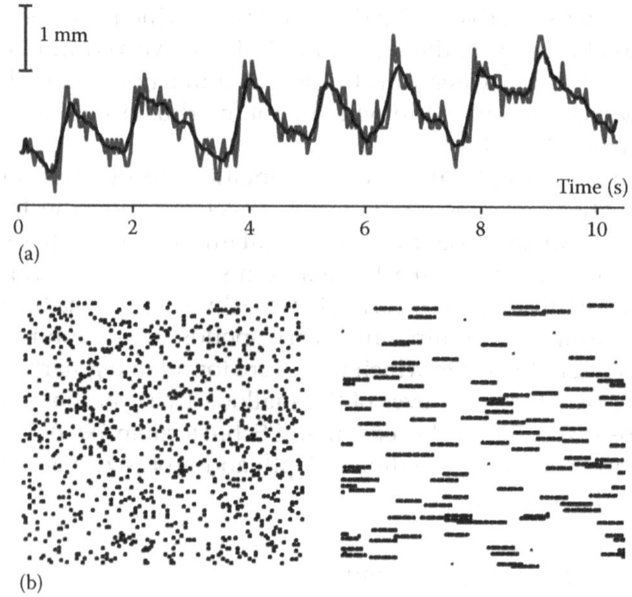

Figure 12.3 Basis for optic flow perception. (a) Characteristic movement of the eye in the head over time during optokinetic stimulation. (Modified from Hanke, F.D. et al., *Vis. Res.*, 48(2), 304, 2008.) The eye makes a following movement to the right (downward movement in trace), during which the external motion is optimally nullified, succeeded by a fast saccade in the opposite direction to re-center the eye in the orbit. The gray line represents the changes in position of one point on the eye over time and noise is due to errors in tracing this specific point. The dark line was obtained by a moving average filter in order to approach the real movement of the eye. Scale 1 mm. (b) Random dot stimulus used during the assessment of the harbor seal's sensitivity to coherent motion. (Modified from Weiffen, M. et al., *SpringerPlus*, 3, 688, 2014.) In the display on the right side, all dots move into the same direction which means that they move at 100% coherence. In contrast, in the display on the left side, all dots move randomly at 0% coherence. Noise was programmed as random position noise in the final phase of the experiment in order to reduce local motion events (Scase et al. 1996). For the visualization of dot movement, the movement of the dots was averaged over 10 frames.

This coherence threshold is comparable to most other terrestrial species tested so far (see for example, Bischof et al. 1999; Douglas et al. 2006). It remains to be determined if, with training, harbor seals could also achieve higher motion sensitivity under the second experimental conditions. Nevertheless, the suite of experiments showed that harbor seals can achieve motion sensitivity at least as high as most vertebrates. Moreover, they seem to possess a global motion integration system which means that they can integrate local motion events such as the motion of single dots into a global percept. Global motion stimuli are most likely ecologically relevant to harbor seals. In the context of foraging, harbor seals could benefit from a global motion integration system and sensitivity to coherent motion when hunting pelagic as well as benthic fish. When hunting pelagic fish, such as a school of herring, sensitivity to coherent global motion would enable the seals to assess the motion direction of a school or the opposite to assess fish moving in a deviant manner that, consequently, might be easy prey objects. Furthermore, when hunting benthic fish such as flatfish, harbor seals could identify the cryptically hiding fish by the coherent movement of their body parts (Lui et al. 2012).

12.2.3 The vibrissal system: Adaptations to the aquatic medium

Vibrissae in the facial region can be found in many marine mammals. In sirenians, vibrissae additionally cover the whole body (Bryden et al. 1978; Reep et al. 2002, 2011), and in baleen whales, vibrissae can also be found around the blowholes (Japha 1907, 1912; Nakai and Shida 1948; Ling 1977). Toothed whales possess vibrissae on the jaw, however, they lose them already prenatally or shortly after birth (Ling 1977). Amazon River dolphins are an exception as, in this species, adults still have bristles on the jaw (Kükenthal 1909; Layne and Caldwell 1964). However, these hairs more resemble *guard hairs* that are found in the body fur of mammals. The river dolphin's bristles might also be richly innervated but lack the *follicle-sinus complex* of vibrissae. In some odontocetes, the follicle crypts that remain after the vibrissae are lost are large in diameter, and recent investigations in the Guiana dolphin (*Sotalia guianensis*) have described these empty crypts as functional electrosensitive organs (Mauck et al. 2000; Czech-Damal et al. 2011).

Vibrissal hairs are anchored in follicles. The whole unit is usually referred to as vibrissal follicle-sinus complex (F-SC; Rice et al. 1986). Due to this feature, vibrissae are also called sinus-hairs. Detailed anatomical investigations of the F-SCs of marine mammals are available for some pinnipeds (Stevens et al. 1973; Hyvärinen and Katajisto 1984; Hyvärinen 1989; Marshall et al. 2006; Hyvärinen et al. 2009) and sirenians (Reep et al. 2001) and revealed specific adaptations which can be linked to their function in the aquatic environment. In comparison to terrestrial species such as the cat (*Felis catus*; Ebara et al. 2002), the mystacial F-SCs of pinnipeds are much larger as their F-SC possess an additional blood sinus, the upper cavernous sinus. Its length accounts for 60% of the total F-SC length. The existence of an additional cavernous sinus probably leads to elevated surface temperatures measured at the vibrissae of, for example, harbor seals (Mauck et al. 2000) and maybe also in Cape fur seals (*Arctocephallus pusillus pusillus*) (Erdsack et al. 2014). This way the mechanoreceptors at the ring sinus are thermally shielded from low external temperatures, and, consequently, the vibrissae can retain high tactile sensitivity even in cold waters (Dehnhardt et al. 1998). In cold waters, low-melting-point monoenoic fatty acids in the adipose tissue around the vibrissae render the tissue very flexible ensuring high vibrissal mobility (Käkelä and Hyvärinen 1993, 1996). Furthermore, the innervation of the F-SCs of pinnipeds and sirenians by the deep vibrissal nerve is much higher in comparison to terrestrial species which might point to the significance of the vibrissal system in the aquatic medium. In ringed, bearded seals (*Erignathus barbatus*) and northern elephant seals, for example, 1000–1600 axons transmit the sensory information from one F-SC to the brain (Hyvärinen and Katajisto 1984; Hyvärinen 1989; Marshall et al. 2006; Hyvärinen et al. 2009; McGovern et al. 2015), thus the innervation density of these pinnipeds is 10 times higher as in cats or rats (Ebara et al. 2002).

12.2.4 Basis for hydrodynamic flow perception

Hydrodynamic flow perception investigated in harbor seals, the California sea lion, and the Florida manatee was preceded by a number of studies examining the haptic abilities of pinnipeds and sireneans whereas comparable studies in cetaceans are still missing. Dykes (1975) first concluded from his single unit recordings from the infraorbital branch of the Nervus trigeminus of harbor seals and gray seals (*Halichoerus grypus*) that the vibrissal function is active touch. Research on the haptic abilities of the walrus (*Odobenus rosmarus divergens*; Kastelein and van Gaalen 1988; Kastelein et al. 1990), the California sea lion (Dehnhardt 1990, 1994; Dehnhardt and Dücker 1996), and the manatee (*Trichechus manatus latirostris*; Bachteler and Dehnhardt 1999; Bauer et al. 2012) revealed that these animals are able to discriminate objects on the basis of their form, texture or size only by means of their vibrissae. Harbor seals can perform discriminations of size and texture equivalent to the other marine mammals, and their vibrissal system functions in air and underwater with the same precision and irrespective of ambient temperature (Dehnhardt and Kaminski 1995; Dehnhardt et al. 1997, 1998; Grant et al. 2013). Altogether these studies reveal abilities in marine mammals that compare to, for example, the tactile abilities of terrestrial species.

As movement between the touch organ and the inspected object is characteristic for any active touch process, pinnipeds usually perform head movements in order to judge object parameters while their vibrissae are protruded to the most frontal position but remain motionless (Dehnhardt 1994). The vibrissae of harbor seals also take this frontal and motionless position during hydrodynamic trail-following (Hanke et al. 2010). This surprises as hydrodynamic vortices should shed from the vibrissae thus leading to vibrations. However, measurements with particle image velocimetry (see Section 12.3.3) and *numerical simulations* revealed that, in contrast to the vibrissae of a sea lion, the vortices shedding from the vibrissae of a harbor seal are directly destroyed and thus fluctuating lift and drag forces acting on the vibrissae are maximally reduced (Figure 12.4a; Hanke et al. 2010). This effect is caused by the different morphology of the vibrissae. The vibrissae of most Phocidae are flattened in one direction and undulated in the other (Figure 12.4b; Watkins and Wartzok 1985; Hyvärinen 1989; Dehnhardt and Kaminski 1995; Ginter et al. 2012). Exceptions hereby are the vibrissae of the bearded seal and the monk seals (Monachinae). These species as well as all eared seals (Otariidae) and walruses possess vibrissae that are oval in diameter and smooth in outline. The undulated shape of, for example, harbor seal vibrissae seems to lead to an almost motionless sensor even when the seal is swimming fast. Thus, the external hydrodynamic event

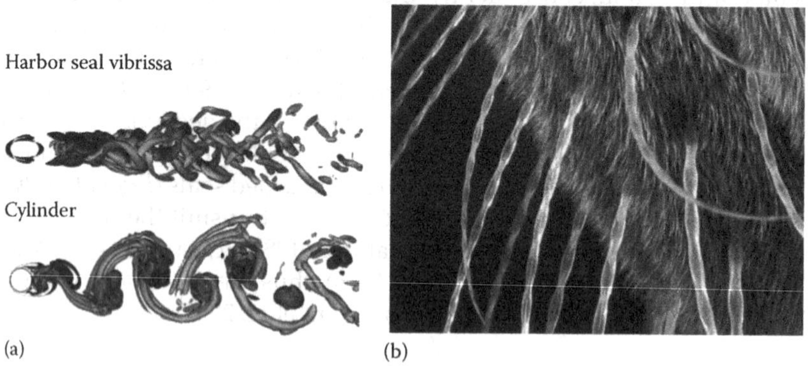

Figure 12.4 **(See color insert.)** Morphology of harbor seal vibrissae and their behavior in flow. (a) Wake flow behind a vibrissae of a harbor seal and behind a circular cylinder as obtained from numerical simulations. (Modified after Hanke, W. et al., *J. Exp. Biol.*, 213, 2665, 2010.) Behind a vibrissa, a complex 3D vortex structure generates downstream from the vibrissa leaving a gap between the vibrissa and the region with fluctuating vortex flows. Furthermore, the complex vortex structure is not stable over time. In contrast, behind a circular cylinder, primary vortices regularly shed directly from the cylinder (Kármán street) and largely persist over time. (b) Close-up view of a snout of a harbor seal showing the characteristic morphology of the vibrissae of most phocids. Most phocids possess vibrissae that are flattened and possess an undulatory shape.

can be directly measured without the need of extracting the signal from noise (Miersch et al. 2011). In contrast, noise is considerable when the vibrissae of sea lions detect water movements. However, the noise signal contains a dominant frequency that is modulated by a hydrodynamic event. As the dominant frequency was found to correlate with flow velocity, the sea lion can extract flow velocity from the dominant frequency contained in the signal measured by the vibrissae (Miersch et al. 2011). Consequently, as already shown (Gläser et al. 2011), sea lions are able to detect and follow hydrodynamic trails. However, their performance when encountering trials that have aged might be worse in comparison to harbor seals due to differences inherent in the two types of hydrodynamic sensors.

12.3 Toolbox

Sensory research of the last decades involving marine mammals has shown that there are still many discoveries to be made. The basis for these discoveries is in our opinion one central and important method which is "seeing/observing-perceiving-thinking." Thus we would like to encourage everybody to use this central method in order to develop and finally work on ideas that even seem to be "crazy" at first thought (see e.g., Czech-Damal et al. 2011).

12.3.1 Operant conditioning

Sensory abilities are often analyzed in behavioral experiments for which the experimental subjects need to learn specific tasks. Here fore, the subjects are trained with the help of *operant conditioning* (Skinner 1938). The basis for operant conditioning is that the experimental subject is behaviorally active and that it learns from the consequences its behavior led to. Thus upon showing a specific behavior, which initially might only happen incidentally, the specific behavior is reinforced whereas others have no consequences or are negatively reinforced or even punished. Over time, the frequency with which a subject shows the desired behavior will increase, the animal learns, alters its behavior from the consequences of its behaviors.

When training mammals such as seals, *positive reinforcement* is sufficient for successful learning (Pryor 1984; Ramirez 1999). Positive reinforcement usually consists of a *primary reinforcer*, a piece of food, and a *secondary reinforcer*, such as a whistle. The application of a secondary reinforcer is essential for the precise timing of the reinforcement which might be difficult with a primary reinforcer under some conditions, and furthermore bridges the gap until the primary reinforcer can be delivered. A secondary reinforcer needs to be established using *classical conditioning* (Pavlov 1902).

12.3.2 Psychophysics

Sensory abilities of an organism are mainly investigated in psychophysical experiments. The aim of psychophysics is to link sensation and perception with physical stimuli (Gescheider 1997). The physical stimuli possess parameters that can/have to be measured and systematically changed. Thus a good access to physics and measurement techniques is a prerequisite for a good psychophysicist. The well-controllable and measurable stimuli allow to assess sensory thresholds that can be determined as *absolute thresholds* defined as the smallest amount of stimulus energy necessary to produce a sensation or as difference threshold defined as the amount of change in a stimulus required to produce a just noticeable difference in sensation (Gescheider 1997).

In order to obtain a sensory threshold, two main experimental methods are usually applied in marine mammal research, the *staircase method/method of limits* or the *method of constant stimuli* (Gescheider 1997; Ehrenstein and Ehrenstein 1999). Using the staircase method, the intensity of the stimulus is decreased step by step by one unit until the experimental subject ceases to respond correctly. Whenever the subject answers incorrectly, defined as a transition point, the stimulus intensity is again increased by one unit. Thresholds are then calculated by averaging over approximately six transition points. A threshold can also be approached starting with subthreshold intensity values. However, with respect to the test subject's motivation a descending series offers the advantage that the session starts with suprathreshold intensity values that are easily detectable. The staircase method allows determining a threshold within one session which is very effective and sometimes even necessary if, for example, changes in the environment need to be considered (see for example, Weiffen et al. 2006). However, the thresholds obtained are often not the lowest as the experimental subjects rapidly learn that if they respond incorrectly, the task will be easier in the following trial.

Using the method of constant stimuli, a set of stimuli is selected before data collection with stimulus intensities well above and below the putative threshold. This set of stimuli is repeatedly presented to the experimental subject and its response is noted. Finally, the threshold can be calculated from the *psychometric function* that depicts the performance (in %) of the subject averaged over all presentations as a function of the stimulus intensity. Threshold estimation with the method of constant stimuli are more time-consuming than the estimation with a staircase method as several sessions are needed before data collection can be completed. However, the thresholds are usually lower in comparison to those obtained with a staircase method as the subject cannot develop expectancies.

In psychophysical experiments, two experimental paradigms are generally used to present the stimuli, a go/no go-paradigm or a *two-alternative-forced-choice-paradigm*. In a *go/no go-experiment*, two stimulus categories are defined, the go-stimulus and the no go-stimulus. When confronted with the go-stimulus, the experimental subject has to give an answer by, for example, moving its head to a response target; when confronted with the no go-stimulus, the subject needs to stay stationary for a predefined time interval. Stimulus presentation is thus successively, and four responses are possible: a hit (go-stimulus and go-response), a miss (go-stimulus but no go-response), a correct rejection (no go-stimulus and no go-response), and a false alarm (no go-stimulus but go-response). As a high hit rate can be achieved by always answering with a go-response, during analysis, hits and false alarms need to be considered, and it is required to document a low false alarm rate to make the data included in the hit rate valid. Using a go/no go-paradigm additionally requires a perfect control of the time interval as some subjects tend to give a go-response in the final milliseconds of stimulus presentation of the no go-time interval when stimulus intensity is close to threshold.

In a two-alternative-forced-choice (2AFC) paradigm, the experimental subject always has to discriminate between two stimuli which can be presented either simultaneously or successively. In each trial, the subject is forced to give a predefined response such as indicating the position of the positive stimulus (S+), which is followed by reinforcement, whereas choosing the negative stimulus (S−) is not reinforced. In a 2AFC experiment, side preferences need to be documented and counteracted by an equal number of trials with the S+ on either side (Gellermann 1933). Generally, the 2AFC procedure faster leads to a threshold in comparison to the go/no go-paradigm as in the latter only the performance in respect to one stimulus category, usually the go-stimuli, is analyzed to determine a threshold. However, double the number of trials is needed including the no go-stimuli as go- and no go-stimuli need to be balanced in order to be able to approach the threshold.

Please note that psychophysical experiments alone cannot describe the sensory abilities of an organism in total. Additional application experiments such as the hydrodynamic trail-following experiment (Dehnhardt et al. 2001) need to be conducted to show what the animals use their sense for and how they use it in their environment.

12.3.3 Particle image velocimetry

Particle image velocimetry (PIV) measurements are fundamental to all hydrodynamic experiments in order to control qualitative and quantitative aspects of the water movements used as stimuli. With the help of PIV, water movements can be visualized and water velocities can be measured, also over time (Figure 12.2c). During PIV measurements, tracer particles are added to the water. These particles are neutrally buyont and thus keep position unless a water movement is generated. A fanned-out laser is illuminating the particles in one layer. A top view video camera is filming the light reflected from particles in the region of interest (ROI) first without (background flow) and then with hydrodynamic event such as induced by, for example, a paddle that is moved from left to right through the ROI. The video raw material can then be analyzed offline in the PIV software DaVis 7.2 (LaVision GmbH, Göttingen, Germany). With this software, velocity vectors are calculated from correlating the displacement of tracer particles in subsequent images (Keane and Adrian 1992).

12.3.4 Challenges associated with the investigation of senses underwater

A big challenge when investigating sensory abilities of (semi)aquatic mammals is that often experiments need to be conducted underwater. Consequently, the equipment needs to be water-proof which renders every purchase cost-intensive and often setups need to be custom built as off-the-shelf setups are not available. Furthermore, setups usually have to be quite large when working with marine mammals thus space might be a limiting factor.

The presentation of optic flow stimuli underwater requires large projection screens or monitors. Generally, for the presentation of high-quality stimuli, stimulus presentation has to be in a shaded area, ideally in an experimental chamber (Hanke et al. 2008; Gläser et al. 2014; Weiffen et al. 2014). For using underwater back projections on a projection screen, the projector needs to be lowered in a water-proof housing. Alternatively, the beamer stays in air but then the water surface needs to be calmed in order to avoid distortion of the image by surface waves. Still the stimuli might be affected by water quality, even if experiments are conducted in an aquarium setting, as, for example, dissolved particles can lower contrast and intensity. This phenomenon can be reduced to some extent when monitors are applied as then light is only attenuated between experimental subject and monitor and not additionally in between projection screen and beamer.

Testing underwater usually requires one or multiple cameras to record the animal's behavior. For optic flow experiments a camera is needed as stimulus presentation needs to take place in a shaded area. In hydrodynamic experiments, especially in the application experiments, a top view camera is essential to be able to analyze the animal's and the trail generator's trails. However, as a large area needs to be overseen the camera has to be installed high above the experimental area which is challenging.

Alternatively, trails from trail generator and experimental subject need to be recorded with tracking devices attached to the generator or subject. Generally with these tracking devices, experimenters face the problem that usually these devices only send signals when the subject or trail generator surfaces. Thus, reliable tracking instruments need to be developed that can operate even when attached to an object fully and constantly submerged.

12.4 Unsolved mysteries

Most experiments reported in this chapter have been conducted with harbor seals as model organism. Thus, the question arises if other marine mammals possess comparable sensory abilities. However, the generalization from harbor seals to other marine mammals even to closely related species is complicated by the fact that all species differ in many aspects and show specific adaptations to their environment and lifestyle. To give an example, the vibrissae of harbor seals are undulated in shape whereas otariids and even some phocids possess smooth vibrissae (see for example, Hanke et al. 2010). Or, even though a number of studies have analyzed the vibrissal follicles in various species many structures within the follicle vary interspecifically and often their function is unknown. In conclusion, visual as well as hydrodynamic experiments comparing the performance of other marine mammals to the documented performance of harbor seals in visual and hydrodynamic flow perception experiments are required.

The finding of optic flow perception in harbor seals will influence research in the field of marine mammal science and underwater locomotion and orientation. It forces researchers to rethink underwater locomotion and orientation in general as optic flow is, for example, induced by movement through dissolved particles that have previously been listed in support of the view that underwater visual orientation is often restricted if not impossible. Current research has focused on optic flow induced by movement through particles. In the future, it awaits determination if seals are also able to use the optic flow induced by movement over the ground or underneath the water surface. And, as the first optic flow experiment was assessing optic flow sensitivity in response to a simulation, it needs to be determined how and what for optic flow is used in marine mammals.

Research on underwater optic flow perception will significantly enhance vision science generally since specifics of the underwater situation (Gläser et al. 2014) are likely to require novel strategies for optic flow analysis. In addition to a pure visual solution of these problems, an analysis of the integration of information from *hydrodynamics* and vision can bring the field forwards. In detail, one challenge is that harbor seals have to cope with optic flow induced by external water movements such as currents. The question thus is if harbor seals are able to analyze optic flow fields in the presence of drift. If harbor seals were not able to solve the task using vision alone, they might be able to sense the external water movement with the help of their vibrissae. By integrating visual and tactile flow information they might then be able to subtract the optic flow induced by external water movements from the overall flow pattern. Generally, the vibrissal system faces a comparable problem; a trail generated by a fish is masked by medium movement noise (Dehnhardt and Mauck 2008). This problem will also have to be addressed in future experiments.

Experimental results from optic flow as well as from hydrodynamic experiments will and already have an impact on underwater robotics, for example, for controlling movement in three-dimensional flow of remotely operating vehicles (ROVs).

Glossary

Absolute threshold: The smallest amount of stimulus energy to produce a sensation.

Active touch: Active tactile exploration of objects. Tactile perception is mediated by pressure, skin stretch, vibration, and temperature sensors.

Benthic prey: Prey that lives close to, on, or in the seabed.

Bioluminescence: Some organisms are able to produce and emit light either themselves or with the help of bioluminescent bacteria.

Breathing current: A breathing current is generated by water being expelled from the gill opening of a fish.

Burst-and-glide swimming: A type of swimming in which undulatory bursts are alternating with periods in which the body is held straight and the organism is gliding forward.

Ciliary body: The ciliary muscle and epithelium are called the ciliary body. The ciliary muscle is connected via the zonular fibers to the lens and mediates accommodation. The ciliary body is furthermore involved in the production and resorption of the aqueous humor.

Classical conditioning: As a result of classical conditioning, a previously neutral stimulus (e.g., light) elicits the response (salivation) associated with an unconditioned stimulus (food). This is achieved by repeatedly pairing the neutral stimulus with the unconditioned stimulus.

Coherent motion: When objects move coherently, they move into the same direction. Coherence can vary between 100% meaning that all objects in a display move into the same direction or 0% meaning that all objects move randomly.

Dark adaptation: Dark adaptation is the ability of the eye to adjust its sensitivity to dim light conditions.

Difference threshold: The amount of change in a stimulus required to produce a just noticeable difference in stimuli.

Focus of expansion: The singular point from which all visual motion seems to emanate during movement of an observer. In pure translations, the heading corresponds to the focus of expansion.

Follicle sinus complex: The unit of the vibrissa and its follicle is called follicle sinus complex.

Go/no go-experiment: In a go/no go-experiment, the animal has to wait for a specific time interval when confronted with the no go-stimulus, but it has to give a response within the respective time interval when confronted with the go-stimulus.

Guard hairs: Guard hairs form one layer of the fur. Guard hairs are long hairs that stick out of the undercoat, the bottom layer of the fur.

Hydrodynamic trail: The water disturbances caused by every object that is pulled or actively swimming through the water.

Hydrodynamics: Hydrodynamics is the ability of an organism to detect water movements.

Lateral line: A sensory organ of aquatic vertebrates that is able to detect water movements and vibrations.

Method of constant stimuli: A predefined set of stimuli spanning the threshold is repeatedly presented to the experimental subject.

Numerical simulation: Numerical simulation is a technique of computational fluid dynamics.

Operant conditioning: Operant conditioning describes a method of learning during which an organism learns through the consequences of its own behavior.

Optic flow: The pattern of visual motion elicited on the retina of a moving observer.

Optokinetic eye movements: The optokinetic nystagmus is a basic motion stabilizing reflex. During optokinetic stimulation, the eye makes a pursuit eye movement in direction of the motion stimulus during which the external motion is ideally nullified. The pursuit eye movement is followed by a fast saccade against stimulus direction in order to re-center the eye in the orbit.

Particle image velocimetry: Particle image velocimetry is a technique to visualize water movements and to measure water velocities.

Path integration: Path integration describes a process during which an organism is continuously documenting distances traveled and directions taken during a trip which provides an estimate of its current position relative to a start position and allows integrating a homing vector which will lead the organism back to its starting point.

Pelagic prey: Prey that lives in the free water column.

Positive reinforcement: An event that increases the probability of a response to reoccur in the future.

Primary reinforcer: A stimulus that is inherently reinforcing as it satisfies biological needs such as hunger.

Psychometric function: A psychometric function is plotting the correct choices (mostly in percent) as a function of the stimulus parameter that is varied during testing.

Psychophysics: Psychophysics is the discipline that links sensation and perception to physical stimuli.

Pupil/pupillary opening: The pupil is the aperture in the iris that allows the light to enter the eye.

Pursuit eye movements: Pursuit eye movements function to closely follow a moving object.

Random dot display: In a random dot display, a specific number of dots is displayed that can either move completely randomly or with a certain percentage of coherence.

Retina: Photosensitive layer in the backside of the eye which transforms light into nerve impulses.

Rhodopsin: Rhodopsin is one of the visual pigments within the photoreceptor cells of the retina.

Rod, cone: Rods and cones are photoreceptors found within the retina that absorb photons. Rods are very sensitive and function in dim light (scotopic vision), whereas cones are less sensitive and function in bright light (photobic vision). Usually color vision is based on the presence of different cone types.

Saccade: A fast (up to 1000°/s) eye movement.

Secondary reinforcer: A stimulus that is associated with the primary reinforcer and thus acquires reinforcing properties.

Staircase method/method of limits: In a staircase method, the intensity of the stimulus is decreased step by step by one unit until the experimental subject ceases to respond correctly. At this point, the transition point, stimulus intensity is again increased by one unit until the experimental subject starts to answer correctly.

Tapetum: Reflective layer behind the retina that reflects photons that were not absorbed by the retina on the first passage back enabling absorption during the second passage of the retina. The tapetum thus increases the sensitivity of the eye.

Two-alternative-forced choice experiment: In a two-alternative forced choice experiment, the experimental subject has to discriminate between two stimuli. Its task is to always choose the positive stimulus but reject/ignore the negative stimulus.

Vision: Vision is mediated by the eyes and describes an organism's ability to process the information contained in light.

References

Bachteler, J. and G. Dehnhardt. 1999. Active touch performance in the Antillean manatee: Evidence for a functional differentiation of facial tactile hairs. *Zoology (Jena)* 102:61–69.

Bauer, G.B., J.C. Gaspard, D.E. Colbert et al. 2012. Tactile discrimination of textures by Florida manatees (*Trichechus manatus latirostris*). *Marine Mammal Science* 28 (4):E456–E471.

Bischof, W.F., S.L. Reid, D.R.W. Wylie, and M.L. Spetch. 1999. Perception of coherent motion in random dot displays by pigeons and humans. *Perception & Psychophysics* 61 (6):1089–1101.

Blake, R.W. 1983. Functional design and burst-and-coast swimming in fishes. *Canadian Journal of Zoology* 61:2491–2494.

Bleckmann, H. 1994. Reception of hydrodynamic stimuli in aquatic and semi aquatic animals. In *Progress in Zoology*, ed. W. Rathmayer. Stuttgart, Germany: Gustav Fischer Verlag.

Bowen, W.D., D. Tully, D.J. Boness, B.M. Bulheier, and G.J. Marshall. 2002. Prey-dependent foraging tactics and prey profitability in a marine mammal. *Marine Ecology Progress Series* 244:235–245.

Britten, K.H. and R.J.A. van Wezel. 1998. Electrical microstimulation of cortical area MST biases heading perception in monkeys. *Nature Neuroscience* 1 (1):59–63.

Britten, K.H., and R.J.A. van Wezel. 2002. Area MST and heading perception in macaque monkeys. *Cerebral Cortex* 12:692–701.

Bryden, M.M., H. Marsh, and B.W. MacDonald. 1978. The skin and hair of the dugong, *Dugong dugong*. *Journal of Anatomy* 126:637–638.

Bublitz, A. 2010. Wasserbewegungen von stationären Fischen und ihre mögliche Bedeutung für fischfressende Tiere. Diploma thesis, Sensory and cognitive ecology. Rostock, Germany: University of Rostock.

Cutting, J.E., K. Springer, P.A. Braren, and S.H. Johnson. 1992. Wayfinding on foot from information in retinal, not optical, flow. *Journal of Experimental Psychology (General)* 121:41–72.

Czech-Damal, N., A. Liebschner, L. Miersch et al. 2011. Electroreception in the Guiana dolphin (*Sotalia guianensis*). *Proceedings of the Royal Society B: Biological Sciences* 279(1729):663–668. doi: 10.1098/rspb.2011.1127.

Dawson, W.W. 1980. The cetacean eye. In *Cetacean behavior: mechanisms and processes*, ed. L.M. Herman. New York: Wiley Interscience.

Dawson, W.W. 1980a. The cetacean eye. In *Cetacean Behavior: Mechanisms and Processes*, ed. L.M. Herman. New York: Academic Press.

Dawson, W.W. 1980b. The cetacean eye. In *Cetacean Behavior, Mechanisms and Function*, ed. L.M. Herman. New York: Wiley Interscience.

Dawson, W.W., M.N. Hawthorne, R.L. Jenkins, and R.T. Goldston. 1982. Giant neural systems in the inner retina and optic nerve of small whales. *Journal of Comparative Neurology* 205:1–7.

Dawson, W.W., J.P. Schroeder, and J.F. Dawson. 1987. The ocular fundus of two cetaceans. *Marine Mammal Science* 3(1):1–13.

Debey, L.B. and N.D. Pyenson. 2013. Osteological correlates and phylogenetic analysis of deep diving in living and extinct pinnipeds: What good are big eyes? *Marine Mammal Science* 29 (1):48–83.

de Gurjao, L.M., M.A. de Antrade Furtado Neto, R.A. dos Santos, and P. Cascon. 2003. Feeding habits of marine tucuxi, *Sotalia fluviatilis*, at Ceara State, northeastern Brazil. *Latin American Journal of Aquatic Mammals* 2:117–122.

Dehnhardt, G. 1990. Preliminary results from psychophysical studies on the tactile sensitivity in marine mammals. In *Sensory Abilities of Cetaceans*, eds. J.A. Thomas and R. Kastelein. New York: Plenum Press.

Dehnhardt, G. 1994. Tactile size discrimination by a California sea lion (*Zalophus californianus*) using its mystacial vibrissae. *Journal of Comparative Physiology A* 175:791–800.

Dehnhardt, G. 2002. Sensory systems. In *Marine Mammal Biology: An Evolutionary Approach*, ed. A.R. Hoelzel. Oxford, UK: Blackwell.

Dehnhardt, G. and G. Dücker. 1996. Tactual discrimination of size and shape by a California sea lion (*Zalophus californianus*). *Animal Learning & Behavior* 24 (4):366–374.

Dehnhardt, G., W. Hanke, S. Wieskotten, Y. Krüger, and L. Miersch. 2014. Hydrodynamic perception in seals and sea lions. In *Flow Sensing in Air and Water*, ed. H. Bleckmann. Berlin, Germany: Springer.

Dehnhardt, G. and A. Kaminski. 1995. Sensitivity of the mystacial vibrissae of harbour seals (*Phoca vitulina*) for size differences of actively touched objects. *Journal of Experimental Biology* 198:2317–2323.

Dehnhardt, G. and B. Mauck. 2008a. Mechanoreception in secondarily aquatic vertebrates. In *Sensory Evolution on the Threshold: Adaptations in Secondarily aquatic Vertebrates*, eds. J.G.M. Thewissen and S. Nummela. Berkeley, CA: University of California Press.

Dehnhardt, G. and B. Mauck. 2008b. The physics and physiology of mechanoreception. In *Sensory Evolution on the Threshold: Adaptations in Secondarily Aquatic Vertebrates*, eds. J.G.M. Thewissen and S. Nummela. Berkeley, CA: University of California Press.

Dehnhardt, G., B. Mauck, and H. Bleckmann. 1998. Seal whiskers detect water movements. *Nature* 394:235–236.

Dehnhardt, G., B. Mauck, W. Hanke, and H. Bleckmann. 2001. Hydrodynamic trail-following in harbor seals (*Phoca vitulina*). *Science* 293:102–104.

Dehnhardt, G., B. Mauck, and H. Hyvärinen. 1998. Ambient temperature does not affect tactile sensitivity of mystacial vibrissae in harbour seals. *Journal of Experimental Biology* 201:3023–3029.

Dehnhardt, G., M. Sinder, and N. Sachser. 1997. Tactual discrimination of size by means of mystacial vibrissae in harbor seals: In air versus underwater. *Zeitschrift für Säugetierkunde* 62:40–43.

Douglas, R.M., A. Neve, J.P. Qittenbaum, N.M. Alam, and G.T. Prusky. 2006. Perception of visual motion coherence by rats and mice. *Vision Research* 46:2642–2647.

Dykes, R.W. 1975. Afferent fibers from mystacial vibrissae of cats and seals. *Journal of Neurophysiology* 38:650–662.

Ebara, S., K. Kumamoto, T. Matsuura, J.E. Mazurkiewicz, and F. Rice. 2002. Similarities and differences in the innervation of mystacial vibrissa follicle-sinus complexes in the rat and cat: A confocal microscopic study. *Journal of Comparative Neurology* 449:103–119.

Ehrenstein, W.H. and A. Ehrenstein. 1999. Modern techniques in neuroscience research. In *Psychophysical Methods*, eds. U. Windhorst and H. Johansson. Berlin, Germany: Springer Verlag.

Erdsack, N., G. Dehnhardt, and W. Hanke. 2014. Thermoregulation of the vibrissal system in harbor seals (*Phoca vitulina*) and Cape fur seals (*Arctocephalus pusillus pusillus*). *Journal of Experimental Marine Biology and Ecology* 452:111–118.

Fasick, J.I. and P.R. Robinson. 2000. Spectral-tuning mechanisms of marine mammal rhodopsins and correlations with foraging depth. *Visual Neuroscience* 17:781–788.

Gaspard, J.C. III, G.B. Bauer, R.L. Reep, K. Dziuk, L. Read, and D.A. Mann. 2013. Detection of hydrodynamic stimuli by the Florida manatee (*Trichechus manatus latirostris*). *Journal of Comparative Physiology A* 199 (6):441–450.

Gellermann, L.W. 1933. Chance orders of alternating stimuli in visual discrimination experiments. *Journal of Genetic Psychology* 42:206–208.

Gescheider, G.A. 1997. *Psychophysics: The Fundamentals*, 3rd edn. New York: Lawrence Erlbaum Associates.

Gibson, J.J. 1950. *Perception of the Visual World*. Boston, MA: Houghton Mifflin.

Ginter, C.C., T.J. DeWitt, F.E. Fish, and C.D. Marshall. 2012. Fused traditional and geometric morphometrics demonstrate pinniped whisker diversity. *PLoS One* 7 (4):e34481.

Gläser, N., B. Mauck, F. Kandil, M. Lappe, G. Dehnhardt, and F.D. Hanke. 2014. Harbour seals (*Phoca vitulina*) can perceive optic flow underwater. *PloS One* 9 (7):e103555.

Gläser, N., S. Wieskotten, C. Otter, G. Dehnhardt, and W. Hanke. 2011. Hydrodynamic trail following in a California sea lion (*Zalophus californianus*). *Journal of Comparative Physiology A* 197:141–151.

Grant, R., S. Wieskotten, N. Wengst, T. Prescott, and G. Dehnhardt. 2013. Vibrissal touch sensing in the harbor seal (*Phoca vitulina*): How do seals judge size? *Journal of Comparative Physiology A* 199:521–533.

Gu, Y., C.R. Fetsch, B. Adeyemo, G.C. DeAngelia, and D.E. Angelaki. 2010. Decoding of MSTd population activity accounts for variations in the precision of heading perception. *Neuron* 66:596–609.

Hanke, F.D., W. Hanke, K.-P. Hoffmann, and G. Dehnhardt. 2008. Optokinetic nystagmus in harbor seals (*Phoca vitulina*). *Vision Research* 48 (2):304–315.

Hanke, F.D., W. Hanke, C. Scholtyssek, and G. Dehnhardt. 2009. Basic mechanisms in pinniped vision. *Experimental Brain Research* 199 (3):299–311.

Hanke, W. and H. Bleckmann. 2004. The hydrodynamic trails of *Lepomis gibbosus* (Centrarchidae), *Colomesus psittacus* (Tetraodontidae) and *Thysochromis ansorgii* (Cichlidae) investigated with scanning particle image velocimetry. *Journal of Experimental Biology* 207:1585–1596.

Hanke, W., C. Brücker, and H. Bleckmann. 2000. The ageing of the low-frequency water disturbances caused by swimming goldfish and its possible relevance to prey detection. *Journal of Experimental Biology* 203:1193–1200.

Hanke, W., M. Witte, L. Miersch et al. 2010. Harbor seal vibrissa morphology suppresses vortex-induced vibrations. *Journal of Experimental Biology* 213:2665–2672.

Herald, E.S., R.L. Brownell Jr., F.L. Frye, E.J. Morris, W.E. Evans, and A.B. Scott. 1969. Blind river dolphin: First side-swimming cetacean. *Science* 166 (3911):1408–1410.

Hinch, S.G., E.M. Standen, M.C. Healey, and A.P. Farrell. 2002. Swimming patterns and behaviour of upriver-migrating adult pink (*Oncorhynchus gorbuscha*) and sockeye (*O. nerka*) salmon as assessed by EMG telemetry in the Fraser River, British Columbia, Canada. *Hydrobiologia* 483:147–160.

Hines, A.H. and T.R. Loughlin. 1980. Observations of sea otters digging for clams at Monterey Harbor, California. *Fishery Bulletin* 78 (1):159–163.

Hyvärinen, H. 1989. Diving in darkness: Whiskers as sense organs of the ringed seal (*Phoca hispida*). *Journal of Zoology* 218:663–678.

Hyvärinen, H. and H. Katajisto. 1984. Functional structure of the vibrissae of the ringed seal (*Phoca hispida* Schr.). *Acta Zoologica Fennica* 171:27–30.

Hyvärinen, H., A. Palviainen, U. Strandberg, and I.J. Holopainen. 2009. Aquatic environment and differentiation of vibrissae: Comparison of sinus hair systems of ringed seal, otter and pole cat. *Brain, Behavior and Evolution* 74:268–279.

Jamieson, G.S. and H.D. Fisher. 1971. The retina of the harbour seal, *Phoca vitulina*. *Canadian Journal of Zoology* 49:19–23.

Japha, A. 1907. Über die Haut nord-atlantischer Furchenwale. *Zool Jb Anat* 24:1–40.

Japha, A. 1912. Die Haare der Waltiere. *Zool Jb Anat* 32:1–42.

Jerlov, N.G. 1976. *Marine Optics*. Amsterdam, the Netherlands: Elsevier Scientific Publishing.

Johnson, G.L. 1901. Contributions to the comparative anatomy of the mammalian eye, chiefly based on ophthalmoscopic examination. *Philosophical Transactions of the Royal Society of Biological Characters* 194:1–30.

Jonas, J.B., U. Schneider, and G.O. Naumann. 1992. Count and density of human retinal photoreceptors. *Graefe's Archive for Clinical and Experimental Ophthalmology* 230 (6):505–510.

Käkelä, R. and H. Hyvärinen. 1993. Fatty acid composition of fats around the mystacial and superciliary vibrissae differs from that of blubber in the Saimaa ringed seal (*Phoca hispida saimensis*). *Comparative Biochemistry and Physiology* 105:547–552.

Käkelä, R. and H. Hyvärinen. 1996. Site-specific fatty acid composition in adipose tissues of several northern aqautic and terrestrial mammals. *Comparative Biochemistry and Physiology* 115:501–514.

Kastelein, R.A., S. Stevens, and P. Mosterd. 1990. The tactile sensitivity of the mystacial vibrissae of a Pacific walrus (*Odobenus rosmarus divergens*). Part 2: Masking. *Aquatic Mammals* 16 (2):8–87.

Kastelein, R.A. and M.A. van Gaalen. 1988. The sensitivity of the vibrissae of a Pacific walrus (*Odobenus rosamarus divergens*) Part 1. *Aquatic Mammals* 14:123–133.

Keane, R.D. and R.J. Adrian. 1992. Theory of cross-correlation analysis of PIV images. *Applied Science Research* 49:191–215.

Kilian, M. 2010. Hydrodynamische Spurverfolgung beim Seehund (*Phoca vitulina*). Diploma thesis, Sensory and cognitive ecology. Rostock, Germany: University of Rostock.

Kröger, R.H.H. and G. Katzir. 2008. Comparative anatomy and physiology of vision in aquatic tetrapods. In *Sensory Evolution on the Threshold—Adaptations in Secondarily Aquatic Vertebrates*, eds. J.G.M. Thewissen and S. Nummela. Berkeley, CA: University of California Press.

Krüger, Y., S. Wieskotten, L. Miersch, G. Dehnhardt, and W. Hanke. 2014. Size discrimination of single vortex rings by stationary harbor seals (*Phoca vitulina*). Poster presented at *107th Annual meeting of the German Zoological Society*, at Göttingen, Germany.

Kükenthal, W. 1909. Untersuchungen an Walen. *Jena Ztschr Naturw* 45:544–588.

Land, M.F. and D.-E. Nilsson. 2002. *Animal Eyes*. Oxford, UK: Oxford University Press.

Lappe, M., M. Jenkin, and L.R. Harris. 2007. Travel distance estimation from visual motion in leaky path integration. *Experimental Brain Research* 180:35–48.

Layne, J.N. and D.K. Caldwell. 1964. Behavior of the amazon dolphin, *Inia geoffrensis* (Blainville), in captivity. *Zoologica: New York Zoological Society* 49 (5):81–112.

Le Boeuf, B.J. and R.M. Laws. 1994. *Elephant Seals: Population, Ecology, Behavior and Physiology*. Berkley, CA: University of California Press.

Levenson, D.H., P.J. Ponganis, M.A. Crognale, J.F. Deegan II, A. Dizon, and G.H. Jacobs. 2006. Visual pigments of marine carnivores: Pinnipeds, polar bear, and sea otter. *Journal of Comparative Physiology A* 192 (8):833–843.

Levenson, D.H. and R.J. Schusterman. 1997. Pupillometry in seals and sea lions: Ecological implications. *Canadian Journal of Zoology* 75:2050–2057.

Levenson, D.H. and R.J. Schusterman. 1999. Dark adaptation and visual sensitivity in shallow and deep-diving pinnipeds. *Marine Mammal Science* 15 (4):1303–1313.

Ling, J.K. 1977. Vibrissae of marine mammals. In *Functional Anatomy of Marine Mammals*, ed. R.J. Harrison. London, UK: Academic Press.

Lui, L.L., A.E. Dobiecki, J.A. Bourne, and M.G.P. Rosa. 2012. Breaking camouflage: Responses of neurons in the middle temporal area to stimuli defined by coherent motion. *European Journal of Neuroscience* 36:2063–2076.

Lythgoe, J.N. and H.J.A. Dartnall. 1970. A "deep sea rhodopsin" in a marine mammal. *Nature* 227:995–996.

Mann, J., B.L. Sargeant, J.J. Watson-Capps et al. 2008. Why do dolphins carry sponges? *PloS One* 3 (12):e3868.

Marshall, C.D., H. Amin, K.M. Kovacs, and C. Lydersen. 2006. Microstructure and innervation of the vibrissal follicle-sinus complex in the bearded seal, *Erignathus barbatus* (Pinnipedia: Phocidae). *Anatomical Record* 288A:13–25.

Mass, A.M. and A.Y. Supin. 1989. Distribution of ganglion cells in the retina of an amazon river dolphin *Inia geoffrensis*. *Aquatic Mammals* 15 (2):49–56.

Mass, A.M. and A.Y. Supin. 2007. Adaptive features of aquatic mammals' eyes. *The Anatomical Record* 290:701–715.

Mass, A.M. and A.Y. Supin. 2009. Vision. In *Encyclopedia of Marine Mammals*. London, UK: Academic Press.

Mauck, B., U. Eysel, and G. Dehnhardt. 2000. Selective heating of vibrissal follicles in seals (*Phoca vitulina*) and dolphins (*Sotalia fluviatilis guianensis*). *Journal of Experimental Biology* 203:2125–2131.

McFarland, W.N. 1971. Cetacean visual pigments. *Vision Research* 11:1065–1076.

McGovern, K.A., C.D. Marshall, and R.W. Davis. 2015. Are vibrissae viable sensory structures for prey capture in northern elephant seals, *Mirounga angustirostris? Anatomical Record* 298:750–760.

Miersch, L., W. Hanke, S. Wieskotten et al. 2011. Flow sensing in pinniped whiskers. *Philosophical Transactions of the Royal Society of London B: Biological Sciences* 366:3077–3084.

Mills, F.H.J. and D. Renouf. 1986. Determination of the vibration sensitivity of harbor seal *Phoca vitulina* (L.) vibrissae. *Journal of Experimental Marine Biology and Ecology* 100:3–9.

Mobley, C.D. 1994. *Light and Water*. San Diego, CA: Academic Press.

Morgane, P.J. and M. Jacobs. 1972. Comparative anatomy of the cetacean nervous system. In *Functional Anatomy of Marine Mammals*, ed. R.J. Harrison. New York: Academic Press.

Nagy, A.R. and K. Ronald. 1970. The harp seal, *Pagophilus groenlandicus (Erxleben, 1777)*. VI. Structure of retina. *Canadian Journal of Zoology* 48:367–370.

Nakai, J. and T. Shida. 1948. Sinus hairs of the sei-whale (*Balaenoptera borealis*). *The Scientific Reports of the Whale Research Institute* 1:41–47.

Newsome, W.T. and E.B. Paré. 1988. A selective impairment of motion perception following lesions of the middle temporal visual area (MT). *Journal of Neuroscience* 8 (6):2201–2211.

Niesterok, B., S. Wieskotten, Y. Krüger, G. Dehnhardt, and W. Hanke. 2014. Localization of artificial breathing currents by harbor seals (*Phoca vitulina*). Paper read at *107th Annual Meeting of the German Zoological Society*, Göttingen, Germany.

Pavlov, I.P. 1902. *The Work of the Digestive Glands*. London, UK: Griffin.

Peichl, L. and K. Moutairou. 1998. Absence of short-wavelength sensitive cones in the retinae of seals (*Carnivora*) and African giant rats (*Rodentia*). *European Journal of Neuroscience* 10:2586–2594.

Pilleri, G. 1974. Side-swimming, vision and sense of touch in *Platanista indi* (Cetacea, Platanistidae). *Experientia* 30 (1):100–104.
Pryor, K. 1984. *Don't Shoot the Dog*. New York: Bentam Books.
Pütter, A. 1903. Die Augen der Wassersäugethiere. *Zoologische Jahrbuecher Abteilung fuer Anatomie und Ontogenie der Tiere* 17:99–402.
Ramirez, K. 1999. *Animal Training—Successful Animal Management through Positive Reinforcement*. Chicago, IL: John G. Shedd Aquarium.
Reep, R.L., J.C. III Gaspard, D.K. Sarko, F.L. Rice, D.A. Mann, and G.B. Bauer. 2011. Manatee vibrissae: Evidence for a "lateral line" function. In *New Perspectives on Neurobehavioral Evolution*, eds. J.I. Johnson, H.P. Zeigler, and P.R. Hof. Boston, MA: Blackwell.
Reep, R.L., C.D. Marshall, and M.L. Stoll. 2002. Tactile hairs on the postcranial body of the Florida manatees: A mammalian lateral line? *Brain Behavior and Evolution* 58:141–154.
Reep, R.L., M.L. Stoll, C.D. Marshall, B.L. Homer, and D.A. Samuelson. 2001. Microanatomy of facial vibrissae in the *Florida manatee*: The basis of specialized sensory function and oripulation. *Brain Behavior and Evolution* 58:1–14.
Regan, D. and R. Gray. 2000. Visually guided collision avoidance and collision achievement. *Trends in Cognitive Science* 4 (3):99–107.
Renouf, D. 1979. Preliminary measurements of the sensitivity of the vibrissae of harbor seals (*Phoca vitulina*) to low frequency vibrations. *Journal of Zoology (London)* 188:443–450.
Reuter, T. and L. Peichl. 2008. Structure and function of the retina in aquatic tetrapods. In *Sensory Evolution on the Threshold—Adaptations in Secondarily Aquatic Vertebrates*, eds. J.G.M. Thewissen and S. Nummela. Berkeley, CA: University of California Press.
Rice, F.L., A. Mance, and B.L. Munger. 1986. A comparative light microscopic analysis of the sensory innervation of the mystacial pad. I. Innveration of vibrissal follicle-sinus complexes. *Journal of Comparative Neurology* 252:154–174.
Rogers, K., ed. 2011. *The Eye: The Physiology of Human Perception*. New York: Britannica Educational Publishing and Rosen Educational Services.
Rossbach, K.A. and D.L. Herzing. 1997. Underwater observations of benthic-feeding bottlenose dolphins (*Tursiops truncatus*) near Grand Bahama Island, Bahamas. *Marine Mammal Science* 13 (3):498–504.
Scase, M.O., O. Braddick, and J.E. Raymond. 1996. What is noise for the motion system? *Vision Research* 36:2579–2586.
Schulte-Pelkum, N., S. Wieskotten, W. Hanke, G. Dehnhardt, and B. Mauck. 2007. Tracking of biogenic hydrodynamic trails in harbour seals (*Phoca vitulina*). *Journal of Experimental Biology* 210:781–787.
Skinner, B.F. 1938. *The Behavior of Organisms: An Experimental Analysis*. New York: Appleton Century.
Smolker, R., A. Richards, R. Connor, J. Mann, and P. Berggren. 1997. Sponge carrying by dolphins (Delphinidae, *Tursiops* sp.): A foraging specialization involving tool use? *Ethology* 103 (6):454–465.
Southall, K.D., G.W. Oliver, J.W. Lewis, and B.J. LeBoeuf. 2002. Visual pigment sensitivity in three deep diving marine mammals. *Marine Mammal Science* 18:275–281.
Standen, E.M., S.G. Hinch, and P.S. Rand. 2004. Influence of river speed on path selection by migrating adult sockeye salmon (*Oncorhynchus nerka*). *Canadian Journal of Fisheries and Aquatic Sciences* 61:905–912.
Stevens, R.J., I.J. Beebe, and T.C. Poulter. 1973. Innervation of the vibrissae of the California sea lion, *Zalophus californianus*. *Anatomical Record* 176:421–442.
Tautz, J. 1989. *Medienbewegung in der Sinneswelt der Arthropoden—Fallstudien zu einer Sinnesökologie*. Stuttgart, Germany: Gutsav Fischer Verlag.
Videler, J.J. and D. Weihs. 1982. Energetic advantages of burst-and-coast swimming of fish at high speeds. *Journal of Experimental Biology* 97:169–178.
Waller, G.N.H. 1982. Retinal ultrastructure of the Amazon River dolphin (*Inia geoffrensis*). *Aquatic Mammals* 9:17–28.
Waller, G.N.H. 1983. Is the blind river dolphin sightless? *Aquatic Mammals* 10 (3):106–108.
Walls, G.L. 1942. *The Vertebrate Eye and Its Adaptive Radiation*. New York: Hafner Press.
Warrant, E.J. and N.A. Locker. 2004. Vision in the deep sea. *Biological Review* 79:671–712.

Warren, W.H. and D.J. Hannon. 1988. Direction of self-motion is perceived from optical flow. *Nature* 336:162–163.

Warren Jr., W.H., B.A. Kay, W.D. Zosh, A.P. Duchon, and S. Sahuc. 2001. Optic flow is used to control human walking. *Nature Neuroscience* 4 (2):213–216.

Warren, W.H., M.W. Morris, and M. Kalish. 1988. Perception of translational heading from optical flow. *Journal of Experimental Psychology* 14 (4):646–660.

Watkins, W.A. and D. Wartzok. 1985. Sensory biophysics of marine mammals. *Marine Mammal Science* 1:219–260.

Weiffen, M., B. Mauck, G. Dehnhardt, and F.D. Hanke. 2014. Sensitivity of a harbor seal (*Phoca vitulina*) to coherent visual motion in random dot displays. *SpringerPlus* 3:688.

Weiffen, M., B. Möller, B. Mauck, and G. Dehnhardt. 2006. Effect of water turbidity on the visual acuity of harbor seals (*Phoca vitulina*). *Vision Research* 46:1777–1783.

Wieskotten, S., G. Dehnhardt, B. Mauck, L. Miersch, and W. Hanke. 2010a. The impact of glide phases on the trackability of hydrodynamic trails in harbour seals (*Phoca vitulina*). *Journal of Experimental Biology* 213:3734–3740.

Wieskotten, S., G. Dehnhardt, B. Mauck, L. Miersch, and W. Hanke. 2010b. Hydrodynamic determination of the moving direction of an artificial fin by a harbour seal (*Phoca vitulina*). *Journal of Experimental Biology* 213:2194–2200.

Wieskotten, S., B. Mauck, L. Miersch, G. Dehnhardt, and W. Hanke. 2011. Hydrodynamic discrimination of wakes caused by objects of different size or shape in a harbour seal (*Phoca vitulina*). *Journal of Experimental Biology* 214:1922–1930.

Young, N., G.M. Hope, W.W. Dawson, and R.L. Jenkins. 1988. The tapetum fibrosum in the eyes of two small whales. *Marine Mammal Science* 4 (4):281–290.

section five

Environmental interactions

chapter thirteen

Disease

Shawn P. Johnson and Claire A. Simeone

Contents

13.1 Introduction

Marine mammals are exquisitely adapted to occupy the niches that they do. Manatees (*Trichechus* sp.) have a specialized digestive tract to break down the sea grass they feed on in both fresh and saltwater. Many species of *pinnipeds,* which include seals, sea lions, and walruses, exhibit the unique behavior of spending much of their time at sea, often migrating long distances, and then all returning to the same site each year to breed on land.

Some marine mammal species are apex predators, meaning that as healthy adults they are not usually preyed upon by other species in the wild. Polar bears (*Ursus maritimus*), for example, feed almost exclusively on seals and other marine mammals without fear of other predators. Blue whales (*Balaenoptera musculus*) only feed on planktonic krill, but they are considered apex predators as well, as they are at the top of the food chain in their ecosystem. Both marine mammal species help to keep prey populations balanced and keep the ecosystem healthy.

Certain marine mammals go even further to keep the ecosystem healthy, and their presence plays a critical role in maintaining the ecological community, far beyond what would be expected for their population size. Named a *keystone species*, sea otters (*Enhydra lutris*) are a well-known example. Their high metabolism requires them to eat a large amount of invertebrates each day, including mussels, clams, and urchins. Their presence keeps the sea urchin population in check. In places where sea otters are no longer a part of the community, sea urchins quickly overpopulate and they destroy entire kelp forests by eating the base of kelp fronds. The entire ecosystem of the kelp forest, and all of the organisms that are a part of it, rely on the sea otters to keep this balance in place.

These animals are highly specialized to their environment, and if there are any changes to that environment, or to the animals themselves, they may be predisposed to disease. Humans are changing the environment—degradation and development of the coast allows for run-off of contaminants, infectious organisms, or nutrients that may trigger a toxic algal bloom. New industrial chemicals are biomagnified in the food chain, and end up being stored in marine mammal blubber—just as environmental concentrations of old, outlawed chemicals such as DDT are starting to decrease.

Marine mammals have a variety of stressors to contend with in this changing environment. In this chapter, we will explore how the physiological adaptations that they have developed to live in a marine environment can either protect or predispose them to disease.

13.2 Cetaceans

13.2.1 Physiologic adaptation: Respiratory system

Members of the order Cetacea are one of the most radically adapted groups to the aquatic environment. They are obligate aquatic species and cannot survive outside of the water due to extreme anatomical and physiological changes. Adaptations to the respiratory and dermal system necessary for aquatic life also play a significant role in maintaining health in these species.

The respiratory system of *cetaceans* is highly adapted for swimming and diving. Their *nares* are positioned at the top of the head to allow for inspiration at the water's surface. The lungs of cetaceans are highly compliant, which help protect them from decompression sickness during repeated deep diving. Within the lung there is very little lymphoid tissue and the bronchial epithelium and glands contain almost no mucoid-producing cells, which typically play a role in trapping foreign particles (Goudappel and Slijper 1958). The alveoli contain no epithelial lining, which typically provides a last layer of protection against invaders. This directly exposes the capillary network to respiratory gases, allowing for rapid exchange of oxygen and carbon dioxide (Haynes and Laurie 1937). It is hypothesized that these protective anatomical respiratory structures are unnecessary to cetaceans that live in a mostly dust-free environment. However, this leaves the cetacean respiratory system susceptible to insult from a variety of external factors and prone to developing pneumonia. Bottlenose dolphins (*Tursiops truncatus*) exposed to volatile hydrocarbons following the *Deepwater Horizon* oil spill in the Gulf of Mexico in 2010 were five times more likely to have moderate to severe lung disease than counterparts at a reference site in Florida (Schwacke et al. 2014).

Immediately following an oil spill, volatile hydrocarbon vapors are released from the oil. Cetaceans appear to be able to detect oil slicks on water but do not necessarily avoid them (Gubbay and Earll 2000). Therefore, dolphins swimming through an oiled area may be exposed through direct contact with the oil in the water column; through ingestion of contaminated water and prey; and through inhalation of petroleum vapors.

Inhalation of petroleum vapors has been linked to decreased lung function, chronic bronchitis, and airway inflammation in humans (Sekkal et al. 2012). Ultrasonographic pulmonary changes seen in dolphins in the spill area following the *Deepwater Horizon* spill were similar to the findings associated with these pulmonary diseases. In addition to pulmonary disease, affected dolphins also showed evidence of adrenal suppression, liver damage, and tooth loss (Schwacke et al. 2014).

13.2.2 *Physiologic adaptation: Skin*

Cetacean skin has undergone significant changes in order to maintain homeostasis in the marine environment. The skin is hairless, smooth, and lacks any cutaneous appendages or glands. In order to protect against the osmotic challenges of living in water, the epidermis provides a protective layer that is impermeable to water. To maintain this protection, the epidermal turnover rate is 8.5 times higher than human skin (Hicks et al. 1985). Cetaceans have also developed an alternative method of repairing skin damage because they are unable to form a transition scab in the water (Bruce-Allen and Geraci 1985). A breach in the dermis allows seawater contact with epidermal cells causing degeneration and swelling of these cells in order to provide a protective barrier to the deeper regenerating cells. As the skin wound heals from below, the degenerate cells are sloughed off as the wound closes. In bottlenose dolphins, the skin heals at the same rate as humans; however, in other cetaceans, dermal wound healing is longer and affected by dermal thickness and environmental conditions such as water temperature and salinity. In the beluga whale (*Delphinapterus leucas*), skin wounds heal in 30–40 days, five times longer than in bottlenose dolphins primarily due to the fact that beluga skin is five times thicker than dolphin skin (Geraci and Bruce-Allen 1987). Belugas likely have thicker skin because they live in substantially colder water and are exposed to water with low salinity when they inhabit river estuaries during the summer months.

The dermis is a primary protective barrier against environmental stressors in cetaceans, but it is highly susceptible to water-quality abnormalities like changes in salinity and increases in pollutants and nutrients in the water caused by local freshwater and storm water run-off. For example, 30% of dolphins living in the southern Indian River Lagoon in Florida have developed lacaziosis (lobomycosis), a chronic mycotic disease of the skin and subdermal tissues caused by a yeast-like organism known as *Lacazia loboi* (Reif et al. 2006). Water quality of the Indian River Lagoon has dramatically decreased over the past 50 years due to changes in watershed drainage and land development, leaving the local dolphin population susceptible to this naturally occurring fungus. There is also evidence that the population of dolphins affected by lacaziosis may have immunosuppression, showing depressed lymphocyte proliferation, which plays a role in responding to external pathogens.

Dolphins that are infected with lacaziosis may have extensive areas of affected skin that cover large areas of the body. Lesions may be associated with sites of previous trauma such as shark bite scars, where the protective mechanisms of the skin may be reduced. Raised, bumpy areas of skin may turn white or discolored, and are often found on the dorsal fin, flukes, or head. As lacaziosis is a *zoonotic disease*, humans are also susceptible to infection by *L. loboi*, and care should be taken when handling animals with visible skin lesions. The fungus is typically spread through breaks or cuts in the skin (Figure 13.1).

Figure 13.1 **(See color insert.)** Bottlenose dolphin (*Tursiops truncatus*) affected by lacaziosis in Sarasota Bay, Florida. Note the raised, bumpy lesions that may affect large areas of the body. (Photo courtesy of Sarasota Dolphin Research Program, obtained under National Marine Fisheries Service Scientific Research Permit No. 15543.)

13.3 Pinnipeds

13.3.1 Physiologic adaptation: Haul-out behavior

Pinnipeds include seals, sea lions, and walruses. These families each have unique habitats and express unique behaviors, but one behavior that they share in common is a tendency to haul out, or congregate in large groups on land or ice. Nearly all California sea lions (*Zalophus californianus*), for example, return to islands off the coast of southern California and Mexico each year to give birth and reproduce. Animals are packed along the beaches and rocks, and these dense groups are ideal for disease transfer (Figure 13.2).

Leptospirosis is a disease caused by the bacteria *Leptospira* sp., characterized by kidney failure and causing large, cyclical outbreaks of mortality in California sea lions (Lloyd-Smith et al. 2007). The bacteria are excreted in the urine, and infected animals may shed bacteria even after they have recovered from illness. Just a few sick animals can infect hundreds of animals in close physical proximity.

Many parasites rely on this haul-out behavior in order to complete their life cycle in a marine mammal definitive host. Hookworms (*Uncinaria* sp.) are acquired by the northern fur seal (*Callorhinus ursinus*) and California sea lion dams, when the worm buries through the skin and migrates into the mammary glands. Larvae are transferred to pups through the milk, where they mature into adult worms in the pup's intestines. As the hookworms actually bury into the intestinal lining and feed on the pup's blood, high hookworm burdens can cause anemia through blood loss, intestinal perforation, and death. Nearly 100% of pups can be infested in a single year, and mortality can exceed 70% (Lyons et al. 2005). The pups that survive will clear the infection, but the eggs that they shed in their feces will survive in the sand until the next time the mothers come to the beaches to pup. The life-history behaviors of returning to the same site year after year to pup, and sharing that beach with hundreds of other animals, encourage the hookworms to persist in this species.

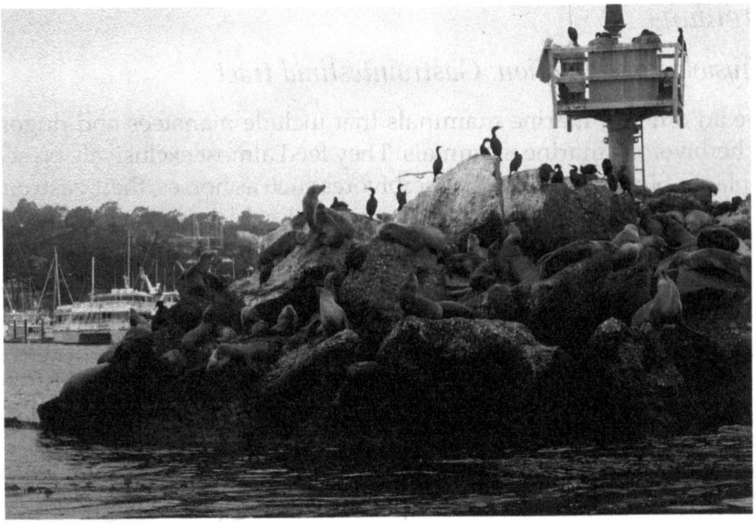

Figure 13.2 **(See color insert.)** California sea lions (*Zalophus californianus*) haul out, or congregate on land, in large groups on a jetty in Monterey Bay, California. (Photo courtesy of The Marine Mammal Center, Sausalito, California.)

13.4 Sea otters

13.4.1 Physiologic adaptation: Metabolism

Sea otters are one of the most recent mammals to have re-entered the marine environment, and their physiological adaptations are different from cetaceans and pinnipeds. Instead of relying on a thick, internalized blubber layer for insulation, sea otters prevent heat loss through an air layer trapped against the skin by an exceptionally dense fur (Williams et al. 1992). As an otter dives, the air layer is compressed, which reduces the insulating quality of fur at depth. This elevated thermal energetic cost, along with a Resting Metabolic Rate three times the rates observed for terrestrial mammals of a similar size, results in large food requirements of roughly 20%–25% of body mass in prey items per day (Costa and Kooyman 1982; Yeates et al. 2007).

Sea otters consume a variety of marine invertebrates, including clams, mussels, and snails. As they filter seawater or scrape biofilm off of kelp, these prey items may collect and concentrate infectious oocysts (Lindsay et al. 2001). Ingestion of infectious *Sarcocystis neurona* or *Toxoplasma gondii* oocysts may cause severe neurological signs such as seizures, coma, and death in otters that consume the contaminated invertebrates (Thomas et al. 2007; Miller et al. 2010). Both protozoal diseases have terrestrial definitive hosts: *Sarcocystis neurona* is spread by the opossum, while *Toxoplasma gondii* is spread to sea otters mainly by cats (Conrad et al. 2005). Wetlands typically filter out a large amount of these contaminants, and prevent them from entering the sea, but in areas where wetlands have become degraded, high levels of protozoal cysts can enter the water. Because sea otters have a relatively restricted territory, and live in kelp forests close to the coast, these environmental changes, along with the otters' unique physiology and increased metabolism, have made them susceptible to protozoal diseases as they flow from land to sea.

13.5 Sirenians

13.5.1 Physiologic adaptation: Gastrointestinal tract

Sirenians are an order of marine mammals that include manatees and dugongs, and they are the only herbivorous marine mammals. They feed almost exclusively on sea grasses and other water plants. Similar to terrestrial herbivores such as horses, their gastrointestinal tract is specially formatted to break down plant material, which is done in an elongated large intestine. In addition to the structures similarly found in the horse, manatees have several specialized glands in their stomach that are suspected to regulate water–salt balance as they forage for plants in both saltwater and freshwater (Reynolds and Rommel 1996).

Because of their specialized diet, manatees are susceptible to diseases that may be associated with the plants they eat. Biotoxins are toxic compounds that are produced by living organisms, and marine algal blooms can produce a variety of biotoxins. In Florida, where one endangered species of manatee lives, more than 70 potentially harmful algal species have been identified, producing toxins such as brevetoxin, domoic acid, okadaic acid, and saxitoxin (Abbott et al. 2009). As an example, brevetoxin (produced by the diatom *Karenia brevis*) is lipophilic, and will concentrate in sea grass blades and rhizomes, and in the epiphytes living on the surface of the grass. As manatees ingest the sea grass they receive a high dose of the toxin, which causes listlessness and neurological signs (O'Shea et al. 1991). In addition, if the toxin is inhaled, pneumonia and bronchitis may develop. While some manatees have been rehabilitated and released into the wild, large toxic blooms have caused fatalities of more than 100 individuals in a single event (Bossart et al. 1998).

Sirenians are not the only marine mammals to be affected by harmful algal blooms (HABs). Toxin-producing HABs are increasing worldwide. One of the toxins that has been most heavily studied in marine mammals is domoic acid, produced by the diatom *Pseudo-nitzschia* sp. While domoic acid is found in oceans across the world, frequent blooms off the coast of California cause widespread morbidity and mortality in the California sea lions there. Domoic acid specifically targets glutamate receptors, where it acts as an excitotoxin, causing massive cellular depolarization and death (Olney et al. 1979). There is a high concentration of glutamate receptors in the hippocampus, a region of the brain that is responsible for spatial processing and memory.

In humans, a syndrome called *Amnesic Shellfish Poisoning* results from ingestion of domoic acid-tainted shellfish, characterized by memory loss, seizures, and death. In sea lions that have consumed domoic acid-tainted fish, similar signs are seen, and seizures, disorientation, and death result from hippocampal necrosis. Interestingly, clinical signs are seen with both acute intoxication as well as chronic, sub-lethal exposure (Goldstein et al. 2008). Chronically exposed animals exhibit complex partial seizures, atrophy of the hippocampus, and involvement of other brain regions—all similar to temporal lobe epilepsy as seen in humans (Figure 13.3).

Biotoxins pose a large risk to marine mammal health. The specializations in prey selection, and the relatively small geographic range that pinnipeds and sirenians use to forage make them susceptible to large algal blooms, and the toxins they produce.

13.6 Polar bears

13.6.1 Physiologic adaptation: Prey sources

Polar bears are apex predators, feeding mainly on marine mammals such as ringed seals (*Phoca hispida*), bearded seals (*Erignathus barbatus*), and hooded seals (*Cystophora cristata*).

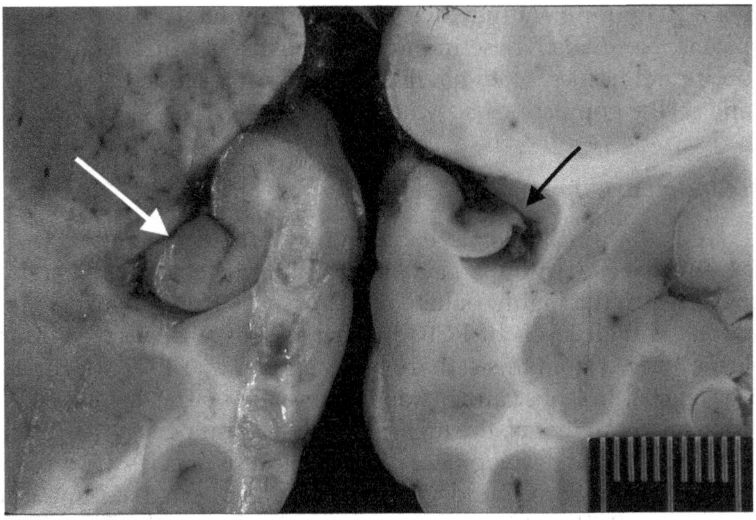

Figure 13.3 **(See color insert.)** Domoic acid specifically targets the hippocampus in the brain. Left: A normal California sea lion brain section (the hippocampus is noted with the white arrow). Right: A California sea lion's brain affected by domoic acid exposure. Note the shrunken hippocampus (black arrow). (Photo courtesy of The Marine Mammal Center, Sausalito, California.)

As all of these seal species have a thick blubber layer as insulation, the polar bear diet is high in fat. Lipophilic contaminants, such as PCBs, DDT, and a variety of flame retardants are stored in fat over the lifetime of the animal. Thus, the polar bear diet is also high in contaminants.

Many contaminants can have effects on reproduction by disrupting hormone pathways and can cause immunosuppression, leaving animals with an increased susceptibility to infectious diseases or certain types of cancer. Studies over three decades have tracked levels of lipophilic contaminants in polar bear fat. Nearly all of the legacy organochlorine contaminants (which have been banned or highly regulated) showed a statistically significant decline, suggesting that banning these compounds will lead to them slowly disappearing from the environment (Dietz et al. 2013a). However, relatively high levels still persist in polar bear tissues, at levels that continue to have the potential to cause health changes.

Unfortunately, when studying contaminants for which no international bans currently exist, the trends are reversed. Brominated flame retardant (BFR) concentrations more than doubled on average across the same study period (Dietz et al. 2013b). In addition to the effects described above, BFRs can also cause neurotoxicity, and may affect the liver and thyroid. Due to their unique diet, polar bears are exposed to high levels of lipophilic contaminants. This is only one health risk that these animals are exposed to as they live in a changing Arctic.

13.7 Marine mammal populations

As we have illustrated, individual marine mammals are susceptible to disease due to a variety of physiological and anatomical adaptations that allow them to inhabit an aquatic environment. However, disease can also have a dramatic negative effect on health of a population of marine mammals and is a major risk to small isolated populations. Many threats from infectious agents to marine mammal populations are a direct consequence of degradation of the environment they live. Many contaminants bioaccumulate in the food sources of marine mammals. These immunosuppressive chemicals are stored in marine mammal blubber and tissues, often having a negative impact on the overall health of the individual, and on the population as a whole.

13.7.1 Population example: Neoplasia

Neoplasia is rare in marine mammals, except for two distinct populations: beluga whales living in St. Lawrence estuary in Canada, and California sea lions along the California coast. Belugas living in the highly polluted St. Lawrence Estuary have a high prevalence of cancer, as well as high levels of PCBs in their tissues. Tumors were found in 27% of the dead belugas studied from this estuary between 1983 and 1999 (Martineau et al. 2002). In California sea lions, 18% of stranded adults along the central California coast are diagnosed with urogenital carcinoma, the highest among a pinniped species. Three factors are known to play a role in the development of cancer in this species: (1) presence of otarine herpesvirus-1, which has been found in 100% of tumors tested; (2) immunosuppressive contaminants such as PCBs, which are frequently elevated in tissues from animals with urogenital carcinoma; and (3) a genetic predisposition, as sea lions with cancer appear to be more genetically similar than those without cancer (Gulland et al. 1996; Ylitalo et al. 2005).

13.7.2 Population example: Viral infections

Morbillivirus is a genus that includes measles, canine distemper virus, rinderpest virus, and marine morbilliviruses. Marine morbilliviruses, including phocine distemper virus (PDV) and the cetacean morbilliviruses (CeMV), cause explosive outbreaks of disease (Kennedy 1990). Morbilliviruses cause immunosuppression, and after they debilitate an animal, secondary infections such as fungal or bacterial disease are often seen. In addition, high contaminant levels have been associated with animals infected with morbillivirus, suggesting that the immunosuppression caused by a high contaminant load could make animals even more susceptible to infection (Hall et al. 1992). Animals that survive the outbreak do have immunity for the virus, but it does not last for the lifetime of the animal. As immunity wanes, and younger animals that are unexposed enter the population, a threshold is reached that allows the population to be susceptible to a large outbreak.

Outbreaks of marine morbilliviruses killed an estimated 18,000 harbor seals in Europe in 1988, several thousand striped dolphins (*Stenella coeruleoalba*) in the Mediterranean from 1990 to 1992, and two outbreaks in 1987 to 1988 and 2013 to 2014 on the U.S. Atlantic coast have killed more than 2500 bottlenose dolphins combined (Kennedy 1990; Aguilar and Raga 1993; Lipscomb et al. 1994). Such large numbers have the potential to severely affect a population, especially if that population is small and relatively isolated.

13.8 Difficulties of diagnosis

Veterinarians use many of the same techniques to diagnose disease in marine mammals as they would in dogs or cats, or as a physician would diagnose disease in a human. Sterile swabs are used to collect samples for bacterial culture; tests such as polymerase chain reaction (PCR) and enzyme-linked immunosorbent assays (ELISAs) are used to screen for diseases; blood tubes are used in the analysis of white and red blood cells.

These tests have caveats when used on marine mammal patients, though. Many marine organisms are difficult to recognize, especially in a laboratory that is used to working primarily with terrestrial pathogens. *Listeria* sp., a bacterium that causes food poisoning in humans, was originally reported in abscesses in pinnipeds, based on the biochemical reactions of the bacteria. However, genetic tests revealed the organism to be *Arcanobacterium*, a marine bacterium that does not cause disease in humans (Johnson et al. 2003).

Tests such as PCR and ELISA are often developed for terrestrial diseases. They may cross-react with marine organisms but may make it difficult to tell from where the disease originated. For instance, seals are susceptible to phocine distemper (PDV), a virus that causes pneumonia and encephalitis, and is closely related to canine distemper virus (CDV). Initial PCR testing showed these seals to be positive to CDV. While there is evidence that some outbreaks of seal distemper may have come from sled dogs with canine distemper, genetic testing revealed that PDV is genetically distinct from CDV (Mahy et al. 1988). Tuberculosis has a similar story. The causative agent, *Mycobacterium* sp., can cause abscesses in the lungs, lymph nodes, and throughout the body, and is zoonotic, meaning it can be spread from humans to animals and back. Original tests of tuberculosis in sea lions described it as *Mycobacterium bovis*, which often infects cows. It was not until genetic testing described a new strain, *Mycobacterium pinnipedii*, that it was understood that the tuberculosis strain was unique to pinnipeds—although it could still spread to other species from them (Cousins et al. 2003).

We have to take marine mammals' unique physiology into account when analyzing changes to blood work. Deep-diving cetaceans or pinnipeds may have a high percentage of red blood cells (described as their hematocrit) in their blood, to carry extra oxygen as they dive. If we compared their hematocrit to the normal values for a dog or a human, we would think a healthy marine mammal might be dehydrated or have an abnormally high hematocrit. It is important to know what normal blood values are for marine mammals, so that we can correctly assess changes in animals with disease.

If a marine mammal is not hauled out on land, or stranded on a beach, it can be difficult to collect samples to diagnose disease. Lipophilic contaminants are stored in the blubber layer, and a skin sample can give us information about the genetics of that animal, and what population it belongs to. Collecting a skin or blubber sample from a whale at sea involves using a dart gun to perform a remote biopsy, which collects a small tissue core sample that can be analyzed. Imagine that, in addition to a moving animal target, you are also bobbing up and down in a small boat.

Many of the medications used to treat marine mammal diseases are the same drugs as those used in humans or other animals. Antibiotics are used to fight bacterial infections, anti-inflammatories control inflammation and associated pain, and anesthetics provide sedation for surgical procedures. However, unique adaptations to marine mammal physiology may change the way drugs are metabolized in the body, translating to altered doses required for treatment. For instance meloxicam, a non-steroidal anti-inflammatory drug (NSAID), is typically excreted from the body in roughly 24 hours in dogs and humans. A study in dolphins showed elevated levels of meloxicam in the blood for more than 7 days, meaning that a once-weekly dose likely provides the same level of inflammatory relief in dolphins as a daily dose in humans (Simeone et al. 2014). Dolphins are also more sensitive to the sedative effects of some drugs, and benzodiazepines such as midazolam cause sedation at doses less than half that required to sedate other terrestrial species (Dold 2014).

13.9 Lingering mysteries

Much of what is known with regard to the diagnosis and treatment of diseases in marine mammals is due to studies that investigate the unique ways that marine mammals respond to disease. However, many questions remain.

Little is known, for instance, about the way diseases are transmitted and persist in a marine environment. Marine morbilliviruses, including PDV and CeMV, have caused explosive outbreaks of disease characterized by high mortality specifically in the Atlantic Ocean. In the Pacific, morbilliviruses have been documented in marine mammals, but have not yet caused widespread mortality (Goldstein et al. 2009). Further research is needed to understand why there is such extensive variation in host susceptibility to marine morbilliviruses, and whether this is driven primarily by host differences, or geographic differences in the viral strains (Grenfell et al. 2014).

Many questions remain about leptospirosis, the zoonotic bacterial infection that causes kidney damage in California sea lions. Cyclical outbreaks of disease are seen, but it is unknown whether the sea lion population continues to become re-infected by an unknown terrestrial reservoir host, or if the disease persists in the sea lion population itself (Lloyd-Smith et al. 2007). Widespread sampling of the population during and outside of epidemics is currently underway to better understand the disease.

The role that biotoxins play in marine mammal disease is continuing to be elucidated. Sampling of stranded animals' body fluids and tissues for the detection of various biotoxins often produces a positive result, but the significance of this positive test may not be clear. The level of toxin required to cause disease varies with type of toxin and is unknown for nearly all biotoxins. Because these are naturally occurring algal species, animals frequently ingest low levels of the toxins in their prey, and positive tests do not necessarily mean that intoxication is the cause of illness or death. A suite of clinical signs in live animals, or certain pathological findings in dead animals, may strongly suggest that intoxication has occurred, especially if the animal strands in a location and time in which a known algal bloom is occurring. However, more research is needed to correlate clinical findings with biotoxin concentrations, in order to determine the concentrations necessary to cause either acute or chronic disease.

As the environment changes from climate change and human pressures, we have seen marine mammals adapt in a variety of ways. northern elephant seals (*Mirounga angustirostris*) are expanding their range as their population grows, resting on beaches where they had never been documented before. The critically endangered Hawaiian monk seal (*Neomonachus schauinslandi*) reproduces largely on sandy atolls, which are disappearing with sea level rise, forcing them to choose new sites. And in years when fish stocks move offshore, California sea lion dams are forced to travel further from rookery sites to feed—meaning they may not produce as much milk to nurse a pup.

It is unknown exactly what diseases will be favored as marine mammals change their behavior. Increased concentrations of animals may allow an introduced disease to spread through a naïve population quickly. Animals already stressed with fewer food resources may be susceptible to common pathogens. Certain species may actually benefit from these changes, if their behavior changes result in less interaction with other individuals. Scientists must respond to a changing environment by altering the questions they ask—in order to understand how diseases affect marine mammals in this new world.

Glossary

Cetacean: A marine mammal of the order *Cetacea,* including whales, dolphins, and porpoises.

ELISA: Enzyme-linked immunosorbent assay.

Keystone species: A species that has a disproportionately large effect on its environment relative to its abundance, such that if it were removed, the ecosystem would change drastically.

Lipophilic: *Fat-loving* or stored in fat.
Nares: Nostrils.
PCR: Polymerase chain reaction.
Pinniped: A marine mammal of the order *Pinnipedia,* including seals, sea lions, and walruses. Derived from Latin meaning *fin-footed.*
Urogenital carcinoma: A type of cancer affecting California sea lions that originates from the cervix or penis and affects the genitourinary tracts.
Zoonosis (Zoonotic disease): A disease that can be transmitted between humans and animals.

References

Abbott, G.M., J.H. Landsberg, A.R. Reich, K.A. Steidinger, S. Ketchen, and C. Blackmore. 2009. Resource guide for public health response to harmful algal blooms in Florida. Technical Report TR-14, Fish and Wildlife Research Institute, Tallahassee, FL, pp. viii+132.

Aguilar, A. and J. A. Raga. 1993. The striped dolphin epizootic in the Mediterranean Sea. *Ambio* 22(8): 524–528.

Bossart, G.D., D.G. Baden, R.Y. Ewing, B. Roberts, and S.D. Wright. 1998. Brevetoxicosis in manatees (*Trichechus manatus latirostris*) from the 1996 epizootic: Gross, histologic, and immunohistochemical features. *Toxicologic Pathology* 26(2): 276–282.

Bruce-Allen, L.J. and J.R. Geraci. 1985. Wound healing in the bottlenose dolphin (*Tursiops truncatus*). *Canadian Journal of Fisheries and Aquatic Sciences* 42(2): 216–228.

Conrad, P.A., M.A. Miller, C. Kreuder, E.R. James, J. Mazet, H. Dabritz, D.A. Jessup, F. Gulland, and M.E. Grigg. 2005. Transmission of *Toxoplasma*: Clues from the study of sea otters as sentinels of *Toxoplasma gondii* flow into the marine environment. *International Journal for Parasitology* 35: 1155–1168.

Cousins, D.V., R. Bastida, A. Cataldi, V. Quse, S. Redrobe, S. Dow, P. Duignan et al. 2003. Tuberculosis in seals caused by a novel member of the *Mycobacterium tuberculosis* complex: *Mycobacterium pinnipedii* sp. nov. *International Journal of Systematic and Evolutionary Microbiology* 53: 1305–1314.

Costa, D.P. and G.L. Kooyman. 1982. Oxygen consumption, thermoregulation, and the effect of fur oiling and washing on the sea otter, *Enhydra lutris. Canadian Journal of Zoology* 60(11): 2761–2767.

Dietz, R., F.F. Riget, C. Sonne, E.W. Born, T. Bechshoft, M.A. McKinney, and R.J. Letcher. 2013a. Three decades (1983–2010) of contaminant trends in East Greenland polar bears (*Ursus maritimus*). Part 1: Legacy organochlorine contaminants. *Environmental International* 59: 485–493.

Dietz, R., F.F. Riget, C. Sonne, E.W. Born, T. Bechshoft, M.A. McKinney, and J. Letcher. 2013b. Three decades (1983–2010) of contaminant trends in East Greenland polar bears (*Ursus maritimus*). Part 2: Brominated flame retardants. *Environmental International* 59: 494–500.

Dold, C. 2014. Cetacea (whales, dolphins, porpoises). In R.E. Miller, M.E. Fowler (eds.), *Fowler's Zoo and Wild Animal Medicine.* Elsevier Saunders, St. Louis, MO, Vol. 8, pp. 422–436.

Geraci, J.R. and L.J. Bruce-Allen. 1987. Slow process of wound repair in beluga whales, *Delphinapterus leucas. Canadian Journal of Fisheries and Aquatic Sciences* 44(9): 1661–1665.

Goldstein, T., J.A.K. Mazet, V.A. Gill, A.M. Doroff, K.A. Burek, and J.A. Hammond. 2009. Phocine distemper virus in northern sea otters in the Pacific Ocean, Alaska, USA. *Emerging Infectious Diseases* 15(6): 925–927.

Goldstein, T., J.A.K. Mazet, T.S. Zabka, G. Langlois, K.M. Colegrove, M. Silver, S. Bargu et al. 2008. Novel symptomatology and changing epidemiology of domoic acid toxicosis in California sea lions (*Zalophus californianus*): An increasing risk to marine mammal health. *Proceedings of the Royal Society B* 275: 267–276.

Goudappel, J.R. and E.J. Slijper. 1958. Microscopic structure of the lungs of the bottlenose whale. *Nature* 182: 479.

Grenfell, B., F. Gulland, and T. Rowles. 2014. Report on RAPID marine mammal morbillivirus workshop. Princeton University, Princeton, NJ, 11pp.

Gulland, F.M.D., J.G. Trupkiewicz, T.R. Spraker, and L.J. Lowenstine. 1996. Metastatic carcinoma of probable transitional cell origin in 6 free-living California sea lions (*Zalophus californianus*), 1979 to 1994. *Journal of Wildlife Diseases* 32(2): 250–258.

Gubbay, S. and R. Earll. 2000. Review of literature on the effects of oil spills on cetaceans. Scottish Natural Heritage Review No. 3. Scottish Natural Heritage, Publications Section Advisory Services, Edinburgh, UK.

Hall, A.J., R.J. Law, D.E. Wells, J. Harwood, H.M. Ross, S. Kennedy, C.R. Allchin, L.A. Campbell, and P.P. Pomeroy. 1992. Organochlorine levels in common seals (*Phoca vitulina*) which were victims and survivors of the 1988 phocine distemper epizootic. *Science of the Total Environment* 115: 145–162.

Haynes, F. and A.H. Laurie. 1937. *On the Histological Structure of Cetacean Lungs*. University Press, Cambridge, UK.

Hicks, B.D., D.J.S. Aubin, J.R. Geraci, and W.R. Brown. 1985. Epidermal growth in the bottlenose dolphin, *Tursiops truncatus*. *Journal of Investigative Dermatology* 85(1): 60–63.

Johnson, S.P., S. Jang, F.M.D. Gulland, M.A. Miller, D.R. Casper, J. Lawrence, and J. Herrera. 2003. Characterization and clinical manifestations of *Arcanobacterium phocae* infections in marine mammals stranded along the central California coast. *Journal of Wildlife Diseases* 39(1): 136–144.

Kennedy, S. 1990. A review of the 1988 European seal morbillivirus epizootic. *The Veterinary Record* 127(23): 563–567.

Lindsay, D.S., K.K. Phelps, S.A. Smith, G. Flick, S.S. Sumner, and J.P. Dubey. 2001. Removal of *Toxoplasma gondii* oocysts from seawater by eastern oysters (*Crassostrea virginica*). *Journal of Eukaryotic Microbiology* 48(s1): 197s–198s.

Lipscomb, T.P., F.Y. Schulman, D. Moffett, and S. Kennedy. 1994. Morbilliviral disease in Atlantic bottlenose dolphins (*Tursiops truncatus*) from the 1987–1988 epizootic. *Journal of Wildlife Diseases* 30(4): 567–571.

Lloyd-Smith, J.O., D.J. Greig, S. Hietala, G.S. Ghneim, L. Palmer, J. St Leger, B.T. Grenfell, and F.M.D. Gulland. 2007. Cyclical changes in seroprevalence of leptospirosis in California sea lions: Endemic and epidemic disease in one host species? *BMC Infectious Diseases* 7:125.

Lyons, E.T., R.L. DeLong, T.R. Spraker, S.R. Melin, J.L. Laake, and S.C. Tolliver. 2005. Seasonal prevalence and intensity of hookworms (*Uncinaria* spp.) in California sea lion (*Zalophus californianus*) pups born in 2002 on San Miguel Island, California. *Parasitology Research* 96: 127–132.

Mahy, B.W.J., T. Barrett, S. Evans, E.C. Anderson, and C.J. Bostock. 1988. Characterization of a seal *morbillivirus*. *Nature* 336(6195): 115–116.

Martineau, D., K. Lemberger, A. Dallaire, P. Labelle, T.P. Lipscomb, P. Michel, and I. Mikaelian. 2002. Cancer in wildlife, a case study: Beluga from the St. Lawrence Estuary, Quebec, Canada. *Environmental Health Perspectives* 110: 285–292.

Miller, M.A., P.A. Conrad, M. Harris, B. Hatfield, G. Langlois, D.A. Jessup, S.L. Magargal et al. 2010. A protozoal-associated epizootic impacting marine wildlife: Mass-mortality of southern sea otters (*Enhydra lutris nereis*) due to *Sarcocystis neurona* infection. *Veterinary Parasitology* 172: 183–194.

Olney, J.W., T. Fuller, and T. de Gubareff. 1979. Acute dendrotoxic changes in the hippocampus of kainite treated rats. *Brain Research* 176: 91–100.

O'Shea, T.J., G.B. Rathbun, R.K. Bonde, C.D. Bergelt, and D.K. Odell. 1991. An epizootic of Florida manatees associated with a dinoflagellate bloom. *Marine Mammal Science* 7(2): 165–179.

Reif, J.S., M.S. Mazzoil, S.D. McCulloch, R.A. Varela, J.D. Goldstein, P.A. Fair, and G.D. Bossart. 2006. Lobomycosis in Atlantic bottlenose dolphins from the Indian River Lagoon, Florida. *Journal of the American Veterinary Medical Association* 228(1): 104–108.

Reynolds, J.E. and S.A. Rommel. 1996. Structure and function of the gastrointestinal tract of the Florida manatee, *Trichechus manatus latirostris*. *The Anatomical Record* 245: 539–558.

Schwacke, L.H., C.R. Smith, F.I. Townsend, R.S. Wells, L.B. Hart, B.C. Balmer, T.K. Collier et al. 2014. Health of common bottlenose dolphins (*Tursiops truncatus*) in Barataria Bay, Louisiana, following the *Deepwater Horizon* oil spill. *Environmental Science and Technology* 48: 93–103.

Sekkal, S., N. Haddam, H. Scheers, K.L. Poels, L. Bouhacina, T.S. Nawrot, H.A. Veulemans, A. Taleb, and B. Nemery. 2012. Occupational exposure to petroleum products and respiratory health: A cross-sectional study from Algeria. *Journal of Occupational Environmental Medicine* 54(11): 1382–1388.

Simeone, C.A., H.H. Nollens, J.M. Meegan, T.L. Schmitt, E.D. Jensen, M.G. Papich, and C.R. Smith. 2014. Pharmacokinetics of single dose oral meloxicam in bottlenose dolphins (*Tursiops truncatus*). *Journal of Zoo and Wildlife Medicine* 45(3): 594–599.

Thomas, N.J., J.P. Dubey, D.S. Lindsay, R.A. Cole, and C.U. Meteyer. 2007. Protozoal meningoencephalitis in sea otters (*Enhydra lutris*): A histopathological and immunohistochemical study of naturally occurring cases. *Journal of Comparative Pathology* 137: 102–121.

Williams, T.D., D.D. Allen, J.M. Groff, and R.L. Glass. 1992. An analysis of California sea otter (*Enhydra lutris*) pelage and integument. *Marine Mammal Science* 8(1): 1–18.

Yeates, L.C., T.M. Williams, and T.L. Fink. 2007. Diving and foraging energetics of the smallest marine mammal, the sea otter (*Enhydra lutris*). *Journal of Experimental Biology* 210: 1960–1970.

Ylitalo, G.M., J.E. Stein, T. Hom, L.L. Johnson, K.L. Tilbury, A.J. Hall, T. Rowles, D. Greig, L.J. Lowenstine, and F.M.D. Gulland. 2005. The role of organochlorines in cancer-associated mortality in California sea lions (*Zalophus californianus*). *Marine Pollution Bulletin* 50(1): 30–39.

chapter fourteen

Toxicology and poisons

John R. Harley and Todd M. O'Hara

Contents

14.1 Introduction

All marine mammals must deal with many classes of poisons, both natural and anthropogenic, via numerous routes of exposure (Figure 14.1). Both marine-based toxins (i.e., domoic acid) and terrestrially originated (intentional and unintentional releases) poisons (i.e., dichlorodiphenyltrichlorethane, DDT) are present in appreciable amounts in the marine environment. Many of the industrial sourced compounds are biologically novel (xenobiotic) on millennial or evolutionary time scales. However, there are a number of marine biotoxins and elements (mercury, cadmium, etc.), including some of the most potent poisons known (e.g., ciguatoxin, brevitoxin, and tetradotoxin) that are produced as secondary metabolites by marine organisms and have likely been present throughout the evolution of marine mammal species (Fowler 1993; Fire and Van Dolah 2012). These poisons have been implicated in a number of acute disease events as well as mass die-offs and likely contribute significantly to the health and management status of some marine mammal populations. For instance, of the 60 identified marine mammal unusual mortality events (UMEs) from 1991 to 2013, the largest proportion of etiologically determined events

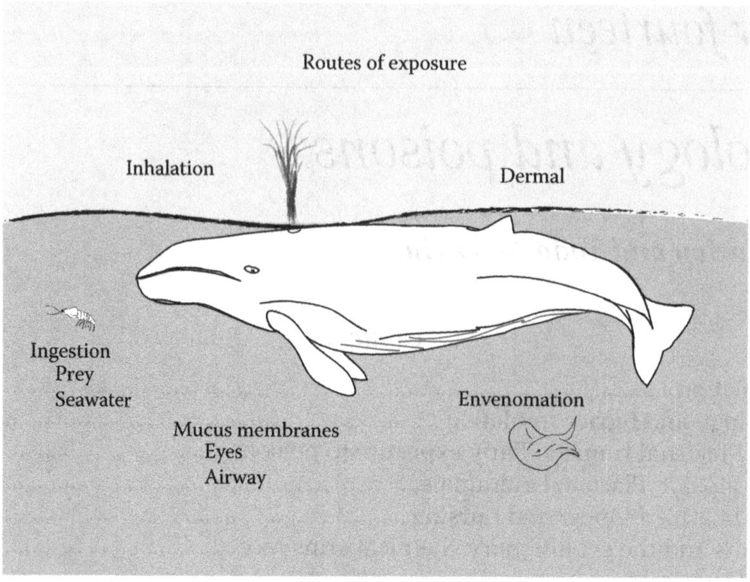

Figure 14.1 **(See color insert.)** A conceptual figure displaying the potential routes of exposure of marine mammals to poisons. Some of these pathways are often overlooked, including the exposure of mucosal membranes to poisons such as brevitoxins or volatile organic compounds. Envenomation may be rare, although some marine mammals are known to feed on venomous animals (i.e., killer whales and stingrays) it has been hypothesized as a potential cause of mortality in some marine mammals. (From Duignan, P.J. et al., *Aquat. Mamm.*, 17, 143, 2000.)

are attributed to biotoxins (see Table 14.1). In fact, several biotoxin-related UMEs, such as domoic acid in California sea lions (*Zalophus californianus*) and brevitoxin in Florida manatees (*Trichechus manatus*) now cause strandings and die-offs on a more-or-less predictable basis, leading to their reclassification as repeated mortality events, as they are no longer, in fact, unusual (F. Gulland, 2014, pers. comm.).

The occurrence of these toxins has likely co-evolved along with host defense systems in the canonical *evolutionary chemical arms race* that has been described in a number of species and ecosystems, and perhaps has contributed to their overt toxicity (Dawkins 1987). Despite an undoubtedly long history of exposure to marine biotoxins, the interaction

Table 14.1 Marine mammal unusual mortality events (UMEs) recorded by the National Marine Fisheries Service (NMFS) from when the UME program was initiated in 1991–2013

Groups affected	Count	Suspected etiology	Count
Cetacean	30	Biotoxin	12
Pinniped	15	Infectious agent	8
Manatee	8	Ecologic factors	6
Sea otter	2	Human interaction	3
Multi-species	5	Undetermined	31
Total	60		60

Source: The table presented here is summarized from data available on the National Marine Fisheries Service, Marine mammal unusual mortality events, National Oceanic and Atmospheric Administration website: http://www.nmfs.noaa.gov/pr/health/mmume/, last updated August 2013.

Note: UMEs have been recorded in the Pacific, Atlantic, and the Gulf of Mexico. Etiology is not necessarily causally determined in every case, but suspected or implicated.

between toxin-producing organisms and marine mammals cannot be said to exist in static isolation; rather, the spatial, temporal, and geographic range of both these groups of organisms are influenced by environmental conditions and are constantly shifting. With climate change altering the seasonal and geographic ranges of (among others) biotoxin-producing harmful algal blooms (HABs), climate variability is an important consideration in the poison exposure paradigm (Moore et al. 2008).

In addition to traditional poisons, marine mammals must also cope with other stressors that result from their unique and often highly specialized physiology, behavior, and diet that are inextricably linked. All marine mammals are air-breathing and, with a few exceptions, forage underwater. During dives, marine mammals deplete their on-board oxygen supplies (hypoxia) which promotes the formation of acutely toxic by-products such as lactic acid, hydrogen ions, and radical oxygen species (Berta et al. 2006). While a more thorough review of the adaptations of diving physiology are presented elsewhere in this textbook (Chapters 2 through 4), it is important to consider that due to their rapid bursts of apneic activity marine mammals must tolerate anaerobic cellular conditions and the generation of a number of toxic by-products using a battery of defense mechanisms, usually antioxidant in nature. These processes inherently alter the response of diving mammals to other toxicants.

While marine mammals display tolerance and resistance to a number of poisons, some physiologic adaptations (populations) and/or acclimations (individuals) of marine mammals to diving may have predisposed some species to be more susceptible to certain chemicals, while the opposite may occur in different host species or chemical classes. For instance, being endothermic mammals in often cold water conditions, most marine mammals have developed thick layers of blubber for insulation and energy stores. However, this blubber layer also tends to accumulate lipophilic compounds such as polychlorinated biphenyls (PCBs) as well as limit the amount of partitionable room for water-soluble compounds since most blubber has low percent water composition (Hoekstra et al. 2002; Van Dolah et al. 2003). Furthermore, during dives blood is directed to the heart and brain to increase oxygen delivery to these highly aerobic tissues (Chapter 2). However, in doing so poisons in the bloodstream are similarly delivered at a greater magnitude per unit of time and tissue mass to these sensitive tissues, thus limiting their biotransformation and elimination in so-called filtering organs such as the liver or kidneys (Geraci et al. 1989).

It is critical to not extrapolate with respect to a chemical's adverse effect on *marine mammals* as this is a non-monophyletic group with a large diversity of lineages and only grouped together based on generalized habitat relationships (i.e., lives in or feeds in the ocean). In the same vein, it is imprudent to generalize regarding the physiology or life-history characteristics of marine mammals, however, many marine mammal clades share traits either conserved from previous common ancestors or co-evolved that allow us to make comparisons. Mammals have evolved complex and often elegant detoxification and sequestering mechanisms for poisons, and adaptive machinery that is present in marine mammals is often found across mammalian lineages. While particular xenobiotic biotransformation pathways may not be unique to marine mammals, it is important to discuss them in the context of adaptation since marine mammals are exposed to different classes of poisons and at varying concentrations than their terrestrial counterparts. As an example, consider the detoxification and sequestration process of a potent neurotoxin, monomethylmercury ($MeHg^+$), to an insoluble crystal complex with selenium (Se) and other components which occurs in the livers of marine mammals (Koeman et al. 1975; Lailson-Brito et al. 2012). While this process is also known to occur in some terrestrial animals, the extent of $MeHg^+$ exposure in fish-consuming mammals in the marine environment is often much higher than in terrestrial systems, leading to hypotheses involving detoxification thresholds, efficiencies, and potential limits in marine mammal species. Furthermore, some of these physiologic and cellular defenses may utilize one or more micronutrients or cofactors, such as Se, which is present in relatively higher concentrations in prey species of marine environment with respect to similar terrestrial trophic levels (Figure 14.2). In fact, Se species (in aquatic organisms) often exist in concentrations at or above thresholds for chronic toxicity for some domesticated terrestrial species (Puls 1994) (Dietz et al. 1996), leading to the paradoxical inference that Hg can reduce Se toxicity (Figure 14.3, Khan and Wang 2009; Lemire et al. 2012). For instance, Dietz et al. (1996) found concentrations of

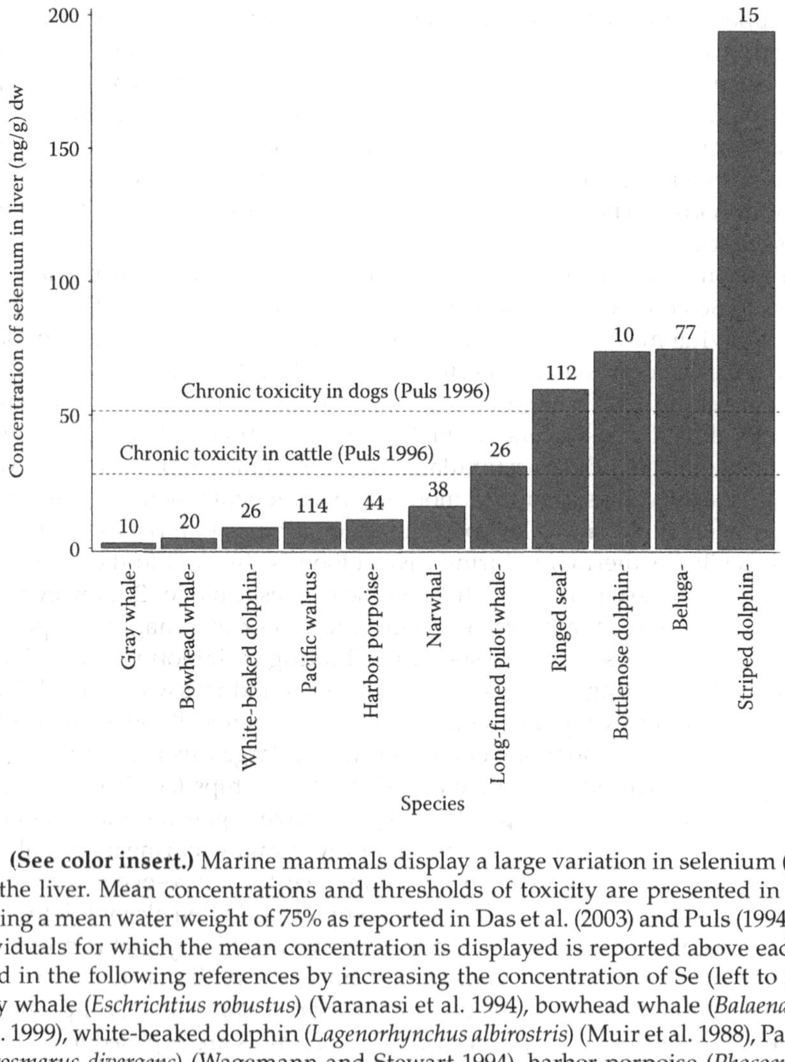

Figure 14.2 **(See color insert.)** Marine mammals display a large variation in selenium (Se) concentrations in the liver. Mean concentrations and thresholds of toxicity are presented in dry weight (dw) assuming a mean water weight of 75% as reported in Das et al. (2003) and Puls (1994). The number of individuals for which the mean concentration is displayed is reported above each bar. Data are reported in the following references by increasing the concentration of Se (left to right in the figure); gray whale (*Eschrichtius robustus*) (Varanasi et al. 1994), bowhead whale (*Balaena mysticetus*) (Krone et al. 1999), white-beaked dolphin (*Lagenorhynchus albirostris*) (Muir et al. 1988), Pacific walrus (*Odobenus rosmarus divergens*) (Wagemann and Stewart 1994), harbor porpoise (*Phocoena phocoena*) (Paludan-Müller et al. 1993), harbor seal (*Phoca vitulina*), narwhal (*Monodon monoceros*) (Wagemann et al. 1983), long-finned pilot whale (*Globicephala melaena*) (Muir et al. 1988), ringed seal (*Pusa hispida*) (Smith and Armstrong 1978), bottlenose dolphin (*Tursiops truncatus*) (Kuehl and Haebler 1995), beluga (*Delphinapterus leucas*) (Wagemann et al. 1996), and striped dolphin (*Stenella coeruleoalba*) (Itano et al. 1984). (Adapted from Das, K. et al., Heavy metals in marine mammals, in *Toxicology of Marine Mammals*, Taylor & Francis, New York, 2003, pp. 135–167.)

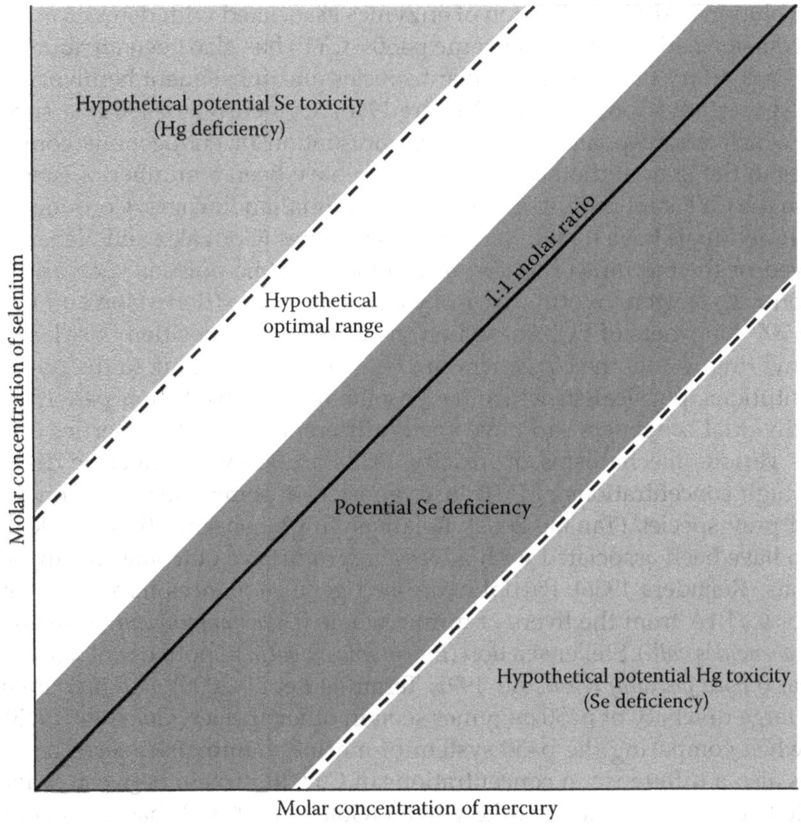

Figure 14.3 **(See color insert.)** Mercury and selenium display a multi-faceted antagonistic relationship in marine mammals, including the formation of an insoluble mineral (HgSe, tiemmannite) occasionally resulting in a near 1:1 molar correlation (i.e., Koeman et al. 1973; Koeman et al. 1975; Brookens et al. 2007). As noted by Khan and Wang (2009) the 1:1 molar ratio can potentially be viewed as Se deficient if Hg–Se is not bioavailable, wherein selenium is unable to participate in Se-dependent detoxification processes (i.e., glutathione peroxidase, selenoproteins). We emphasize that, somewhat counterintuitively, with high concentrations of Se seen in marine mammals (see Figure 14.2) that potential selenosis could occur in marine mammals as a result of Hg deficiency. (Adapted from Khan, M. and Wang, F., *Environ. Toxicol. Chem.*, 28, 1567, 2009.)

Se up to 7.69, 4.99, and 9.09 μg/g in the liver of ringed seals, beluga whales, and polar bears, respectively; which compare favorably to the lower thresholds of chronic toxicity reported in the liver for cattle (1.25–7.0 μg/g), pig (4.0–20.0 μg/g), and rabbit (7.04 μg/g) (Puls 1994). Research into the diversity of selenocysteine (Sec) incorporating proteins (selenoproteomes) found that terrestrial eukaryotes have a smaller and less diverse selenoproteome than aquatic eukaryotes (Lobanov et al. 2007). The authors of the aforementioned study attribute the diversity of selenoproteomes in aquatic species to, at least in part, the abundance of bioavailable selenium (organic selenides) in the marine environment. While terrestrial mammals likely share a number, if not all, of their related marine mammal selenoproteins (a comparative study between mammalian selenoproteomes is not known), it is worth noting that Lobanov et al. (2007) found that the lack of diversity of terrestrial selenoproteomes was due to several independent losses, thus, it is conceivable that marine mammal selenoproteomes differ from their terrestrial counterparts.

The evolution and diversification of enzymes associated with detoxification and xenobiotic biotransformation (i.e., cytochrome p450s; CYP) has also been attributed to the production of secondary metabolites by plant species and subsequent herbivory, an example of chemical warfare (Gonzalez and Nebert 1990). Cytochrome P450 is a superfamily of enzymes which are responsible for biotransformation of endogenous compounds and xenobiotics in the liver and other tissues. There have been a number of isozymes identified within the CYP superfamily in nearly all mammalian lineages. Consequently, species and even individuals have high diversity of p450 genes (Gonzalez and Nebert 1990). CYPs are involved in the biotransformation of complex organic poisons such as PCBs via the conjugation with oxygen to form a hydroxylated metabolite (Robertson and Hansen 2001). There are 209 congeners of PCBs based on the number and position of chlorine atoms on the biphenyl ring. While the congeners can be grouped according to the number of chlorine substitutions, physical structure (i.e., coplanar), or substitution pattern (para, meta, ortho), individual congeners can have vastly different physical properties and toxicities, as well as various mechanisms of toxicity. PCBs are known endocrine disrupters, and relatively high concentrations of PCBs are reported in some marine mammals and their associated prey species (Tanabe 2002). Relatively high concentrations of PCB congeners of concern have been associated with adverse reproductive outcomes in ringed seals and harbor seals (Reijnders 1986). Partial expressed gene sequences have been obtained via RT-PCR for CYP1A from the livers of minke whale (*Balaenoptera acutorostrata*), Dall's porpoise (*Phocoenoides dalli*), Steller sea lion (*Eumetopias jubatus*), spotted seal (*Phoca largha*), and ribbon seal (*Phoca fasciata*) (Goksøyr 1995; Teramitsu et al. 2000). Perhaps unsurprisingly, given the large diversity of p450 enzymes seen in other species, Goksøyr (1995) found differences when comparing the p450 system of marine mammals to terrestrial mammals. There was also a difference in concentrations of CYP2B protein between seals (harp seal and hooded seal) and whales (minke whale) (Goksøyr 1995). It has been suggested that small cetaceans have a limited capacity to biotransform certain PCB congeners compared to terrestrial mammals, perhaps due to limited activity of phenobarbital or methylcholanthrene-dependent microsomal systems (Tanabe et al. 1988). It is conceivable that, due to their ancient divergence from terrestrial herbivorous ancestors, some cetaceans have lost certain p450 enzymes as compared to terrestrial mammals. In humans, the mechanism of PCB biotransformation appears to be dependent on the formation of an arene oxide intermediate (Tanabe et al. 1988; Robertson and Hansen 2001) and Tanabe et al. (1988) suggested the lack of observed biotransformation of certain isomers of PCBs in small cetaceans was a result of the inability to form this hydroxylated intermediate.

What follows is a discussion of several driving forces of physiologic adaptation of marine mammals to poisons as well as discussion of individual organism acclimation. Attention will be given to traditional poisons (chemicals), and physical forms of *contamination* such as debris, external oiling and sound pollution will be only briefly discussed. Specific examples will be provided by order in subsequent sections based on the specific process of interest.

14.2 Physiology and bioaccumulation

Some marine mammals have been advocated as ecosystem sentinels for environmental poisons due to their unique physiology and ecology (Ross 2000) that integrates multiple components of an ecosystem. Unfortunately, many of the characteristics that cause them to be effective sentinels also result in higher toxicant concentrations than relevant prey species (biomagnification). Marine mammals are generally long-lived, and most occupy high trophic levels (strict predators). Many studies have reported concentrations of some poisons positively correlated with age, although these relationships are often complicated by growth/dilution dynamics and reproductive status (Dietz et al. 1996; Knott et al. 2011). For instance, Ross et al. (2004) found increasing concentrations of persistent organic pollutants (POPs, i.e., PCBs, PCDEs) in older male harbor seals, but found decreasing concentrations in females associated with parity and lactation (i.e., transfer to pup). The effect of reproduction on poison accumulation and offloading in marine mammals is an important driver of observed concentrations in adult females and represents a significant exposure route *in utero* and to nursing pups. These dynamics will be discussed further in subsequent sections.

Some lipophilic poisons can increase in concentration with increasing trophic level due to biomagnification (van de Vijver et al. 2003; Coelho et al 2013, numerous others). Since many marine mammals are strict carnivores, they often assume high trophic positions, especially among marine mammal species that prey upon other marine mammals such as killer whales (Saulitis et al. 2000), polar bears (Gormezano and Rockwell 2013), leopard seals (Boveng et al. 1998), and walruses (Lowry and Fay 1984). Trophic transfer of poisons is often complicated by varying abilities of taxonomic groups, or even individual species to sequester or biotransform certain poisons (Tanabe et al. 1988). In some instances, biotransformation at certain trophic positions can greatly alter the potency of a particular poison. For instance, ciguatoxin, which is a potent neurologic biotoxin associated with the dinoflagellate *Gambierdiscus toxicus*, enters lower trophic positions as gambiertoxins which are produced by *G. toxicus*. As the gambiertoxins are biotransformed by herbivorous and predatory fish, they are oxidized into more potent ciguatoxins (Lewis and Holmes 1993; Van Egmond et al. 2004).

14.2.1 Maternal transfer: Early exposure to poisons

14.2.1.1 Placental mammals

For many species, especially placental mammals, one of the most sensitive cohorts to poisoning is the fetus and neonate (transplacental and lactational transfer). As opposed to oviparous animals, in which embryos do not develop within the mother, marine mammal fetuses are exposed to maternal toxicants both *in utero* (for compounds which can cross the placental barrier; transplacental) as well as during nursing (trans-mammary). For instance methylmercury is transferred to the developing fetus through the placenta, and while studies of embryonic development in marine mammals are rare, methylmercury has been shown to impair development and cause adverse health outcomes in fish eating humans and wildlife (Grandjean et al. 1994; Debes et al. 2006; Johansson et al. 2007; Tonk et al. 2010). Some of these studies and criteria (oral intake and tissue concentrations) have been used to extrapolate potential adverse effects in marine mammals.

Interestingly, one of the main excretory routes for methylmercury, besides excretion in urine and bile, is through hair growth (Clarkson and Magos 2006). Mammalian hair is composed of large amounts of the protein keratin, which contains a large number of cysteine residues which form the disulfide bridges necessary for its rigidity. Methylmercury binds tightly to the sulfur-containing cysteine residues, and thus hair concentrations of methylmercury are often several-fold higher than blood or other tissues (e.g., Rea et al. 2013). Both pinniped and cetacean fetuses develop hair *in utero* (lanugo), which in pinnipeds can persist for several weeks post-partum before it is shed and a new coat develops. Most mammals, including humans, display signs of lanugo development *in utero*, which is likely a conserved trait harkening back to common ancestors which maintained full coats of hair throughout their life histories, and therefore cannot be easily implicated as a direct adaptation of the developing fetus to excreting toxicants. However, it is interesting to speculate that the shedding of contaminants *in utero* via lanugo perhaps persists as one of the selective pressures to protect from neurotoxicity in situations where mercury exposure may be high.

14.2.1.2 Blubber and nursing

The term *blubber* is often misrepresented in both popular culture and occasionally scientific literature, so we will use the definition provided by Reeb et al. (2007) where blubber is integument layers below the epidermis (dermis and hypodermis). As sizeable endotherms that make their living in the marine environment, many marine mammals utilize extensive lipid-rich blubber tissue for thermal insulation and other needs (store nutrients, enhance hydrodynamics, protection from predators, etc.). In migratory species, such as gray whales (*Eschrichtius robustus*), not only are individual animals often subject to large changes in ambient water temperature, but migratory periods are often marked by little or no foraging activity (fasting). The laying down and mobilization of energy stored in adipose tissue is truly remarkable in marine mammals. While duration of nursing is often radically different between marine mammal species (on scale of a few weeks to several years), in some species with short nursing periods, such as the northern elephant seal (*Mirounga angustirostris*), females can lose up to 42% of their initial body mass during nursing (58% of adipose mass) (Costa et al. 1986). The energetics of nursing is discussed in Chapters 8 and 10, however, the transfer of toxicants during lactation is interesting to consider from the perspective of both the lactating female and the nursing pup. While the nursing young is exposed to a milieu of potential toxicants during a sensitive developmental stage, it does represent a significant excretory route for reproductive females (Addison and Brodie 1987; Debier et al. 2006; Frouin et al. 2012). A decrease in maternal contaminant concentrations following parity has been described in sea otters (Jessup et al. 2010). In particular, primiparous females (females giving birth to their first offspring) often transfer a majority of their blubber (lipid store)-based contaminant load to their firstborn. Indeed, Beckmen et al. (2003) found increased blood concentrations of PCB congeners as well as *p,p*-DDE in pups born to young dams compared to pups born to older, presumably multiparous dams. Similarly, Ylitalo et al. (2001) found higher concentrations of OCs in first-recruited resident male Alaskan killer whales as compared to second or later recruitment.

14.3 Biotoxins and the evolutionary chemical arms race

Toxins that are produced by prey species either as defensive mechanisms or as secondary metabolites play a role in the predator–prey relationship. While examples exist of specific adaptations to poisons in predator–prey relationships (i.e., the skin of the rough-skinned newt *Taricha granulosa* produces tetradotoxin to which the common garter snake *Thamnophis sirtalis* demonstrates resistance), it is often difficult to assign a *quid-pro-quo* physiologic or behavioral adaptation of a species or taxonomic group as a result of exposure to a poisonous prey (Brodie and Brodie 1990).

A number of marine organisms have been identified that produce compounds that are toxic to marine mammals such as dinoflagellates, diatoms, and prokaryotes (Valério et al. 2010) and there are a number of fish, mollusk, and cnidarian species that are poisonous or venomous to marine mammals. These poisons are often encountered by marine mammals through diet. However, saxitoxin produced by the dinoflagellate *Gonyaulax catenella* (responsible for some *red tides* along with *Karenia brevis*) can be aerosolized by cell lysis via wave/wind action and has been known to cause respiratory problems in humans and other mammalian models (Franz and LeClaire 1989; Kirkpatrick et al. 2004). The saxitoxins have been identified as the toxins associated with paralytic shellfish poisoning (PSP) which has been linked to episodic mass poisonings of humans, marine mammals, and birds (Van Dolah et al. 2003).

While the co-evolution of chemical warfare between predator and prey has undoubtedly been occurring for millennia, it is perhaps surprising that cetaceans (and pinnipeds) do not display more resistance to marine biotoxins. Historically, diagnostic methods have limited the ability to identify or detectably measure algal biotoxins, and thus the etiology of mass-strandings and die-offs was uncertain. However, with increasingly sensitive and precise assays, there is now good evidence to suggest that algal biotoxins are at least partially responsible for previously unexplained unusual mortality events (UMEs) (Scholin et al. 2000; Van Dolah et al. 2003). The unique nature of HABs, including non-species-specific mortality, has led some to speculate about evidence for mass-strandings in fossilized remains from the late Miocene epoch (approximately 7 Mya) (Pyenson et al. 2014). At a site in the Atacama desert of Chile, Pyenson et al. (2014) found four layers of multi-species assemblages stratified in the sedimentary substrate, including species of Balaenopteridae, Phocidae, and an aquatic sloth (*Thalassocnus natans*). The unique orientation (supine) and proximity of the remains led the authors to conclude that the individuals died rapidly at sea and were promptly washed ashore (spatial focusing) by the unique structure and currents of the supratidal flat (Pyenson et al. 2014). While pathology of 6–9 Myr old fossils is undoubtedly filled with uncertainty, it is worth noting that HABs have likely contributed to the evolution of marine mammals and have even been hypothesized to have played a role in Phanerozoic mass extinctions (Castle and Rodgers 2009).

Marine mammal mortality events have been associated with other marine biotoxins such as domoic acid (Scholin et al. 2000) and brevetoxin (Flewelling et al. 2005). The task of attributing mass-strandings and die-offs to marine biotoxins is often complicated by other factors such as viruses, bacteria, and other pathogens as well as nutrition and immunologic status of the individuals and populations. Rather than being pesky confounding variables, however, these factors likely contribute to the overall health picture of the animal, and the balance of two or more factors may well tip the scales from health to outright disease. With respect to some mortality events documented in the USA, there is very strong diagnostic evidence that some HAB-related mortality events for certain species are now considered endemic (e.g., California sea lions and domoic acid, manatees and brevetoxin; F. Gulland, pers. comm.).

Since most marine mammal species are not directly consuming the toxin-producing organisms, it is important to consider the trophic transfer and biotransformation of biotoxins. For instance, Bricelj et al. (2005) found that genetic mutations in the sodium channel pore region of wild softshell clams (*Mya arenaria*) decreased affinity of the receptors to PSTs (saxitoxin and tetrodotoxin). These *resistant* clams were able to tolerate much higher concentrations of STX than their *sensitive* wild-type counterparts, which is presumably favorable for the clams but rather unpropitious for the species which consume them. Saxitoxin is a fascinating chemical, not only because it is a Schedule I Chemical Warfare Agent but because it is produced by two kingdoms of life (cyanobacteria and dinoflagellates; Cusick and Sayler 2013). The two synthesis pathways appear to have evolved independently, albeit based on conserved ancestral proteins (Hackett et al. 2013). Several species of dinoflagellates that do not produce saxitoxin share homologs to toxic species, suggesting that the evolution of the synthesis system was not exclusively driven toward toxin production (Cusick and Sayler 2013; Hackett et al. 2013). Indeed, the main site of the saxitoxin toxicity in mammals (sodium channels) had not evolved when genes associated with saxitoxin production (*sxt* cluster) appeared in cyanobacteria (Murray et al. 2011).

14.3.1 Cetaceans

Large whales and other cetaceans are exposed to toxins present in HABs in areas where their temporal and spatial ranges overlap. Evidence of PSTs and DA in fecal samples of North Atlantic right whales (*Eubalaena glacialis*) suggests that individuals are exposed to these algal toxins on a regular basis (70%–80% contained traces of PST, 25%–30% contained traces of DA, and 22% of the fecal samples collected showed exposure to both DA and PST, Doucette et al. 2012). Interestingly, while PST is often recognized accumulating in shellfish, the fact that it can accumulate in the tissues of highly motile fish such as cod raises some interesting ecologic considerations. For instance, during the fall of 1987, 14 humpback whales (*Megaptera novaengliae*) died in Cape Cod from apparent saxitoxin poisoning (Geraci et al. 1989). While plankton and shellfish in the immediate vicinity did not contain traceable concentrations of saxitoxin, stomach content analysis of the whales found a high prevalence of mackerel, which spawn in the waters of the Gulf of St. Lawrence. The concentrations of STX in mackerel liver were 154 μg/100 g tissue, which is nearly double the threshold for human consumption of mussels of 80 μg/100 g mussel (Geraci et al. 1989). It is tempting to view HABs as limited by the spatial and temporal lifespan of the toxin-producing algae, however, when considering the lifespan and motility of fish consumers able to accumulate the toxin, the overall scope of the effects of HABs increases dramatically.

14.3.2 Pinnipeds

Saxitoxin has been implicated in the mortality of a number of Hawaiin monk seals (*Monachus schauinslandi*) due to the presence of the toxin in seal tissues (mean 12.55 ± 10.09 μg dcSTX per 100 g tissue, $n = 7$) and prey species, as well as pulmonary trauma consistent with drowning due to paralysis (Hernández et al. 1998). However, the authors conclude that, as with most cases of algal toxin poisoning, without reference concentrations, it is difficult to establish the etiology of mortality. There is some speculation that ciguatoxin, a toxin present in reef fish, has contributed to the decline of Hawaiian monk seal populations, although currently this remains rather speculative (Van Dolah et al. 2003). Recent work has shown detectable levels of ciguatoxin activity in blood samples from Hawaiian monk seals (0.43–5.49 pg/ml P-CTX-1 equivalent), affirming prior speculation that individuals were indeed exposed to the toxin and transfer to marine mammals is possible (Bottein et al. 2011).

Although the impact of biotoxins on Hawaiin monk seals is concerning due to their conservation status, the power of these poisons to influence pinniped health on a population scale is perhaps better demonstrated in the dense and closely monitored colonies of California sea lions (*Zalophus californianus*) and harbor seals (*Phoca vitulina*) along the California coastline. In 1998, a UME of California sea lions was convincingly linked through multiple lines of evidence to domoic acid—a neurotoxin produced by diatoms of the *Pseudo-nitzschia* genus (Scholin et al. 2000). Mammalian exposure is due primarily to dietary consumption of contaminated prey species (Lefebvre and Robertson 2010), which in the case of California sea lions appears to be largely attributed to the filter feeding anchovy (Lefebvre et al. 1999; Scholin et al. 2000). Domoic acid is structurally analogous to glutamate, a neurotransmitter, which is an excitatory amino acid (Pulido 2008). Domoic acid activates ionotropic glutamate receptors which overstimulates the neuron by allowing high levels of calcium to enter the cell and also prevents rapid desensitization in a mechanism consistent with excitotoxicity (Jeffery et al. 2004; Ramsdell and Stafstrom 2009).

Intriguingly, although several isomers of domoic acid have been identified it appears that deleterious effects are caused by a single toxic species, the same chemical species that is found accumulating in high concentrations in shellfish (Iverson et al. 1989). This would appear to suggest that the absorbed molecule does not undergo any biotransformation processes, as opposed to other compounds such as gambiertoxin/ciguatoxin. Furthermore, Suzuki and Hierlihy (1993) found that in rats nearly 100% of administered domoic acid was recovered in the urine, suggesting that biotransformation and other forms of elimination play little to no role in the toxicodynamics of domoic acid in mammals.

Concentrations of domoic acid of 96.8 μg/g in feces were found in California sea lions exhibiting acute signs of neurotoxicity including seizures, erratic head movements, and ataxia (Gulland et al. 2002; Bargu et al. 2012). Though catastrophic neurologic effects have been well documented, vexing results in rats suggest that the blood–brain barrier is effective in limiting the distribution of the toxin into the brain (Preston and Hynie 1991). However, the excitatory nature of domoic acid is consistent with the idea that acute neurologic effects can be induced by relatively small concentrations in the brain tissue. Recent evidence has suggested a host of deleterious effects on a variety of tissues, including myocardial lesions in California sea lions (*Zalophus californianus*) which may explain rapid lethality following acute exposure (Gulland et al. 2002; Pulido 2008).

Linking disease of stranded California sea lions to blooms of a toxic diatom can be difficult, especially considering confounding considerations such as the patchiness of diatom blooms, wind or current transport of blooms as well as lag time associated with disease progression (Goldstein et al. 2008). For instance, Goldstein et al. (2008) found clinical signs of both chronic and acute domoic acid toxicity associated with abnormal epileptiform discharges, but high incidence of chronic neurologic cases lagged behind cases of acute toxicity by 4 months. It is conceivable that the observed lag time in chronic toxicity is caused by the progression of toxicosis due to low level exposure, however, another more feasible explanation is that signs of chronic toxicity are merely a progression of an acute sub-lethal toxic response.

There are a several interesting observations that arise from studying stranding data from California sea lions from the last 15 years (Goldstein et al. 2008; Bargu et al. 2012). Strandings associated with acute and chronic domoic acid toxicity were reported in every year from 1998 to 2006 except 1999 (Bejarano et al. 2008). This may be explained by an increased stranding response effort or improved diagnostics and surveillance, or both. However, there is also evidence that blooms of *Pseudo-nitzschia* have co-occurred during periods of increased domoic acid poisoning frequency (Scholin et al. 2000; Bejarano et al. 2008). The majority of strandings associated with domoic acid appear to be adult females (68%) as opposed to non-domoic acid cases which are dominated by pups (59%) (Bejarano et al. 2008). While this difference could be associated with sex-based differences in toxicodynamics or activity of domoic acid, it seems a more plausible explanation is differences in exposure related to foraging ecology. For instance, the breeding season for California sea lions occurs during the summer months in California (May–August), which generally coincides with blooms of *Pseudo-nitzschia*. Adult females spend a large portion of the breeding season foraging at sea, while the adult males are generally hauled out defending territory (Francis and Heath 1991; Bejarano et al. 2008). The high incidence of adult female poisonings during this critical time period is especially concerning from an epidemiologic standpoint when considering the potential role of domoic acid in abortion and adverse birthing outcomes in California sea lions (Goldstein et al. 2009).

14.3.3 Other

Other marine mammal clades exhibit a wide range of responses to biotoxins, and in particular those associated with HABs. An unexplained die-off of Alaskan sea otters (*Enhydra lutris*) at Kodiak Islands, AK in 1987 may have been associated with saxitoxin, after investigators reported finding concentrations of PSP toxin in blue mussels (*Mytilus edulis*) 50 times above the upper threshold for human consumption in Alaska (>5800 μg/100 g mussel as compared to the threshold of 80 μg/100 g mussel, Degange and Vacca 1989). In a controlled setting, sea otters appear to reduce their feeding rates when presented with clams containing high concentrations (226 ± 96 μg/g STX) versus clams with lower concentrations (37 ± 9 μg/g STX) of saxitoxin. In this study, the otters also avoided the siphons, gills, and pericardial tissue of clams, which contain higher concentrations of the toxins compared to the other clam tissues (60%–80% of total STX sequestered by clam). The method of presumed chemodetection is unknown, however, the authors speculate that the otters may be responding to a tingling sensation reported in humans exposed to PST, or they may have STX-specific gustatory receptors such as those found in fish (Yamamori et al. 1988).

14.4 Toxic oxygen: Adaptations to anaerobic diving

While cellular and physiologic adaptations to diving will be discussed elsewhere in this volume, it is worth noting that some of the most potent poisons that marine mammals are exposed to come not from environmental exposure but from their own physiology. As air-breathing mammals that forage underwater, marine mammals have developed a unique physiology that allows them to cope with extended periods of anaerobic activity.

Molecular oxygen (O_2) is often thought of as an incontestably wholesome chemical, given its necessity for cellular respiration and the swift and severe consequences that arise in the absence thereof. However, the atom O is quite damaging in many forms that are produced via normal physiologic processes and/or as a consequence of some toxins. An array of O by-products are known to cause damage to macromolecules, some changes being irreversible. In particular, the apneic behavior of many marine mammals (diving, sleeping) creates risk of a number of oxygen-related injuries such as ischemia–reperfusion injury and the generation of reactive oxygen species (ROS). Voluntary breath-holds in northern elephant seal pups have been shown to increase the activity of hypoxanthine (which generates ROS upon reperfusion) in plasma, but without observable effects to biomarkers of oxidative stress (4-hydroxynonenal and 8-iso prostaglandin F2α) (Vázquez-Medina et al. 2006). Similarly, basal concentrations of CO and carboxyhemoglobin (COHb) in Weddell seals and northern elephant seals are several times higher than in human smokers (Pugh 1959; Tift et al. 2014). Both Tift et al. (2014) and Vázquez-Medina et al. (2006) suggest that endogenous production of CO and XO confers protective effects against oxidative damage caused by apnea. Carbon monoxide (CO) is widely considered to be a toxic gas and has been known to cause hypoxia and death in humans due to its affinity for heme protein which prevents oxygen from appropriately binding to hemoglobin, lowering blood O_2 saturation levels. However, CO is produced endogenously, largely through the breakdown of heme, which also produces biliverdin. While biliverdin is water soluble and rapidly excreted, biliverdin reductase (BVRA) reduces biliverdin to bilirubin which is not water soluble and a well-known neurotoxicant (Ostrow et al. 2004). Barañano et al. (2002), among others, noted the ostensive paradox of an energy-requiring process evolving which converts a non-toxic product into a toxic compound. However, CO, as well as other products of the breakdown of heme protein (such as bilirubin and biliverdin) are known to have antioxidant properties as well as protect against ischemia–reperfusion injury (Stocker et al. 1987; Barañano et al. 2002). Barañano et al. (2002) found that relatively low concentrations of bilirubin have cytoprotective effects through redox cycling with bilverdin reductase (BVRA), and that BVRA might be more important than traditional antioxidants such as glutathione in oxidative cytoprotection. There is some evidence to suggest that marine mammals have higher rates of endogenous production of antioxidants (including glutathione, superoxide dismutase, and glutathione peroxidase) compared to terrestrial mammals, presumably as an adaptation or acclimation to dealing with periods of apnea (Wilhelm Filho et al. 2002). This will alter their response to environmental poisons unrelated to molecular oxygen products and by-products of cellular respiration.

14.5 Metals and detoxification

High trophic level feeding marine mammals (fish and marine mammal eating) often have high concentrations of some elements (Hg, Se) compared to many terrestrial mammals, while cadmium (Cd) is often associated with consumption of invertebrates (Dehn et al. 2005). Despite reports of concentrations above levels of concern in other mammalian species (Dietz et al. 1998, 2013; Rea et al. 2013) there have been few reports of poisoning events (signs of toxicosis) associated with heavy metals, although this may be due to the difficulties associated with performing detailed clinical assessments and necropsies on marine mammals of suitable condition.

There has been speculation that Arctic marine mammals have adapted to high levels of cadmium (Dietz et al. 1998). Temporal trends assessments for Cd in the Arctic environment in the last few decades do not yield themselves to clear interpretation (Riget and Dietz 2000), and it is conceivable that Cd has been at appreciable levels at least since the fifteenth century (Hansen et al. 1989). The toxic effects of Cd have been known for several decades, the most infamous poisoning event occurring in Japan's Toyama Prefecture in the early 1900s, although the link between the "itai-itai" (*ouch-ouch*) disease and Cd was not reported until 1968. Cadmium accumulates in the kidney and the liver and is a well-known nephrotoxin, where it causes glomerular and tubular damage (for review of toxicity in humans see Godt et al. 2006). Chronic Cd poisoning is associated with a number of bone maladies, including osteoporosis and increased occurrence of fractures, both of which appear to be due to decreased rates of bone mineralization. This is likely related to renal dysfunction (Berglund et al. 2000), although the exact mechanisms for altered mineralization due to Cd toxicity in unknown.

Metallothioneins (MTs) are a conserved family of proteins among mammals, with isoforms identified in a number of species of marine mammals (for review see Das et al. 2000). The element-binding metallothioneins function in uptake, transfer, and excretion of both essential (such as Cu and Zn) and nonessential metals (such as Ag, Hg, Cd). The majority of metal interactions (binding sites) take place on thiol groups of cysteine residues, which can account for up to 30% of MT residues. Metallothioneins have been postulated to be involved in defending organisms against heavy metals such as Hg and Cd, where metals bound to MT are less toxic than free metals. The relationship between MTs and Hg is not clear—binding of Hg to MTs in the kidney and the liver of rats has been observed for several decades (Wiśniewska et al. 1970; Trojanowska and Sapota 1974), and species of Hg have been clearly demonstrated to induce MTs in other mammalian models. Indeed, it was assumed that sequestration by MT is a main detoxification function for mercury (Roesijadi 1992). However, studies in marine mammal species have indicated that a relatively small percentage of mercury is bound to MTs in the liver and kidneys. For instance, one study found that the percentage of tissue Hg bound to metallothioneins ranged from 14.5% to 20.0% in the kidney and 2.0% to 3.2% in the liver of California sea lions (*Zalophus californianus*) (Lee et al. 1977). Similarly, low percentages of Hg bound to MT have been found in narwhals (5% in the liver, 10% in the kidney, Wagemann et al. 1984), and pilot whales (79% of THg in the insoluble fraction, Caurant et al. 1996). It is possible that MTs represent an acute response to intracellular $MeHg^{+}$ or Hg^{2+}, and that transient sequestration via binding to MTs facilitates transport to other organs or eventual biotransformation of metals into inactive forms (e.g., tiemmanite).

The low percentages of Hg bound to MTs are intriguing because it has been proposed that toxic effects of heavy metals occur when free metals exceed the binding capacity of MTs, or other high-affinity scavenging proteins. The *spillover hypothesis* was first proposed by Winge et al. (1974) who showed that pathologic effects of Cd were only seen when free Cd entered other protein fractions. The situation becomes more complex, however, when considering that there is more than one metal present and these metals have different concentrations and binding affinities for the numerous isoforms of MT proteins (Hamilton and Mehrle 1986).

Interestingly, the expression of MT mRNA has been shown to increase following exposure to mercury chloride (Hg^{2+}), accompanied by an increase in the MT concentration (Reus et al. 2003). The authors speculate that MTs bind Hg due to the high concentration of sulfhydryl groups (–SH) on cysteine residues, and that the MT–Hg complex enters the nucleus where it disassociates and attaches the metal response element of the MT gene, resulting in an increase in the MT expression. The mammalian MT gene has several 5′ upstream elements that allow it to responds to stimuli, one of them being the metal response element (MRE). The MREs, in the presence of heavy metals such as Zn and Cd, respond to metal-regulatory transcription factors such as MT1-F, which has been described in a number of vertebrate species (Andrews 2000). Another explanation for the induction of MTs is the role they play in mitigating cellular oxidative stress. Metallothionein mRNA expression has been shown to be induced not only by heavy metals, but by strongly oxidizing agents such as hydrogen peroxide (Andrews 2000). Mercury is known to cause oxidative stress, thus the transcriptional response of cells to Hg might be occurring in response to oxidative damage rather than Hg bound to MT, or it could be a combination of the two mechanisms.

Glutathione peroxidase is an endogenous antioxidant which prevents both lipid peroxidation as well as catalyzes hydrogen peroxide into water. Other antioxidant enzymes (and cofactors) have been shown to *rescue* rats from lipid peroxidation following chronic Cd exposure (in this case *N*-acetylcysteine or Vitamin E were co-delivered with Cd), resulting in protection from renal toxicity (Shaikh et al. 1999). Diving mammals have been shown to have higher activities of GPx compared to terrestrial mammals (Wilhelm Filho et al. 2002). Similar results were demonstrated with higher GPx activity in ringed seals versus domestic pig in the heart, the lung, and the muscle tissue (Vázquez-Medina et al. 2006). The higher activity of antioxidant defenses as an adaptation to diving physiology has been discussed earlier in this chapter, however, there is no reason not to suspect that higher activities of the enzymes could not provide, at least in part, protective effects against heavy metal-induced oxidative stress as well. In fact, it is more difficult to speculate that increased GPx activity would be independent of some role in poison detoxification or biotransformation.

14.5.1 Cetaceans

$MeHg^+$ biomagnifies through trophic levels and is present in high concentrations in tissues such as the liver, the kidney, and the pelage of marine mammals (for review of concentrations in Arctic marine mammals see Dietz et al. 2013). Mercury has both natural and anthropogenic sources, although temporal trends of Hg in ice cores indicate that concentrations increased markedly following industrialization of the early twentieth century. Temporal trends of Hg concentrations in biota are scant and difficult to compare, and analyses are confounded by well described spatial trends and difficulties in comparing different age, sex, and tissues (Braune et al. 2005). Some species/locations show increasing concentrations of Hg across multiple decades, and for some species where preindustrial archived tissues exist (generally hard tissues, i.e., Dietz et al. 2009).

The relationship between Hg and Se has been known for several decades (Koeman et al. 1975) although the mechanisms of demethylation and subsequent detoxification are only recently beginning to be described (Khan and Wang 2009). Tiemmanite crystals (mercury selenide; HgSe) have been reported in the livers of Guiana dolphins (Lailson-Brito et al. 2012), pilot whales (Caurant et al. 1996), and in the lung tissue of Atlantic bottlenose dolphins and short finned pilot whales (*Globicephala macrorhynchus*) (Rawson et al. 1995). Similarly, using micro-x-ray fluorescence, Nakazawa et al. (2011) found evidence for tiemmanite molecules in the liver, the kidney, the lung, the spleen, the pancreas, the muscle, and the brain of Striped dolphin (*Stenella coeruleoalba*). It is unclear whether the tiemmanite crystals found in the lung and other tissues of the latter study were present due to localized demethylation and conjugation, sequestration from other tissues, or, in the case of the lungs, inhalation of the particles directly.

Cultured cells selected for Cd resistance have been shown to have elevated levels of MT protein compared to non-resistant cells (Durnam and Palmiter 1987). In fact, the induction of MT by various compounds has been shown to protect against Cd-induced toxicity (for review see Klaassen et al. 1999). Interestingly, it has been shown in rats that immature individuals have higher circulating MT concentrations compared to adults (Goering and Klaassen 1984), and similar trends have been described in Franciscana dolphins (*Pontoporia blainvillei*) where concentrations of MT were higher in juveniles than adults (Polizzi et al. 2014). Although the differences in the aforementioned study were not statistically significant, the relationship of higher MT in immature > mature animals has been described in gray seals (*Halichoerus grypus*) as well as in mice (Andrews et al. 1984; Teigen and Andersen 1999). Maternal milk has been suggested to be a significant source of dietary Cd exposure during nursing, and studies investigating mineral contents of harp seal milk have found concentrations of 0.059–0.101 μg/g (Webb et al. 1984) and 0.021–0.089 μg/g (Wagemann and Stewart 1988). Whether the elevated levels of MT in immature mammals are due to induction via maternal contaminant transfer or differences in developmental physiology is unknown (perhaps a higher demand for Cu or Zn), however, high concentrations of MT in fetal tissues must certainly confer protections against Cd toxicity.

Similar to Hg, reports of concentrations of Cd bound to MT are varied among marine mammals. Amiard-Triquet and Caurant (1997) found relatively high concentrations of Cd in the livers of pilot whales, and found that concentrations of MT-like proteins were correlated with Cd concentrations in adults. However, they found that in the summer (July) only 51% of Cd was bound to MTs, and in November the percent bound was only 6%. This is in contrast to Wagemann et al. (1984), who found 77% of cytosolic Cd bound to MTs in the livers of narwhals. Das et al. (2000) suggested that this might be evidence of an adaptation of Arctic narwhals to high concentrations of Cd in prey, since environmental concentrations of Cd appear not to have fluctuated strongly in Northern latitudes (this is in contrast to Hg which has shown an increase since post-industrialization) (Fitzgerald et al. 2005). However, given the large variation seen in pilot whales, it is likely that there are other physiologic and molecular modulators of Cd bound to MT, both as total amount and proportion of total MT. These include the presence of other metal ions in diet which could compete for MT binding, the induction of hormone systems which are known to impact MT synthesis, or seasonal variation in physiology which could drive transport or sequestration of heavy metals (i.e., nursing, lipid mobilization, etc.).

14.5.2 Pinnipeds

In addition to Hg and Se, some pinnipeds have high concentrations of Cd (Dietz et al. 1996, 1998). Unlike other heavy metals such as lead and mercury, which have shown a marked increase in the Arctic since the industrial revolution, Cd levels appear to have been consistent since the fifteenth century based on hair samples from Greenland (Hansen et al. 1989). Using Greenland ringed seals (*Phoca hispida*) as an example, Dietz et al. (1998) reported that based on previous studies cadmium levels found in the kidneys of ringed seals exceeded thresholds associated with renal damage in other mammals (200 μg/g ww, as determined by WHO 1992). In the kidneys with very high concentrations of cadmium (up to 726 μg/g Cd ww in the kidney cortex), Dietz et al. (1998) found no significant differences in renal structure or necrosis was observed compared to the kidneys with low (1.63–5.19 μg/g Cd) and intermediate (86.5–91.3 μg/g Cd) concentrations. The concentrations of Cd in the livers of bowhead whales (*Balaena mysticetus*) have been also found that are associated with toxic thresholds in domestic animals (Woshner et al. 2001). Rosa et al. (2008) found moderate to severe thickening of the Bownan's capsule in the kidneys of bowhead whales associated with age and Cd concentrations, however, there was no evidence of renal dysfunction. This supports the conclusion of Dietz et al. (1998) that given historically high Cd concentrations in the marine environment (as compared to terrestrial, especially Arctic regions) marine mammals may have evolved to tolerate or detoxify relatively high concentrations of Cd.

Besides genetic adaptation of phenotypic plasticity, there is another mechanism which could help explain how pinnipeds, and indeed other species of marine mammals, appear to tolerate concentrations of heavy metals often without gross injury. A proposed mechanism for Cd-induced hepatotoxicity is oxidative stress (Shaikh et al. 1999), which occurs when MT synthesis cannot effectively sequester free Cd. Methylmercury is known to similarly generate oxidative stress, although often in different tissues (for review see Farina et al. 2011). The more infamous heavy metals of concern in the marine environment do not exist in isolation—marine mammals are exposed to these metals through a variety of routes but generally diet, which for high trophic level marine mammals such as pinnipeds and odontecetes means a diet of fish, invertebrates, and possibly other marine mammals (for instance orcas and walruses consume seals and sea lions). Other nutritional components of a fish-based diet, such as selenium (Se), may counter some of the potential negative effects of heavy metal intoxication in general. While the mercury–selenium relationship is well publicized and is discussed elsewhere in this chapter, selenium is also a key component of the glutathione-peroxidase (GPx) family of enzymes. Proteins with selenium containing active sites (selenoproteins) often contain selenocysteine active sites, such as human GPx1, in which selenium takes the place of sulfur in the cysteine residue forming a selanol group. Interestingly, the selenocysteine amino acid is not coded for by the standard genetic alphabet, rather, it is synthesized into the growing polypeptide via the traditional stop codon TGA (UGA in mRNA) (Chambers et al. 1986; Böck et al. 1991). The translation of the stop codon into a functional protein requires a specific sequence in the 3′ untranslated region (Berry et al. 1991) as well as specialized tRNA and translation factors (Leinfelder et al. 1988; Forchhammer et al. 1989). We emphasize the role of Se in both toxic metal detoxification as well as defense against oxidative stress through (1) direct interactions between Se and other metals and (2) indirect actions of Se cofactor requiring proteins.

14.6 Other poisons

Following catastrophic anthropogenic release of crude oil, such as the Exxon Valdez oil spill (1989) or the Deepwater Horizon (2010), there is a large amount of public and scientific concern devoted to the health and status of local cetacean populations. Species that depend on fur or feathers for insulation are at risk of pelage contamination and subsequent dysfunction, which can result in both poor temperature regulation, transfer to young (nursing), and oral exposure to crude hydrocarbons following grooming/preening. However, cetaceans generally have very thick skin (odontocetes epidermal layer is 10–20 times thicker than terrestrial mammals) and deep blubber tissue (hypodermal) (Geraci et al. 1986; Reeb et al. 2007). Thus the main threat to cetaceans following oil spills may in fact be inhalation of toxic fumes or irritation of mucous membranes (eyes, blowhole) (Chapter 13, Geraci 1990; Smultea and Würsig 1995). The vapors of volatile hydrocarbons contain a number of harmful toxicants which can cause localized inflammation of lung tissue, and some compounds can accumulate in blood and filtering organs causing hepatoxicity and neurotoxicity (Geraci and St. Aubin 1982). Polycyclic aromatic hydrocarbons (PAHs), chemical constituents of crude and refined oil, have been found in a number of marine mammal tissues (Carvan and Busbee 2003; Kannan and Perrotta 2008) and are known to be toxic through the production of reactive intermediates during biotransformation. However, there are limited studies examining the effects of these compounds in marine mammals (Carvan and Busbee 2003). These compounds are known to have endocrine disrupting effects, although these effects are generally seen in acute exposure scenarios, and there is a dearth of information concerning chronic low level exposure, as one might hypothesize in marine mammals.

There is some evidence to suggest that bottlenose dolphins (*Tursiops truncatus*) are able to detect oil slicks, however reports of avoidance behavior are mixed and may depend on environmental conditions and the nature of the oil product involved. We caution that generalizations should not be made. In an experimental oil spill, three female dolphins appeared to avoid surfacing in oil slicks following initial contact, however in observational studies, such as those following the *Mega Borg* oil spill in the Gulf of Mexico in 1990, nine groups of bottlenose dolphins did not show consistent avoidance of most types of oil slicks (Smith et al. 1983; Smultea and Würsig 1995). The mechanism of chemodetection of oil slicks is unknown, and may be related to visual signaling, echolocation, tactile or olfactory sensing or combinations of these.

14.7 Emerging tools

The emergence of next-generation sequencing tools (DNA and RNA) has undoubtedly advanced the field of toxicology, and a number of intrepid researchers have attempted to apply these tools to non-model organisms such as marine mammals. Edwards et al. (2013) developed a workflow incorporating undergraduate students and novel coursework and sequenced metagenomes of the California sea lion, proving that next-generation sequencing tools are accessible and validated for work with marine mammals. At the time of writing, there are seven genomes available on NCBIs website including the minke whale, bottlenose dolphin, polar bear, killer whale, Weddell seal, Pacific walrus, and Florida manatee. The generation of whole genome sequence data has the potential to elucidate a number of questions associated with marine mammal physiology and adaptation; with direct relevance to toxicology. Because of the polyphyletic nature of marine mammals, many of the phenotypic adaptations present in numerous orders provide excellent examples of convergent evolution. Recent work to examine loci that are currently under selective pressure in three orders of marine mammals (represented by walrus, manatee, and killer whale genomes) found evidence that genes associated with the glutathione metabolism pathway (ANPEP and GCLC) were under positive selection (Foote et al., 2015). It is likely that the selective pressures acting on these loci are based on hypoxia and diving physiology; however, given the importance of glutathione in detoxification pathways, it is certainly intriguing to consider the adaptation of the glutathione system in the context of toxicology. Many unexplored gene–environment (toxicants) interactions and responses of marine mammals (e.g., gene expression) to toxicants are now open to investigators with the courage and imagination to venture into these unknown waters.

High-throughput gene chips, such as cDNA microarrays, capable of measuring the expression levels of thousands of genes (increase, decrease, or no change in specific mRNA) have been utilized to examine the gene status of a number of marine mammal species in response to contaminants and disease (Mancia et al. 2012, 2014). The microarray, while still being utilized and customized to particular marine mammal species and gene groups (i.e., immune function genes) may have already peaked in usefulness with the advancement of high-throughput sequencing technologies such as RNA-seq (Shendure 2008). Incorporating additional omics tools (transcriptomics, metabolomics, proteomics) have the ability to push marine mammal toxicology research to the forefront of emerging technologies. High-throughput mass-spectroscopy techniques for identifying proteins have been utilized to study the effects of MeHg on human lymphocytes at environmentally relevant concentrations (i.e., concentrations found in harbor seals) (Das et al. 2008).

Novel methods for collecting cetacean saliva and blow, including using radio-controlled unmanned aerial vehicles, can potentially provide insight into the respiratory condition and health status of marine mammals (Acevedo-Whitehouse et al. 2010). Bacteria species present in blow and respiratory mucosa can be analyzed to investigate disease status and prevalence (Acevedo-Whitehouse et al. 2010), while other researchers have examined the utility of blows to measure levels of hormones (Hogg et al. 2009) or the detection of volatile organics (Cumeras et al. 2014).

In addition to direct sampling of marine mammals, there have been a number of developments in the detection and prediction of HABs including the genomic identification of toxin-producing bacteria species as well as satellite imaging in order to detect concentrations of algae in seawater (for review see Anderson et al. 2012). There are several challenges to detecting algal toxins in seawater or biologic tissue, including occasionally very low concentrations. However, the development of liquid chromatography coupled with mass spectroscopy (LC-MS) has provided the ability to simultaneously identify low concentration toxins as well as novel marine phycotoxins (Hummert et al. 2002; Zhang and Zhang 2015).

14.8 Lingering questions

As described above, the rapid evolution of omics technologies and increasingly sensitive/high-throughput mass spectroscopy has advanced our capability to examine in ultra-fine scale both the presence of potentially deleterious compounds as well as their negative effects. However, connecting the presence of toxins or toxicants to individual health status is still challenging, and extrapolation to local or global population level effects is extremely difficult (Fedorenkova et al. 2010; Van Straalen et al. 2010). It is important to reiterate that not all responses to the presence of a toxicant are adverse, and in many cases are essential or beneficial (such as protective hermetic effects). There are a number of populations of marine mammals that have been shown to have concentrations of contaminants over established traditional thresholds of concern (i.e., Basu et al. 2009; Doucette et al. 2012; Rea et al. 2013). However, from a population perspective, the ramifications of high concentrations of contaminants observed are unclear. Additionally, as has been emphasized in this chapter and elsewhere, toxic thresholds which have been established in model organisms (i.e., humans, rats, mink) might not be relevant to marine mammals, which have evolved in a drastically different environment than their terrestrial counterparts.

In some instances where the etiology of mass die-offs is deduced (i.e., UMEs due to HABs) the pathology of toxins may be known from laboratory animal models or observational studies (Silvagni et al. 2005), however, observations of intoxication are often limited to dead/moribund or severely intoxicated animals. Pathology associated with chronic low level or sub-lethal exposure to poisons is poorly understood, and observing changes in complex traits such as foraging ecology and behavior are often difficult and dependent on low numbers of observations (Kvitek et al. 1991; Cook et al. 2011). The study of cognitive deficits as a result of exposure to particular poisons in humans is often subtle and dependent on baseline markers of typical development. In some cases, associations between an adverse outcome (neurologic disease) and a toxicant (mercury) are noted but cannot be causally determined due to the observational nature or lack of controlled study design (Van Hoomissen et al. 2015). It is unclear whether these poisons have the same effects on marine mammals as for non-marine mammals, or furthermore if cognitive deficits have significant effects on the behavior or fitness of marine mammals. For instance, concentrations of THg in the brainstems of polar bears were associated with decreased *N*-methyl-D-aspartate (NMDA) receptor levels, however, it is unclear if these subclinical effects are associated with any gross pathologies or observable behavioral changes (Basu et al. 2009). Linking an observed biochemical or physiologic change to an adverse effect can be rather elusive and frustrating for marine mammal toxicologists. However, progress is being made.

The study of marine mammal toxicology and their unique adaptations to dealing with both new and old poisons is inexorably linked to other fields of study. Nutrition, physiology, ecology, and genetics are all components of toxicology, and developments in these fields shape the future advancement of toxicologic research. The interaction between environmental, wildlife, and human health is beginning to be appreciated with the promotion and implementation of the One Health Initiative, and the field of marine mammal toxicology can undoubtedly benefit from a holistic approach to examining the effects of poisons on marine mammal health.

Glossary

Acclimation: The ability of an individual organism to respond to changes in environmental or physiologic conditions.

Adaptation: An evolved phenotypic trait which has arisen through natural selection.

Bioaccumulation: Net accumulation of a chemical compound in an organism from all sources (air, diet, water) resulting from a higher rate of absorption than excretion.

Biomagnification: An increase in contaminant concentration from one trophic level to the next attributable to accumulation from diet.

Biotoxin: A poison which is produced by a living organism (Fowler 1993).
Biotransformation: Metabolism or modification of chemical compounds by an organism using biochemical processes.
Blubber: Integument layers below the epidermis (dermis and hypodermis) (Reeb et al. 2007).
Epidermis: The outermost, often pigmented, integument layer (Reeb et al. 2007).
Essential element: A chemical element that is required for normal physiologic function.
Maternal transfer (transplacental in mammals): Chemicals that are transferred from the mother to the developing embryo/fetus.
Parity: The amount of pregnancies an organism has carried to gestation, or viable gestational condition.
Poison: A substance which disrupts normal physiologic function via chemical interaction (Fowler 1993).
Pollutant: A poisonous substance that is derived and or synthesized via anthropogenic activities and is not found naturally in the environment.
Toxicology: The study of poisons.

References

Acevedo-Whitehouse, K., A. Rocha-Gosselin, and D. Gendron. 2010. A novel non-invasive tool for disease surveillance of free-ranging whales and its relevance to conservation programs. *Animal Conservation* 13:217–225. doi: 10.1111/j.1469-1795.2009.00326.x.

Addison, R.F. and P.F. Brodie. 1987. Transfer of organochlorine residues from blubber through the circulatory system to milk in the lactating grey seal *Halichoerus grypus*. *Canadian Journal of Fisheries and Aquatic Science* 44:782–786. doi: 10.1139/f87-095.

Amiard-Triquet, C. and F. Caurant. 1997. Adaptation of the delphinid *Globicephala melas* (Traill, 1809) to cadmium contamination. *Bulletin de la Société Zoologique de France* 122:127–136.

Anderson, D.M., A.D. Cembella, and G.M. Hallegraeff. 2012. Progress in understanding harmful algal blooms: Paradigm shifts and new technologies for research, monitoring, and management. *Annual Review of Marine Science* 4:143–176. doi: 10.1146/annurev-marine-120308-081121.

Andrews, G.K. 2000. Regulation of metallothionein gene expression by oxidative stress and metal ions. *Biochemical Pharmacology* 59:95–104.

Andrews, G.K., E.D. Adamson, and L. Gedamu. 1984. The ontogeny of expression of murine metallothionein: Comparison with the alpha-fetoprotein gene. *Developmental Biology* 103:294–303.

Barañano, D., M. Rao, C.D. Ferris, and S.H. Snyder. 2002. Biliverdin reductase: A major physiologic cytoprotectant. *Proceedings of the National Academy of Sciences* 99:16093–16098.

Bargu, S., T. Goldstein, K. Roberts, C. Li, and F. Gulland. 2012. Pseudo-nitzschia blooms, domoic acid, and related California sea lion strandings in Monterey Bay, California. *Marine Mammal Science* 28:237–253. doi: 10.1111/j.1748-7692.2011.00480.x.

Basu, N., A.M. Scheuhammer, C. Sonne, R.J. Letcher, E.W. Born, and R. Dietz. 2009. Is dietary mercury of neurotoxicological concern to wild polar bears (*Ursus maritimus*)? *Environmental Toxicology and Chemistry* 28:133–140. doi: 10.1897/08-251.1.

Beckmen, K.B., J.E. Blake, G.M. Ylitalo, J.L. Stott, and T.M. O'Hara. 2003. Organochlorine contaminant exposure and associations with hematological and humoral immune functional assays with dam age as a factor in free-ranging northern fur seal pups (*Callorhinus ursinus*). *Marine Pollution Bulletin* 46:594–606. doi: 10.1016/S0025-326X(03)00039-0.

Bejarano, A.C., F.M. Gulland, T. Goldstein, J. St Leger, M. Hunter, L.H. Schwacke, F.M. Van Dolah, and T.K. Rowles. 2008. Demographics and spatio-temporal signature of the biotoxin domoic acid in California sea lion (*Zalophus californianus*) stranding records. *Marine Mammal Science* 24:899–912.

Berglund, M., A. Akesson, P. Bjellerup, and M. Vahter. 2000. Metal-bone interactions. *Toxicology Letters* 112–113:219–225.

Berry, M.J., L. Banu, Y. Chen, S.J. Mandel, J.D. Kiefer, J.W. Harney, and P.R. Larsen. 1991. Recognition of UGA as a selenocysteine codon in Type I deiodinase requires sequences in the 3′ untranslated region. *Nature* 353:273–276.
Berta, A., J.L. Sumich, and K.M. Kovacs. 2006. *Marine Mammals: Evolutionary Biology*, 2nd edn. Academic Press, Burlington, MD.
Böck, A., K. Forchhammer, J. Heider, and C. Baron. 1991. Selenoprotein synthesis: An expansion of the genetic code. *Trends in Biochemical Sciences* 16:463–467.
Bottein, M.-Y.D., L. Kashinsky, Z. Wang, C. Littnan, and J.S. Ramsdell. 2011. Identification of ciguatoxins in Hawaiian monk seals *Monachus schauinslandi* from the Northwestern and main Hawaiian Islands. *Environmental Science and Technology* 45:5403–5409. doi: 10.1021/es2002887.
Boveng, P.L., L.M. Hiruki, M.K. Schwartz, and J.L. Bengtson. 1998. Population growth of Antarctic fur seals: Limitation by a top predator, the leopard seal? *Ecology* 79:2863–2877. doi: 10.1890/0012-9658(1998)079[2863:PGOAFS]2.0.CO;2.
Braune, B.M., P.M. Outridge, A.T. Fisk, D.C.G. Muir, P.A. Helm, K. Hobbs, P.F. Hoekstra et al. 2005. Persistent organic pollutants and mercury in marine biota of the Canadian Arctic: An overview of spatial and temporal trends. *Science of the Total Environment* 351–352:4–56. doi: 10.1016/j.scitotenv.2004.10.034.
Bricelj, V.M., L. Connell, K. Konoki, S.P. MacQuarrie, T. Scheuer, W.A. Catterall, and V.L. Trainer. 2005. Sodium channel mutation leading to saxitoxin resistance in clams increases risk of PSP. *Nature* 434:763–767. doi: 10.1038/nature03415.
Brodie, E.D.I. and E.D. Brodie Jr. 1990. Tetrodotoxin resistance in garter snakes: An evolutionary response of predators to dangerous prey. *Evolution* 44:651–659. doi: 10.2307/2409442.
Brookens, T.J., J.T. Harvey, and T.M. O'Hara. 2007. Trace element concentrations in the Pacific harbor seal (*Phoca vitulina richardii*) in central and northern California. *Science of the Total Environment* 372:676–692. doi: http://dx.doi.org/10.1016/j.scitotenv.2006.10.006.
Carvan III, M.J. and D.L. Busbee. 2003. Mechanisms of aromatic hydrocarbon toxicity: Implications for cetacean morbidity and mortality. In *Toxicology of Marine Mammals, New Perspectives: Toxicology of the Marine Environment*, eds. J.G. Vos, G. Bossart, M. Fournier, and T. O'Shea. Taylor & Francis, London, UK, pp. 429–457.
Castle, J.W. and J.H. Rodgers. 2009. Hypothesis for the role of toxin-producing algae in Phanerozoic mass extinctions based on evidence from the geologic record and modern environments. *Environmental Geosciences* 16:1–23. doi: 10.1306/eg.08110808003.
Caurant, F., M. Navarro, and J.C. Amiard. 1996. Mercury in pilot whales: Possible limits to the detoxification process. *Science of the Total Environment* 186:95–104.
Chambers, I., J. Frampton, P. Goldfarb, N. Affara, W. McBain, and P.R. Harrison. 1986. The structure of the mouse glutathione peroxidase gene: The selenocysteine in the active site is encoded by the "termination" codon, TGA. *EMBO Journal* 5 (6):1221–1227.
Clarkson, T.W. and L. Magos. 2006. The toxicology of mercury and its chemical compounds. *Critical Reviews in Toxicology* 36 (8):609–662. doi: 10.1080/10408440600845619.
Coelho, J.P., C.L. Mieiro, E. Pereira, A.C. Duarte, and M.A. Pardal. 2013. Mercury biomagnification in a contaminated estuary food web: Effects of age and trophic position using stable isotope analyses. *Marine Pollution Bulletin* 69 (1–6):110–115. doi: 10.1016/j.marpolbul.2013.01.021.
Cook, P., C. Reichmuth, and F. Gulland. 2011. Rapid behavioural diagnosis of domoic acid toxicosis in California sea lions. *Biology Letters* 7:536–538. doi: 10.1098/rsbl.2011.0127.
Costa, D.P., B.J. LeBoeuf, A.C. Huntley, and C.L. Ortiz. 1986. The energetics of lactation in the northern elephant seal, *Mirounga angustirostris*. *Journal of Zoology* 209:21–33. doi: 10.1111/j.1469-7998.1986.tb03563.x.
Cumeras, R., W.H.K. Cheung, F. Gulland, D. Goley, and C.E. Davis. 2014. Chemical analysis of whale breath volatiles: A case study for non-invasive field health diagnostics of marine mammals. *Metabolites* 4:790–806. doi: 10.3390/metabo4030790.
Cusick, K.D. and G.S. Sayler. 2013. An overview on the marine neurotoxin, saxitoxin: Genetics, molecular targets, methods of detection and ecological functions. *Marine Drugs* 11:991–1018. doi: 10.3390/md11040991.
Das, K., U. Siebert, A. Gillet, A. Dupont, C. Di-Poï, S. Fonfara, Z. Mazzucchelli, and E. De Pauw-Gillet. 2008. Mercury immune toxicity in harbour seals: Links to in vitro toxicity. *Environmental Health* 7:52. doi: 10.1186/1476-069X-7-52.

Das, K., V. Debacker, and J.M. Bouquegneau. 2000. Metallothioneins in marine mammals: A review. *Cellular and Molecular Biology* 46:283–294.

Das, K., V. Debacker, S. Pillet, and J.M. Bouquegneau. 2003. Heavy metals in marine mammals. In *Toxicology of Marine Mammals*. eds. J.G. Vos, G.D. Bossart, M. Fournier, and T.J. O'Shea. New York, Taylor & Francis, pp. 135–167.

Dawkins, R. 1987. *The Blind Watchmaker. Why the Evidence of Evolution Reveals a Universe without Design*. Norton, New York.

Debes, F., E. Budtz-Jørgensen, P. Weihe, R.F. White, and P. Grandjean. 2006. Impact of prenatal methylmercury exposure on neurobehavioral function at age 14 years. *Neurotoxicology and Teratology* 28:363–375. doi: 10.1016/j.ntt.2006.02.004.

Debier, C., C. Chalon, B.J. Le Boeuf, T. de Tillesse, Y. Larondelle, and J.-P. Thomé. 2006. Mobilization of PCBs from blubber to blood in northern elephant seals (*Mirounga angustirostris*) during the post-weaning fast. *Aquatic Toxicology* 80 (2):149–157. doi: 10.1016/j.aquatox.2006.08.002.

Degange, A.R. and M.M. Vacca. 1989. Sea otter mortality at Kodiak Island, Alaska, during summer 1987. *Journal of Mammalogy* 70:836–838.

Dehn, L.-A., G.G. Sheffield, E.H. Follmann, L.K. Duffy, D.L. Thomas, G.R. Bratton, R.J. Taylor, and T.M. O'Hara. 2005. Trace elements in tissues of phocid seals harvested in the Alaskan and Canadian Arctic: Influence of age and feeding ecology. *Canadian Journal of Zoology* 83:726–746. doi: 10.1139/z05-053.

Dietz, R., J. Nørgaardt, and J.C. Hansen. 1998. Have Arctic marine mammals adapted to high cadmium levels? *Marine Pollution Bulletin* 36 (6):490–492.

Dietz, R., P.M. Outridge, and K.A. Hobson. 2009. Anthropogenic contributions to mercury levels in present-day Arctic animals—A review. *Science of the Total Environment* 407:6120–6131. doi: http://dx.doi.org/10.1016/j.scitotenv.2009.08.036.

Dietz, R., F. Riget, and P. Johansen. 1996. Lead, cadmium, mercury and selenium in Greenland marine animals. *Science of the Total Environment* 186:67–93.

Dietz, R., C. Sonne, N. Basu, B. Braune, T. O'Hara, R.J. Letcher, T. Scheuhammer et al. 2013. What are the toxicological effects of mercury in Arctic biota? *Science of the Total Environment* 443:775–790. doi: 10.1016/j.scitotenv.2012.11.046.

Doucette, G.J., C.M. Mikulski, K.L. King, P.R. Roth, Z. Wang, L.F. Leandro, S.L. DeGrasse et al. 2012. Endangered North Atlantic right whales (*Eubalaena glacialis*) experience repeated, concurrent exposure to multiple environmental neurotoxins produced by marine algae. *Environmental Research* 112:67–76. doi: 10.1016/j.envres.2011.09.010.

Duignan, P.J., J.E.B. Hunter, I.N. Visser, G.W. Jones, and A. Nutman. 2000. Stingray spines: A potential cause of killer whale mortality in New Zealand. *Aquatic Mammals* 17:143–147.

Durnam, D.M. and R.D. Palmiter. 1987. Analysis of the detoxification of heavy metal ions by mouse metallothionein. *Experientia Supplementum: Metallothionein II* 52:457–463.

Edwards, R.A., J.M. Haggerty, N. Cassman, J.C. Busch, K. Aguinaldo, S. Chinta, M. Houle Vaughn et al. 2013. Microbes, metagenomes and marine mammals: Enabling the next generation of scientist to enter the genomic era. *BMC Genomics* 14:600. doi: 10.1186/1471-2164-14-600.

Farina, M., M. Aschner, and J.B.T. Rocha. 2011. Oxidative stress in MeHg-induced neurotoxicity. *Toxicology and Applied Pharmacology* 256:405–417. doi: 10.1016/j.taap.2011.05.001.

Fedorenkova, A., J.A. Vonk, H.J.R. Lenders, N.J. Ouborg, A.M. Breure, and A.J. Hendriks. 2010. Ecotoxicogenomics: Bridging the gap between genes and populations. *Environmental Science and Technology* 44:4328–4333.

Fire, S.E. and F.M. Van Dolah. 2012. Marine biotoxins: Emergence of harmful algal blooms as health threats to marine wildlife. In *New Directions in Conservation Medicine: Applied Cases of Ecological Health*, eds. A. Aguire, R. Ostfeld, and P. Daszak. Oxford University Press, New York, pp. 374–389.

Fitzgerald, W.F., D.R. Engstrom, C.H. Lamborg, C.-M. Tsang, P.H. Balcom, and C.R. Hammerschmidt. 2005. Modern and historic atmospheric mercury fluxes in northern Alaska: Global sources and arctic depletion. *Environmental Science and Technology* 39:557–568. doi: 10.1021/es049128x.

Flewelling, L.J., J.P. Naar, J.P. Abbott, D.G. Baden, N.B. Barros, G.D. Bossart, M.-Y.D. Bottein et al. 2005. Brevetoxicosis red tides and marine mammal mortalities. *Nature* 435:755–756.

Foote, A.D., Y. Liu, G.W.C. Thomas, T. Vinar, J. Alfoldi, J. Deng, S. Dugan et al. 2015. Convergent evolution of the genomes of marine mammals. *Nature Genetics* 47:272–275.

Forchhammer, K., W. Leinfelder, and A. Bock. 1989. Identification of a novel translation factor necessary for the incorporation of selenocysteine into protein. *Nature* 342:453456.

Fowler, M. 1993. *Veterinary Zootoxicology*. CRC Press Inc., Boca Raton, FL.

Francis, J.M. and C.B. Heath. 1991. Population abundance, pup mortality, and copulation frequency in the California sea lion in relation to the 1983 El Niño on San Nicolas Island. In *Pinnipeds and El Niño*, eds. F. Trillmich and K.A. Ono. Ecological Studies. Springer-Verlag, Berlin, Heidelberg, pp. 119–128.

Franz, D.R. and R.D. LeClaire. 1989. Respiratory effects of brevetoxin and saxitoxin in awake guinea pigs. *Toxicon* 27:647–654. doi: http://dx.doi.org/10.1016/0041-0101(89)90015-9.

Frouin, H., M. Lebeuf, M. Hammill, and M. Fournier. 2012. Transfer of PBDEs and chlorinated POPs from mother to pup during lactation in harp seals *Phoca groenlandica*. *Science of the Total Environment* 417–418:98–107. doi: http://dx.doi.org/10.1016/j.scitotenv.2011.11.084.

Geraci, J.R.1990. Physiologic and toxic effects on cetaceans. In *Sea Mammals and Oil: Confronting the Risks*, eds. J.R. Geraci and D.J. St. Aubin. Academic Press, San Diego, CA, pp. 167–198.

Geraci, J.R., D.M. Anderson, R.J. Timperi, D.J. St. Aubin, G.A. Early, J.H. Prescott, and C.A. Mayo. 1989. Humpback whales (*Megaptera novaeangliae*) fatally poisoned by dinoflagellate toxin. *Canadian Journal of Fisheries and Aquatic Sciences* 46:1895–1898. doi: 10.1139/f89-238.

Geraci, J.R. and D.J. St. Aubin. 1982. Study of the effects of oil on cetaceans. Final report, Prepared for U.S. Department of the Interior, Bureau of Land Management. Washington, DC.

Geraci, J.R., D.J. St. Aubin, and B.D. Hicks. 1986. The epidermis of odontocetes: A view from within. In *Research on Dolphins*, eds. M.M. Bryden and R. Harrison. Oxford University Press, Oxford, UK, pp. 3–21.

Godt, J., F. Scheidig, C. Grosse-Siestrup, V. Esche, P. Brandenburg, A. Reich, and D.A. Groneberg. 2006. The toxicity of cadmium and resulting hazards for human health. *Journal of Occupational Medicine and Toxicology* 1:22. doi: 10.1186/1745-6673-1-22.

Goering, P.L. and C.D. Klaassen. 1984. Resistance to cadmium-induced hepatotoxicity in immature rats. *Toxicology and Applied Pharmacology* 74:321–329. doi: http://dx.doi.org/10.1016/0041-008X(84)90285-0.

Goksøyr, A. 1995. Cytochrome P450 in marine mammals: Isozyme forms, catalytic functions, and physiological regulations. In *Whales, Seals, Fish and Man. Proceedings of the International Symposium on the Biology of Marine Mammals in the North East Atlantic*, eds. A.S. Blix, L. Walløe, and Ø. Ulltang. November 29–December 1, 1994. Elsevier Science, Tromsø, Norway, pp 629–639. http://www.sciencedirect.com/science/article/pii/S0163699506800624.

Goldstein, T., J.A.K. Mazet, T.S. Zabka, G. Langlois, K.M. Colgrove, M. Silver, S. Bargu et al. 2008. Novel symptomatology and changing epidemiology of domoic acid toxicosis in California sea lions (*Zalophus californianus*): An increasing risk to marine mammal health. *Proceedings of the Royal Society C* 275:267–276. doi: 10.1098/rspb.2007.1221.

Goldstein, T., T.S. Zabka, R.L. DeLong, E.A. Wheeler, G. Ylitalo, S. Bargu, T. Leighfield et al. 2009. The role of domoic acid in abortion and premature parturition of California sea lions (*Zalophus californianus*) on San Miguel Island, California. *Journal of Wildlife Diseases* 45:91–108. doi: 10.7589/0090-3558-45.1.91.

Gonzalez, F.J. and D.W. Nebert. 1990. Evolution of the P450 gene superfamily: Animal-plant "warfare", molecular drive and human genetic differences in drug oxidation. *Trends in Genetics* 6:182–186.

Gormezano, L.J. and R.F. Rockwell. 2013. Dietary composition and spatial patterns of polar bear foraging on land in western Hudson Bay. *BMC Ecology* 13:51. doi: 10.1186/1472-6785-13-51.

Grandjean, P., P. Weihe, and J.B. Nielsen. 1994. Methylmercury: Significance of intrauterine and postnatal exposures. *Clinical Chemistry* 40:1395–1400.

Gulland, F.M.D., M. Haulena, D. Fauquier, M.E. Lander, T. Zabka, R. Duerr, and G. Lanlois. 2002. Domoic acid toxicity in Californian sea lions (*Zalophus californianus*): Clinical signs, treatment and survival. *Veterinary Record* 150:475–480. doi: 10.1136/vr.150.15.475.

Hackett, J.D., J.H. Wisecaver, M.L. Brosnahan, D.M. Kulis, D.M. Anderson, D. Bhattacharya, F.G. Plumley, and D.L. Erdner. 2013. Evolution of saxitoxin synthesis in cyanobacteria and dinoflagellates. *Molecular Biology and Evolution* 30:70–78. doi: 10.1093/molbev/mss142.

Hamilton, S.J. and P.M. Mehrle. 1986. Metallothionein in fish: Review of its importance in assessing stress from metal contaminants. *Transactions of the American Fisheries Society* 115:596–601.

Hansen, J.C., T.Y. Toribara, and A.G. Muhs. 1989. Trace metals in human and animal hair from the 15th century graves at Qilakitsoq compared with recent samples. *Meddelelser om Grønland, Man and Society* 12:161–167.

Hernández, M., I. Robinson, A. Aguilar, L.M. Gonzalez, L.F. Lopes-Jurado, M.I. Reyero, E. Cacho et al. 1998. Did algal toxins cause monk seal mortality? *Nature* 393:28–29. doi: 10.1038/29906.

Hoekstra, P.F., T.M. O'Hara, S.J. Pallant, K.R. Solomon, and D.C.G. Muir. 2002. Bioaccumulation of organochlorine contaminants in bowhead whales (*Balaena mysticetus*) from Barrow, Alaska. *Archives of Environmental Contamination and Toxicology* 42:497–507. doi: 10.1007/s00244-001-0046-x.

Hogg, C.J., T.L. Rogers, A. Shorter, K. Barton, P.J.O. Miller, and D. Nowacek. 2009. Determination of steroid hormones in whale blow: It is possible. *Marine Mammal Science* 25:605–618. doi: 10.1111/j.1748-7692.2008.00277.x.

Hummert, C., A. Rühl, K. Reinhardt, G. Gerdts, and B. Luckas. 2002. Simultaneous analysis of different algal toxins by LC-MS. *Chromatographia* 55:673–680.

Itano, K., S. Kawai, N. Miyazaki, R. Tatsukawa, and T. Fujiyama. 1984. Mercury and selenium levels in striped dolphins caught off the pacific coast of Japan. *Agricultural and Biological Chemistry* 48:1109–1116. doi: 10.1080/00021369.1984.10866282.

Iverson, F., J. Truelove, E. Nera, L. Tryphonas, J. Campbell, and E. Lok. 1989. Domoic acid poisoning and mussel-associated intoxication: Preliminary investigations into the response of mice and rats to toxic mussel extract. *Food and Chemical Toxicology* 27:377–384. doi: http://dx.doi.org/10.1016/0278-6915(89)90143-9.

Jeffery, B., T. Barlow, K. Moizer, S. Paul, and C. Boyle. 2004. Amnesic shellfish poison. *Food and Chemical Toxicology* 42 (4):545–557. doi: http://dx.doi.org/10.1016/j.fct.2003.11.010.

Jessup, D.A., C.K. Johnson, J. Estes, D. Carlson-Bremer, W.M. Jarman, S. Reese, E. Dodd et al. 2010. Persistent organic pollutants in the blood of free-ranging sea otters (*Enhydra lutris* ssp.) in Alaska and California. *Journal of Wildlife Diseases* 46 (4):1214–1233.

Johansson, C., A.F. Castoldi, N. Onishchenko, L. Manzo, M. Vahter, and S. Ceccatelli. 2007. Neurobehavioural and molecular changes induced by methylmercury exposure during development. *Neurotoxicity Research* 11 (3–4):241–260.

Kannan, K. and E. Perrotta. 2008. Polycyclic aromatic hydrocarbons (PAHs) in livers of California sea otters. *Chemosphere* 71:649–655. doi: http://dx.doi.org/10.1016/j.chemosphere.2007.11.043.

Khan, M. and F. Wang. 2009. Mercury-selenium compounds and their toxicological significance: Toward a molecular understanding of the mercury-selenium antagonism. *Environmental Toxicology and Chemistry* 28:1567–1577.

Kirkpatrick, B., L.E. Fleming, D. Squicciarini, L.C. Backer, R. Clark, W. Abraham, J. Benson et al. 2004. Literature review of Florida red tide: Implications for human health effects. *Harmful Algae* 3 (2):99–115. doi: http://dx.doi.org/10.1016/j.hal.2003.08.005.

Klaassen, C.D., J. Liu, and S. Choudhuri. 1999. Metallothionein: An intracellular protein to protect against cadmium toxicity. *Annual Review of Pharmacology and Toxicology* 39:267–94. doi: 10.1146/annurev.pharmtox.39.1.267.

Knott, K.K., P. Schenk, S. Beyerlein, D. Boyd, G.M. Ylitalo, and T.M. O'Hara. 2011. Blood-based biomarkers of selenium and thyroid status indicate possible adverse biological effects of mercury and polychlorinated biphenyls in Southern Beaufort Sea polar bears. *Environmental Research* 111:1124–1136. doi: 10.1016/j.envres.2011.08.009.

Koeman, J.H., W.H. Peeters, C.H.M. Koudstaal-Hol, P.S. Tjioe, and J.J.M. De Goeij. 1973. Mercury-selenium correlations in marine mammals. *Nature* 245:385–386. doi: 10.1038/245385a0.

Koeman, J.H., W.S.M. van de Ven, J.J.M. de Goeij, P.S. Tjioe, and J.L. van Haaften. 1975. Mercury and selenium in marine mammals and birds. *Science of the Total Environment* 3:279–287. doi: http://dx.doi.org/10.1016/0048-9697(75)90052-2.

Krone, C.A., P.A. Robisch, K.L. Tilbury, J.E. Stein, E.A. Mackey, and P.R. Becker. 1999. Elements in liver tissues of bowhead whales (*Balaena mysticetus*). *Marine Mammal Science* 15:123–142. doi: 10.1111/j.1748-7692.1999.tb00785.x.

Kuehl, D.W. and R. Haebler. 1995. Organochlorine, organobromine, metal, and selenium residues in bottlenose dolphins (*Tursiops truncatus*) collected during an unusual mortality event in the Gulf of Mexico, 1990. *Archives of Environmental Contamination and Toxicology* 28:494–499. doi: 10.1007/BF00211632.

Kvitek, R.G., A.R. DeGange, and M.K. Beitler. 1991. Paralytic shellfish poisoning toxins mediate feeding behavior of sea otters. *Limnology and Oceanography* 36 (2):393–404.

Lailson-Brito, J., R. Cruz, P.R. Dorneles, L. Andrade, A. de Freitas Azevedo, A.B. Fragoso, L. Gama Vidal et al. 2012. Mercury-selenium relationships in liver of Guiana dolphin: The possible role of Kupffer cells in the detoxification process by tiemannite formation. *PLoS One* 7 (7):e42162. doi: 10.1371/journal.pone.0042162.

Lee, S.S., B.R. Mate, K.T. von der Trenck, R.A. Rimerman, and D.R. Buhler. 1977. Metallothionein and the subcellular localization of mercury and cadmium in the California sea lion. *Comparative Biochemistry and Physiology Part C: Comparative Pharmacology* 57 (1):45–53.

Lefebvre, K.A., C.L. Powell, M. Busman, G.J. Doucette, P.D.R. Moeller, J.B. Silver, P.E. Miller et al. 1999. Detection of domoic acid in northern anchovies and California sea lions associated with an unusual mortality event. *Natural Toxins* 7:85–92.

Lefebvre, K.A. and A. Robertson. 2010. Domoic acid and human exposure risks: A review. *Toxicon* 56:218–230.

Leinfelder, W., E. Zehelein, M. Mandrand Berthelot, and A. Bock. 1988. Gene for a novel tRNA species that accepts L-serine and cotranslationally inserts selenocysteine. *Nature* 331:723–725.

Lemire, M., A. Philibert, M. Fillion, C.J. Passos, J.R. Guimarães, F. Barbosa Jr., and D. Mergler. 2012. No evidence of selenosis from a selenium-rich diet in the Brazilian Amazon. *Environment International* 40:128–136. doi: 10.1016/j.envint.2011.07.005.

Lewis, R.J. and M.J. Holmes. 1993. Origin and transfer of toxins involved in ciguatera. *Comparative Biochemistry and Physiology Part C: Comparative Pharmacology* 106:615–628.

Lobanov, A.V., D.E. Fomenko, Y. Zhang, A. Sengupta, D.L. Hatfield, and V.N. Gladyshev. 2007. Evolutionary dynamics of eukaryotic selenoproteomes: Large selenoproteomes may associate with aquatic life and small with terrestrial life. *Genome Biology* 8:R198. doi: 10.1186/gb-2007-8-9-r198.

Lowry, L.F. and F.H. Fay. 1984. Seal eating by walruses in the Bering and Chukchi Seas. *Polar Biology* 3:11–18. doi: 10.1007/BF00265562.

Mancia, A., L. Abelli, J.R. Kucklick, T.K. Rowles, R.S. Wells, B.C. Balmer, A.A. Hohn, J.E. Baatz, and J.C. Ryan. 2014. Microarray applications to understand the impact of exposure to environmental contaminants in wild dolphins (*Tursiops truncatus*). *Marine Genomics* 19:47–57. doi: 10.1016/j.margen.2014.11.002.

Mancia, A., J.C. Ryan, R.W. Chapman, G.W. Warr, F.M. Gulland, and F.M. Van Dolah. 2012. Health status, infection and disease in California sea lions (*Zalophus californianus*) studied using a canine microarray platform and machine-learning approaches. *Developmental and Comparative Immunology* 36 (4):629–637.

Moore, S.K., V.L. Trainer, N.J. Mantua, M.S. Parker, E.A. Laws, L.C. Backer, and L.E. Fleming. 2008. Impacts of climate variability and future climate change on harmful algal blooms and human health. *Environmental Health* 7 (Suppl. 2):S4. doi: 10.1186/1476-069X-7-S2-S4.

Muir, D.C.G., R. Wagemann, N.P. Grift, R.J. Norstrom, M. Simon, and J. Lien. 1988. Organochlorine chemical and heavy metal contaminants in white-beaked dolphins (*Lagenorhynchus albirostris*) and pilot whales (*Globicephala melaena*) from the coast of Newfoundland, Canada. *Archives of Environmental Contamination and Toxicology* 17:613–629. doi: 10.1007/BF01055830.

Murray, S.A., T.K. Mihali, and B.A. Neilan. 2011. Extraordinary conservation, gene loss, and positive selection in the evolution of an ancient neurotoxin. *Molecular Biology and Evolution* 28:1173–1182. doi: 10.1093/molbev/msq295.

Nakazawa, E., T. Ikemoto, A. Hokura, Y. Terada, T. Kunito, S. Tanabe, and I. Nikai. 2011. The presence of mercury selenide in various tissues of the striped dolphin: Evidence from μ-XRF-XRD and XAFS analyses. *Metallomics* 3:719–725. doi: 10.1039/C0MT00106F.

National Marine Fisheries Service, Marine mammal unusual mortality events, National Oceanic and Atmospheric Administration website: http://www.nmfs.noaa.gov/pr/health/mmume/, last updated August 2013.

Ostrow, J.D., L. Pascolo, D. Brites, and C. Tiribelli. 2004. Molecular basis of bilirubin-induced neurotoxicity. *Trends in Molecular Medicine* 10 (2):65–70. doi: 10.1016/j.molmed.2003.12.003.

Paludan-Müller, P., C. Agger, R. Dietz, and C. Kinze. 1993. Mercury, cadmium, zinc, copper and selenium in harbour porpoise (*Phocoena phocoena*) from West Greenland. *Polar Biology* 13:311–320. doi: 10.1007/BF00238358.

Polizzi, P.S., M.B. Romero, L.N. Chiodi Boudet, K. Das, P.E. Denuncio, D.H. Rodriguez, and M.S. Gerpe. 2014. Metallothioneins pattern during ontogeny of coastal dolphin, *Pontoporia blainvillei*, from Argentina. *Marine Pollution Bulletin* 80:275–281. doi: 10.1016/j.marpolbul.2013.10.037.

Preston, E. and I. Hynie. 1991. Transfer constants for blood-brain barrier permeation of the neuroexcitatory shellfish toxin, domoic acid. *Canadian Journal of Neurological Sciences* 18:39–44.

Pugh, L.G. 1959. Carbon monoxide content of the blood and other observations on Weddell seals. *Nature* 183:74.

Pulido, O.M. 2008. Domoic acid toxicologic pathology: A review. *Marine Drugs* 6:180–219.

Puls, R. 1994. *Mineral Levels in Animal Health*, 2nd edn. Sherpa International, Clearbrook, British Columbia, Canada.

Pyenson, N.D., C.S. Gutstein, J.F. Parham, J.P Le Roux, C. Carreño Chavarria, H. Little, A. Metallo et al. 2014. Repeated mass strandings of Miocene marine mammals from Atacama Region of Chile point to sudden death at sea. *Proceedings of the Royal Society B* 281(1781). doi: 10.1098/rspb.2013.3316. http://rspb.royalsocietypublishing.org/content/281/1781/20133316.

Ramsdell, J.S. and C.E Stafstrom. 2009. Rat kainic acid model provides unexpected insight into an emerging epilepsy syndrome in sea lions. *Epilepsy Currents* 9:142–143. doi: 10.1111/j.1535-7511.2009.01321.x.

Rawson, A.J., T.P. Bradley, A. Teetsov, S.B. Rice, E.M. Haller, and G.W. Patton. 1995. A role for airborne particulates in high mercury levels of some cetaceans. *Ecotoxicology and Environmental Safety* 30:309–314. doi: http://dx.doi.org/10.1006/eesa.1995.1035.

Rea, L.D., J.M. Castellini, L. Correa, B.S. Fadely, and T.M. O'Hara. 2013. Maternal steller sea lion diets elevate fetal mercury concentrations in an area of population decline. *Science of the Total Environment* 454–455:277–282. doi: 10.1016/j.scitotenv.2013.02.095.

Reeb, D., P.B. Best, and S.H. Kidson. 2007. Structure of the integument of southern right whales, *Eubalaena australis*. *Anatomical Record* 290:596–613. doi: 10.1002/ar.20535.

Reijnders, P.J.H. 1986. Reproductive failure in common seals feeding on fish from polluted coastal waters. *Nature* 324:456–457.

Reus, I.S., I. Bando, D. Andrés, and M. Cascales. 2003. Relationship between expression of HSP70 and metallothionein and oxidative stress during mercury chloride induced acute liver injury in rats. *Journal of Biochemical and Molecular Toxicology* 17:161–168. doi: 10.1002/jbt.10074.

Riget, F. and R. Dietz. 2000. Temporal trends of cadmium and mercury in Greenland marine biota. *Science of the Total Environment* 245:49–60. doi: http://dx.doi.org/10.1016/S0048-9697(99)00432-5.

Robertson, L.W. and L.G. Hansen. 2001. *PCBs: Recent Advances in Environmental Toxicology and Health Effects*. University Press of Kentucky, Lexington, KY.

Roesijadi, G. 1992. Metallothioneins in metal regulation and toxicity in aquatic animals. *Aquatic Toxicology* 22:81–113. doi: http://dx.doi.org/10.1016/0166-445X(92)90026-J.

Rosa, C., J.E. Blake, G.R. Bratton, L.-A. Dehn, M.J. Gray, and T.M. O'Hara. 2008. Heavy metal and mineral concentrations and their relationship to histopathological findings in the bowhead whale (*Balaena mysticetus*). *Science of the Total Environment* 399:165–178. doi: 10.1016/j.scitotenv.2008.01.062.

Ross, P.S. 2000. Marine mammals as sentinels in ecological risk assessment. *Human and Ecological Risk Assessment* 6:2946.

Ross, P.S., S.J. Jeffries, M.B. Yunker, Addison, R.F., M.G. Ikonomou, and J.C. Calambokidis. 2004. Harbor seals (*Phoca vitulina*) in British Columbia, Canada, and Washington State, USA, reveal a combination of local and global polychlorinated biphenyl, dioxin, and furan signals. *Environmental Toxicology and Chemistry* 23:157–165.

Saulitis, E., C. Matkin, L. Barrett-Lennard, K. Heise, and G. Ellis. 2000. Foraging strategies of sympatric killer whale (*Orcinus orca*) populations in Prince William Sound, Alaska. *Marine Mammal Science* 16 (1):94–109.

Scholin, C.A., F. Gulland, G.J. Doucette, S. Benson, M. Busman, F.P. Chavez, J. Cordaro et al. 2000. Mortality of sea lions along the central California coast linked to a toxic diatom bloom. *Nature* 403:8084.

Shaikh, Z.A., T.T. Vu, and K. Zaman. 1999. Oxidative stress as a mechanism of chronic cadmium-induced hepatotoxicity and renal toxicity and protection by antioxidants. *Toxicology and Applied Pharmacology* 154:256–263. doi: 10.1006/taap.1998.8586.

Shendure, J. 2008. The beginning of the end for microarrays? *Nature Methods* 5:585–587.

Silvagni, P.A., L.J. Lowenstine, T. Spraker, T.P. Lipscomb, and F.M.D. Gulland. 2005. Pathology of domoic acid toxicity in California sea lions (*Zalophus californianus*). *Veterinary Pathology* 42 (2):184–191. doi: 10.1354/vp.42-2-184.

Smith, T.G. and F.A.J. Armstrong. 1978. Mercury and selenium in ringed and bearded seal tissues from Arctic Canada. *Arctic* 31:75–84. doi: 10.2307/40508886.

Smith, T.G., J.R. Geraci, and D.J. St. Aubin. 1983. Reaction of bottlenose dolphins, *Tursiops truncatus*, to a controlled oil spill. *Canadian Journal of Fisheries and Aquatic Sciences* 40 (9):1522–1525.

Smultea, M.A. and B. Würsig. 1995. Behavioral reactions of bottlenose dolphins to the Mega Borg oil spill, Gulf of Mexico 1990. *Aquatic Mammals* 21 (3):171–181.

Stocker, R., Y. Yamamoto, A.F. McDonagh, A.N. Glazer, and B.N. Ames. 1987. Bilirubin is an antioxidant of possible physiological importance. *Science* 235:1043–1046. doi: 10.1126/science.3029864.

Suzuki, C.A.M. and S.L. Hierlihy. 1993. Renal clearance of domoic acid in the rat. *Food and Chemical Toxicology* 31:701–706.

Tanabe, S. 2002. Contamination and toxic effects of persistent endocrine disrupters in marine mammals and birds. *Marine Pollution Bulletin* 45:69–77. doi: http://dx.doi.org/10.1016/S0025-326X(02)00175-3.

Tanabe, S., S. Watanabe, H. Kan, and R. Tatsukawa. 1988. Capacity and mode of PCB metabolism in small cetaceans. *Marine Mammal Science* 4:103–124.

Teigen, S.W., R.A. Andersen, H.L. Daae, and J.U. Skaare. 1999. Heavy metal content in liver and kidneys of grey seals (*Halichoerus grypus*) in various life stages correlated with metallothionein levels: Some metal—Binding characteristics of this protein. *Environmental Toxicology and Chemistry* 18 (10):2364–2369. http://doi.org/10.1002/etc.5620181034.

Teramitsu, I., Y. Yamamoto, I. Chiba, H. Iwata, S. Tanabe, Y. Fujise, A. Kajusaka, F. Akahori, and S. Fujita. 2000. Identification of novel cytochrome P450 1A genes from five marine mammal species. *Aquatic Toxicology* 51:145–153.

Tift, M.S., P.J. Ponganis, and D.E. Crocker. 2014. Elevated carboxyhemoglobin in a marine mammal, the northern elephant seal. *Journal of Experimental Biology* 217:1752–1757. doi: 10.1242/jeb.100677.

Tonk, E.C.M., D.M.G. de Groot, A.H. Penninks, I.D.H. Waalkens-Berendsen, A.P.M. Wolterbeek, W. Slob, A. H. Piersma, and H.V. Loveren. 2010. Developmental immunotoxicity of methylmercury: The relative sensitivity of developmental and immune parameters. *Toxicological Sciences* 117 (2):325–335.

Trojanowska, B. and A. Sapota. 1974. Binding of cadmium and mercury by metallothionein in the kidneys and liver of rats following repeated administration. *Archives of Toxicology* 32(4):351–360.

Valério, E., S. Chaves, and R. Tenreiro. 2010. Diversity and impact of prokaryotic toxins on aquatic environments: A review. *Toxins* 2 (10): 2359–2410.

Van de Vijver, K.I., P.T. Hoff, K. Das, W.V. Dongen, E.L. Esmans, T. Jauniaux, J.M. Bouquegneau, R. Blust, and W. de Coen. 2003. Perfluorinated chemicals infiltrate ocean waters: Link between exposure levels and stable isotope ratios in marine mammals. *Environmental Science & Technology* 37(24): 5545–5550.

Van Dolah, F.M., G.J. Doucette, F.M.D. Gulland, T.L. Rowles, and G.D. Bossart. 2003. 10 impacts of algal toxins on marine mammals. In *Toxicology of Marine Mammals, New Perspectives: Toxicology and the Environment*, Vol. 3, eds. J.G. Vos, G.D. Bossart, M. Fournier, and T.J. O'Shea. Taylor & Francis, New York, pp. 247–269.

Van Egmond, H.P., M.E. Van Apeldoom, and G.J.A. Speijers. 2004. Marine biotoxins. FAO Food and Nutrition Paper 80. UN Food and Agriculture Organization (FAO), Rome, Italy.

Van Hoomissen, S., F.M.D. Gulland, D.J. Greig, J.M. Castellini, and T.M. O'Hara. 2015. Blood and hair mercury concentrations in the Pacific Harbor Seal (*Phoca vitulina richardii*) pup: Associations with neurodevelopmental outcomes. *EcoHealth*. http://link.springer.com/article/10.1007%2Fs10393-015-1021-8.

Van Straalen, N.M., D. Roelofs, C.A.M. Van Gestel, and T.E. De Boer. 2010. Comment on Ecotoxicogenomics: Bridging the gap between genes and populations. *Environmental Science and Technology* 44(23):9239–9240.

Varanasi, U., J.E. Stein, K.L. Tilbury, J.P. Meador, C.A. Sloan, R.C. Clark, and S.L. Chan. 1994. Chemical contaminants in gray whales (*Eschrichtius robustus*) stranded along the west coast of North America. *Science of the Total Environment* 145(1):29–53.

Vázquez-Medina, J.P., T. Zenteno-Savín, and R. Elsner. 2006. Antioxidant enzymes in ringed seal tissues: Potential protection against dive-associated ischemia/reperfusion. *Comparative Biochemistry and Physiology Part C: Toxicology & Pharmacology* 142(3):198–204.

Wagemann R., R. Hunt, and J.F. Klaverkamp. 1984. Subcellular distribution of heavy metals in liver and kidney of a narwhal whale (*Monodon monoceros*): An evaluation for the presence of metallothionein. *Comparative Biochemistry and Physiology* C 78:301–307.

Wagemann R., S. Innes, and P.R. Richard. 1996. Overview and regional and temporal differences of heavy metals in Arctic whales and ringed seals in the Canadian Arctic. *Science of the Total Environment* 186:41–66. doi: 10.1016/0048-9697(96)05085-1.

Wagemann, R., N.B. Snow, A. Lutz, and D.P. Scott. 1983. Heavy metals in tissues and organs of the narwhal (*Monodon monoceros*). *Canadian Journal of Fisheries and Aquatic Sciences* 40(S2): s206–s214.

Wagemann, R. and R.E.A. Stewart. 1994. Concentrations of heavy metals and selenium in tissues and some foods of walrus (*Odobenus rosmarus rosmarus*) from the Eastern Canadian Arctic and Sub-Arctic, and associations between metals, age, and gender. *Canadian Journal of Fisheries and Aquatic Sciences*. 51:426–436. doi: 10.1139/f94-044.

Wagemann, R., R.E.A. Stewart, W.L. Lockhart, B.E. Stewart, and M. Povoledo. 1988. Trace metals and methyl mercury: Associations and transfer in harp seal (*Phoca groenlandica*) mothers and their pups. *Marine Mammal Science* 4(4):339–355. http://doi.org/10.1111/j.1748-7692.1988.tb00542.x

Webb, B.E, R.E.A. Stewart, and D.M. Lavigne. 1984. Mineral constituents of harp seal milk. *Canadian Journal of Zoology* 62:831–833.

Wilhelm Filho, D., F. Sell, L. Ribeiro, M. Ghislandi, F. Carrasquedo, C.G. Fraga, J.P. Wallauer, P.C. Simões-Lopes, and M.M. Uhart. 2002. Comparison between the antioxidant status of terrestrial and diving mammals. *Comparative Biochemistry and Physiology Part A: Molecular & Integrative Physiology* 133 (3):885–892.

Winge, D., J. Krasno, and A.V. Colucci. 1974. Cadmium accumulation in rat liver: Correlation between bound metals and pathology. In *Trace Element Metabolism in Animals*, eds. W.G. Hoekstra, J.W. Suttie, H.E. Ganther, and W. Mertz. University Park Press, Baltimore, MD, pp. 500–502.

Wiśniewska, J.M., B. Trojanowska, J. Piotrowski, and M. Jakubowski. 1970. Binding of mercury in the rat kidney by metallothionein. *Toxicology and Applied Pharmacology* 16 (3):754–763.

World Health Organization. 1992. Cadmium. Environmental Health Criteria 134. World Health Organization, Geneva, Switzerland.

Woshner, V.M., T.M. O'Hara, G.R. Bratton, R.S. Suydam, and V. R. Beasley. 2001. Concentrations and interactions of selected essential and non-essential elements in bowhead and beluga whales of Arctic Alaska. *Journal of Wildlife Diseases* 37 (4): 693–710.

Yamamori, K., M. Nakamura, T. Matsui, and T.J. Hara. 1988. Gustatory responses to tetrodotoxin and saxitoxin in fish: A possible mechanism for avoiding marine toxins. *Canadian Journal of Fisheries and Aquatic Sciences* 45 (12): 2182–2186.

Ylitalo, G.M., C.O. Matkin, J. Buzitis, M.M. Krahn, L.L. Jones, T. Rowles, and J.E. Stein. 2001. Influence of life-history parameters on organochlorine concentrations in free-ranging killer whales (*Orcinus orca*) from Prince William Sound, AK. *Science of the Total Environment* 281 (1):183–203.

Zhang, C. and J. Zhang. 2015. Environmental analytical chemistry current techniques for detecting and monitoring algal toxins and causative harmful algal blooms. *Journal of Environmental and Analytical Chemistry* 2:1–12. doi: 10.4172/JREAC.1000123.

chapter fifteen

Conclusions and questions

Michael A. Castellini

Contents

15.1 Introduction

During the process of writing this book, the authors often encountered some fact we did not know, or an idea that came from reading or editing the chapters. For example, there are some blood oxygen data that I collected almost 35 years ago from diving seals that I still do not understand. The authors from the pressure chapter (Chapter 4) discussed a new concept in their writing that might help explain those findings, and we are going to re-examine those data again using their new ideas. Even with multiple decades of experience working with marine mammals, we all found out *something new* or were exposed to a new and interesting interpretation.

Our concept for this book was to explore the physiological, anatomical, and biochemical adaptations that must be considered when asking how marine mammals live in the ocean. By now, you should have a good understanding of the core issues that face these groups in their marine existence: From hydrodynamic forces, to diving, staying warm, obtaining water, sensory systems, fasting, and disease, there are many basic mammalian systems that must be *tweaked* for life in the sea. Yet, there are many more physiological concepts that we could not cover in the space of one book, a few of which should be discussed before we conclude.

15.2 Behavior in a physiological box

You have learned about species of marine mammals that can dive for over an hour or forage for food at over 1000 m depth. Just as with humans, these may be the extremes of their performance, and daily routines do not include such amazing physiological feats. Almost every chapter referred to behavioral patterns in marine mammals, and this is why the *physiological box* concept becomes important. Biochemistry, physics, and physiology define the limits of how a marine mammal can dive, but it is behavior that places an animal within that box. An animal will dive to hunt for food, to move from one area to another, to socialize, to breed, to communicate, or to get away from a predator. All of those behaviors will determine the pattern of the dives. Physiology and biochemistry set what the animal is capable of, not what it does from moment-to-moment. For example, when we were first working on the *aerobic dive limit* concept many years ago, we discovered that over 95% of the free dives of Weddell seals were under the ADL limit by putting dive recorders on dozens of seals and then turning them loose to discover their "natural diving patterns."

The sea otter is another example we have covered that applies to the *physiological box.* The daily behavior of sea otters is driven by a high metabolism, the need to obtain great amounts of food, and to keep their fur coat waterproof. These are the physiological and biochemical *facts* about sea otters. However, the facts do not provide much information on whether the otter is going to forage at any particular location, or whether it is going to capture a crab or an urchin. To some extent, the otter is going to capture what it can to eat, depending on where it is diving. While the physiological box constrains the otter (it cannot hold its breath long enough to go below 30–40 m), what it does inside that constraint is not really limited by its physiology. What it does inside that box is more determined by the concept of *physiological ecology.*

15.3 Physiological ecology

Physiological ecology is at the interface of the energetic needs of the animal and the ecological niche of its environment. As you might imagine, these questions are not easily answered, particularly for our marine mammal species that spend most or all of their lives outside of our ability to observe them.

For example, how would you determine how much a blue whale consumes in krill each day, or how much milk it feeds its calf? Despite the difficulty of finding these answers, many are at the core of ecological overlap in habitat or resource use by humans. The field of physiological ecology is essential to most management, Endangered Species Act findings, and other policy determinations for marine mammals. It is a fascinating area of study, but one that relies heavily on fundamental limits dictated by physiological and biochemical rules, combined with *physiology in a box* theories and knowledge.

Several sections in this book discussed the obvious, and not obvious, conflicts between fishing and marine mammals. In Alaska, populations of Steller sea lions have been declining for decades. "Is it food?" is not only an important question but has large economic and social impacts for humans given the billion-dollar fishing industry that is subject to marine mammal protection regulations. These regulations are based on sometimes very limited data due to the difficulty of working with these species. Should fisheries around sea lion rookeries be closed to fishing at the times of year when mothers and pups are nearby? To obtain part of that answer, scientists deploy dive recorders with satellite tracking so that they can find out the distribution and depth of where the female sea lions forage. But, how do you know if they are foraging while diving in a certain area, or just passing through? While stomach temperature sensors are a start, they only can tell when the first fish enters the stomach, not an actual measure of how much is consumed, or what species is ingested. Sea lions also tend to regurgitate, or eliminate, these sensors in a matter of mere hours. Other clever approaches have used sensors to measure the jaw angle, allowing you to know when an animal opens its mouth to catch prey. But then, more information is needed to confirm that the animal was successful. Having read this textbook, you can likely now better understand the numerous pieces of information, all that require different measurement techniques, that have to come together to answer the seemingly simplest of questions.

15.4 Other physiological questions

There are multiple further physiological and biochemical studies of marine mammals that occur at the next deeper levels of study and form many research theses and postdoctoral studies. Many of the authors in this textbook were chosen specifically for their expertise relating to their specific chapters, but also because they are on the forefront of research and teaching in diverse areas of marine mammal physiology, including the particularly challenging topics we will just briefly mention.

15.4.1 Pre-partum physiology

Pinnipeds and otters exhibit a phenomenon known as embryonic diapause. Within a very short time period after giving birth (roughly 4–10 days), the female mates again. The egg is fertilized immediately and divides to about the 16-cell stage. At this point, it enters embryonic diapause, and sits dormant for about 3 months. The embryo is reactivated to result in a pregnancy period of 9 months, which results in a predictable, annual cycle that can take advantage of prime periods of prey abundance, pupping substrate (e.g., pagophilic seals), or breeding opportunities. For a physiologist studying these species, it leads to challenges in the standard approach of using blood hormones to detect early pregnancy. The female may very well be carrying a fertilized egg, but the effects of the diapause period will prevent the appearance of the fairly standard hormone profile found in other mammals.

The period of embryonic diapause also comes into play as trial period for the female. There is evidence that if the female does not accumulate sufficient energy reserves to support the pregnancy for whatever reason, the pregnancy may terminate and she re-absorbs the embryo. This results in a divergence in adult females during the breeding season. Those females who have had a successful year will be present at the rookeries to give birth to their pups, and to breed. Those females who did not maintain their pregnancy may also come to the rookery solely to breed, or they may not be present at all, opting to wait a year to improve their body condition before they breed again. For a review of marine mammal reproductive biology (see Schroeder 1990; Stewart and Stewart 2003).

15.4.2 Digestive physiology

It is very easy to measure how many fish, squid, or other food items a marine mammal consumes per day in a controlled setting. The difficulty is in measuring how efficiently those food items are digested and converted into energy, muscle mass, etc. If a small dolphin consumes 10,000 cal/day of raw fish, what is the digestive and assimilation efficiency of those 10,000 cal? It turns out that digestive efficiency is roughly 90%, but varies with numerous variables including prey species and time of year. The field of digestive physiology focuses on assimilation and digestive efficiency by using a suite of metabolic tracers in food items, combined with analyzing calories ingested versus the calories remaining in fecal material. These tracers include protein and amino acid markers to estimate protein metabolism among a host of other methods for specific tissues or types of metabolism. This returns us to the question of physiological ecology and policy management in the wild, particularly as it relates to the overlap of fisheries and marine mammal prey. While it may be simple enough to determine the species and general caloric value of a prey species, it only provides the roughest approximation of what is available to the marine mammal in question, which may or may not be even remotely close to the true energetic value to the consumer. This was an essential component of the "Is it food" policy debate during the listing of the Steller sea lion under the Endangered Species Act. One major theory during this process was that despite their energetic content, some fish species should be considered *junk food* (Rosen and Trites 2000), precluding their need to be included in fishing regulations. As techniques for monitoring animals in the wild grow ever more detailed and comprehensive, we will hopefully be in a better position to answer these sometimes politically charged questions with solid scientific data. For an in-depth review of marine mammal digestive physiology (see Worthy 1990).

15.4.3 Sleep physiology

Dive recorders have shown that elephant seals dive repetitively and without stopping for months at sea. When do they sleep if they are diving hundreds of times per day with no long surface periods? It turns out that many species of phocids are able to sleep while underwater. Some cetaceans remarkably sleep with one hemisphere of their brain while the other half maintains vigilance. Sleeping and diving have become so physiologically intertwined in phocids that seals will hold their breath while sleeping on dry land. This phenomenon enables us to study many of the same attributes of breath-holding seen in diving (e.g., bradycardia, hematocrit regulation) but without the conflict of exercise, as noted in Chapter 3. Electroencephalograms of brain activity show that seals will stay sleeping on the beach, as they come in and out of breath-hold periods. In the case of elephant seals, this effect of sleep apnea can be as long as 20 min. Comparisons to estimates of diving metabolic rate in Weddell seals suggested that it was only slightly lower than during sleeping. For a review of sleep apnea physiology (see Castellini 1996).

15.4.4 Tissue level biochemistry and molecular biology

The field of molecular medicine and biochemistry is touched on many times in this book, but overall, this research area is far behind that of laboratory-based medical research on terrestrial mammals. The difficulty lies in obtaining samples in sufficiently large numbers for marine mammal species. Genetic analysis has been essential for some Endangered Species Act determinations, and cell culture has been essential to the fields of contaminant chemistry and responses to external stressors. Many of the basic questions on cellular metabolic pathway regulation, impacts of temperature, pressure, and pH on cell function and enzymatic reactions remain at the very edge of marine mammal biochemistry. New work on oxygen free radical damage and oxidative stress is important and is an area that touches diving physiology and other medical aspects of marine mammal biology (e.g., Hindle et al. 2009).

15.5 Lessons from teaching physiology of marine mammals

Most of the authors of this book have taught classes in marine mammal physiology, biochemistry, and anatomy, which have featured prominently in the way we have addressed many of our chapter topics. There are some questions students ask consistently that we are still lacking the tools and information to answer in detail and with authority.

15.5.1 Can marine mammals be hurt by loud underwater noise?

Chapter 11 explains many of the issues surrounding marine noise and marine mammal responses. Increasing background noise in the ocean due to human maritime activity makes this a particularly contentious area of research. As the Arctic Ocean warms and human enterprise takes advantage of reduced sea ice, massive efforts are underway to measure sound levels in an attempt to predict what may cause whales or seals to move or alter their migration routes. The controversy around Navy SONAR and potential for damage to whales has been decades-long and is at the basis of a great deal of current work on acoustics.

15.5.2 How do whales find their way during migrations?

The search continues for how whales navigate across great stretches of open ocean. Theories include navigation by stars, polarized sunlight angles in the water, magnetic sensing, and acoustics. Luckily, advances in custom telemetry are allowing for dive recorders that are smaller, more hydrodynamic, and increasingly powerful with additional sensors to describe the water salinity and temperature. These transmitters are our window into the whale's movements up and down coastlines, allowing new analyses of where and when they spend most of their time, with important information about linkages to prevailing ocean conditions.

15.5.3 What causes whales to strand?

Every year there are many cases of good intentioned humans attempting to rescue stranded dolphins or small whales. In most cases, even if they are successful in dislodging the animals from the beach, the animals turn back to shore only to strand again. Sadly, very few of these rescue attempts are successful. Sound pollution is routinely implicated in stranding events, however animals have beached for eons, long before human noise in the ocean was a rising concern. Alternative theories include the apparent strong social adhesion of cetaceans, such that one individual strands due to illness, only to be followed by other members of their group. Both natural algal bloom events and anthropogenic chemical pollution have been suggested as disorienting factors. The details of each stranding case vary so widely that we are still many years from identifying the underlying conditions, and there very well may not be a single unifying principle.

15.5.4 Can species interbreed?

This issue comes up particularly in the context of the changing ice patterns in the Arctic. It is predicted that polar bears will be forced into ice refugia in the North Eastern Canadian Arctic within the next century. As they are forced to spend more time on land, there are suggestions that they may begin to breed with brown bears. While this may occur to some extent, and indeed may have already occurred in a few isolated cases, polar bears in their current evolutionary state do not forage well on land. They will likely remain closely associated with whatever ice remains in their regions. Similarly, it has been questioned whether Arctic seals will interbreed if they become more closely associated with each other under changing ice conditions? Perhaps, but the larger concern is whether this closer association may accelerate disease transmission between previously isolated populations.

15.6 Requisites for ocean living

The goal of this book was to answer many of the questions that surround the enigmatic marine mammals. Our ultimate hope was that we have stimulated you to ask even more questions about the physiology of marine mammals and their life in the sea. There is one constant among this widely diverse group of animals. For every adaptation that solves a physiological challenge arising from life in the ocean, there are a suite of cascading and complex physiological and behavioral tweaks required to integrate effectively into the numerous and varied species that we identify collectively as the *marine mammals*.

Now, it is our turn to close by coming full circle. How would *you* design a marine mammal?

References

Castellini, M.A. 1996. Dreaming about diving: Sleep apnea in seals. *News in Physiological Sciences* 11: 208–214.

Hindle, A., M. Horning, J. Mellish, and J. Lawler. 2009. Diving into old age: Muscular senescence in a large-bodied, long-lived marine mammal, the Weddell seal (*Leptonychotes weddellii*). *Journal of Experimental Biology* 212: 790–796.

Rosen D.A.S. and A.W. Trites. 2000. Pollock and the decline of the Steller sea lions: Testing the junk food hypothesis. *Canadian Journal of Zoology* 78: 1243–1250.

Schroeder, J.P. 1990. Reproductive aspects of marine mammals. In *CRC Handbook of Marine Mammal Medicine: Health, Disease, and Rehabilitation*, CRC Press, Boca Raton, FL, pp. 353–369.

Stewart, R.E.A. and B.E. Stewart. 2003. Female reproductive systems. In *Encyclopaedia of Marine Mammals*. Academic Press, London, UK, pp. 422–428.

Worthy, G.A.J. 1990. Nutritional energetics of marine mammals. In *CRC Handbook of Marine Mammal Medicine: Health, Disease and Rehabilitation*, CRC Press, Boca Raton, FL, pp. 489–520.

Index

G

9781482242676